Theory Of Oscillations

SERIES ON OPTIMIZATION

Published

Vol. 2   Differential Games of Pursuit
         by L. A. Petrosjan

Vol. 3   Game Theory
         by L. A. Petrosjan and N. A. Zenkevich

Vol. 4   Theory of Oscillations
         by V. I. Zubov

SERIES ON
OPTIMIZATION
VOL. 4

THEORY OF OSCILLATIONS

Vladimir I Zubov
Faculty of Applied Mathematics
St. Petersburg State University
Russia

Translated by

J M Donetz

World Scientific
Singapore • New Jersey • London • Hong Kong

Published by

World Scientific Publishing Co. Pte. Ltd.
P O Box 128, Farrer Road, Singapore 912805
USA office: Suite 1B, 1060 Main Street, River Edge, NJ 07661
UK office: 57 Shelton Street, Covent Garden, London WC2H 9HE

Library of Congress Cataloging-in-Publication Data
Theory of oscillations / Vladimir I. Zubov ; English translation by J. M.
　　Donetz.
　　　　p.　cm. -- (Series on optimization; vol. 4)
　　Includes bibliographical references (pp. 393–395) and index.
　　ISBN: 9810209789 (alk. paper)
　　1. Oscillations.
QA865.Z83 1998
531'.11--dc21　　　　　　　　　　　　　　　　　　　　97-47135
　　　　　　　　　　　　　　　　　　　　　　　　　　　　CIP

British Library Cataloguing-in-Publication Data
A catalogue record for this book is available from the British Library.

Copyright © 1999 by World Scientific Publishing Co. Pte. Ltd.

All rights reserved. This book, or parts thereof, may not be reproduced in any form or by any means, electronic or mechanical, including photocopying, recording or any information storage and retrieval system now known or to be invented, without written permission from the Publisher.

For photocopying of material in this volume, please pay a copying fee through the Copyright Clearance Center, Inc., 222 Rosewood Drive, Danvers, MA 01923, USA. In this case permission to photocopy is not required from the publisher.

This book is printed on acid-free paper.

Printed in Singapore by Uto-Print

Preface

Physical phenomena related to process dynamics are described in many cases by nonlinear systems of ordinary differential equations, in other words such descriptions may eventually reduce to these systems. For this reason, oscillations in mechanical, electric and radio engineering systems are investigated with the use of mathematical tools involving nonlinear equations. Development of oscillations in controlled and uncontrolled systems is largely governed by their stationary states and behavior in the neighborhood of such stationary states.

This book also concentrates on the behavior of integral curves of systems of differential equations in the neighborhood of rest points, periodic orbits, and almost periodic solutions of these systems. This investigation leads, in some cases, to construction of a family of integral curves coming arbitrarily close to the stationary states involved. In modern mathematics, existence theory and search methods for stationary states (or modes) are far from being complete and, for this reason, are discussed in detail. We make use of the perturbation method, perturbation theory of stationary modes, perturbation theory of periodic orbits, and other methods. Consideration of controlled systems raises specific issues are also mentioned.

Conceptually, this book derives, directly or indirectly, from Zubov's "Oscillations in Nonlinear and Controlled Systems" and "Analytic Dynamics of Systems". It should be noted that Section 5.2 and the Appendix of this book were written by S.V. Zubov.

The author would like to thank A.A. Ovseenko for typesetting the manuscript, and J.M. Donetz for translating this book.

Contents

1 Preliminary Representations and Analyses of Motion Family Behavior   **1**
   1.1 Basic Properties of Functions Defined by Ordinary Differential Equations   1
   1.2 Properties of Solutions to Linear Systems 12
   1.3 Behavior of Integral Curves of a Nonlinear System of ODE's with Unbounded Increase in Time 31
   1.4 Behavior of Integral Curves in the Neighborhood of a Singular Point . 46
   1.5 Lyapunov Stability of Nonperturbed Motions 60

2 On Behavior of Trajectories in the Neighborhood of a Periodic Orbit   **69**
   2.1 Autonomous Dynamical Systems 69
   2.2 Preliminary Studies on Properties of the Neighborhood of a Periodic Orbit 77
   2.3 Unity Roots of Characteristic Equation 87
   2.4 The Case of Several Complex Roots with Unity Moduli 100
   2.5 Perturbation Theory of Periodic Orbits 111

3 Natural and Forced Oscillations in Systems with Many Degrees of Freedom   **131**
   3.1 Basic Properties of Periodic Dynamical Systems 131
   3.2 Small Parameter Method 146
   3.3 Natural Oscillations of Autonomous Systems in a Neighborhood of an Equilibrium 161
   3.4 Forced Periodic and Almost Periodic Oscillations 175
   3.5 Influence of External Perturbations on Stationary Modes 187

4 Methods for Investigation and Construction of Stationary Modes   **199**
   4.1 Elements of Recurrent Function Theory 199
   4.2 Forced Multifrequency Oscillations 221
   4.3 Motions of Nonlinear Systems Determined by Boundary Conditions . 231
   4.4 Application of Method of Successive Approximations for Finding Stationary Modes 247
   4.5 Quantitative Stability Criteria for Stationary Modes in Automatic Control Systems 257

5 Oscillations in Nonlinear and Controlled Systems   **265**
   5.1 Construction and Investigation of Stationary Self-oscillations in Automatic Control Systems 265
   5.2 Self-oscillations in Hysteresis Systems 278

5.3	Controlled Motion of Charged Particles in Magnetic Field	293
5.4	Phase Plane	309
5.5	Oscillations in Autonomous System	315
5.6	Analytical Test for Qualitative Behavior of Integral Curves on a Plane in the Neighborhood of Periodic Motion	344

Appendix. Theory of Rated Stability   357

1	Introduction and Basic Definitions	357
2	An Investigation of Rated Stability by Lyapunov's Direct Method	360
3	Analysis of Rated Stability by Lyapunov's First Method	380

Bibliography   393

Index   397

Theory Of Oscillations

Chapter 1

Preliminary Representations and Analyses of Motion Family Behavior

1.1 Basic Properties of Functions Defined by Ordinary Differential Equations

Idealization of many physical phenomena, in particular, those of oscillatory nature, in mechanical, electrical, electromechanical and other systems, both controlled and uncontrolled, calls for mathematical description which can be provided by a system of differential equations. This generates a need for analysis of general properties of functions defined by ordinary differential equations without any reference to a physical phenomenon.

1. Consider the set of all continuous vector-valued functions $x(t) = (x_1(t), \ldots, x_n(t))$ given on the same interval $[a, b]$, and denote this set by $C[a, b]$. In other words, $C[a, b]$ is the collection of all sets, each containing n functions that are real and continuous in $[a, b]$.

Definition 1. The set K of functions $x \in C[a, b]$ is called equicontinuous if for $\epsilon > 0$ there is $\delta > 0$ such that $\|x(t') - x(t'')\| < \epsilon$ for $|t' - t''| < \delta$, $t', t'' \in [a, b]$ for all $x \in K$. Here

$$\|x\| = \sqrt{\sum_{i=1}^{n} x_i^2}, \quad \|x(t') - x(t'')\| = \sqrt{\sum_{i=1}^{n} (x_i(t') - x_i(t''))^2}.$$

Definition 2. The set K of functions $x \in C[a, b]$ is uniformly bounded if there exists a number $M < +\infty$ such that $\|x\| \leq M$ for all $x \in K$.

Theorem 1 (Arzela theorem) *From any infinite set K of functions, which is equicontinuous and uniformly bounded, it is possible to select an infinite sequence $x^{(1)}, \ldots, x^{(p)}, \ldots$, uniformly converging on $[a, b]$ to function $x \in C[a, b]$.*

Proof: Take an arbitrary infinite sequence $x^{(1)}, \ldots, x^{(p)}, \ldots$, belonging to K, and show that we can select from there a sequence uniformly converging on $[a, b]$ to the

function $x \in C[a,b]$. To do this, we will consider the set of all rational numbers $R \subset [a,b]$. Since R is countable, all of its elements can be designated by numbers in such a way that
$$R = \{r_i\}, \ j = 1, 2, \ldots.$$
The numerical sequence $\{x^{(p)}(r_1)\}, p = 1, 2, \ldots$, is norm-bounded by the number M, therefore we can choose a convergent sequence $x_1^{(1)}(r_1), \ldots, x_1^{(p)}(r_1), \ldots$. The numerical sequence $\{x_1^{(p)}(r_2)\}$ is also norm-bounded by the magnitude of M, which offers a means of selecting a convergent subsequence $x_2^{(1)}(r_2), \ldots, x_2^{(p)}(r_2), \ldots$. Continuing this process we obtain a subsequence $\{x_q^{(p)}\}$ that converges in a point $r_q, q = 1, 2, \ldots$. Let $\tilde{x}_q = x_q^{(q)}$. We will show that the sequence $\{\tilde{x}_q\}$ is uniformly converging on $[a,b]$ to the function $x \in C[a,b]$. Basically, the sequence $\{\tilde{x}_q\}$ converges in any point of the set R by construction. In order to establish its convergence at each point of $[a,b]$, it is sufficient to show that with t fixed, $\{\tilde{x}_q(t)\}$ converges in itself. For this purpose, let $\epsilon > 0$. By equicontinuity, there exists $\delta > 0$ such that for $|t - t'| < \delta$, t and $t' \in [a,b]$ there is $\|\tilde{x}_q(t) - \tilde{x}_q(t')\| < \epsilon$, $q = 1, 2, \ldots$. Choose $r_j \in R$ such that $|t - r_j| < \delta$; then we get $\|\tilde{x}_q(t) - \tilde{x}_q(r_j)\| < \epsilon$.

Next, the sequence $\{\tilde{x}_q(r_j)\}$ converges in itself; hence there is a number p_0 such that $\|\tilde{x}_p(r_j) - \tilde{x}_q(r_j)\| < \epsilon$ for p and $q > p_0$. On this basis, for $p > p_0, q > p_0$ we obtain

$$\|\tilde{x}_p(t) - \tilde{x}_q(t)\| \leq \|\tilde{x}_p(t) - \tilde{x}_p(r_j)\| + \|\tilde{x}_q(t) - \tilde{x}_q(r_j)\| + \|\tilde{x}_p(r_j) - \tilde{x}_q(r_j)\| < 3\epsilon.$$

Thus, the sequence $\{\tilde{x}_q(t)\}$ converges at each $t \in [a,b]$. Denote by $x(t)$ the limit of this sequence at each point $t \in [a,b]$.

It remains to prove that the sequence $\{\tilde{x}_q(t)\}$ converges uniformly on $[a,b]$, then $x(t)$ is a continuous function and, therefore, $x(t) \in C[a,b]$. For this purpose, we will choose $\epsilon > 0$. Then, by equicontinuity, it is in agreement with the corresponding $\delta > 0$. Cover the interval $[a,b]$ with the line segments δ, the number of which equals l. In each of these, we choose a rational number, say, $r_1, \ldots, r_l$. By the convergence of the sequence $\{\tilde{x}_q\}$ at each such point, we may choose an integer $p_0 > 0$ such that

$$\|\tilde{x}_p(r_j) - \tilde{x}_q(r_j)\| < \epsilon; \ p > p_0, \ q > p_0, \ j = 1, \ldots, l.$$

Evaluate the expression $\|\tilde{x}_p(t) - \tilde{x}_q(t)\|$. We have

$$\|\tilde{x}_p(t) - \tilde{x}_q(t)\| \leq \|\tilde{x}_p(t) - \tilde{x}_p(r_j)\| + \|\tilde{x}_q(t) - \tilde{x}_q(r_j)\| + \|\tilde{x}_p(r_j) - \tilde{x}_q(r_j)\| < 3\epsilon, \quad (*)$$

where j is selected so that r_j belongs to the same line segment δ as the number t. This inequality implies the inequality $\|\tilde{x}_p(t) - x(t)\| \leq 3\epsilon$ with $p \geq p_0$ for all $t \in [a,b]$, if, within the inequality $(*)$, passage is made as $q \to \infty$. This means that the sequence $\{x_q(t)\}$ converges uniformly on $[a,b]$. ∎

2. Arzela theorem gives the proof of the solution existence theorem of initial value problem for the system of ordinary differential equations proposed by Peano.

Consider the system of ordinary differential equations

$$\frac{dx_s}{dt} = f_s(t, x_1, \ldots, x_n), \ s = 1, \ldots, n, \quad (1.1.1)$$

Basic Properties of Functions Defined by Ordinary Differential Equations

for which the functions $f_s(t, x_1, \ldots, x_n)$ are assumed to be given for $t \in (-\infty, +\infty)$ and $x \in G$, where G is the domain of the n-dimensional Euclidean space E_n, the functions being real and continuous therein.

Theorem 2 *If* 1) *the point* $x_0 = (x_1^{(0)}, \ldots, x_n^{(0)}) \in G$,

2) *the two numbers* $a > 0$ *and* $b > 0$, *are given such that all of the points* x *satisfying the inequality* $\mid x_s - x_s^{(0)} \mid \leq b$, $s = 1, \ldots, n$, *are members of* G,

3)
$$M = \sup_{\substack{s=1,\ldots,n, \\ |x_s - x_s^{(0)}| \leq b, \\ |t - t_0| \leq a}} |f_s(t, x_1, \ldots, x_n)|,$$

then there exist n *continuously differentiable functions* $x_1(t), \ldots, x_n(t)$, *given for* $|t - t_0| \leq h$ *and satisfy* (1.1.1) *with initial conditions*

$$x_s(t_0) = x_s^{(0)}, \ s = 1, \ldots, n;$$

here
$$h = \min(a, b/M).$$

Proof: Break the interval $[t_0 - h, t_0 + h]$ into $2q$ equal parts with points $t_0 \pm \frac{lh}{q}$, $l = 0, \ldots, q$, and construct the set of points by

$$x^{(l+1)} = x^{(l)} + \frac{h}{q} f\left(t_0 + \frac{hl}{q}, x^{(l)}\right), \quad (1.1.2)$$

where $f = (f_1, \ldots, f_n), l = 0, \ldots, q - 1$. Join by means of straight line segments the points $x^{(l)}$ and $x^{(l+1)}$ constructed by (1.1.2). The resulting polygonal line is called the Euler polygonal line. By changing $q = 1, 2, \ldots$, we obtain an infinite set of Euler polygonal lines.

It will be shown below that this set is equipotentially bounded and, by Theorem 1, provides a means of selecting the subsequence which converges to a continuous function $x(t)$. Find that this function is the desired solution of the initial value Cauchy problem for equation (1.1.1).

Proof will be furnished for one half of the segment $[t_0 - h, t_0 + h]$, viz. for $[t_0, t_0 + h]$. Each Euler polygonal line has the derivative bounded with respect to t, since on the interval $t \in \left[t_0 + \frac{lh}{q}, t_0 + \frac{(l+1)h}{q}\right]$ the Euler polygonal line is given by the equation

$$x_q(t) = x^{(l)} + \left(x^{(l+1)} - x^{(l)}\right)\left(t - t_0 - \frac{hl}{q}\right)\frac{q}{h}. \quad (1.1.3)$$

Relation (1.1.3) implies the inequality $\mid \frac{d}{dt} x_{sq}(t) \mid \leq M$ from which it follows that the set of continuous curves $x_q(t), q = 1, 2, \ldots$, is equicontinuous. The polygonal line $x_q(t)$ consists of q segments. An increment can be estimated on each segment from the formula $\mid x_s^{(l+1)} - x_s^{(l)} \mid \leq \frac{h}{q} M$. Therefore,

$$\mid x_{qs}(t) - x_s^{(0)} \mid \leq Mh \leq b \text{ for all } t \in [t_0, t_0 + h].$$

Thus, the set of Euler polygonal lines is uniformly bounded for $q = 1, 2, \ldots$, and, additionally, is enclosed in the cube $|x_s - x_s^{(0)}| \leq b$ which is completely immersed in the domain G.

Based on Theorem 1, from the set of polygonal lines we select a subsequence converging to the continuous function $x(t)$ and, for simplicity of notation, denote it by $x^{(q)}$. It will be shown that the function $x(t)$ we have constructed is the required solution of the initial value problem for system (1.1.1). Introduce two functions $y^{(q)}$ and $t^{(q)}$ defined by the formulas

$$y^{(q)}(t) = x^{(q)}\left(\frac{lh}{q}\right); \tag{1.1.4}$$

$$t^{(q)}(t) = t_0 + \frac{lh}{q} \quad \text{when} \quad t \in \left[\frac{lh}{q}, \frac{(l+1)h}{q}\right]. \tag{1.1.5}$$

Evidently, $t^{(q)} \to t$, $y^{(q)} \to x(t)$ as $q \to \infty$ uniformly on $[t_0, t_0 + h]$. From (1.1.4) it follows that

$$|y_s^{(q)}(t) - x_s^{(0)}| \leq b, \tag{1.1.6}$$

and (1.1.5) suggests that

$$|t^{(q)}(t) - t_0| \leq h. \tag{1.1.7}$$

From (1.1.6), (1.1.7), and from the uniform convergence of $y^{(q)} \to x(t)$ and $t^{(q)} \to t$ we find that $f(t^{(q)}, y^{(q)}) \to f(t, x(t))$ as $q \to \infty$ uniformly on $[t_0, t_0 + h]$. On the other hand, we have

$$x^{(q)}(t) = x^{(0)} + \int_{t_0}^{t} f(t^{(q)}, y^{(q)}) d\tau. \tag{1.1.8}$$

When passing to the limit in (1.1.8) as $q \to \infty$ for each $t \in [t_0, t_0 + h]$, we get

$$x(t) = x^{(0)} + \int_{t_0}^{t} f(\tau, x(\tau)) d\tau. \tag{1.1.9}$$

Differentiating both sides with respect to t, we find from (1.1.9) that $x(t)$ is the solution of system (1.1.1). Substituting $t = t_0$ in (1.1.9), we find that $x(t_0) = x^{(0)}$. ∎

Remark. Theorem 2 does not guarantee the uniqueness of a continuously differentiable solution satisfying the initially given $x(t_0) = x^{(0)}$, since the possibility of selecting various subsequences from the sequence of polygonal lines implies nonuniqueness.

3. If more rigid conditions are imposed on the right-hand sides of system (1.1.1), then it becomes possible to prove, along with the existence of solutions, their uniqueness in the same interval.

Theorem 3 *If 1) all the conditions in Theorem 2 are satisfied, 2) the functions $f_s(t, x_1, \ldots, x_n)$ satisfy the Lipschitz condition*

$$|f_s(t, \bar{x}_1, \ldots, \bar{x}_n) - f_s(t, x_1, \ldots, x_n)| \leq L \sum_{i=1}^{n} |\bar{x}_i - x_i|,$$

$$s = 1, \ldots, n \quad \text{for} \quad |t - t_0| \leq a_1, \ |\bar{x}_s - x_s^{(0)}| \leq b, \tag{1.1.10}$$

Basic Properties of Functions Defined by Ordinary Differential Equations

then there exists a unique, continuously differentiable solution of the initial value problem for system (1.1.1), $x_s = x_s(t), s = 1, \ldots, n$, which satisfies the initial condition $x_s(t_0) = x_s^{(0)}$ and is defined for $|t - t_0| \leq h, h = \min(a, b/M)$.

Proof: Theorem 2 has established that if condition 1) of Theorem 3 is satisfied, then there exists the solution of system (1.1.1) $x_s = x_s(t), s = 1, \ldots, n$, satisfying the following conditions:

1) the functions $x_s(t)$ are given for $|t - t_0| \leq h$ and are continuously differentiable;
2) $x_s(t_0) = x_s^{(0)}, s = 1, \ldots, n$.

We will show that this solution is unique. That is, if there is the alternative solution of system (1.1.1)

$$x_s = \bar{x}_s(t), \quad s = 1, \ldots, n,$$

satisfying conditions 1) and 2), then

$$x_s(t) \equiv \bar{x}_s(t), \quad s = 1, \ldots, n \text{ for } |t - t_0| \leq h.$$

Suppose this is not so. Substituting the solutions x and $\bar{x}$ into system (1.1.1) and integrating both sides of the identities obtained between the limits t_0 and t, we have

$$\bar{x}(t) = x^{(0)} + \int_{t_0}^{t} f(\tau, \bar{x}(\tau)) d\tau; \qquad (1.1.11)$$

$$x(t) = x^{(0)} + \int_{t_0}^{t} f(\tau, x(\tau)) d\tau. \qquad (1.1.12)$$

Subtracting termwise one identity from another and taking the norms of both sides, we get

$$\|x(t) - \bar{x}(t)\| = \left\| \int_{t_0}^{t} (f(\tau, x(\tau)) - f(\tau, \bar{x}(\tau))) d\tau \right\|$$

$$\leq \left| \int_{t_0}^{t} \|f(\tau, x(\tau)) - f(\tau, \bar{x}(\tau))\| d\tau \right| \leq \left| \int_{t_0}^{t} nL \|x(\tau) - \bar{x}(\tau)\| d\tau \right|. \qquad (1.1.13)$$

The last inequality utilizes the relation

$$\sum_{j=1}^{n} |x_j(t) - \bar{x}_j(t)| \leq \sqrt{n} \|x - \bar{x}\|.$$

For definiteness, it will be assumed that in (1.1.13) $t \geq t_0$. Using inequality (1.1.13) p times, we obtain

$$\|x(t) - \bar{x}(t)\| \leq (nL)^p \int_{t_0}^{t} \int_{t_0}^{\tau_1} \ldots \int_{t_0}^{\tau_{p-1}} \|x(\tau_p) - \bar{x}(\tau_p)\| d\tau_p \ldots d\tau_1$$

$$= \frac{(nL)^p}{(p-1)!} \int_{t_0}^{t} (t - \tau)^{p-1} \|x(\tau) - \bar{x}(\tau)\| d\tau. \qquad (1.1.14)$$

Introduce the notation: $M_1 = \max_{t - t_0 \leq h} \|\bar{x}(t) - x(t)\|$. Then from (1.1.14) we have

$$M_1 \leq \frac{(nL)^p}{(p-1)!} \int_{t_0}^{t_0 + h} (t_0 + h - \tau)^{p-1} M_1 d\tau = \frac{(nL)^p}{p!} h^p M_1. \qquad (1.1.15)$$

The right-hand side of (1.1.15) contains the factor $\frac{(nL)^p}{p!}h^p$ which is the p-th term of a convergent series for e^{nLh}. Consequently, there is p_0 such that for $p > p_0$ this term is less than one. If $M_1 \neq 0$, then cancellation of both sides in (1.1.15) by M_1 would result in the fulfilment of the inverse inequality $\frac{(nL)^p}{p!}h^p \geq 1$ for all $p \geq 1$, which is impossible. ■

Although this theorem is popularly known as the Picard theorem, in actual Picard's is responsible for the method of successive approximations which is rather common in the existence proof of solutions without using Theorem 2. This method involves construction of the sequence of functions $x^{(p)}(t)$ which converges to the solution $x(t)$ as $p \to \infty$. The sequence of functions is constructed from the formulas

$$x^{(0)}(t) = x^{(0)}, \ldots, x^{(p+1)}(t) = x^{(0)} + \int_{t_0}^{t} f(\tau, x^{(p)}(\tau))d\tau.$$

4. In parallel with system (1.1.1), we will consider a system of differential equations

$$\frac{dx_s}{dt} = f_s(t, x_1, \ldots, x_n) + r_s(t, x_1, \ldots, x_n), \quad s = 1, \ldots, n. \tag{1.1.16}$$

Assume that the functions $r_s(t,x)$ possess the same properties as the functions $f_s(t,x)$ of system (1.1.1).

Next, compare the solutions of systems (1.1.1) and (1.1.16).

Theorem 4 *If 1) on the segment $|t - t_0| \leq h_1$ there exist the solution of system (1.1.1) $x = x(t)$ and the solution of system (1.1.16) $x = y(t)$, $y(t_0) = y^{(0)}$,*

2) both solutions for $|t - t_0| \leq h_1$, are contained in the n-dimensional cube $|x_s - x_s^{(0)}| \leq b$, $s = 1, 2, \ldots, n$, belonging to G,

3) the function $r(t, x_1, \ldots, x_n)$, calculated from the solution $x = y(t)$ of system (1.1.16), satisfies the inequality $\|r(t, y(t))\| \leq \delta_2$, $|t - t_0| \leq h_1$ and the initial conditions adhere to the inequality $\|x^{(0)} - y^{(0)}\| \leq \delta_1$,

4) the functions $f_s(t, x_1, \ldots, x_n)$ satisfy the Lipschitz condition

$$|f_s(t, \bar{x}_1, \ldots, \bar{x}_n) - f_s(t, \bar{\bar{x}}_1, \ldots \bar{\bar{x}}_n)| \leq L \sum_{i=1}^{n} |\bar{x}_i - \bar{\bar{x}}_i|,$$

$$|\bar{x}_s^{(0)} - x_s^{(0)}| \leq b, \ |\bar{\bar{x}}_s - \bar{x}_s^0| \leq b, \ |t - t_0| \leq a, \ a \geq h_1, \ s = 1, \ldots, n,$$

then the following inequality holds

$$\|x(t) - y(t)\| \leq \delta_1 e^{nL|t-t_0|} + \frac{\delta_2}{nL}(e^{nL|t-t_0|} - 1), \ |t - t_0| \leq h_1. \tag{1.1.17}$$

Proof: Substituting the solutions $x = x(t)$ and $x = y(t)$ into equations (1.1.1) and (1.1.16), respectively, we obtain two identities:

$$\frac{dx(t)}{dt} = f(t, x(t)), \tag{1.1.18}$$

Basic Properties of Functions Defined by Ordinary Differential Equations

$$\frac{dy(t)}{dt} = f(t, y(t)) + r(t, y(t)). \qquad (1.1.19)$$

Introduce the vector-valued function $z(t) = x(t) - y(t)$. Subtracting termwise the identity (1.1.19) from the identity (1.1.18), we find

$$\frac{dz}{dt} = f(t, z(t) + y(t)) - f(t, y(t)) - r(t, y(t)). \qquad (1.1.20)$$

Scalar multiplying both sides of (1.1.20) by the vector $z(t)$ yields

$$\frac{1}{2}\frac{d}{dt}\|z\|^2 = [f(t, z+y) - f(t, y)]z - r(t, y)z. \qquad (1.1.21)$$

Evaluating the right-hand side of (1.1.21) with respect to the absolute value, we define

$$|(f(t, z+y) - f(t, y))z - rz| \leq |(f(t, z+y) - f(t, y))z|$$
$$+ |rz| \leq \|f(t, z+y) - f(t, y)\| \cdot \|z\| + \|z\| \cdot \|r\|. \qquad (1.1.22)$$

In order to obtain (1.1.22), we employ the Cauchy-Bunyakovsky inequality

$$\left|\sum_{j=1}^n x_j y_j\right| \leq \sqrt{\sum_{j=1}^n x_j^2} \sqrt{\sum_{j=1}^n y_j^2}.$$

Further, using the Lipschitz condition, from (1.1.22) we find

$$\frac{1}{2}\frac{d}{dt}\|z\|^2 \leq nL\|z\|^2 + \|r\| \cdot \|z\|.$$

From cancellation by $\|z\|$ we obtain

$$\frac{d}{dt}\|z\| \leq nL\|z\| + \|r\|. \qquad (1.1.23)$$

Multiplying both sides of inequality (1.1.23) by the factor $e^{-nL(t-t_0)}$ and integrating with respect to t between t_0 and t, we find

$$\|z\|e^{-nL(t-t_0)} - \|z^{(0)}\| \leq \int_{t_0}^t e^{-nL(\tau-t_0)}\|r\|d\tau. \qquad (1.1.24)$$

From (1.1.24) we obtain

$$\|z\| \leq \delta_1 e^{nL(t-t_0)} + \frac{\delta_2}{nL}(e^{nL(t-t_0)} - 1). \qquad (1.1.25)$$

In deriving (1.1.25), it was assumed that $t \geq t_0$.

Combining both cases $t \geq t_0$ and $t \leq t_0$ into one case, we find that for $|t-t_0| \leq h_1$ there exists the inequality

$$\|x(t) - y(t)\| \leq \delta_1 e^{nL|t-t_0|} + \frac{\delta_2}{nL}(e^{nL|t-t_0|} - 1). \qquad (1.1.26) \blacksquare$$

Corollary 1. If the conditions of Theorem 3 are satisfied, the solution of system (1.1.1) is a set of n functions dependent on the values of t, t_0, $x_1^0, \ldots, x_n^{(0)}$, each function being continuous in all these values.

Indeed, consider two solutions of system (1.1.1):

$$x^{(1)} = x(t, x_1^{(0)}, t_{01}), \quad x^{(2)} = x(t, x_2^{(0)}, t_{02}).$$

Substitute them into system (1.1.1) and integrate between various limits:

$$x(t_1, x_1^{(0)}, t_{01}) = x_1^{(0)} + \int_{t_{01}}^{t_1} f(\tau, x^{(1)}(\tau))d\tau, \tag{1.1.27}$$

$$x(t_2, x_2^{(0)}, t_{02}) = x_2^{(0)} + \int_{t_{02}}^{t_2} f(\tau, x^{(2)}(\tau))d\tau. \tag{1.1.28}$$

Denote one of the t_{0i} values by t_0, and one of the t_i values by t. Next, denote by $z(t)$ the vector-valued function

$$x(t, \bar{x}_1^{(0)}, t_0) - x(t, \bar{x}_2^{(0)}, t_0) = z(t). \tag{1.1.29}$$

Subtracting termwise the identity (1.1.28) from the identity (1.1.27), we obtain

$$x(t_1, x_1^{(0)}, t_{01}) - x(t_2, x_2^{(0)}, t_{02}) = x_1^{(0)} - x_2^{(0)} + \int_{t_0}^{t} f(\tau, x(\tau, \bar{x}_1^{(0)}, t_0))d\tau$$

$$- \int_{t_0}^{t} f(\tau, x(\tau, \bar{x}_2^{(0)}, t_0))d\tau + \int_{t_{01}}^{t_0} f(\tau, x^{(1)}(\tau))d\tau$$

$$+ \int_{t}^{t_1} f(\tau, x^{(1)}(\tau))d\tau - \int_{t_{02}}^{t_0} f(\tau, x^{(2)}(\tau))d\tau - \int_{t}^{t_2} f(\tau, x^{(2)}(\tau))d\tau, \tag{1.1.30}$$

$$\bar{x}_1^{(0)} = x_1(t_0, x_1^{(0)}, t_{01}), \quad \bar{x}_2^{(0)} = x_2(t_0, x_2^{(0)}, t_{02}). \tag{1.1.31}$$

From the equality (1.1.31) it follows, by the continuity of solutions to system (1.1.1) in t, that $\|\bar{x}_1^{(0)} - x_1^{(0)}\|$, $\|\bar{x}_2^{(0)} - x_2^{(0)}\|$ will be arbitrarily small if $|t_0 - t_{01}|$ and $|t_0 - t_{02}|$ are sufficiently small. Hence, if the value of $\|x_1^{(0)} - x_2^{(0)}\|$ is sufficiently small, then the value of $\|\bar{x}_2^{(0)} - \bar{x}_1^{(0)}\|$ will be arbitrarily small.

We now use Theorem 4 to evaluate the integral

$$\int_{t_0}^{t} (f(\tau, x(\tau, \bar{x}^{(0)}, t_0)) - f(\tau, x(\tau, \bar{x}_2^{(0)}, t_0)))d\tau.$$

Using inequality (1.1.26) and the Lipschitz condition, we obtain

$$\left\|\int_{t_0}^{t} (f(\tau, x(\tau, \bar{x}_1^{(0)}, t_0)) - f(\tau, x(\tau, \bar{x}_2^{(0)}, t_0)))d\tau\right\| \leq \left|\int_{t_0}^{t} nL\|z(\tau)\|d\tau\right|$$

$$\leq nL\delta_1 |t - t_0| e^{nLh_1}. \tag{1.1.32}$$

Relations (1.1.32) implies the inequality

$$\left\|x(t_1, x_1^{(0)}, t_{01}) - x(t_2, x_2^{(0)}, t_{02})\right\| \leq \left\|x_1^{(0)} - x_2^{(0)}\right\|$$

$$+nL\delta_1 \mid t-t_0 \mid e^{nLh_1} + \mid t_0 - t_{01} \mid M\sqrt{n} + \mid t_1 - t \mid M\sqrt{n}$$
$$+ \mid t_0 - t_{02} \mid M\sqrt{n} + \mid t_2 - t \mid M\sqrt{n}. \tag{1.1.33}$$

Since the right-hand side of (1.1.33) can be made arbitrarily small for sufficiently small $\mid t_1 - t_2 \mid$, $\mid t_{01} - t_{02} \mid$, $\|x_1^{(0)} - x_2^{(0)}\|$, the left-hand side of (1.1.33) will also be arbitrarily small. Note that the number M in (1.1.33) is determined by the equality

$$M = \max_{\substack{|t-t_0|\le a, \\ |x_s-x_s^{(0)}|\le b, \\ s=1,\ldots,n,}} |f_s(t,x_1,\ldots,x_n)|, \quad h_1 < h = \min\left(a, \frac{b}{M}\right). \quad \blacksquare$$

Corollary 2. Assume that the right-hand sides of system (1.1.1) depend on m parameters $\alpha_1, \ldots, \alpha_m$ so that

$$\frac{dx_s}{dt} = f_s(t, x_1, \ldots, x_n, \alpha_1, \ldots, \alpha_m), \quad s = 1, 2, \ldots, n. \tag{1.1.34}$$

Let the functions f_s satisfy the conditions: 1) the functions f_s given for $t \in (-\infty, +\infty)$, $x \in G$, $\alpha \in F$, $\alpha = (\alpha_1, \ldots, \alpha_m)$ are real and continuous therein. Here G and F are the open sets located respectively in the n-dimensional and m-dimensional spaces; 2) for every $t_0 \in (-\infty, +\infty)$, $x^{(0)} \in G$, $\alpha^{(0)} \in F$ we can select numbers a, b, c and L such that the n-dimensional cube $\mid x_s - x_s^{(0)} \mid \le b$, $s = 1, \ldots, n$, is contained in G, the m-dimensional cube $\mid \alpha_j - \alpha_j^{(0)} \mid \le c$, $j = 1, \ldots, m$, is contained in F and the Lipschitz condition holds for the variables $(x_1, \ldots, x_n)$:

$$\mid f_s(t, \bar{x}_1, \ldots, \bar{x}_n, \alpha_1, \ldots, \alpha_m) - f_s(t, \bar{\bar{x}}_1, \ldots, \bar{\bar{x}}_n, \alpha_1, \ldots, \alpha_m) \mid$$

$$\le L \sum_{s=1}^{n} \mid \bar{x}_s - \bar{\bar{x}}_s \mid \quad \text{with} \quad \mid t - t_0 \mid \le a, \; \mid \bar{x}_s - x_s^{(0)} \mid \le b,$$

$$\mid \bar{\bar{x}}_s - x_s^{(0)} \mid \le b, \; \mid \alpha_j - \alpha_j^{(0)} \mid \le c, \; s = 1, \ldots, n, \; j = 1, \ldots, m.$$

The solution of system (1.1.34) $x_s = x_s(t, t_0, x_1^{(0)}, \ldots, x_n^{(0)}, \alpha_1, \ldots, \alpha_m)$, is then defined for $\mid t - t_0 \mid \le h$, where $h = \min\left(a, \frac{b}{M}\right)$, and

$$M = \max_{\substack{|t-t_0|\le a, \\ |x_s-x_s^{(0)}|\le b, \\ |\alpha_j-\alpha_j^{(0)}|\le c, \\ s=1,\ldots,n, \; j=1,\ldots,n}} \mid f_s(t, x_1, \ldots, x_n, \alpha_1, \ldots, \alpha_m) \mid,$$

satisfies the condition $x_s = x_s^{(0)}$ for $t = t_0$ and is a continuous function of the parameters $\alpha_1, \ldots, \alpha_m$ in the closed domain

$$\mid \alpha_j - \alpha_j^{(0)} \mid \le c, \; j = 1, \ldots, m. \tag{1.1.35}$$

Indeed, the solution of system (1.1.34) with the initial condition $x_s = x_s^{(0)}$ at $t = t_0$ for any parameters $\alpha_1, \ldots, \alpha_m$ will be defined with $\mid t - t_0 \mid \le h$ since Theorem 2 is applicable in this case to all the parameter values from the closed domain (1.1.35).

Suppose $x(t)$ is a solution to system (1.1.34) with the initial condition $x = x^{(0)}$ at $t = t_0$ corresponding to the parameter values $\bar{\alpha}_1, \ldots, \bar{\alpha}_m$, and $y(t)$ is a solution to the same system with the same initial conditions corresponding to the other values of parameters $\bar{\bar{\alpha}}_1, \ldots, \bar{\bar{\alpha}}_m$. We say that $\bar{\alpha}$ and $\bar{\bar{\alpha}}$ are the points of the set (1.1.35). Denote by $z(t)$ the vector $z(t) = x(t) - y(t)$. Then it will satisfy the identity

$$\frac{dz}{dt} = f(t, z+y, \bar{\alpha}) - f(t, y, \bar{\alpha}) + R, \qquad (1.1.36)$$

where $R = f(t, y, \bar{\alpha}) - f(t, y, \bar{\bar{\alpha}})$. Also, consider the system of equations

$$\frac{dz}{dt} = f(t, z+y, \bar{\alpha}) - f(t, y, \bar{\alpha}). \qquad (1.1.37)$$

System (1.1.37) has the solution $z \equiv 0$ with the initial condition $z = 0$ at $t = t_0$. Applying Theorem 4 to systems (1.1.36) and (1.1.37), we find that

$$\|z\| \leq \frac{\delta_2}{nL}(e^{nL|t-t_0|} - 1), \ |t-t_0| \leq h, \ \delta_2 = \max_{|t-t_0|\leq h} \|R\|.$$

Because of the uniform continuity of the function $f(t, x, \alpha)$ in α in the closed domain (1.1.35) relative to x and t satisfying the inequalities $|t-t_0| \leq a, |x_s - x_s^{(0)}| \leq b, s = 1, \ldots, n$, δ_2 will be as small as desired if $\|\bar{\alpha} - \bar{\bar{\alpha}}\|$ is sufficiently small. ∎

As in Corollary 1, it can be shown that the above solution to system (1.1.34) will be a continuous function in all its arguments.

To be noted is that from fulfilment of the Lipschitz conditions with respect to the variables $x_1, \ldots, x_n$ in the neighborhood of each point $t_0 \in (-\infty, +\infty)$, $x^{(0)} \in G$, it follows that a solution to system (1.1.34) will be the function which is given for $t_0 \in (-\infty, +\infty), \alpha \in F, |t-t_0| \leq h, h = h(t_0, x_0, \alpha)$ and is continuous in all its arguments.

5. Consider the problem of differentiability solution for the system (1.1.34) with respect to parameters and initial conditions.

Theorem 5 *If the right-hand sides of system (1.1.34) satisfy the conditions: 1) the functions f_s given for $t \in (-\infty, +\infty), x \in G, \alpha \in F, \alpha = (\alpha_1, \ldots, \alpha_m)$, are real and continuous there,*

2) for each $t_0 \in (-\infty, +\infty), x^{(0)} \in G, \alpha^{(0)} \in F$ it is possible to select numbers a, b, c, L such that the closed domain

$$|x_s - x_s^{(0)}| \leq b, \ s = 1, \ldots, n, \qquad (1.1.38)$$

is contained in G, the closed domain

$$|\alpha_j - \alpha_j^{(0)}| \leq c, \ j = 1, \ldots, m, \qquad (1.1.39)$$

is contained in F, and the functions f_s are p times continuously differentiable with respect to $x_1, \ldots, x_n, \alpha_1, \ldots, \alpha_m$ in (1.1.38) – (1.1.39), then the system (1.1.34) has the solution

$$x_s = x_s(t, t_0, x^{(0)}, \alpha), \qquad (1.1.40)$$

Basic Properties of Functions Defined by Ordinary Differential Equations

which is defined for $|t-t_0| \leq h$ and is p times continuously differentiable with respect to $\alpha_1, \ldots, \alpha_m$ in the closed domain (1.1.39).

Proof: Denote the solution (1.1.40) computed with the parameter values $\alpha = (\alpha_1, \ldots, \alpha_m)$ by $x(t)$, and the solution (1.1.40) corresponding to the parameter values $\alpha_\delta = (\alpha_1 + \delta, \alpha_2, \ldots, \alpha_m)$ by $y(t)$. Let $z(t) = \frac{y(t) - x(t)}{\delta}$. Henceforth it will be assumed that the points α and α_δ are contained in (1.1.39), α_δ being considered with sufficiently small δ (δ can assume either the values of an arbitrary sign, if α is an interior point, or those of the same sign, if α is a boundary point). The vector z satisfies the identity

$$\frac{dz}{dt} = \frac{1}{\delta}(f(t, z(t)\delta + x(t), \alpha_\delta) - f(t, x(t), \alpha)). \tag{1.1.41}$$

Relationship (1.1.41) will be regarded as a system of differential equations defining the z function with the initial condition $z = 0$ for $t = 0$. The right-hand sides of this system depend continuously on the unique parameter δ, in particular for $\delta = 0$, therefore from Corollary 2 and Theorem 4, $z(t)$ will also be a continuous function of this parameter for $\delta = 0$. Thus it is established that there exists $\lim\limits_{\delta \to 0} z(t)$.

Because of the available uniqueness, $z(t)$ necessarily satisfies the identity

$$z(t) = \frac{y(t) - x(t)}{\delta}$$

$$= \frac{x(t, t_0, x^{(0)}, \alpha_1 + \delta, \alpha_2, \ldots, \alpha_m) - x(t, t_0, x^{(0)}, \alpha_1, \ldots, \alpha_m)}{\delta}.$$

This means that the solution to system (1.1.34) is continuously differentiable with respect to α_1.

Similarly, continuous differentiability is established up to the order p with respect to all parameters.

To complete the proof, it remains to show that the right-hand side of system (1.1.41) is continuous for $\delta = 0$. The right-hand side of system (1.1.41) can be represented as

$$\frac{f(t, z\delta + x, \alpha_\delta) - f(t, x, \alpha)}{\delta} = \frac{1}{\delta}\int_0^\delta \frac{\partial f(t, z\tau + x(t), \alpha_\tau)}{\partial \tau} d\tau. \tag{1.1.42}$$

From the form of the right-hand side of (1.1.42) it follows that it is continuous in δ for $\delta \neq 0$ and has the limit as $\delta \to 0$. ∎

Corollary. From Theorem 5 it follows that the solution of (1.1.41) is also continuously differentiable up to the order p with respect to the initial conditions $x_1^{(0)}, \ldots, x_n^{(0)}$.

This statement becomes evident if in the system (1.1.34) the required functions are replaced by the formula $x = \bar{x} + x^{(0)}$, where $\bar{x}$ is the new required function with the initial condition $\bar{x} = 0$ for $t = t_0$. Then $\bar{x}$ will satisfy the system

$$\frac{d\bar{x}}{dt} = f(t, \bar{x} + x^{(0)}, \alpha). \tag{1.1.43}$$

That is, the variables $x_1^{(0)}, \ldots, x_n^{(0)}$ have become parameters. ∎

1.2 Properties of Solutions to Linear Systems

1. Consider the system
$$\frac{dx}{dt} = \mathbf{A}x, \tag{1.2.1}$$
where $\{\mathbf{A}\}_{ik} = a_{ik}$ are constant real numbers ($i, k = 1, \ldots, n$). The fundamental matrix for system (1.2.1) is given by
$$\mathbf{X}(t) = e^{\mathbf{A}t} \tag{1.2.2}$$
and the solution of system (1.2.1) satisfying the condition $x(t_0) = x_0$ is
$$x(t) = e^{\mathbf{A}(t-t_0)} x_0. \tag{1.2.3}$$
Indeed, differentiating (1.2.2), we have
$$\frac{d}{dt} e^{\mathbf{A}t} = e^{\mathbf{A}t} \mathbf{A} = \mathbf{A} e^{\mathbf{A}t},$$
and $\mathbf{X}(0) = \mathbf{E}$.

All the fundamental matrices are contained in the formula
$$\mathbf{X}_1(t) = \mathbf{X}(t) \mathbf{C} = e^{\mathbf{A}t} \mathbf{C}, \tag{1.2.4}$$
where $\mathbf{C}$ is an arbitrary constant nonsingular matrix. Substituting $t = t_0$ in (1.2.4), we obtain
$$\mathbf{C} = e^{-\mathbf{A}t_0} \mathbf{X}_1(t_0) = e^{-\mathbf{A}t_0} \mathbf{X}_0, \text{ i.e. } \mathbf{X}_1(t) = e^{\mathbf{A}(t-t_0)} \mathbf{X}_0.$$

The general solution of the nonhomogeneous system
$$\frac{dy}{dt} = \mathbf{A}y + f(t) \tag{1.2.5}$$
can be obtained from the formula
$$y(t) = e^{\mathbf{A}t} \left(\int_{t_0}^{t} e^{-\mathbf{A}\tau} f(\tau) d\tau + c \right). \tag{1.2.6}$$

The solution of equation (1.2.5) satisfying the condition $y(t_0) = y_0$ is obtained from (1.2.6) for $c = e^{-\mathbf{A}t_0} y_0$, i.e.
$$y(t) = e^{\mathbf{A}(t-t_0)} y_0 + \int_{t_0}^{t} e^{\mathbf{A}(t-\tau)} f(\tau) d\tau. \tag{1.2.7}$$

Consider now the structure of the fundamental matrix $e^{\mathbf{A}t}$. Let $\mathbf{S}$ be a constant nonsingular matrix such that $\mathbf{S}^{-1} \mathbf{A} \mathbf{S} = \Xi$ has the canonical form
$$\Xi = \begin{pmatrix} \Xi_0 & O & O & \cdots & O \\ O & \Xi_1 & O & \cdots & O \\ \cdots & \cdots & \cdots & \cdots & \cdots \\ O & O & O & \cdots & \Xi_s \end{pmatrix},$$

Properties of Solutions to Linear Systems

where Ξ_0 is a diagonal matrix with elements $\lambda_1, \lambda_2, \ldots, \lambda_q$, and p_i is the order of the matrix

$$\Xi_i = \begin{pmatrix} \lambda_{q+i} & 1 & 0 & \cdots & 0 \\ 0 & \lambda_{q+i} & 1 & \cdots & 0 \\ \cdots & \cdots & \cdots & \cdots & \cdots \\ 0 & 0 & 0 & \cdots & \lambda_{q+i} \end{pmatrix}.$$

Here λ_j, $j = 1, \ldots, q+s$ are characteristic numbers or eigenvalues of the matrix **A**, and are not necessarily different. If λ_j is a prime eigenvalue, then it is a member of Ξ_0. Hence, if all the eigenvalues are different, **A** is similar to the diagonal matrix

$$\Xi = \begin{pmatrix} \lambda_1 & 0 & \cdots & 0 \\ 0 & \lambda_2 & \cdots & 0 \\ \cdots & \cdots & \cdots & \cdots \\ 0 & 0 & \cdots & \lambda_n \end{pmatrix}.$$

The elementary divisors of the matrix **A**

$$\lambda - \lambda_1, \lambda - \lambda_2, \ldots, \lambda - \lambda_q, (\lambda - \lambda_{q+1})^{\rho_1}, \ldots, (\lambda - \lambda_{q+s})^{\rho_s}$$

and $q + \sum_{i=1}^{s} \rho_i = n$, then

$$e^{\mathbf{A}t} = e^{\mathbf{S}\Xi\mathbf{S}^{-1}t} = \mathbf{S}e^{\Xi t}\mathbf{S}^{-1}.$$

It is well known that

$$e^{\Xi t} = \begin{pmatrix} e^{\Xi_0 t} & \mathbf{O} & \mathbf{O} & \cdots & \mathbf{O} \\ \mathbf{O} & e^{\Xi_1 t} & \mathbf{O} & \cdots & \mathbf{O} \\ \cdots & \cdots & \cdots & \cdots & \cdots \\ \mathbf{O} & \mathbf{O} & \mathbf{O} & \cdots & e^{\Xi_s t} \end{pmatrix},$$

where

$$e^{\Xi_0 t} = \begin{pmatrix} e^{t\lambda_1} & 0 & \cdots & 0 \\ 0 & e^{t\lambda_2} & \cdots & 0 \\ 0 & 0 & \cdots & e^{t\lambda_q} \end{pmatrix},$$

$$e^{\Xi_i t} = e^{t\lambda_{q+i}} \begin{pmatrix} 1 & t & \frac{t^2}{2!} & \cdots & \frac{t^{\rho_i - 1}}{(\rho_i - 1)!} \\ 0 & 1 & t & \cdots & \frac{t^{\rho_i - 2}}{(\rho_i - 2)!} \\ \cdots & \cdots & \cdots & \cdots & \cdots \\ 0 & 0 & 0 & \cdots & 1 \end{pmatrix}.$$

Note that the matrix

$$\mathbf{Y}(t) = e^{\mathbf{A}t}\mathbf{S} = \mathbf{S}e^{\Xi t} \tag{1.2.8}$$

will also be fundamental.

2. Suppose **S** has vectors $S_1, \ldots, S_n$ as its columns. Denote the matrix columns $\mathbf{Y}(t)$ by $y_1(t), \ldots, y_n(t)$. They form a set of n linearly independent solutions to system (1.2.1). From equality (1.2.8) we get:

$$y_1(t) = e^{t\lambda_1} S_1, \; y_2(t) = e^{t\lambda_2} S_2, \; \ldots, \; y_q(t) = e^{t\lambda_q} S_q,$$

$$y_{q+1}(t) = e^{t\lambda_{q+1}} S_{q+1},$$
$$y_{q+2}(t) = e^{t\lambda_{q+1}} (tS_{q+1} + S_{q+2}),$$
$$\ldots$$
$$y_{q+\rho_1}(t) = e^{t\lambda_{q+1}} \left(\frac{t^{\rho_1-1}}{(\rho_1-1)!} S_{q+1} + \ldots + S_{q+\rho_1} \right),$$
$$\ldots$$
$$y_{n-\rho_s+1}(t) = e^{t\lambda_{q+s}} S_{n-\rho_s+1},$$
$$\ldots$$
$$y_n(t) = e^{t\lambda_{q+s}} \left(\frac{t^{\rho_s-1}}{(\rho_s-1)!} \right) S_{n-\rho_s+1} + \ldots + tS_{n-1} + S_n.$$

To summarize the above discussion, we can state that the linear system with constant coefficients (1.2.1) has n linearly independent solutions $y_1(t), \ldots, y_n(t)$ of the form

$$y_j(t) = p_j(t)e^{\lambda_j t}, \qquad (1.2.9)$$

where $p_j(t)$ represents polynomials in t, whose degree is less than the multiplicity of the root of λ_j, the polynomial coefficients $p_j(t)$ being constant vectors.

From (1.2.9) it follows that the zero solution to system (1.2.1) is asymptotically Lyapunov stable if and only if the conditions

$$\text{Re}(\lambda_j) < 0, \ j = 1, \ldots, n, \qquad (1.2.10)$$

are satisfied. In other words, condition (1.2.10) is necessary and sufficient for the natural oscillations of system (1.2.1) to be damped.

If among the eigenvalues $\lambda_1, \ldots, \lambda_n$ there are only k values satisfying the condition

$$\text{Re}(\lambda_j) < 0, \ j = 1, \ldots, k, \qquad (1.2.11)$$

then it is stated that there exists the k-dimensional subspace whose special feature is that all the natural motions of system (1.2.1) originating therein tend to a zero solution as $t \to +\infty$. In fact, any solution

$$x = \sum_{j=1}^{k} p_j(t)e^{\lambda_j t} c_j, \qquad (1.2.12)$$

where $c_1, \ldots, c_k$ are arbitrary constants, has the property $x \to 0$ as $t \to +\infty$. This implies the above statement.

If among the eigenvalues λ_j there are numbers with zero real parts, then two cases are possible here. In the first case, where there are no purely imaginary eigenvalues, the system (1.2.1) has the whole family of equilibria which fills the l-dimensional subspace, where l is the number of all zero roots, considering their multiplicities only when prime elementary divisors correspond to these zero roots. If (1.2.11) holds, then each of the above equilibria determines the k-dimensional subspace filled with the motions of system (1.2.1) asymptotically approaching it.

Properties of Solutions to Linear Systems

If there are pure imaginary eigenvalues, then among the natural motions of system (1.2.1) there exist periodic and even almost periodic motions. The latter appears where there are at least two purely imaginary eigenvalues λ_s and λ_k such that the real numbers $i\lambda_s$ and $i\lambda_k$ are incommensurable. Periodic and almost periodic motions fill the s-dimensional linear subspace where s is the number of purely imaginary roots, their multiplicities being considered only when prime elementary divisors correspond to them. If the condition (1.2.11) is satisfied, each periodic and almost periodic mode of system (1.2.1) is associated with the k-dimensional linear subspace completely filled with the motions asymptotically approaching the natural continuous oscillations of system (1.2.1).

Such is the general outline of development of oscillations in system (1.2.1). In order to prove the above statements, it suffices to consider the structure of solutions to system (1.2.1). The phase space related to the original system (1.2.1) can be represented as a homeomorphic image of the Cartesian product of phase spaces in several systems of differential equations obtained from (1.2.1) by nonsingular linear transformation in order to put the matrix $\mathbf{A}$ into canonical Jordan form (see above).

3. Let us now focus on the oscillations of a linear system induced by some applied force. To this end, consider a linear nonhomogeneous system of the form

$$\frac{dx}{dt} = \mathbf{A}x + f(t), \qquad (1.2.13)$$

where the elements of the one-column matrix $f(t)$ are continuous t-periodic functions with the same period ω, or almost periodic functions of t. We get:

$$x(t) = e^{\mathbf{A}t}x_0 + \int_0^t e^{\mathbf{A}(t-\tau)} f(\tau) d\tau. \qquad (1.2.14)$$

It will be shown below that (1.2.14) implies the basic property of system (1.2.13) which is as follows. If the natural motions of system (1.2.13) do not involve stationary oscillations - periodic motions and equilibria other than $x = 0$ in system (1.2.1), then in system (1.2.13) there arise the forced motions, periodic or almost periodic, to suit the type of a disturbing influence, be it periodic or almost periodic. In this case the periodic or almost periodic motion is unique.

In fact, the system does not involve stationary natural oscillations if and only if all the eigenvalues $\lambda_j, j = 1, \ldots, n$, have nonzero real part. With this in mind, to put the matrix $\mathbf{A}$ into canonical Jordan form, we have to carry out in system (1.2.13) the nonsingular linear transformation of the required functions

$$x = \mathbf{S}y, \qquad (1.2.15)$$

where y is the new desired vector function, and $\mathbf{S}$ is a constant nonsingular matrix; we obtain

$$\frac{dy}{dt} = \mathbf{S}^{-1}\mathbf{A}\mathbf{S}y + \mathbf{S}^{-1}f(t). \qquad (1.2.16)$$

System (1.2.16) can be represented as a collection of more than two unrelated systems of the form

$$\frac{dy_1}{dt} = \lambda_1 y_1 + f_1(t),$$

$$\frac{dy_2}{dt} = y_1 + \lambda_1 y_2 + f_2(t), \tag{1.2.17}$$

$$\ldots$$

$$\frac{dy_k}{dt} = y_{k-1} + \lambda_1 y_k + f_k(t)$$

and q first order equations of the form

$$\frac{dy_j}{dt} = \lambda_j y_j + f_j(t), \tag{1.2.18}$$

where $f_1, \ldots, f_k, f_j$ are components of the vector $\mathbf{S}^{-1} f(t)$.

For definiteness, let $\mathrm{Re}(\lambda_j) < 0$ in (1.2.18); then the function

$$\varphi_j(t) = e^{\lambda_j t} \int_{-\infty}^{t} e^{-\lambda_j \tau} f_j(\tau) d\tau \tag{1.2.19}$$

will be a unique periodic or almost periodic solution to (1.2.18). It will be shown that the function (1.2.19) satisfies equation (1.2.18), which is established by direct differentiation:

$$\frac{d\varphi_j(t)}{dt} = \lambda_j e^{\lambda_j t} \int_{-\infty}^{t} e^{-\lambda_j \tau} f_j(\tau) d\tau + e^{\lambda_j t} e^{-\lambda_j t} f_j(t) = \lambda_j \varphi_j + f_j.$$

We shall now find a forced oscillation in system (1.2.17). For definiteness, let $\mathrm{Re}(\lambda_1) < 0$. Then, by the proved statement, the function

$$\varphi_1(t) = e^{\lambda_1 t} \int_{-\infty}^{t} e^{-\lambda_1 \tau} f_1(\tau) d\tau$$

will be a solution to the first equation of system (1.2.17) which has the same property in the sense of periodicity and almost periodicity as the function $f_1(t)$. The functions $\varphi_2(t), \ldots, \varphi_k(t)$ are defined by

$$\varphi_2(t) = e^{\lambda_1 t} \int_{-\infty}^{t} e^{-\lambda_1 \tau} (f_2(\tau) + \varphi_1(\tau)) d\tau,$$

$$\ldots \tag{1.2.20}$$

$$\varphi_k(t) = e^{\lambda_1 t} \int_{-\infty}^{t} e^{-\lambda_1 \tau} (f_k(\tau) + \varphi_{k-1}(\tau)) d\tau.$$

From (1.2.19), (1.2.20) it follows that the system (1.2.13) has a periodic or almost periodic solution $x(t) = \varphi(t)$ depending on whether the function $f(t)$ is periodic or almost periodic. This solution is obtained by making up the vector of the functions $\varphi_1, \ldots, \varphi_k, \varphi_j$ and in addition multiplying by the matrix $\mathbf{S}$ on the left. Replacing $x = z + \varphi$, we get for the required function z a homogeneous system of the form (1.2.1). Since the system obtained after replacement has no natural stationary oscillations, the system (1.2.1) has neither periodic modes nor equilibria, as distinct from $x = 0$. So, the resulting stationary forced oscillation of system (1.2.13), $x = \varphi(t)$, is a unique

Properties of Solutions to Linear Systems

oscillation for a system of this type. This case is called non-resonance, and the system (1.2.1) is referred to as non-resonance.

4. It now remains to discuss what is called a resonance case, where among the proper motions of the system are found stationary oscillations, i.e., where among the eigenvalues $\lambda_1, \ldots, \lambda_n$ there are numbers with zero real parts. A special feature of this case is that all of the external disturbing influences fall into two classes. One of these comprises the influences initiating in the system stationary forced oscillations, and the other the oscillations producing in the system a resonance ("nonstationary" forced oscillations containing secular components, or oscillations with the unbounded amplitude of a more complicated nature). Analytical description of the propositions will be given below.

We first assume that the external disturbing influence $f(t)$ is a continuous ω-periodic vector-valued function of t.

From (1.2.14) it follows that a necessary and sufficient condition for the availability of an ω-periodic solution to system (1.2.13) is provided by satisfying the equality $x(\omega) = x(0)$. In other words, for a vector x_0 there must be

$$x_0 = e^{\mathbf{A}\omega} x_0 + \int_0^\omega e^{\mathbf{A}(\omega - \tau)} f(\tau) d\tau. \tag{1.2.21}$$

From this it follows that the initial condition x_0 determining the ω-periodic solution satisfies a linear system of the form

$$\left(\mathbf{E} - e^{\mathbf{A}}\omega\right) x_0 = \int_0^\omega e^{\mathbf{A}(\omega - \tau)} f(\tau) d\tau. \tag{1.2.22}$$

System (1.2.22) has a unique solution for any choice of vector $f(t)$ if and only if

$$\det \left(\mathbf{E} - e^{\mathbf{A}\omega}\right) \neq 0. \tag{1.2.23}$$

This determinant is nonzero if and only if among the eigenvalues there are no values of the form $\frac{i2k\pi}{\omega}$, $k = 0, \pm 1, \pm 2, \ldots$, i.e., among natural oscillations there are no oscillations whose frequencies coincide with the frequency of one of the harmonics appearing in a Fourier series of the vector function $f(t)$.

If it is considered that the system (1.2.1) experiences only the external disturbing periodic influence $f(t)$, then the resonance-type case is the one where eigenvalues include magnitudes of the form $\frac{i2k\pi}{\omega}$, where k is an integer. A special feature of the non-resonance-type case is that the eigenvalues do not include magnitudes of such a form. In this case, there exists a unique ω-periodic solution to system (1.2.13) given by a formula

$$x(t) = e^{\mathbf{A}t} \left(\mathbf{E} - e^{\mathbf{A}\omega}\right)^{-1} \int_0^\omega e^{\mathbf{A}(\omega - \tau)} f(\tau) d\tau + \int_0^t e^{\mathbf{A}(t - \tau)} f(\tau) d\tau. \tag{1.2.24}$$

5. Consider the case where the determinant (1.2.23) goes to zero; that is, the eigenvalues $\lambda_1, \ldots, \lambda_n$ involve numbers of the form $\frac{i2k\pi}{\omega}$, $k = 0, \pm 1, \ldots$. Derive the conditions for the vector function $f(t)$ which are necessary and sufficient to give rise

to ω-periodic oscillations in system (1.2.13) under the external disturbing force $f(t)$. To this end, multiply both sides of equation (1.2.22) by the matrix $e^{-\mathbf{A}\omega}$ and obtain

$$\left(e^{-\mathbf{A}\omega} - \mathbf{E}\right)x_0 = \int_0^\omega e^{-\mathbf{A}\tau}f(\tau)d\tau. \tag{1.2.25}$$

From algebra it is known that for the system (1.2.25) to have a solution, it is necessary and sufficient that its right-hand side be orthogonal to all the solutions of the adjoint homogeneous system

$$\left(e^{-\mathbf{A}\omega} - \mathbf{E}\right)^* y_0 = 0. \tag{1.2.26}$$

Thus, for the external disturbing influence to induce the forced ω-periodic oscillations in system (1.2.13), it is necessary and sufficient that the condition

$$\int_0^\omega y_0^* e^{-\mathbf{A}\tau} f(\tau) d\tau = 0 \tag{1.2.27}$$

hold for any solution y_0 of system (1.2.26). Since for all such y_0 the vector function

$$y(t) = e^{-\mathbf{A}^* t} y_0$$

is an ω-periodic solution of the adjoint system of differential equations

$$\frac{dy}{dt} = -\mathbf{A}^* y, \tag{1.2.28}$$

the condition (1.2.27) becomes

$$\int_0^\omega y^*(\tau) f(\tau) d\tau = 0, \tag{1.2.29}$$

where $y(t)$ runs through the entire set of ω-periodic solutions to system (1.2.28). Any solution y_0 of the system

$$\left(-\mathbf{A}^* + \mathbf{E}\frac{i2k\pi}{\omega}\right) y_0 = 0 \tag{1.2.30}$$

is a solution to the adjoint system (1.2.26) because the equation (1.2.26) is believed to be derived from the ω-periodicity condition for solutions to the adjoint system of differential equations (1.2.28), and the system (1.2.30) can be regarded as a result of defining an ω-periodic solution of system (1.2.28) to be

$$y = e^{-\frac{i2k\pi t}{\omega}} y_0.$$

Next, from equation (1.2.30) it follows that

$$(-\mathbf{A}^*)^n y_0 = \left(-\frac{i2k\pi}{\omega}\right)^n y_0,$$

whence follows the relation

$$e^{-\tau \mathbf{A}^*} y_0 = e^{\frac{-i2k\pi\tau}{\omega}} y_0,$$

Properties of Solutions to Linear Systems

by which (1.2.27) becomes

$$\int_0^\omega y^*(\tau)f(\tau)d\tau = 0$$

for all $y(t) = e^{-\frac{i2k\pi t}{\omega}} y_0$ that are solutions to the adjoint system (1.2.28). Thus, if the condition (1.2.29) holds, in system (1.2.13) we observe the entire family of ω-periodic oscillations, which is a superposition of the oscillation induced by the external force $f(t)$, and the family of ω-periodic natural oscillations which depends on m arbitrary constants, where m is the number of eigenvalues of the form $\frac{i2k\pi}{\omega}$ (the multiplicities of eigenvalues are considered if the prime elementary divisors correspond to them).

6. It remains to discuss the issue of the forced oscillations under resonance, induced in system (1.2.13) by an almost periodic external force $f(t)$. In this case, we may also select the conditions for the vector function $f(t)$ that are necessary and sufficient to set up almost periodic forced oscillations in system (1.2.13). However, we will discuss here a system of the form (1.2.17) and (1.2.18) without writing down these conditions direct for the vector function $f(t)$ appearing in system (1.2.13). Suppose $\mathrm{Re}(\lambda_j) = 0$ in equation (1.2.18). In order to provide an almost periodic solution for equation (1.2.18), it is necessary and sufficient that the following conditions be satisfied:

$$M[e^{-\lambda_j t}f_j(t)] = \lim_{T \to \infty} \frac{1}{2T}\int_{-T}^T e^{-\lambda_j \tau}f_j(\tau)d\tau = 0, \quad (1.2.31)$$

$$\left|\int_0^T e^{-\lambda_j \tau}f_j(\tau)d\tau\right| < M' < +\infty. \quad (1.2.32)$$

Conditions (1.2.31) and (1.2.32) are satisfied, say, if Fourier indices of the function f_j are not in a small, but fixed neighborhood of the point $i\lambda_j$. Suppose the conditions (1.2.31) and (1.2.32) hold for the function f_1 and the number λ_1, and $\mathrm{Re}(\lambda_1) = 0$. Then the first equation of system (1.2.17) has the solution

$$y_1 = c_1 e^{\lambda_1 t} + e^{\lambda_1 t}\int_0^t e^{-\lambda_1 \tau}f_1(\tau)d\tau, \quad (1.2.33)$$

which is almost periodic for any choice of c_1. Substitute this solution in the second equation of (1.2.17) and choose a number c_1 so that the mean value of the function $M[e^{-\lambda_1 t}y_1(t) + e^{-\lambda_1 t}f_2(t)]$ is zero. If the condition similar to (1.2.32)

$$\left|\int_0^T e^{-\lambda_1 \tau}f_2(\tau)d\tau + \int_0^T \left(c_1 + \int_0^\tau e^{-\lambda_1 \theta}f_1(\theta)d\theta\right)d\tau\right| < M' < \infty$$

is satisfied, then we can claim that the second equation of system (1.2.17) under such a choice of the function y_1 has a family of almost periodic solutions, etc. All of these conditions give the necessary and sufficient conditions for the vector function $f(t)$ under which the system (1.2.13) has an almost periodic forced oscillation. This forced oscillation, combined with the natural oscillations of the system forms the family of almost periodic oscillations of the system (1.2.13) determined by m arbitrary constants, where m is the number of all eigenvalues with zero real parts (the multiplicities of eigenvalues are considered only if the prime elementary divisors correspond to them).

An external almost periodic perturbation may be of the form

$$f(t) = \sum_{j=1}^{N} c_j(t) e^{i\mu_j(t)}, \qquad (1.2.34)$$

where $c_j(t)$ are continuous periodic vector functions with the same period ω. In the case of (1.2.34), where there are no eigenvalues of the form

$$i(\frac{2k\pi}{\omega} + \mu j), \ j = 1, \ldots, N, \ k = 0, \pm 1, \ldots, \qquad (1.2.35)$$

almost periodic oscillations of the same form as the function in (1.2.34) will arise in system (1.2.13). If among $\lambda_1, \ldots, \lambda_n$ there are eigenvalues as in (1.2.35), then the necessary and sufficient conditions for the function (1.2.34), whose fulfilment gives rise to a forced oscillation in system (1.2.13), has the same form (1.2.29) if $y(t)$ represent periodic solutions to the adjoint system (1.2.28) corresponding to eigenvalues as in (1.2.35).

Similarly, we can study the case where a free member of $f(t)$ in system (1.2.13) is a finite linear combination with constant coefficients of the functions as in (1.2.34) corresponding to various periods ω.

In the above cases of oscillation development in system (1.2.13), where r eigenvalues $\lambda_j, j = 1, \ldots, r$, with negative real parts are given, there is a family of integral curves dependent on r arbitrary constants which are coming arbitrarily close to the original stationary mode.

7. We shall now consider a linear system with periodic coefficients

$$\frac{dx_s}{dt} = p_{s1}x_1 + \ldots + p_{sn}x_n, \ s = 1, \ldots, n, \qquad (1.2.36)$$

or, in matrix form,

$$\frac{dx}{dt} = \mathbf{P}(t)x, \qquad (1.2.37)$$

where x is a one-column matrix, and the matrix elements $\{\mathbf{P}(t)\}_{ik} = p_{ik}(t)$ are continuous ω-periodic functions of t. Let $\mathbf{X}(t)$ be a fundamental matrix satisfying the condition $\mathbf{X}(0)=\mathbf{E}$. Then any other fundamental matrix $\mathbf{X}_1(t)$ satisfying the system (1.2.37) can be represented as

$$\mathbf{X}_1(t) = \mathbf{X}(t)\mathbf{C}, \qquad (1.2.38)$$

where $\mathbf{C}$ is a nonsingular constant matrix.

Show that $\mathbf{X}(t + \omega)$ is a solution to system (1.2.37). Replace t in (1.2.37) by $(t + \omega)$:

$$\frac{d}{dt}\mathbf{X}(t+\omega) = \mathbf{P}(t+\omega)\mathbf{X}(t+\omega) = \mathbf{P}(t)\mathbf{X}(t+\omega).$$

Properties of Solutions to Linear Systems 21

By (1.2.38), we have $\mathbf{X}(t+\omega) = \mathbf{X}(t)\mathbf{C}$. Setting $t = 0$, we obtain $\mathbf{C} = \mathbf{X}(\omega)$, i.e. $\mathbf{X}(t+\omega) = \mathbf{X}(t)\mathbf{X}(\omega)$.

The equation
$$|\mathbf{X}(\omega) - \rho \mathbf{E}| = 0 \tag{1.2.39}$$
is called the characteristic equation of system (1.2.37). It is central to the theory of linear equations with periodic coefficients and has the following properties.

1^0. The characteristic equation is independent of the chosen fundamental system. Indeed, instead of the fundamental system $\mathbf{X}$, we choose another system, $\mathbf{Y}$. Then $\mathbf{Y}(t+\omega) = \mathbf{Y}(t)\mathbf{C}_1$. It will be shown that the roots of the characteristic equation $|\mathbf{C}_1 - \rho \mathbf{E}| = 0$ coincide with those of equation (1.2.39). By (1.2.38), we have
$$\mathbf{Y}(t) = \mathbf{X}(t)\mathbf{C}_2 \text{ and } \mathbf{X}(t) = \mathbf{Y}(t)\mathbf{C}_2^{-1},$$
i.e.,
$$\mathbf{Y}(t+\omega) = \mathbf{X}(t+\omega)\mathbf{C}_2 = \mathbf{X}(t)\mathbf{X}(\omega)\mathbf{C}_2 = \mathbf{Y}(t)\mathbf{C}_2^{-1}\mathbf{X}(\omega)\mathbf{C}_2.$$
Hence,
$$\mathbf{C}_1 = \mathbf{C}_2^{-1}\mathbf{X}(\omega)\mathbf{C}_2,$$
$$|\mathbf{C}_1 - \rho \mathbf{E}| = \left|\mathbf{C}_2^{-1}\mathbf{X}(\omega)\mathbf{C}_2 - \rho \mathbf{E}\right| = \left|\mathbf{C}_2^{-1}(\mathbf{X}(\omega) - \rho \mathbf{E})\mathbf{C}_2\right|$$
$$= \left|\mathbf{C}_2^{-1}\right| |\mathbf{X}(\omega) - \rho \mathbf{E}| |\mathbf{C}_2| = |\mathbf{X}(\omega) - \rho \mathbf{E}|.$$

2^0. The characteristic equation will not change if system (1.2.37) is subject to nonsingular linear transformation with periodic coefficients of period ω. Indeed, let us transform (1.2.37) by substituting $\mathbf{Y}(t) = \mathbf{Q}(t)\mathbf{X}(t)$, i.e. $\mathbf{X}(t) = \mathbf{Q}^{-1}(t)\mathbf{Y}(t)$, hence
$$\mathbf{Y}(t+\omega) = \mathbf{Q}(t+\omega)\mathbf{X}(t+\omega) = \mathbf{Q}(t)\mathbf{X}(t)\mathbf{X}(\omega)$$
$$= \mathbf{Q}(t)\mathbf{Q}^{-1}(t)\mathbf{Y}(t)\mathbf{X}(\omega) = \mathbf{Y}(t)\mathbf{X}(\omega).$$
Thus, the characteristic equation of the transformed system coincides with that of the original system.

It is well known that
$$\det(\mathbf{X}(\omega)) = e^{\int_0^\omega \sum_{i=1}^n p_{ii}(t)dt} > 0.$$

Find the form of a fundamental matrix for system (1.2.37). Note that the matrix-valued function
$$\varphi(t) = \mathbf{X}(t)e^{-\frac{t}{\omega}\ln \mathbf{X}(\omega)} \tag{1.2.40}$$
is continuously differentiable, ω-periodic in t and nonsingular. In fact,
$$\varphi(t+\omega) = \mathbf{X}(t)\mathbf{X}(\omega)e^{-\frac{t+\omega}{\omega}\ln \mathbf{X}(\omega)} = \varphi(t).$$
From (1.2.40) we have $\mathbf{X}(t) = \varphi(t)e^{\frac{t}{\omega}\ln \mathbf{X}(\omega)}$. Let $\mathbf{A} = \frac{1}{\omega}\ln \mathbf{X}(\omega)$, then
$$\mathbf{X}(t) = \varphi(t)e^{\mathbf{A}t}. \tag{1.2.41}$$

Hence, the matrix of the fundamental system can be represented as a product of the nonsingular ω-periodic matrix and the matrix e^{At} discarding the factor $\mathbf{X}(\omega)$ after increasing t by the period ω.

We will clarify the matrix structure of the fundamental system of solutions for system (1.2.37). Let $\mathbf{S}$ be a constant nonsingular matrix such that $\mathbf{S}^{-1}\mathbf{AS} = \Xi$ has the canonical form

$$\Xi = \begin{pmatrix} \Xi_0 & O & O & \cdots & O \\ O & \Xi_1 & O & \cdots & O \\ \cdots & \cdots & \cdots & \cdots & \cdots \\ O & O & O & \cdots & \Xi_s \end{pmatrix}$$

where Ξ_0 is a diagonal matrix with elements $\lambda_1, \ldots, \lambda_k$, and ρ_i is the order of the matrix

$$\Xi_i = \begin{pmatrix} \lambda_{k+i} & 1 & 0 & \cdots & 0 & 0 \\ 0 & \lambda_{k+i} & 1 & \cdots & 0 & 0 \\ \cdots & \cdots & \cdots & \cdots & \cdots & \cdots \\ 0 & 0 & 0 & \cdots & \lambda_{k+i} & 1 \\ 0 & 0 & 0 & \cdots & 0 & \lambda_{k+i} \end{pmatrix}$$

where $\lambda_j, j = 1, \ldots, k+s$ are the eigenvalues of the matrix $\mathbf{A}$ that are not necessarily distinct, i.e., the elementary divisors of the matrix $\mathbf{A}$ are: $\lambda - \lambda_1, \lambda - \lambda_2, \ldots, \lambda - \lambda_k, (\lambda - \lambda_{k+1})^{\rho_1}, \ldots, (\lambda - \lambda_{k+s})^{\rho_s}$; $k + \sum_{i=1}^{s} \rho_i = n$. Let $\mathbf{X}^{(1)}(t) = \mathbf{X}(t)\mathbf{S}, \tilde{\varphi}(t) = \varphi(t)\mathbf{S}$. From (1.2.41) we then have

$$\mathbf{X}^{(1)}(t) = \varphi(t)\mathbf{S}\left(\mathbf{S}^{-1}e^{\mathbf{A}t}\mathbf{S}\right) = \tilde{\varphi}(t)e^{\Xi t}, \qquad (1.2.42)$$

where $\tilde{\varphi}(t+\omega) = \tilde{\varphi}(t)$

It is known that

$$e^{\Xi t} = \begin{pmatrix} e^{\Xi_0 t} & O & \cdots & O \\ O & e^{\Xi_1 t} & \cdots & O \\ \cdots & \cdots & \cdots & \cdots \\ O & O & \cdots & e^{\Xi_s t} \end{pmatrix}$$

where

$$e^{\Xi_0 t} = \begin{pmatrix} e^{t\lambda_1} & 0 & \cdots & 0 \\ 0 & e^{t\lambda_2} & \cdots & 0 \\ \cdots & \cdots & \cdots & \cdots \\ 0 & 0 & \cdots & e^{t\lambda_k} \end{pmatrix}$$

and

$$e^{\Xi_i t} = e^{t\lambda_{k+1}} \begin{pmatrix} 1 & t & \cdots & \frac{t^{\rho_i - 1}}{(\rho_i - 1)!} \\ 0 & 1 & \cdots & \frac{t^{\rho_i - 2}}{(\rho_i - 2)!} \\ \cdots & \cdots & \cdots & \cdots \\ 0 & 0 & \cdots & 1 \end{pmatrix}.$$

The magnitudes $\lambda_k = \frac{1}{\omega} \ln \rho_k$, where ρ_k are the roots of the characteristic equation (1.2.39), are referred to as the characteristic indices of system (1.2.37).

Properties of Solutions to Linear Systems

Let $\varphi_1(t), \ldots, \varphi_n(t)$ be the periodic vector columns of matrix $\tilde{\varphi}(t)$. From (1.2.42) we find that the column-solutions $x_1^{(1)}(t), \ldots, x_n^{(1)}(t)$ of the fundamental matrix $\mathbf{X}^{(1)}(t)$ are of the form

$$x_1^{(1)}(t) = e^{\lambda_1 t}\varphi_1(t),$$

$$\ldots$$

$$x_k^{(1)}(t) = e^{\lambda_k t}\varphi_k(t),$$
$$x_{k+1}^{(1)}(t) = e^{\lambda_{k+1} t}\varphi_{k+1}(t),$$
$$x_{k+2}^{(1)}(t) = e^{\lambda_{k+1} t}(t\varphi_{k+1}(t) + \varphi_{k+2}(t)),$$

$$\ldots$$

$$x_{k+\rho_1}^{(1)}(t) = e^{\lambda_{k+1} t}\left(\frac{t^{\rho_1-1}}{(\rho_1-1)!}\varphi_{k+1}(t) + \ldots + t\varphi_{k+\rho_1-1}(t) + \varphi_{k+\rho_1}(t)\right), \tag{1.2.43}$$

$$\ldots$$

$$x_{n-\rho_s+1}^{(1)}(t) = e^{\lambda_{k+s} t}\varphi_{n-\rho_s+1}(t),$$

$$\ldots$$

$$x_n^{(1)}(t) = e^{\lambda_{k+s} t}\left(\frac{t^{\rho_s-1}}{(\rho_s-1)!}\varphi_{n-\rho_s+1}(t) + \ldots + t\varphi_{n-1}(t) + \varphi_n(t)\right).$$

From (1.2.43) it is seen that if equation (1.2.37) is to have an ω-periodic solution, then it is necessary and sufficient that $\rho = 1$ be the root of the characteristic equation, in other words the characteristic index is zero. From (1.2.43) it also follows that if all the characteristic indices of system (1.2.37) have negative real parts, i.e. the root moduli of the characteristic equation is less than unit, then all the solutions tend to zero as t increases without bound. Hence, the nonperturbed motion $x = 0$ is asymptotically stable. If at least one characteristic index of the system has positive real part, i.e. the modulus of one of roots ρ_i is greater than unit, then the solutions of system (1.2.37) corresponding to this index are unbounded and the nonperturbed motion is unstable. If the system (1.2.37) has no characteristic indices with positive real parts, but has those with zero real parts, then the solutions corresponding to these indices can be both bounded and unbounded. If the characteristic index with zero real part is prime or multiple, but has its associated prime elementary divisors, then all these solutions will be bounded; otherwise for the characteristic index $i\lambda$ under consideration, there will be an unbounded solution of the form

$$x_s(t) = e^{i\lambda t}(t^p\varphi_{s0}(t) + \ldots + t\varphi_{sp-1}(t) + \varphi_{sp}(t)),$$

where φ_{sj} are periodic functions, $j = 0, \ldots, p$, $s = 1, \ldots, n$. By substituting $y = \varphi^{-1}(t)x$, the system (1.2.37) is reduced to a system with constant coefficients $\frac{dy}{dt} = \mathbf{A}y$ [Lyapunov (1950)].

Consider a nonhomogeneous system of linear equations

$$\frac{dy}{dt} = \mathbf{P}(t)y + f(t), \tag{1.2.44}$$

where the elements of the column matrix f are the continuous functions of t. It is well known that if $\mathbf{X}(t)$ is a fundamental system of solutions to (1.2.37), then the function

$$y(t) = \mathbf{X}(t)\left(\int_{t_0}^t \mathbf{X}^{-1}(\tau)f(\tau)d\tau + \mathbf{X}^{-1}(t_0)y_0\right) \tag{1.2.45}$$

is a solution to system (1.2.44) satisfying the condition $y(t_0) = y_0$. Since $\mathbf{X}(t) = \varphi(t)e^{\mathbf{A}t}$, the formula (1.2.45) becomes

$$y(t) = \varphi(t)e^{\mathbf{A}t}\left(\int_{t_0}^t e^{-\mathbf{A}\tau}\varphi^{-1}(\tau)f(\tau)d\tau + e^{-\mathbf{A}t_0}\varphi^{-1}(t_0)y_0\right). \tag{1.2.46}$$

8. We now turn our attention to the discussion of solutions to the linear systems with periodic coefficients from the standpoint of development of the natural and forced oscillations therein. If the characteristic indices include k indices with zero real parts, then the system (1.2.37) contains a family of stationary natural oscillations that are periodic or, in the general case, almost periodic. The latter is found only where there is incommensurability of the number $2\pi/\omega$ and the imaginary part of at least one of the characteristic indices with zero real part. Such a family of natural stationary oscillations depends on as many arbitrary constants as there are characteristic indices with zero real parts, their multiplicities being considered only where prime elementary divisors correspond to them. Each stationary oscillation appearing in this family is conditionally and asymptotically stable in the presence of quantities with negative real parts among the characteristic indices. This implies the existence of a family of integral curves in system (1.2.37) approaching the above stationary mode arbitrarily closely. The family of integral curves depends on the number of arbitrary constants which coincides with the number of characteristic indices with negative real parts. If for any $j = 1, \ldots, n$

$$\operatorname{Re}(\lambda_j) \neq 0, \tag{1.2.47}$$

then system (1.2.37) has a unique stationary mode $x = 0$.

All the above assertions are based on the assumption that the general solution of system (1.2.37) is

$$x = \sum_{j=1}^n c_j p_j(t)e^{\lambda_j t}, \tag{1.2.48}$$

where $p_j(t)$, in the general case, are polynomials with respect to t, whose degrees are less than the multiplicities of the corresponding characteristic indices λ_j and whose coefficients are t-periodic vector-valued functions with period ω.

9. We now focus on the forced oscillations under the acting periodic or almost periodic influence. If system (1.2.37) has no natural stationary oscillations, except for $x = 0$, i.e., provided that the condition (1.2.47) holds, then a periodic unique forced oscillation is set up in system (1.2.44) on condition that $f(t)$ is a $p\omega/q$-periodic function, while the almost periodic one is set up if $f(t)$ is an almost periodic function or the function of the period ω_1 incommensurable with ω, where p and q are natural numbers such that p/q is the irreducible fraction. If among $\lambda_1, \ldots, \lambda_n$ there are eigenvalues with zero real parts, then resonance is possible when the forced oscillations

Properties of Solutions to Linear Systems 25

set up in system (1.2.44) under the external acting force $f(t)$, which is periodic or almost periodic, will have the "amplitude" unboundedly increasing with time. If $f(t)$ is the $p\omega/q$-periodic function and the imaginary parts of characteristic indices with zero real parts are not coincident with the magnitudes of the form $2k\pi/p\omega, k = 0, \pm 1, \ldots$, then a periodic forced oscillation of period $p\omega$ is set up in system (1.2.44). If, however, the imaginary parts of characteristic indices with zero real parts include the indices equal to $2k\pi/p\omega, k = 0, \pm 1, \ldots$, then the forced oscillations will be periodic only when the function $f(t)$ is subject to the condition

$$\int_0^\omega y^*(\tau)f(\tau)d\tau = 0, \tag{1.2.49}$$

where $y(t)$ is any of the periodic solutions to the system

$$\frac{dy}{dt} = -\mathbf{P}^*(t)y, \tag{1.2.50}$$

corresponding to the characteristic indices of the form $i2k\pi/p\omega$.

We will show the necessity of this condition. Let $x(t)$ be the $p\omega$-periodic forced solution of system (1.2.44), and $y(t)$ the periodic solution of system (1.2.50) corresponding to the characteristic index $i2k\pi/p\omega$. Then the function $z = y^*x$ will be $p\omega$-periodic. Calculate $\frac{dz}{dt}$:

$$\frac{dz}{dt} = \frac{dy^*}{dt}x + y^*\frac{dx}{dt} = -y^*\mathbf{P}(t)x + y^*(\mathbf{P}(t)x + f(t)) = y^*f(t).$$

By the periodicity of the function z,

$$\int_0^{p\omega} \frac{dz}{dt}dt = \int_0^{p\omega} y^*(t)f(t)dt = 0.$$

For the sufficiency of condition (1.2.49), it can be established as in (1.2.29) after reducing the system (1.2.44) to the form (1.2.13). If $f(t)$ is an almost periodic function, or a periodic function, whose period is incommensurable with ω, then a forced oscillation is not necessarily almost periodic. However, we can always select necessary and sufficient conditions for the forced oscillation of system (1.2.44) to be almost periodic, as well. Such conditions have a simple form, especially if $f(t)$ can be represented as

$$f(t) = \sum_{j=1}^N c_j(t)e^{i\mu_j t},$$

where $c_j(t)$ are the $p\omega/q$-periodic vector-valued functions. In this case, the forced oscillation assumes the form

$$x = \sum_{j=1}^N \gamma_j(t)e^{i\mu_j t}, \tag{1.2.51}$$

where $\gamma_j(t)$ are the $p\omega$-periodic vector functions if the characteristic indices do not include magnitudes of the form

$$i\left(\frac{2k\pi}{p\omega} + \mu_j\right), \quad j = 1, \ldots, N, k = 0, \pm 1, \ldots. \tag{1.2.52}$$

If this condition is violated, it is possible to select necessary and sufficient conditions on the function $c_j(t)$ under which the forced oscillation in system (1.2.44) again assumes the form (1.2.51). Such conditions are obtained from analysis of the equation system

$$\frac{dx}{dt} = \mathbf{P}(t)x + c_j(t)e^{i\mu_j t}, \qquad (1.2.53)$$

which, by transforming $x = ye^{i\mu_j t}$, is reduced to the above form.

Statements about forced oscillations in system (1.2.44) can be interpreted as a result of reducing system (1.2.44) to system (1.2.13). The general case, where $f(t)$ is the almost periodic function of an arbitrary type, can also be discussed in some detail by reducing (1.2.44) to (1.2.13).

10. We again consider the system

$$\frac{dx}{dt} = \mathbf{P}(t)x, \qquad (1.2.54)$$

where the matrix elements $\{\mathbf{P}(t)\}_{ik} = p_{ik}(t)$, $i,k = 1,\ldots,n$ are real, continuous bounded functions of t defined for all $t \geq 0$. Let

$$x_1(t),\ldots,x_n(t) \qquad (1.2.55)$$

be a solution to system (1.2.54).

The characteristic number $\chi\{f\}$ of the continuous function $f(t)$, defined on the entire real semi-axis $t \geq 0$, is called the number a, satisfying the conditions

$$\overline{\lim_{t\to+\infty}} \, |f(t)|\, e^{(a+\epsilon)t} = +\infty \quad \text{and} \quad \lim_{t\to+\infty} f(t)e^{(a-\epsilon)t} = 0$$

for any $\epsilon > 0$, however small it may be. If for any λ we obtain $\lim_{t\to+\infty} f(t)e^{\lambda t} = 0$, then it is believed that $\chi(f) = +\infty$; if $\lim_{t\to+\infty} |f(t)|\, e^{\lambda t} = +\infty$, then it is believed that $\chi(f) = -\infty$. Under this condition any function $f(t)$ has a finite or infinite characteristic number. Any nonzero constant has a zero characteristic number, and characteristic number of zero equals $+\infty$.

It can readily be seen that $\chi(f) = -\overline{\lim_{\substack{t\to+\infty \\ f\neq 0}}} \ln\frac{|f(t)|}{t}$ [Lyapunov (1950)], [Malkin (1966)]. We will present some properties of characteristic numbers of the functions established by A.M. Lyapunov.

1^0. The characteristic number of the sum of two functions equals the least of the characteristic numbers of these functions when the numbers are distinct, and it is not less than these numbers when they are equal. Indeed, let $\chi(f_1) = \lambda_1, \chi(f_2) = \lambda_2$ and $\lambda_1 \leq \lambda_2$. Then for any $\epsilon > 0$

$$(f_1 + f_2)e^{(\lambda_1-\epsilon)t} = f_1 e^{(\lambda_1-\epsilon)t} + f_2 e^{(\lambda_1-\epsilon)t} \xrightarrow[t\to\infty]{} 0,$$

i.e. $\chi(f_1 + f_2) \geq \lambda_1$. If $\lambda_1 < \lambda_2$, then for $0 < \epsilon < \lambda_2 - \lambda_1$ there will be

$$f_1 e^{(\lambda_1+\epsilon)t} \xrightarrow[t\to\infty]{} \infty, \quad f_2 e^{(\lambda_1+\epsilon)t} \xrightarrow[t\to\infty]{} 0.$$

Hence $(f_1 + f_2)e^{(\lambda_1+\epsilon)t} \xrightarrow[t\to\infty]{} \infty$, i.e. for $\lambda_1 < \lambda_2$ we get $\chi(f_1 + f_2) = \lambda_1$. ∎

2^0. The characteristic number of the product of two functions is not less than the sum of their characteristic numbers. Indeed, let $\chi(f_1) = \lambda_1, \chi(f_2) = \lambda_2$. Then for any $\epsilon > 0$

$$f_1 f_2 e^{(\lambda_1+\lambda_2-\epsilon)t} = f_1 e^{(\lambda_1-\epsilon/2)t} f_2 e^{(\lambda_2-\epsilon/2)t} \xrightarrow[t\to\infty]{} 0,$$

i.e. $\chi(f_1 f_2) \geq \lambda_1 + \lambda_2$. ∎

For example, the characteristic number of the product of two functions $f_1 = e^{t\sin t}$ and $f_2 = e^{-t\sin t}$ is zero, and $\chi(e^{\pm t\sin t}) = -1$, since

$$\overline{\lim_{t\to+\infty}} \frac{\ln e^{\pm t\sin t}}{t} = \overline{\lim_{t\to+\infty}}(\pm \sin t) = +1,$$

i.e. $\chi(f_1 f_2) > \chi(f_1) + \chi(f_2) = -2$.

Corollary. The sum of the characteristic numbers f and $1/f$ is nonpositive: $\chi(f) + \chi(1/f) \leq 0$, since

$$0 = \chi(f \cdot \frac{1}{f}) \geq \chi(f) + \chi(1/f).$$

3^0. For the sum of the characteristic numbers f and $1/f$ to be zero, it is necessary and sufficient that the expression $\frac{1}{t}\ln|f(t)|$ would approach a particular limit as $t \to +\infty$.

4^0. If the sum of the characteristic numbers of f and $1/f$ is zero, then the characteristic number of the product of the function f and any function φ is the sum of the characteristic numbers of the factors.

5^0. The characteristic number of the integral is not less than that of the integrand. Note that for $\chi(f) \leq 0$ there is $\int_a^t f(t)dt$, otherwise $-\int_t^{+\infty} f(t)dt$.

The smallest of the characteristic numbers of the functions $x_s(t), s = 1, \ldots, n$, is called the characteristic number of the solution (1.2.55) involved, and is denoted by $\chi(x_1, \ldots, x_n)$.

We shall now present Lyapunov's assertions about the characteristic numbers of solutions for system (1.2.54).

1^0. Any solution to system (1.2.54), as distinct from zero, has a finite characteristic number.

2^0. Let $\{x_{ij}(t)\}$ be the fundamental system of solutions to equation (1.2.54), λ_j the characteristic number of the solution $x_{1j}, \ldots, x_{nj}, j = 1, \ldots, n$, all $\lambda_1, \ldots, \lambda_n$ being distinct. Then the characteristic number of the solution $x_s = \sum_{l=1}^{k} c_l x_{sl}$ with the constants c_l will be equal to the smallest of the characteristic numbers $\lambda_1, \ldots, \lambda_k$ (by property 1). Therefore, if $\lambda_1, \ldots, \lambda_n$ are distinct, then the characteristic number of any solution to system (1.2.54) will be one of the numbers $\lambda_1, \ldots, \lambda_n$, i.e. the following assertion is valid: the system (1.2.54) cannot have more than n solutions with distinct characteristic numbers.

Assume that among the characteristic numbers $\lambda_1, \ldots, \lambda_n$ of some fundamental system of solutions there are equal numbers. Then the characteristic number of the solution, which is a linear combination of solutions with the same λ_j of the fundamental system, may prove to be greater than the characteristic numbers of the solutions involved in the combination. In this case, the new solution will be incorporated into the fundamental system instead of one of the solutions being combined. If the new fundamental system again has solutions whose combination yields a solution with the characteristic number greater than those of the solutions to be grouped, then the system is transformed as the preceding one. Since the quantity of characteristic numbers is finite, this results in a system of solutions where any linear combination of solutions will possess the characteristic number equal to that of a combined solution. The obtained fundamental system is referred as normal. The characteristic numbers $\lambda_1, \ldots, \lambda_n$ of the normal system of solutions (which may include equal numbers) are called the characteristic numbers of the system of differential equations.

We shall now give Lyapunov's example of the system of equations

$$\frac{dx_1}{dt} = x_1 \cos \ln t + x_2 \sin \ln t,$$

$$\frac{dx_2}{dt} = x_1 \sin \ln t + x_2 \cos \ln t, \tag{1.2.56}$$

whose normal system of solutions

$$x_{11} = e^{t \sin \ln t}, \quad x_{21} = e^{t \sin \ln t}$$

$$x_{12} = e^{t \cos \ln t}, \quad x_{22} = -e^{t \cos \ln t} \tag{1.2.57}$$

has the characteristic number equal to -1.

3^0. If all characteristic numbers $\lambda_1, \ldots, \lambda_n$ of some fundamental systems are distinct, then this system is normal.

4^0. Any fundamental system of solutions, for which the sum of characteristic numbers $\sum_{i=1}^{n} \lambda_i$ of all the solutions involved attains its upper limit, is normal. Suppose this fundamental system is not normal. Then from its solutions it is possible to combine a new solution with a greater characteristic number and to obtain a new fundamental system with the sum of characteristic numbers greater than $\sum_{i=1}^{n} \lambda_i$, which contradicts the assumption.

The $(n \times n)$ matrix $\mathbf{A}(t)$ given for all $t \geq 0$ will be referred to as the Lyapunov type matrix if it is bounded and continuous together with $\frac{d\mathbf{A}(t)}{dt}$ and the value of $\det(\mathbf{A}^{-1}(t))$ is bounded.

In system (1.2.54), we make a replacement

$$y = \mathbf{A}(t)x, \tag{1.2.58}$$

where $\mathbf{A}(t)$ is the Lyapunov type matrix. Denote $\mathbf{A}^{-1}(t) = \mathbf{B}(t)$, then

$$x = \mathbf{B}(t)y. \tag{1.2.59}$$

Properties of Solutions to Linear Systems 29

The system (1.2.54) converts to the system

$$\frac{dy}{dt} = \frac{d\mathbf{A}(t)}{dt}x + \mathbf{A}(t)\frac{dx}{dt} = \left(\frac{d\mathbf{A}(t)}{dt}\mathbf{B}(t) + \mathbf{A}(t)\mathbf{P}(t)\mathbf{B}(t)\right)y.$$

Denoting $\mathbf{Q}(t) = \frac{d\mathbf{A}(t)}{dt}\mathbf{B}(t) + \mathbf{A}(t)\mathbf{P}(t)\mathbf{B}(t)$, we get

$$\frac{dy}{dt} = \mathbf{Q}(t)y. \quad (1.2.60)$$

Note that in the conversion of (1.2.58) the coefficients $\{\mathbf{Q}(t)\}_{ik} = q_{ik}(t)$ will be the bounded functions. The group of characteristic numbers in (1.2.60) is always identical with that in (1.2.54). Indeed, from (1.2.58) it is seen that $\chi(y_1, \ldots, y_n) \geq \chi(x_1, \ldots, x_n)$, and from (1.2.59) we derive the inverse inequality. Therefore, the characteristic number $\chi(y_1, \ldots, y_n)$ of each solution for the system (1.2.60) coincides with the characteristic number $\chi(x_1, \ldots, x_n)$ of an appropriate solution of system (1.2.54). After looking through all the solutions of the normal system for equations (1.2.54), we will prove the assertion made above.

5^0. The sum of characteristic numbers $\sum\limits_{i=1}^{n} \lambda_i$ in system (1.2.54) does not exceed the characteristic number of the function

$$e^{\int_0^t \sum_{s=1}^{n} p_{ss} dt}. \quad (1.2.61)$$

Indeed, let $\mathbf{X}(t) = \{x_{ij}(t)\}$ be a fundamental system of solutions for (1.2.54), then we get

$$\det(\mathbf{X}(t)) = \det(\mathbf{X}(0)) e^{\int_0^t \sum_{s=1}^{n} p_{ss} dt}.$$

By employing the properties relative to the characteristic number of the sum and the product of functions, we find

$$\chi\left(e^{\int_0^t \left(\sum_{s=1}^{n} p_{ss} dt\right)}\right) = \chi(\det \mathbf{X}(t)) \geq \sum_{i=0}^{n} \lambda_i.$$

6^0. If for some fundamental system of solutions the sum of its characteristic numbers

$$\lambda_1 + \ldots + \lambda_n = \chi\left(e^{\int_0^t \sum_{s=1}^{n} p_{ss} dt}\right),$$

then this system is normal, since $\chi\left(e^{\int_0^t \sum_{s=1}^{n} p_{ss} dt}\right)$ is the upper limit for the sum of characteristic numbers in the system (1.2.54). For the system (1.2.56), the sum of characteristic numbers of the normal solution system for (1.2.57), which is -2, is smaller than the characteristic number of function (1.2.61)

$$\chi\left(e^{2\int_0^t \cos\ln t\,dt}\right) = \chi\left(e^{t(\cos\ln t + \sin\ln t)}\right) = -\sqrt{2}.$$

The system (1.2.54) will be called regular if

$$\sum_{i=1}^{n}\lambda_i = -\chi\left(e^{-\int_0^t \sum_{s=1}^{n} p_{ss}\,dt}\right).$$

Note that the system (1.2.54) will always be regular if $\mathbf{P}$ is a constant or periodic matrix. The system (1.2.56) offers an example of irregular system. For regular systems

$$\chi\left(e^{\int_0^t \sum_{s=1}^{n} p_{ss}\,dt}\right) = -\chi\left(e^{-\int_0^t \sum_{s=1}^{n} p_{ss}\,dt}\right).$$

7^0. If the system (1.2.54) is regular, then after transformation (1.2.58) with the Lyapunov type matrix the newly obtained system (1.2.60) will also be regular.

As shown by Lyapunov, if the system (1.2.54) with the triangular matrix $\mathbf{P}(t)$ is to be regular, it is necessary and sufficient that $\frac{1}{t}\int_0^t p_{ss}\,dt$ have definite limits as $t \to +\infty$, $s = 1,\ldots,n$. O.Perron [Perron (1930)] established the following theorem: if the system (1.2.54) is to be regular, it is necessary and sufficient that $\lambda_s = -\mu_s$, $s = 1,\ldots,n$, where μ_s are the characteristic numbers of the adjoint system [Erougin (1946), Lyapunov (1950)]:

$$\frac{dx}{dt} = -\mathbf{P}^*(t)x.$$

The system (1.2.54) is called reducible if there exists the Lyapunov type matrix $\mathbf{Z}$ such that substituting $y = \mathbf{Z}x$ reduces equation (1.2.54) to the equation

$$\frac{dy}{dt} = \mathbf{A}y, \qquad (1.2.62)$$

where $\mathbf{A}$ is a constant matrix. Since all of the systems with constant coefficients are regular, only the regular systems can be reducible. As shown before, the system with periodic coefficients is reduced to system (1.2.62) by substituting $y = \varphi^{-1}(t)x$. A.Lyapunov established that if $\mathbf{X}(t)$ is a normal fundamental system of solutions for (1.2.54) and

$$\frac{e^{-\sum_{s=1}^{n}\lambda_s t}}{\det(\mathbf{X}(t))}, \quad x_{ij}(t)e^{\lambda_j t}, \quad i,j = 1,\ldots,n,$$

are bounded, then, by substituting

$$y = \mathbf{Z}x, \{\mathbf{Z}\}_{ik} = \frac{\Delta_{ki}(t)}{\Delta(t)} e^{-\lambda_i t}, \ i, k = 1, \ldots, n,$$

where $\mathbf{Z}$ is a Lyapunov type matrix, the system (1.2.54) is reduced to the system

$$\frac{dy_s}{dt} = -\lambda_s y_s, \ s = 1, \ldots, n,$$

with constant coefficients.

1.3 Behavior of Integral Curves of a Nonlinear System of ODE's with Unbounded Increase in Time

1. Consider a system of differential equations of the form

$$\frac{dx_s}{dt} = \sum_{j=1}^n p_{sj}(t) x_j + X_s(t, x_1, \ldots, x_n), \ s = 1, \ldots, n. \tag{1.3.1}$$

For the right-hand sides of the system, we will make the following assumptions:
  a) the functions $p_{sj}(t)$, $s, j = 1, \ldots, n$, are given for $t \geq 0$, and are real, continuous and bounded therein;
  b) the functions $X_s(t, x_1, \ldots, x_n)$, $s = 1, \ldots, n$, are given for $t \geq 0$, and are continuous, real, and satisfy the Lipschitz condition

$$| X_s(t, \bar{x}_1, \ldots, \bar{x}_n) - X_s(t, \bar{\bar{x}}_1, \ldots, \bar{\bar{x}}_n) |$$

$$\leq L \left(\sum_{j=1}^n | \bar{x}_j - \bar{\bar{x}}_j | \right) \max \left(\left(\sum_{j=1}^n \bar{x}_j^2 \right)^{m/2}, \left(\sum_{j=1}^n \bar{\bar{x}}_j^2 \right)^{m/2} \right). \tag{1.3.2}$$

Also, $X_s(t, 0, \ldots, 0) \equiv 0$, $s = 1, \ldots, n$, $t \geq 0$. Here m, L and r are positive constants.
  Denote by $\lambda_1, \ldots, \lambda_n$ the collection of characteristic numbers for the linear system

$$\frac{dx_s}{dt} = \sum_{j=1}^n p_{sj}(t) x_j. \tag{1.3.3}$$

Let $\mathbf{Y}(t)$ be a normal fundamental matrix of solutions to equations (1.3.3), whose j-th column is a solution to system (1.3.3) corresponding to the characteristic number λ_j, $j = 1, \ldots, n$. Let y_{sj} be the element of the matrix $\mathbf{Y}(t)$ at row s and column j.
  In this section we will derive conditions, under which the system (1.3.1) has a family of integral curves exhibiting the property $x_s(t) \to 0$ as $t \to +\infty$, $s = 1, \ldots, n$, define the methods for representing such families of integral curves with the use of special type rows, and discuss the structure of the latter.

Theorem 6 *If 1) the right-hand sides of system (1.3.1) satisfy conditions a) and b),*

2) system (1.3.3) is regular,

3) among the characteristic numbers of system (1.3.3) there are k positive numbers, viz. $\lambda_1 > 0, \ldots, \lambda_k > 0$, then system (1.3.1) has a family of integral curves dependent on k arbitrary constants $c_1, \ldots, c_k$,

$$x_s = x_s(t, c_1, \ldots, c_k), \quad s = 1, \ldots, n. \tag{1.3.4}$$

Every integral curve, appearing in the family (1.3.4), satisfies the inequality

$$\left(\sum_{s=1}^{n} x_s^2\right)^{1/2} \leq b \left(\sum_{j=1}^{n} c_j^2\right)^{1/2} e^{-(\bar{\lambda}-\epsilon)t}. \tag{1.3.5}$$

Here $\bar{\lambda} = \min(\lambda_1, \ldots, \lambda_k)$, b and ϵ are positive constants, $\epsilon < \lambda$, and ϵ can be arbitrarily small, while the constant b may depend on ϵ so that, in this case, the b unboundedly increases as $\epsilon \to +0$. The inequality (1.3.5) holds only for the integral curves determined by the values $c_1, \ldots, c_k$ subject to the condition $\left(\sum_{j=1}^{k} c_j^2\right)^{1/2} \leq c$, where c is a sufficiently small positive value.

Before proving this theorem, we will first discuss its design. The basic idea of the proof is that the integral curves of system (1.3.1) satisfying the condition (1.3.5) are regarded as solutions to a system of integral equations, whose solutions, for which the inequality (1.3.5) holds, are in their turn the solutions to system (1.3.1). We first turn to vector-matrix notation. Next, denote by x the vector $(x_1, \ldots, x_n)$, and by $X(t,x)$ the vector $(X_1(t,x), \ldots, X_n(t,x))$. Then system (1.3.1) becomes

$$\frac{dx}{dt} = \mathbf{P}(t)x + X(t,x), \tag{1.3.1'}$$

while system (1.3.3) can be represented as

$$\frac{dx}{dt} = \mathbf{P}(t)x, \tag{1.3.3'}$$

where $\mathbf{P}(t)$ is the square matrix, whose elements are the functions $\mathbf{P}_{sj}(t)$, $s = 1, \ldots, n$. In what follows, as in section 1.1, the norm of vector means its Euclidean norm $\|x\| = \left(\sum_{j=1}^{n} x_j^2\right)^{1/2}$. The norm of the matrix $\mathbf{A}$ will be taken to be

$$\|\mathbf{A}\| = \left(\sum_{s,j=1}^{n} a_{sj}^2\right)^{1/2},$$

where a_{sj}, $s, j = 1, \ldots, n$ are elements of the matrix $\mathbf{A}$.

Consider the system of integral equations

$$x(t) = \int_{t_0}^{+\infty} \mathbf{G}(t,\tau) X(\tau, x(\tau)) d\tau + \mathbf{G}(t, t_0) c_1, \tag{1.3.6}$$

where c_1 is the n-dimensional constant vector column, and $\mathbf{G}(t,\tau)$ is the matrix determined from the formulas

$$\mathbf{G}(t,\tau) = \mathbf{Y}(t)\mathbf{G}_1\mathbf{Y}^{-1}(\tau),\ 0 \leq \tau \leq t,$$

$$\mathbf{G}(t,\tau) = \mathbf{Y}(t)\mathbf{G}_2\mathbf{Y}^{-1}(\tau),\ \tau > t > 0. \tag{1.3.7}$$

In equalities (1.3.7), the matrices $\mathbf{G}_1$ and $\mathbf{G}_2$ are constant and assume the form

$$\mathbf{G}_1 = \begin{pmatrix} \mathbf{I}_k & \mathbf{O} \\ \mathbf{O} & \mathbf{O} \end{pmatrix},$$

$$\mathbf{G}_2 = \begin{pmatrix} \mathbf{O} & \mathbf{O} \\ \mathbf{O} & -\mathbf{I}_{n-k} \end{pmatrix}. \tag{1.3.8}$$

In formulas (1.3.8), $\mathbf{I}_k$ and $\mathbf{I}_{n-k}$ are respectively the unit matrices of orders k and $n-k$, and $\mathbf{O}$ are the matrices, whose elements are zero and of such orders that $\mathbf{G}_1$ and $\mathbf{G}_2$ form the square matrices of order n.

Thus, to prove the theorem we have:

1) to establish the existence of solutions to system (1.3.6) with sufficiently small $\|c_1\|$ satisfying the inequality (1.3.5);

2) to establish that such solutions are simultaneously solutions to system (1.3.1) and, conversely, any solution to system (1.3.1) satisfying inequality (1.3.5), is simultaneously a solution to system (1.3.6).

2. We now turn to implementation of the above plan.

Lemma 1 *The matrix $\mathbf{G}(t,\tau)$ defined by the equalities (1.3.7) has the following properties:*

1) the following inequality holds for any $\delta > 0$

$$\|\mathbf{G}(t,\tau)\| \leq \gamma_1 e^{\delta(t+\tau)} e^{-\bar{\lambda}(t-\tau)}\ \text{with}\ t \geq \tau \geq 0, \tag{1.3.9}$$

$$\|\mathbf{G}(t,\tau)\| \leq \gamma_1 e^{\delta(t+\tau)}\ \text{with}\ \tau > t \geq 0, \tag{1.3.10}$$

where $\gamma_1 > 0$ is constant in t and τ can be a function of δ such that $\gamma_1(\delta) \to +\infty$ as $\delta \to +0$;

2) the following identities hold for $\tau \neq t$

$$\frac{\partial \mathbf{G}(t,\tau)}{\partial \tau} = -\mathbf{G}(t,\tau)\mathbf{P}(\tau), \tag{1.3.11}$$

$$\frac{\partial \mathbf{G}(t,\tau)}{\partial t} = \mathbf{P}(t)\mathbf{G}(t,\tau); \tag{1.3.12}$$

3) the matrix $\mathbf{G}(t,\tau)$ at the point $t = \tau$ satisfies the condition

$$\mathbf{G}(t,t-0) - \mathbf{G}(t,t+0) = \mathbf{I}, \tag{1.3.13}$$

where $\mathbf{I}$ is the n-dimensional unit matrix.

Proof: Denote by Λ a diagonal matrix whose diagonal elements are characteristic numbers of the system $(\lambda_1, \ldots, \lambda_n)$. Each element z_{sj} of the matrix $\mathbf{Z}(t) = \mathbf{Y}(t)e^{\Lambda t}$ has a non-negative characteristic number, since

$$z_{sj} = y_{sj} e^{\lambda_j t}. \tag{1.3.14}$$

From this it follows that $\chi(z_{sj}) \geq \chi(y_{sj}) - \lambda_j \geq 0$. By the regularity of system (1.3.3), the matrix $\mathbf{Z}^{-1}(t)$ has the elements, whose characteristic numbers are also non-negative. In fact,

$$\mathbf{Z}^{-1}(t) = e^{-\Lambda t} \mathbf{Y}^{-1}(t). \tag{1.3.15}$$

Denote by $\bar{y}_{sj}(t)$ the elements of the matrix $\mathbf{Y}^{-1}(t)$, and by $\bar{z}_{sj}(t)$ the elements of the matrix $\mathbf{Z}^{-1}(t)$. Equality (1.3.15) implies $\bar{z}_{sj} = e^{-\lambda_s t} \bar{y}_{sj}(t)$, whence it follows that

$$\chi(\bar{z}_{sj}) \geq \lambda_s + \chi(\bar{y}_{sj}). \tag{1.3.16}$$

By the construction rule of an inverse matrix, $\bar{y}_{sj}(t)$ is determined from the formula

$$\bar{y}_{sj}(t) = \Delta_{sj}(\mathbf{Y})/\Delta(\mathbf{Y}),$$

where Δ_{sj} are cofactors of the element y_{js} of the matrix $\mathbf{Y}$, i.e. its minors obtained by deleting the j-th row and the s-th column and taken with the $(-1)^{s+j}$ sign. Here $\Delta(\mathbf{Y})$ is the determinant of the matrix $\mathbf{Y}$.

Let

$$\mu = \chi(1/\Delta(\mathbf{Y})) = \chi\left(e^{-\int_0^t \sum_{s=1}^n \rho_{ss}(t)dt}\right), \quad \sum_{j=1}^n \lambda_j = S.$$

Considering that the characteristic number of the product of functions is not smaller than the sum of their characteristic numbers and that the characteristic number of the sum of functions is not smaller than the smallest characteristic number of summands, we conclude that

$$\chi(\Delta_{sj}(\mathbf{Y})) \geq S - \lambda_s,$$

whence

$$\chi(\bar{y}_{sj}(t)) \geq \mu + S - \lambda_s.$$

By regularity of system (1.3.3), $\mu + S = 0$. By (1.3.16), we find

$$\chi(\bar{z}_{sj}) \geq 0. \tag{1.3.17}$$

Since the diagonal matrices are permutable, we have the identical equality

$$\mathbf{G}(t,\tau) = \mathbf{Z}(t)\mathbf{G}_1 e^{-\Lambda(t-\tau)} \mathbf{Z}^{-1}(\tau) \quad \text{with} \quad 0 \leq \tau \leq t, \tag{1.3.18}$$

$$\mathbf{G}(t,\tau) = \mathbf{Z}(t)\mathbf{G}_2 e^{-\Lambda(t-\tau)} \mathbf{Z}^{-1}(\tau) \quad \text{with} \quad \tau > t \geq 0. \tag{1.3.19}$$

From (1.3.17) we obtain

$$|\bar{z}_{sj}(\tau)| \leq \bar{c}_{sj} e^{\delta \tau}, \tag{1.3.20}$$

where $\bar{c}_{sj}$ are some positive constants. Similarly, from (1.3.14) we find

$$|z_{sj}(t)| \leq c_{sj} e^{\delta t}. \qquad (1.3.21)$$

Considering that with $0 \leq \tau \leq t$ for $j = 1, \ldots, k$ the following inequality holds

$$e^{-\lambda_j(t-\tau)} \leq e^{-\bar{\lambda}(t-\tau)}, \qquad (1.3.22)$$

and with $\tau > t \geq 0$ for $j = k+1, \ldots, n$ the next inequality holds

$$e^{-\lambda_j(t-\tau)} \leq 1 \ (\lambda_j \leq 0), \qquad (1.3.23)$$

we find that the equality (1.3.18) and inequalities (1.3.20) — (1.3.22) imply (1.3.9), and the equality (1.3.19) and inequalities (1.3.20), (1.3.21), (1.3.23) imply (1.3.10). We have thus established the first property of the matrix $\mathbf{G}(t, \tau)$.

We shall now show that the matrix $\mathbf{G}(t, \tau)$ with $t \neq \tau$ satisfies identities (1.3.11) and (1.3.12). For this purpose, we recall that the derivative of the inverse matrix $\mathbf{Y}^{-1}(t)$ satisfies the relation

$$\frac{d}{dt} \mathbf{Y}^{-1}(t) = -\mathbf{Y}^{-1}(t) \frac{d\mathbf{Y}(t)}{dt} \mathbf{Y}^{-1}(t). \qquad (1.3.24)$$

If $\mathbf{Y}(t)$ is, as before, a normal fundamental system of solutions to equations (1.3.3), it follows that

$$\frac{d\mathbf{Y}(t)}{dt} = \mathbf{P}(t) \mathbf{Y}(t).$$

From this, and from (1.3.24), we find that

$$\frac{d}{dt} \mathbf{Y}^{-1}(t) = -\mathbf{Y}^{-1}(t) \mathbf{P}(t), \qquad (1.3.25)$$

i.e., the inverse matrix to $\mathbf{Y}(t)$ is a solution to a system of differential equations which is adjoint with respect to (1.3.3). With this in mind, by differentiating immediately the matrix $\mathbf{G}(t, \tau)$ with respect to its arguments and taking into account formulas (1.3.7), we obtain identities (1.3.11) and (1.3.12).

It follows from formulas (1.3.8) that the matrices $\mathbf{G}_1$ and $\mathbf{G}_2$ are related by $\mathbf{G}_1 - \mathbf{G}_2 = \mathbf{I}$, whence we have

$$\mathbf{G}(t, t-0) - \mathbf{G}(t, t+0) = \mathbf{Y}(t)(\mathbf{G}_1 - \mathbf{G}_2)\mathbf{Y}^{-1}(t) = \mathbf{I}. \qquad \blacksquare$$

Introduce a set of vector functions $x(t) \in D_{\gamma_0 \epsilon}$, given for $t \geq t_0$, that are continuous and satisfy the inequality

$$\|x\| \leq \gamma_0 e^{-(\lambda - \epsilon)t}, \ t \geq t_0, \qquad (1.3.26)$$

where ϵ and γ_0 are positive constants, while $\epsilon < \bar{\lambda}$ is sufficiently small. On this set of functions we consider the operator H:

$$H(x) = \int_{t_0}^{\infty} \mathbf{G}(t, \tau) X(\tau, x(\tau)) d\tau + \mathbf{G}(t, t_0) c_1, \qquad (1.3.27)$$

where c_1 is the n-dimensional vector whose components are arbitrary constants; and $t_0 \geq 0$ is the number fixed once and for all.

Lemma 2 *For any sufficiently small $\epsilon > 0$ there exist two numbers γ_0 and c_0 such that the operator (1.3.27) transforms the set of functions $D_{\gamma_0\epsilon}$ into itself only if $\|c_1\| \leq c_0$.*

Proof: First choose $\gamma_0 < r$. Then the vector function $X(t,x)$ is defined for $t \geq t_0$ on any vector function $x \in D_{\gamma_0\epsilon}$ and, furthermore, the following inequality holds

$$\|X(t,x)\| \leq L\gamma_0^{1+m} e^{-(1+m)(\bar{\lambda}-\epsilon)t}. \tag{1.3.28}$$

Evaluate the operator $H(x)$ for $x \in D_{\gamma_0\epsilon}$:

$$\|H(x)\| \leq \left\|\int_{t_0}^{+\infty} \mathbf{G}(t,\tau)X(\tau,x(\tau))d\tau\right\| + \|\mathbf{G}(t,t_0)c_1\|. \tag{1.3.29}$$

From (1.3.29) it follows that

$$\|H(x)\| \leq \int_{t_0}^{+\infty} \|\mathbf{G}(t,\tau)\|\,\|X(\tau,x(\tau)\|d\tau + \|\mathbf{G}(t,t_0)\|\,\|c_1\|$$

$$\leq \int_{t_0}^{t} \|\mathbf{G}(t,\tau)\|L\gamma_0^{1+m} e^{-(1+m)(\bar{\lambda}-\epsilon)\tau}d\tau$$

$$+ \int_{t}^{+\infty} \|\mathbf{G}(t,\tau)\|L\gamma_0^{1+m} e^{-(1+m)(\bar{\lambda}-\epsilon)\tau}d\tau + \|\mathbf{G}(t,t_0)\|\,\|c_1\|. \tag{1.3.30}$$

From inequalities (1.3.9), (1.3.10) and (1.3.30) we find

$$\|H(x)\| \leq \gamma_1 \gamma_0^{1+m} L \int_{t_0}^{t} e^{-\bar{\lambda}(t-\tau)} e^{\delta(\tau+t)} e^{-(1+m)(\bar{\lambda}-\epsilon)\tau} d\tau$$

$$+ \gamma_1 \gamma_0^{1+m} L \int_{t}^{\infty} e^{\delta(t+\tau)} e^{-(1+m)(\bar{\lambda}-\epsilon)\tau} d\tau + \|\mathbf{G}(t,t_0)\|\,\|c_1\|. \tag{1.3.31}$$

In these evaluations the number $\delta > 0$ is taken to be so small that infinite integrals converge. For this purpose, it suffices to choose the value of δ as a function of ϵ such that the inequality $\delta < \frac{1}{2}\min(\epsilon, m(\bar{\lambda}-\epsilon))$ holds. Having computed integrals in (1.3.31), we obtain

$$\|H(x)\| \leq \gamma_1 \gamma_0^{1+m} L \left(\frac{1}{\bar{\lambda}+\delta-(1+m)(\bar{\lambda}-\epsilon)}e^{-(\bar{\lambda}-\delta)t}\right.$$

$$\times \left(e^{(\bar{\lambda}+\delta-(1+m)(\bar{\lambda}-\epsilon))t} - e^{(\bar{\lambda}+\delta-(1+m)(\bar{\lambda}-\epsilon))t_0}\right) + \frac{1}{\delta-(1+m)(\bar{\lambda}-\epsilon)}$$

$$\left.\times e^{\delta t}\left(-e^{(\delta-(1+m)(\bar{\lambda}-\epsilon))t}\right)\right) + \gamma_1 e^{-(\bar{\lambda}-\delta)(t-t_0)}\|c_1\|e^{2\delta t_0}. \tag{1.3.32}$$

Setting $\|c_1\| \leq c_0 = \gamma_0^{1+m}$, we finally establish

$$\|H(x)\| \leq \gamma_0^{1+m}\gamma_2 e^{-(\bar{\lambda}-\epsilon)t}, \tag{1.3.33}$$

where γ_2 is the value, which is constant in t and γ_0 and is generally the function of ϵ such that $\gamma_2(\epsilon) \to +\infty$ as $\epsilon \to +0$.

We now choose a number γ_0 such that $\gamma_0^m \gamma_2 < 1$. With the thus chosen numbers γ_0 and $c_0 = \gamma_0^{1+m}$, the values of the operator H given on $D_{\gamma_0 \epsilon}$ again become the elements of the set $D_{\gamma_0 \epsilon}$, that is, $H(D_{\gamma_0 \epsilon}) \subset D_{\gamma_0 \epsilon}$ if $\|c_1\| \leq c_0 = \gamma_0^{1+m}$ and $\gamma_0^m \gamma_2 < 1$. To complete the proof, it remains to establish the continuity of functions $H(x)$ in t. Denote by $y(t)$ the vector function which is the value of the operator H on the element $x \in D_{\gamma_0 \epsilon}$:

$$y(t) = H(x) = \int_{t_0}^{t} \mathbf{G}(t,\tau) X(\tau, x(\tau)) d\tau$$

$$+ \int_{t}^{+\infty} \mathbf{G}(t,\tau) X(\tau, x(\tau)) d\tau + \mathbf{G}(t,t_0) c_1. \qquad (1.3.34)$$

The right-hand side of equality (1.3.34) has a continuous t-derivative. So, the values of operator H are not only continuous, but also differentiable functions of t. ∎

Lemma 3 *If the conditions of Theorem 6 hold for each value of the vector c_1 subject to the inequality $\|c_1\| \leq c_0$, the system of integral equations (1.3.6) has a unique continuously differentiable real solution given for $t \geq t_0$ and satisfying the inequality*

$$\|x(t)\| \leq b\|c_1\| e^{-(\bar{\lambda} - \epsilon)t} \text{ with } \|c_1\| \leq c_0, t \geq t_0. \qquad (1.3.35)$$

Proof: With $\|c_1\| \leq c_0$, the operator H maps the set of functions $D_{\gamma_0 \epsilon}$ into itself, therefore for any initial function $x^{(0)} \in D_{\gamma_0 \epsilon}$ each member of the sequence

$$x^{(p)} = H(x^{p-1}), \ p \geq 1, \qquad (1.3.36)$$

is also an element of $D_{\gamma_0 \epsilon}$. For simplicity, we take $x^{(0)}(t) \equiv 0$ as a zero approximation. Then $x^{(1)} = \mathbf{G}(t, t_0) c_1$. We show that the series

$$\sum_{p=1}^{\infty} \left(x^{(p)}(t) - x^{(p-1)}(t) \right) \qquad (1.3.37)$$

uniformly converges for $t \geq t_0$ to a function $x(t)$ satisfying the system of integral equations (1.3.6), and the estimate (1.3.35), and is a unique solution of this type corresponding to a given value of c_1.

Estimate the vector $x^{(p+1)}(t) - x^{(p)}(t)$ by the norm. Towards this end, introduce the function $z^{(p)}(t)$

$$z^{(p)}(t) = \max_{\theta \geq t} \left\| x^{(p+1)}(\theta) - x^{(p)}(\theta) \right\|. \qquad (1.3.38)$$

From (1.3.36) we have

$$x^{(p+1)}(t) - x^{(p)}(t) = H(x^{(p)}) - H(x^{(p-1)})$$

$$= \int_{t_0}^{+\infty} \mathbf{G}(t,\tau) \left(X(\tau, x^{(p)}(\tau)) - X(\tau, x^{(p-1)}(\tau)) \right) d\tau. \qquad (1.3.39)$$

From (1.3.39) we find

$$z^{(p)}(t) = \max_{\theta \geq t} \left\| \int_{t_0}^{+\infty} \mathbf{G}(\theta, \tau) \left(X(\tau, x^{(p)}(\tau)) - X(\tau, x^{(p-1)}(\tau)) \right) d\tau \right\|$$

$$\leq L \max_{\theta \geq t} \int_{t_0}^{+\infty} \|\mathbf{G}(\theta, \tau)\| z^{(p-1)}(\tau) \max \left(\|x^{(p)}(\tau)\|^m, \|x^{(p-1)}(\tau)\|^m \right) d\tau. \quad (1.3.40)$$

First estimate the function $z^{(1)}(\tau)$. This function is defined by the equality

$$z^{(1)}(t) = \max_{\theta \geq t} \|x^{(1)}(\theta) - x^{(0)}(\theta)\| = \max_{\theta \geq t} \|\mathbf{G}(\theta, t_0) c_1\|$$

$$\leq b_1 \|c_1\| e^{-(\bar{\lambda}-\epsilon)t}. \quad (1.3.41)$$

As noted above, all the successive approximations are in $D_{\gamma_0 \epsilon}$, therefore for each $p \geq 2$ the inequality $\|x^{(p)}\| \leq \gamma_0 e^{-(\bar{\lambda}-\epsilon)t}$ holds. From this inequality, and from (1.3.40), for $p = 2$ we find

$$z^{(2)}(t) \leq L \gamma_0^m b_1 \|c_1\| \max_{\theta \geq t} \int_{t_0}^{+\infty} \|\mathbf{G}(\theta, \tau)\| e^{-(1+m)(\bar{\lambda}-\epsilon)\tau} d\tau. \quad (1.3.42)$$

Utilizing the results obtained in the proof of Lemma 2, we evaluate the integral (1.3.42):

$$z^{(2)}(t) \leq \gamma_2 \gamma_0^m L b_1 \|c_1\| e^{-(\bar{\lambda}-\epsilon)t}. \quad (1.3.43)$$

Similarly, for the function $z^{(p)}(t)$ we have the estimate

$$z^{(p)}(t) \leq L \gamma_0^m \max_{\theta \geq t} \int_{t_0}^{+\infty} \|\mathbf{G}(\theta, \tau)\| e^{-m(\bar{\lambda}-\epsilon)\tau} z^{(p-1)}(\tau) d\tau. \quad (1.3.44)$$

From (1.3.44) we find

$$z^{(p)}(t) \leq b_1 \|c_1\| (L \gamma_2 \gamma_0^m)^{p-1} e^{-(\bar{\lambda}-\epsilon)t} \text{ for } t \geq t_0. \quad (1.3.45)$$

From Lemma 2 it follows that $c_0 = \gamma_0^{1+m}$, therefore if the inequality

$$c_0^{\frac{m}{1+m}} < \frac{1}{L\gamma_2},$$

holds, then the series (1.3.37) proves to be uniformly convergent for $t \geq t_0$ since the series

$$\sum_{p=1}^{\infty} z^{(p)}(t) \leq \frac{b_1 \|c_1\|}{1 - \gamma_2 L c_0^{\frac{m}{1+m}}} e^{-(\bar{\lambda}-\epsilon)t}, \quad (1.3.46)$$

which is majorant for the series made up of the norms of series terms (1.3.37), converges uniformly in $t \geq t_0$. Therefore, for $\|c_1\| \leq c_0$, where c_0 is the constant satisfying the inequality $c_0 \leq (L\gamma_2)^{-\frac{m+1}{m}}$, the series (1.3.37) converges to a continuous function $x(t)$ given for $t \geq t_0$.

Behavior of Integral Curves of a Nonlinear System

We shall now show that $x(t)$ is a unique solution of this form for the system of integral equations (1.3.6) with the fixed value of vector c_1. Towards this end, consider the expression
$$z(t) = x(t) - H(x(t)). \tag{1.3.47}$$
Our claim will be proved if we establish that the right-hand side of (1.3.47) is identically zero. Since $x^{(p+1)}(t) - H(x^{(p)}(t)) \equiv 0$, the equality (1.3.47) can be written as
$$z(t) = x(t) - x^{(p+1)}(t) + H(x^{(p)}) - H(x),$$
whence
$$\|z(t)\| \leq \|x - x^{(p+1)}\| + \|H(x^{(p)}) - H(x)\|$$
$$= \|x - x^{(p+1)}\| + \left\| \int_{t_0}^{+\infty} \mathbf{G}(t,\tau) \left(X(\tau, x^{(p)}(\tau)) - X(\tau, x(\tau)) \right) d\tau \right\|. \tag{1.3.48}$$
Since the integrals
$$\int_{t_0}^{+\infty} \mathbf{G}(t,\tau) X(\tau, x^{(p)}(\tau)) d\tau, \quad \int_{t_0}^{+\infty} \mathbf{G}(t,\tau) X(\tau, x(\tau)) d\tau$$
uniformly converge in p and t, for each $\epsilon_1 > 0$ it is possible to select a number $T > t_0$ such that for all $t \geq t_0$ we obtain
$$\left\| \int_T^{+\infty} \mathbf{G}(t,\tau) X(\tau, x^{(p)}(\tau)) d\tau \right\| < \epsilon_1/5, \quad \left\| \int_T^{+\infty} \mathbf{G}(t,\tau) X(\tau, x(\tau)) d\tau \right\| < \epsilon_1/5. \tag{1.3.49}$$
On the line segment $[t_0, T]$ the sequence $x^{(p)}(t)$ uniformly converges to the function $x(t)$, hence we may select a number p_0 such that for $p \geq p_0$ the following inequalities hold:
$$\|x(t) - x^{(p+1)}(t)\| < \epsilon_1/5, \; t \in [t_0, T],$$
$$\left\| \int_{t_0}^{t} \mathbf{G}(t,\tau) \left(X(\tau, x^{(p)}(\tau)) - X(\tau, x(\tau)) \right) d\tau \right\| < \epsilon_1/5, \tag{1.3.50}$$
$$\left\| \int_{t}^{T} \mathbf{G}(t,\tau) \left(X(\tau, x^{(p)}(\tau)) - X(\tau, x(\tau)) \right) d\tau \right\| < \epsilon_1/5, \; t \in [t_0, T].$$
From (1.3.48), (1.3.49) and (1.3.50) we find that $\|z(t)\| \leq \epsilon_1$ for $p \geq p_0$. Since $z(t)$ is independent of p and the number $\epsilon_1 > 0$ is arbitrarily small, we have $\|z(t)\| \equiv 0$, hence $x(t)$ is a solution to the system of integral equations (1.3.6).

We shall now establish uniqueness of the solution obtained. Suppose there exists an alternative continuous solution to system (1.3.6), $\bar{x}(t) \in D_{\gamma_0 \epsilon}$, corresponding to the same value of vector c_1 as the above solution $x(t)$. Then we have two identities: $x(t) = H(x(t))$ and $\bar{x}(t) = H(\bar{x}(t))$. Subtracting one from the other and estimating by the norm, we find
$$\|x(t) - \bar{x}(t)\| \leq \left\| \int_{t_0}^{+\infty} \mathbf{G}(t,\tau) \left(X(\tau, x(\tau)) - X(\tau, \bar{x}(\tau)) \right) d\tau \right\|$$
$$\leq L\gamma_0^m \int_{t_0}^{+\infty} \|\mathbf{G}(t,\tau)\| e^{-m(\bar{\lambda}-\epsilon)\tau} \|x(\tau) - \bar{x}(\tau)\| d\tau. \tag{1.3.51}$$

In deriving (1.3.51), we utilized the inequalities $\|x\| \leq \gamma_0 e^{-(\bar{\lambda}-\epsilon)t}$, $\|\bar{x}\| \leq \gamma_0 e^{-(\bar{\lambda}-\epsilon)t}$. Let

$$a = \sup_{t \geq t_0} \|x(t) - \bar{x}(t)\|,$$

then we obtain the inequality

$$a \leq a L \gamma_0^m \sup_{t \geq t_0} \int_{t_0}^{+\infty} \|\mathbf{G}(t,\tau)\| e^{-m(\bar{\lambda}-\epsilon)\tau} d\tau < a\gamma_2 L \gamma_0^m. \tag{1.3.52}$$

If it so happened that $a \neq 0$, i.e. if $x(t)$ and $\bar{x}(t)$ did not coincide even at one point, then from the inequality (1.3.52) it would be possible to obtain the inequality $1 \leq \gamma_2 L \gamma_0^m$, but this yields a contradiction to the choice of a number γ_0 from the condition $\gamma_2 L \gamma_0^m < 1$. Therefore, $a = 0$ and $x(t) = \bar{x}(t)$. ∎

3. Proof of Theorem 6: First show that between the solutions to the system of integral equations (1.3.6), constructed in Lemma 3, and the solutions to the original system of differential equations there exists a fairly intimate relation. Suppose that $x(t)$ is a solution to system (1.3.6) mentioned in Lemma 3. By substituting this solution into system (1.3.6), we obtain the identity

$$x(t) = \int_{t_0}^{+\infty} \mathbf{G}(t,\tau) X(\tau, x(\tau)) d\tau + \mathbf{G}(t,t_0) c_1$$

$$= \int_{t_0}^{t} \mathbf{G}(t,\tau) X(\tau, x(\tau)) d\tau + \int_{t}^{+\infty} \mathbf{G}(t,\tau) X(\tau, x(\tau)) d\tau + \mathbf{G}(t,t_0) c_1. \tag{1.3.53}$$

Differentiating both sides of (1.3.53), we find

$$\frac{dx(t)}{dt} = \mathbf{G}(t,t-0) X(t,x(t)) - \mathbf{G}(t,t+0) X(t,x(t))$$

$$+ \int_{t_0}^{t} \mathbf{P}(t) \mathbf{G}(t,\tau) X(\tau, x(\tau)) d\tau + \int_{t}^{+\infty} \mathbf{P}(t) \mathbf{G}(t,\tau) X(\tau, x(\tau)) d\tau$$

$$+ \mathbf{P}(t) \mathbf{G}(t,t_0) c_1 = \mathbf{P}(t) x(t) + X(t,x(t)). \tag{1.3.54}$$

From (1.3.54) it is seen that the function $x(t)$ is a solution to system (1.3.1).

We shall now show that any solution of system (1.3.1), which is an element of the set $D_{\gamma_0 \epsilon}$, is necessarily a solution to system (1.3.6). Let $x(t)$ be such a solution to system (1.3.1), then we obtain the identity

$$\frac{dx(t)}{dt} = \mathbf{P}(t) x(t) + X(t, x(t)). \tag{1.3.55}$$

Identity (1.3.55), considered at the point τ, will be multiplied on the left by the matrix $\mathbf{G}(t, \tau)$ and will be first integrated between the limits t and $+\infty$, and then between t_0 and t. Obtain two identities

$$\int_{t}^{+\infty} \mathbf{G}(t,\tau) \dot{x}(\tau) d\tau = \int_{t}^{+\infty} \mathbf{G}(t,\tau) \mathbf{P}(\tau) x(\tau) d\tau$$

$$+ \int_t^{+\infty} \mathbf{G}(t,\tau)X(\tau,x(\tau))d\tau, \qquad (1.3.56)$$

$$\int_{t_0}^t \mathbf{G}(t,\tau)\dot{x}(\tau)d\tau = \int_{t_0}^t \mathbf{G}(t,\tau)\mathbf{P}(\tau)x(\tau)d\tau$$

$$+ \int_{t_0}^t \mathbf{G}(t,\tau)X(\tau,x(\tau))d\tau. \qquad (1.3.57)$$

Integrating the left-hand side of (1.3.56) and (1.3.57) by parts, we find

$$-\mathbf{G}(t,t+0)x(t) + \int_t^{+\infty} \mathbf{G}(t,\tau)\mathbf{P}(\tau)x(\tau)d\tau$$

$$= \int_t^{+\infty} \mathbf{G}(t,\tau)\mathbf{P}(\tau)x(\tau)d\tau + \int_t^{+\infty} \mathbf{G}(t,\tau)X(\tau,x(\tau))d\tau, \qquad (1.3.58)$$

$$\mathbf{G}(t,t-0)x(t) - \mathbf{G}(t,t_0)x(t_0) + \int_{t_0}^t \mathbf{G}(t,\tau)\mathbf{P}(\tau)x(\tau)d\tau$$

$$= \int_{t_0}^t \mathbf{G}(t,\tau)\mathbf{P}(\tau)x(\tau)d\tau + \int_{t_0}^t \mathbf{G}(t,\tau)X(\tau,x(\tau))d\tau. \qquad (1.3.59)$$

Adding termwise (1.3.58) to (1.3.59) and utilizing the properties of matrix $\mathbf{G}$, we obtain the equality

$$x(t) = \int_{t_0}^{+\infty} \mathbf{G}(t,\tau)X(\tau,x(\tau))d\tau + \mathbf{G}(t,t_0)x(t_0), \qquad (1.3.60)$$

whence it follows that the vector function $x(t)$ is a solution to the system of integral equations (1.3.6) with the condition $c_1 = x(t_0)$.

It remains to show that the system (1.3.1) has the set of solutions possessing the property as in (1.3.5) and depending on k arbitrary constants.

The vector $\mathbf{G}(t,t_0)c_1$ becomes

$$\mathbf{G}(t,t_0)c_1 = \mathbf{Y}(t)\mathbf{G}_1\mathbf{Y}^{-1}(t_0)c_1. \qquad (1.3.61)$$

If we denote by c the vector $\mathbf{Y}^{-1}(t_0)c_1$, then we may assume without loss of generality that its components are arbitrary constants, since the components of the vector c_1 are arbitrary constants and the matrix $\mathbf{Y}(t_0)$ is nonsingular. The vector $\mathbf{G}_1 c$ has only k first components, as distinct from zero $c_1, \ldots, c_k$, while the remaining $n-k$ components are zero. Denote this vector by c_0. Then relation (1.3.61) may be written as

$$\mathbf{G}(t,t_0)c_1 = \mathbf{Y}(t)c_0. \qquad (1.3.62)$$

Solutions to system (1.3.6) are dependent on the values of vector c_0 and, therefore, on k arbitrary constants.

Considering the replacement of vectors and the estimate in (1.3.35), we find that each integral curve of system (1.3.1), which is a member of the obtained family, will satisfy inequality (1.3.5). ∎

Remark 1. In the proof of Lemma 2, we employed the condition for the vector-valued function $X(t,x)$

$$\|X(t,x)\| \leq L \|x\|^{1+m}, \tag{1.3.63}$$

which, in the general case, is weaker than the Lipschitz condition

$$\|X(t,x') - X(t,x'')\| \leq \|x' - x''\| L \max(\|x'\|^m, \|x''\|^m). \tag{1.3.64}$$

However, Lemma 3 was proved by utilizing the Lipschitz condition (1.3.64). When only condition (1.3.63) is fulfilled, the system of integral equations (1.3.6) has the set of solutions with the same estimate for the rate of decrease as in Lemma 3. But in this case it is not possible to guarantee the uniqueness of solutions to system (1.3.1).

We shall now show that if the conditions a) and b) imposed on the right-hand sides of system (1.3.1) are satisfied, and (1.3.64) is replaced by (1.3.63), the system (1.3.6) also has a set of solutions depending on k arbitrary constants, and this set of solutions is contained in the set $D_{\gamma_0 \epsilon}$ if the conditions of Theorem 6 are satisfied. Let us consider the sequence

$$x^{(0)}(t) = 0, \quad x^{(p+1)}(t) = H(x^{(p)}(t)), \quad p = 0, 1, \ldots.$$

By Lemma 2, with $\|c_1\| < c_0$ and $\gamma_0 = c^{\frac{1}{1+m}}$, all successive approximations belong to the collection of functions $D_{\gamma_0 \epsilon}$, i.e., they satisfy the inequality

$$\|x^{(p)}(t)\| \leq \gamma_0 e^{-(\bar{\lambda}-\epsilon)t}, \quad t \geq t_0. \tag{1.3.65}$$

Let us prove that the functions in this sequence are equicontinuous. From differentiation we get

$$\frac{dx^{(p+1)}(t)}{dt} = \mathbf{P}(t)x^{(p+1)}(t) + X(t, x^{(p)}(t)). \tag{1.3.66}$$

From (1.3.65) and (1.3.66) we find the inequality

$$\left\|\frac{d}{dt}x^{(p+1)}(t)\right\| \leq \gamma_3 e^{-(\bar{\lambda}-\epsilon)t}. \tag{1.3.67}$$

From inequality (1.3.67) it follows that the collection of functions $\{x^{(p)}(t)\}$ is equicontinuous. Let us take a sequence of positive numbers $T_p \to +\infty$ as $p \to \infty$

$$T_{p+1} > T_p > \ldots > t_0.$$

The sequence of functions $x^{(p)}(t)$ satisfies the conditions of the Arzela theorem on the closed interval $[t_0, T_1]$, hence there is the sequence $x_1^{(1)}, x_1^{(2)}, \ldots$, uniformly converging in the closed interval $[t_0, T_1]$ to the function $x(t)$. The sequence of functions $x_1^{(p)}$ satisfies the conditions of the Arzela theorem in the closed interval $[t_0, T_2]$, hence it is possible to select the sequence uniformly converging in the closed interval $[t_0, T_2]$ to the function which is again denoted by $x(t)$. This function coincides with the above function in the closed interval $[t_0, T_1]$. Continuing this process unboundedly, we find, in particular, the sequence $x_q^{(p)}$ uniformly converging in the closed interval $[t_0, T_q]$ to the function which is again denoted by $x(t)$.

Denote by $\tilde{x}^{(q)}$ the function $x_q^{(q)}$ and show that the sequence $\tilde{x}^{(q)}$ uniformly converges for $t \geq t_0$ to the function $x(t)$. Note that:

1) the sequence $\tilde{x}^{(q)}(t)$ uniformly converges to $x(t)$ in any finite interval $[t_0, T]$;
2) the function $x(t)$ for each finite $t \geq t_0$ satisfies the inequality $\|x(t)\| \leq \gamma_0 e^{-(\tilde{\lambda}-\epsilon)t}$.

Simultaneous consideration of such properties demonstrates the validity of this proposition.

As in Lemma 3, it may be shown that the constructed function $x(t)$ is a solution to the system of integral equations (1.3.6) and, therefore, to the system (1.3.1). As before, we find that the system (1.3.6) has a set of solutions depending on k arbitrary constants. Hence the system (1.3.1) also has a set of solutions, each solution satisfies inequality (1.3.5) and depends on k arbitrary constants.

In conclusion we will give an example which shows that under conditions as in (1.3.63) the nonuniqueness may occur in the original system of differential equations. Consider the system

$$\frac{dx}{dt} = -x + xy^{1/3}, \quad \frac{dy}{dt} = -|x|y^{1/3}$$

satisfying the condition (1.3.63) for $m = 1/3$, $L = 1$. This system has a solution $x = x_0 e^{-t}, y = 0$. Integrating the second equation, we find $(y^{2/3})' = -\frac{2}{3}|x|$, whence

$$y = \sqrt{\left(y_0^{2/3} - \frac{2}{3}\int_0^t |x(\tau)|\,d\tau\right)^3}.$$

If there exists a number $\bar{t} > 0$ such that the subradical expression goes to zero, then for $t \geq \bar{t}$ there will be $y \equiv 0$. Therefore, an infinite number of integral curves fall on the Ox axis in a finite time and stay there, which implies nonuniqueness. Although more involved situations may arise here, we will further employ Theorem 6 only in its original formulation without giving coverage to special problems of nonuniqueness.

Remark 2. Lemma 3 gives a fairly rough estimation of the series terms $\sum_{p=0}^{+\infty}(x^{(p+1)}(t) - x^{(p)}(t))$. Here we will derive the estimates for these terms characterizing more exactly the rate of their decrease as $t \to +\infty$; we have

$$\left\|x^{(1)}(t)\right\| \leq b_1\|c_1\|e^{-(\tilde{\lambda}-\epsilon)t}, \quad \left\|x^{(2)}(t)\right\| \leq \gamma_0 e^{-(\tilde{\lambda}-\epsilon)t},$$

$$\left\|x^{(2)}(t) - x^{(1)}(t)\right\| \leq \int_{t_0}^{+\infty} \|G(t,\tau)\| \left\|X(\tau, x^{(1)}(\tau))\right\| d\tau$$

$$\leq L\gamma_0^m \int_{t_0}^{+\infty} \|G(t,\tau)\|e^{-m(\tilde{\lambda}-\epsilon)\tau}b_1\|c_1\|e^{-(\tilde{\lambda}-\epsilon)\tau}d\tau$$

$$\leq L\gamma_3\gamma_0^m b_1\|c_1\|e^{-\mu t}e^{-(\tilde{\lambda}-\epsilon)t}. \quad (1.3.68)$$

Next, we get

$$\left\|x^{(p+1)}(t) - x^{(p)}(t)\right\|$$

$$\leq L\gamma_0^m \int_{t_0}^{+\infty} \|\mathbf{G}(t,\tau)\| e^{-m(\bar{\lambda}-\epsilon)\tau} \left\|x^{(p)}(\tau) - x^{(p-1)}(\tau)\right\| d\tau. \tag{1.3.69}$$

In terms of (1.3.68), from (1.3.69) we find

$$\left\|x^{(p+1)}(t) - x^{(p)}(t)\right\| \leq (\gamma_3 L\gamma_0^m)^p b_1 \|c_1\| e^{-p\bar{\mu}t - (\bar{\lambda}-\epsilon)t}, \tag{1.3.70}$$

where μ is a positive constant, which is independent of p and ϵ for sufficiently small values of $\epsilon > 0$. Denoting $L\gamma_3$ by b_2, replacing γ_0 with $c_0^{\frac{1}{m+1}}$, and $\|c_1\|$ with c_0, we finally obtain

$$\left\|x^{(p+1)}(t) - x^{(p)}(t)\right\| \leq b_1 c_0^{1+\frac{pm}{1+m}} b_2^p e^{-(p\bar{\mu}+\bar{\lambda}-\epsilon)t}. \tag{1.3.71}$$

From the estimate (1.3.71) it follows that the series $\sum_{p=0}^{\infty} \left(x^{(p+1)}(t) - x^{(p)}(t)\right)$ will be convergent, and the series of norms for its terms has the majorant

$$b_1 \sum_{p=1}^{\infty} b_2^p c_0^{1+\frac{pm}{1+m}} e^{-(p\bar{\mu}+\bar{\lambda}-\epsilon)t} \tag{1.3.72}$$

and, therefore, converges with any admissible value of $\|c_1\|$ for all $t \geq T(c_1)$.

Remark 3. Analysis of the proof of Theorem 6 shows that it can be extended to the case where system (1.3.3) is irregular

$$\chi\left(e^{-\int_0^t \sum_{s=1}^n p_{ss}(\tau)d\tau}\right) + \sum_{j=1}^n \lambda_j = -\sigma, \tag{1.3.73}$$

where $\sigma > 0$ is the irregularity coefficient. Irregularity primarily implies that the function $\mathbf{G}(t,\tau)$ has the estimates somewhat differing from those derived in Lemma 1. Changes in the estimates are due to the fact that the elements of the matrix $\mathbf{Y}^{-1}(t)$ now have distinct characteristic numbers. It should be recalled that the element $\bar{y}_{sj}$ of this matrix is determined from the formula $\bar{y}_{sj}(t) = \frac{\Delta_{sj}(\mathbf{Y})}{\Delta(\mathbf{Y})}$. Since the characteristic number of the function product is not less than the sum of characteristic numbers of factors and the characteristic number of the function sum is not less than the smallest characteristic number of the summable functions, then in terms of

$$\chi\left(\frac{\Delta_{sj}}{\Delta}\right) \geq \chi\left(\frac{1}{\Delta}\right) + \chi(\Delta_{sj}),$$

we find that

$$\chi(\bar{y}_{sj}) \geq -\lambda_s - \sigma.$$

As a result the factor $e^{\sigma\tau}$ appears in the estimate of $\|\mathbf{G}(t,\tau)\|$, that is:

$$\|\mathbf{G}(t,\tau)\| \leq \gamma_1 e^{-\bar{\lambda}(t-\tau)} e^{(\sigma+\delta)\tau} e^{\delta t}, \ t \geq \tau \geq 0, \tag{1.3.74}$$

and

$$\|\mathbf{G}(t,\tau)\| \leq \gamma_1 e^{(\sigma+\delta)\tau} e^{\delta t} e^{-\sigma(t-\tau)}, \ 0 \leq t < \tau. \tag{1.3.75}$$

Inequalities (1.3.74) — (1.3.75) are derived on the assumption that the characteristic numbers $\lambda_1, \ldots, \lambda_n$ of system (1.3.3) include k positive numbers which exceed σ, while the remaining $n - k$ numbers do not exceed σ, that is:

$$\lambda_j > \sigma, \ j = 1, \ldots, k; \ \lambda_j \leq \sigma, \ j = k+1, \ldots, n. \qquad (1.3.76)$$

In order to pursue the course of reasoning employed in the proof of Lemma 2, it is essential to make the assumption which provides a means to weaken the influence of additional multipliers in (1.3.74) and (1.3.75) on the estimates. Denote by $\bar{\lambda}$ the smallest characteristic number in the group $\lambda_1, \ldots, \lambda_k$; then $\bar{\lambda} > \sigma$. First ensure that the set of functions $D_{\gamma_0\epsilon}$ with sufficiently small $\epsilon > 0$ is mapped into itself with the help of the operator

$$H(x) = \int_{t_0}^{+\infty} \mathbf{G}(t, \tau) X(\tau, x(\tau)) d\tau + \mathbf{G}(t, t_0) c_1.$$

This becomes possible on satisfying the inequality

$$m > \sigma/\bar{\lambda}, \qquad (1.3.77)$$

which is found in the course of reasoning as in Lemma 2, its necessity being open to question, though. From inequality (1.3.77) we have that for all sufficiently small $\epsilon > 0$ the value of $\sigma - m(\bar{\lambda} - \epsilon)$ will be negative. By properly selecting δ, for such values of ϵ, γ_0 and $\|c_1\|$ it is possible to ensure that the requirement of $H(x) \in D_{\gamma_0\epsilon}$, be satisfied if $x \in D_{\gamma_0\epsilon}$.

Based on (1.3.77) and on the earlier discussion, Lemma 3 for the case of the irregular system (1.3.3) is proved by analogy. From this it follows that henceforth the irregularity for system (1.3.3) does not present any difficulty. The changes introduced into the proof of Theorem 6 with due regard to the irregularity of system (1.3.3) show that the following statement is valid.

Theorem 7. *If 1) the right-hand sides of system (1.3.1) satisfy conditions a) and b);*

2) the system (1.3.3) is irregular, i.e.,

$$\sum_{j=1}^{n} \lambda_j + \chi\left(e^{-\int_0^t \sum_{s=1}^{n} p_{ss}(t) dt}\right) = -\sigma, \ \sigma > 0;$$

3) the characteristic numbers $\lambda_1, \ldots, \lambda_n$ of system (1.3.3) do not comprise the k numbers exceeding σ and the $n - k$ numbers not exceeding σ so that $\lambda_j > \sigma, \ j = 1, \ldots, k$, and $\lambda_j \leq \sigma, \ j = k+1, \ldots, n$;

4) the number m from condition b) satisfies the inequality $m > \sigma/\bar{\lambda}$, where $\bar{\lambda} = \min_{j=1,\ldots,k} \lambda_j$, then system (1.3.1) has a family of curves depending on k arbitrary constants $c_1, \ldots, c_k$. In this case, every integral curve of the family satisfies inequality (1.3.5).

Remarks 1 and 2 apply to this statement as much as to Theorem 6 itself.

1.4 Behavior of Integral Curves in the Neighborhood of a Singular Point

1. We shall investigate the behavior of integral curves in the neighborhood of a singular point on the right-hand sides of a system of equations and the analytical representation of the sets of such solutions. First consider the problems for systems of partial differential equations, and then for systems of ordinary equations using the results obtained.

Take the system of partial differential equations

$$\frac{\partial z_j}{\partial t} + \sum_{s=1}^{n} \frac{\partial z_j}{\partial x_s} \left(\sum_{h=1}^{n} p_{sh} x_h + X_s(t, x_1, \ldots, x_n, z_1, \ldots, z_k) \right)$$

$$= \sum_{h=1}^{k} q_{jh} z_h + \sum_{h=1}^{n} r_{jh} x_h + Z_j(t, x_1, \ldots, x_n, z_1, \ldots, z_k), \; j = 1, \ldots, k. \quad (1.4.1)$$

Assume that all the functions appearing in system (1.4.1) satisfy the following conditions:

a) the functions $p_{jh}(t), q_{jh}(t), r_{jh}(t)$ are continuous, given for $t \geq 0$, real and bounded, the functions $X_s(t, x_1, \ldots, x_n, z_1, \ldots, z_k)$ and $Z_j(t, x_1, \ldots, x_n, z_1, \ldots, z_k)$ are given for $t \geq 0, \|x\| \leq r, \|z\| \leq r$, being real and continuous therein;

b) the functions X_s, Z_j are continuously differentiable with respect to $x_1, \ldots, x_n$, $z_1, \ldots, z_k$ and satisfy the Lipschitz condition of the form

$$| X_s(t, \bar{x}_1, \ldots, \bar{x}_n, \bar{z}_1, \ldots, \bar{z}_k) - X_s(t, \bar{\bar{x}}_1, \ldots, \bar{\bar{x}}_n, \bar{\bar{z}}_1, \ldots, \bar{\bar{z}}_k) |$$

$$\leq L \left(\sum_{j=1}^{n} |\bar{x}_j - \bar{\bar{x}}_j| + \sum_{j=1}^{k} |\bar{z}_j - \bar{\bar{z}}_j| \right) \max \left\{ \left(\sum_{s=1}^{n} |\bar{x}_s| + \sum_{j=1}^{k} |\bar{z}_j| \right)^m, \right.$$

$$\left. \left(\sum_{s=1}^{n} |\bar{\bar{x}}_s| + \sum_{j=1}^{k} |\bar{\bar{z}}_j| \right)^m \right\}, \; s = 1, \ldots, n; \quad (1.4.2)$$

$$|Z_j(t, \bar{x}_1, \ldots, \bar{x}_n, \bar{z}_1, \ldots, \bar{z}_k) - Z_j(t, \bar{\bar{x}}_1, \ldots, \bar{\bar{x}}_n, \bar{\bar{z}}_1, \ldots, \bar{\bar{z}}_k)|$$

$$\leq L \left(\sum_{s=1}^{n} |\bar{x}_s - \bar{\bar{x}}_s| + \sum_{j=1}^{k} |\bar{z}_j - \bar{\bar{z}}_j| \right) \max \left\{ \left(\sum_{s=1}^{n} |\bar{x}_s| + \sum_{j=1}^{k} |\bar{z}_j| \right)^m, \right.$$

$$\left. \left(\sum_{s=1}^{n} |\bar{\bar{x}}_s| + \sum_{j=1}^{k} |\bar{\bar{z}}_j| \right)^m \right\}, \; j = 1, \ldots, k, \quad (1.4.3)$$

$$X_s(t, 0, \ldots, 0, 0, \ldots, 0) \equiv Z_j(t, 0, \ldots, 0, 0, \ldots, 0) \equiv 0,$$

$$s = 1, \ldots, n, \; j = 1, \ldots, k. \quad (1.4.4)$$

Next, denote by $\lambda_1, \ldots, \lambda_n$ the characteristic numbers of the system of linear differential equations

$$\frac{dx_s}{dt} = \sum_{j=1}^{n} p_{sj}(t) x_j, \; s = 1, \ldots, n, \quad (1.4.5)$$

Behavior of Integral Curves in the Neighborhood

and by $\mu_1, \ldots, \mu_k$ the characteristic numbers of the system

$$\frac{dz_j}{dt} = \sum_{h=1}^{k} q_{jh}(t) z_h, \quad j = 1, \ldots, k. \tag{1.4.6}$$

Theorem 8 *If system (1.4.1) satisfies the following conditions: 1) constraints a) and b) in this section are valid;*
2) systems (1.4.5) and (1.4.6) are regular;
3) the values $\lambda_1, \ldots, \lambda_n$ are positive, and the values $\mu_1, \ldots, \mu_k$ with $l \leq k$ satisfying the equalities $\lambda_j = \mu_j, j = 1, \ldots, l$, then the system (1.4.1) has a solution set which is dependent on l arbitrary constants

$$z_j = f_j(t, x_1, \ldots, x_n, c_1, \ldots, c_l), \quad j = 1, \ldots, k, \tag{1.4.7}$$

satisfying the system (1.4.1) and the condition

$$f_j(t, 0, \ldots, 0, c_1, \ldots, c_l) \equiv 0, \quad j = 1, \ldots, k.$$

Proof: Consider a system of differential equations that are the equations of characteristics for system (1.4.1):

$$\frac{dx_s}{dt} = \sum_{h=1}^{n} p_{sh} x_h + X_s(t, x_1, \ldots, x_n, z_1, \ldots, z_k);$$

$$\frac{dz_j}{dt} = \sum_{h=1}^{k} q_{jh} z_h + \sum_{h=1}^{n} r_{jh} x_h + Z_j(t, x_1, \ldots, x_n, z_1, \ldots, z_k), \tag{1.4.8}$$

$$s = 1, \ldots, n, \quad j = 1, \ldots, k.$$

It follows from the conditions of Theorem 8 that the system (1.4.8) has a family of integral curves (by Theorem 6) which can be represented as

$$x_s = \bar{\alpha}_s + \bar{f}_s(t, \alpha_1, \ldots, \alpha_n, \beta_1, \ldots, \beta_{l_1}), \quad s = 1, \ldots, n, \tag{1.4.9}$$

$$z_j = \bar{q}_j(t, \alpha_1, \ldots, \alpha_n, \beta_1, \ldots, \beta_{l_1}), \quad j = 1, \ldots, k, \tag{1.4.10}$$

where

$$\bar{\alpha}_s = \sum_{j=1}^{n} y_{sj}(t) \alpha_j, \quad \beta_i > 0, \quad i = 1, \ldots, l_1, \quad l \leq l_1 \leq k.$$

Here $\mathbf{Y}(t) = \{y_{sj}(t)\}$, $s, j = 1, \ldots, n$ is a normal fundamental system of solutions for (1.4.5), the j-th column in the matrix $\mathbf{Y}(t)$ being the solution corresponding to the characteristic number λ_j. The functions $\bar{f}_s$ and $\bar{q}_s$ are continuously differentiable with respect to $\alpha_1, \ldots, \alpha_n, \beta_1, \ldots, \beta_{l_1}$ in a sufficiently small neighborhood of a point $\alpha_1 = \ldots = \alpha_n = \beta_1 = \ldots = \beta_{l_1} = 0$, and, additionally, $\frac{\partial \bar{f}_s}{\partial \alpha_h}$ and $\frac{\partial \bar{f}_s}{\partial \beta_i}$ tend to zero as

$$\|\alpha\| + \|\beta\| \to 0, \quad s = 1, \ldots, n, \quad h = 1, \ldots, n, \quad i = 1, \ldots, l_1.$$

Show this. Construct a normal fundamental system for the system of equations

$$\frac{dx_s}{dt} = \sum_{h=1}^{n} p_{sh} x_h,$$

$$\frac{dz_j}{dt} = \sum_{h=1}^{k} q_{ih} z_h + \sum_{h=1}^{n} r_{jh} x_h, \quad s = 1, \ldots, n, \; j = 1, \ldots, k. \qquad (1.4.11)$$

In this case, we substitute in the first n columns the solutions with characteristic numbers $\lambda_1, \ldots, \lambda_n$, and in the columns ranging from $(n+1)$ to $(n+k)$ the solutions with characteristic numbers $\mu_1, \ldots, \mu_k$. Denote the matrix made up of such solutions by $\hat{\mathbf{Y}}$,

$$\hat{\mathbf{Y}} = \begin{pmatrix} \mathbf{Y} & \mathbf{O} \\ \mathbf{H} & \mathbf{Z} \end{pmatrix}. \qquad (1.4.12)$$

In order to show that the system (1.4.8) has the set of solutions (1.4.9) and (1.4.10), we define the matrix $\hat{\mathbf{G}}(t,\tau)$ in terms of $\hat{\mathbf{Y}}(t)$ just as in the preceding section. Let $\lambda_1, \ldots, \lambda_n, \mu_1, \ldots, \mu_{l_1}$ be the positive characteristic numbers of system (1.4.11), $l \leq l_1 \leq k$. Set

$$\hat{\mathbf{G}}(t,\tau) = \hat{\mathbf{Y}}(t)\mathbf{G}_1\hat{\mathbf{Y}}^{-1}(\tau), \; 0 \leq t \leq \tau,$$

$$\hat{\mathbf{G}}(t,\tau) = \hat{\mathbf{Y}}(t)\mathbf{G}_2\hat{\mathbf{Y}}^{-1}(\tau), \; 0 \leq \tau \leq t,$$

where

$$\mathbf{G}_1 = \begin{pmatrix} \mathbf{I}_{n+l_1} & \mathbf{O} \\ \mathbf{O} & \mathbf{O} \end{pmatrix}, \; \mathbf{G}_2 = \begin{pmatrix} \mathbf{O} & \mathbf{O} \\ \mathbf{O} & -\mathbf{I}_{k-l_1} \end{pmatrix}.$$

Therefore, the system of integral equations (1.3.6) for this case becomes

$$(x;z) = \int_{t_0}^{+\infty} \hat{\mathbf{G}}(t,\tau)(X(\tau,x,z); Z(\tau,x,z))d\tau + \hat{\mathbf{G}}(t,t_0)c_1. \qquad (1.4.13)$$

Here $(x;z)$ means the $(n+k)$-dimensional column vector $(x_1, \ldots, x_n, z_1, \ldots, z_k)$. We can denote by c the $(n+k)$-dimensional vector determined from the formula $c = \mathbf{G}_1\hat{\mathbf{Y}}^{-1}(t_0)c_1$. The system (1.4.13) becomes

$$(x;z) = \int_{t_0}^{\infty} \hat{\mathbf{G}}(t,\tau)(X;Z)d\tau + \hat{\mathbf{Y}}(t)c. \qquad (1.4.14)$$

Denote the first $n+l_1$ components of the vector c by $\alpha_1, \ldots, \alpha_n; \beta_1, \ldots, \beta_{l_1}$; then from (1.4.14) we get:

$$x_s = \bar{\alpha}_s + \int_{t_0}^{+\infty} \left(\sum_{j=1}^{n} \{\hat{\mathbf{G}}(t,\tau)\}_{sj} X_j(\tau,x,z) \right.$$

$$\left. + \sum_{j=n+1}^{n+k} \{\hat{\mathbf{G}}(t,\tau)\}_{sj} Z_j(\tau,x,z) \right) d\tau, \; s = 1, \ldots, n, \qquad (1.4.15)$$

$$z_j = \sum_{h=1}^{n} \{\mathbf{H}\}_{jh} \alpha_h + \sum_{h=1}^{l_1} \{\mathbf{Z}\}_{jh} \beta_h + \int_{t_0}^{+\infty} \left(\sum_{h=1}^{n} \{\hat{\mathbf{G}}(t,\tau)\}_{jh} \right.$$

Behavior of Integral Curves in the Neighborhood

$$\times X_h(\tau, x, z) + \sum_{h=k+1}^{n+k} \{\hat{\mathbf{G}}(t, \tau)\}_{jh} Z_h(\tau, x, z)\right) d\tau, \ j = 1, \ldots, k. \tag{1.4.16}$$

The integrals in (1.4.15) and (1.4.16) are continuously differentiable with respect to $\alpha_1, \ldots, \alpha_n; \beta_1, \ldots, \beta_{l_1}$, as with respect to parameters in a sufficiently small neighborhood of a point $\alpha_1 = \alpha_2 = \ldots = \alpha_n = \beta_1 = \ldots = \beta_{l_1} = 0$, the derivatives of these integrals vanish at $\alpha_1 = \ldots = \alpha_n = \beta_1 = \ldots = \beta_{l_1} = 0$. We now set in relations (1.4.9), (1.4.10) $\beta_j = \alpha_j c_j$, $j = 1, \ldots, l$ and $\beta_j = 0$, $j = l+1, \ldots, l_1$. Solve (1.4.9) with respect to $\alpha_1, \ldots, \alpha_n$ and obtain

$$\alpha_s = \varphi_s(t, x_1, \ldots, x_n, c_1, \ldots, c_l), \ s = 1, \ldots, n. \tag{1.4.17}$$

Here the functions φ_s are continuously differentiable with respect to $x_1, \ldots, x_n$ and have a property such that $\varphi_s(t, 0, \ldots, 0, c_1, \ldots, c_l) = 0$. Using (1.4.17), we eliminate $\alpha_1, \ldots, \alpha_n$ between equalities (1.4.10):

$$z_j = f_j(t, x_1, \ldots, x_n, c_1, \ldots, c_l), \ j = 1, \ldots, k. \tag{1.4.18}$$

The functions f_j are continuously differentiable with respect to $x_1, \ldots, x_n$ and satisfy the system (1.4.1) and the conditions

$$f_j(t, 0, \ldots, 0, c_1, \ldots, c_l) \equiv 0. \qquad \blacksquare$$

Corollary. If the functions X_s and Z_j in system (1.4.1) are holomorphic in a sufficiently small neighborhood of a point $x_1 = \ldots = x_n = z_1 = \ldots = z_k = 0$ equally for all $t \geq 0$ and their expansions into integral non-negative powers of $x_1, \ldots, x_n, z_1, \ldots, z_k$ start at least with second-order terms, then applying Theorem 6, we find that the functions $\bar{f}_s$ in equalities (1.4.9) and $\bar{q}_j$ in equalities (1.4.10) will be holomorphic with respect to $\alpha_1, \ldots, \alpha_n, \beta_1, \ldots, \beta_{l_1}$. Hence, the functions $\varphi_s(t, x_1, \ldots, x_n, c_1, \ldots, c_l)$ on the right-hand sides of (1.4.17) will be holomorphic with respect to $x_1, \ldots, x_n$ for sufficiently small $\|x\|$, so the functions $z_j = f_j(t, x_1, \ldots, x_n, c_1, \ldots, c_l)$ on the right-hand sides of (1.4.18) will also be holomorphic with respect to $x_1, \ldots, x_n$ for sufficiently small $\|x\|$ and, additionally,

$$z_j = \sum_{p=1}^{\infty} z_j^{(p)}, \tag{1.4.19}$$

where $z_j^{(p)}$ are homogeneous forms of the variables $x_1, \ldots, x_n$ of power p with the coefficients which are polynomials in $c_1, \ldots, c_l$ and, simultaneously, continuous functions of t for $t \geq 0$ [Zubov (1964)].

In fact, the expression (1.4.9) can be represented in terms of Theorem 6 as a series arranged, after replacement of $\beta_j = c_j \alpha_j, j = 1, \ldots, l$, in powers of $\alpha_1, \ldots, \alpha_n$ with the coefficients that are polynomials in $c_1, \ldots, c_l$. This suggests that the functions φ_s in (1.4.17), and hence the functions in (1.4.18), will have the same properties. In general, the radius of convergence of series (1.4.19) is the function of t unboundedly decreasing to zero as $t \to +\infty$. In what follows, we will give an example which illustrates the validity of this statement. $\blacksquare$

2. We shall now discuss the structure of coefficients of the forms $z_j^{(p)}$ regarded as the functions of time. In general, where none of the additional assumptions is made of the series coefficients X_s and Z_j, except that these coefficients are the real continuous bounded functions of time given for $t \geq 0$, we may only state that the coefficients of the form $z_j^{(p)}$ are the continuous functions given for $t \geq 0$ and have nonpositive characteristic numbers.

Assume that all the functions p_{sh}, q_{jh}, r_{jh} in system (1.4.1) as well as the coefficients in the expansions of functions X_s, Z_j are constant. Then, it may be said that the coefficients of the form $z_j^{(p)}$ are polynomials in t, whose coefficients, in their turn, are trigonometric polynomials with frequencies $\bar{\lambda}_j - \bar{\mu}_j$, $j = 1,\ldots,l$, where $\bar{\lambda}_j$ and $\bar{\mu}_j$ are real numbers such that the quantities $\kappa_j = -\lambda_j + i\bar{\lambda}_j$, $j = 1,\ldots,n$, and $\bar{\kappa}_j = -\mu_j + i\bar{\mu}_j$, $j = 1,\ldots,k$ are respectively the eigenvalues of the matrices $\mathbf{P} = \{\mathbf{P}_{sh}\}$ and $\mathbf{Q} = \{q_{jh}\}$ [Zubov (1964)]. We will show this. Represent the set of solutions (1.4.9) and (1.4.10) as

$$x_s = \sum_{m_1+\ldots+m_{n+l} \geq 1} L_s^{(m_1,\ldots,m_{n+l})}(t) \alpha_1^{m_1} \ldots \alpha_n^{m_n} \beta_1^{m_{n+1}} \ldots \beta^{m_{n+l}}$$

$$\times e^{t\left(\sum_{j=1}^n \kappa_j m_j + \sum_{j=1}^l \bar{\kappa}_j m_{n+j}\right)}, \quad (1.4.20)$$

$$z_j = \sum_{m_1+\ldots+m_{n+l} \geq 1} K_j^{(m_1,\ldots,m_{n+l})}(t) \alpha_1^{m_1} \ldots \alpha_n^{m_n} \beta_1^{m_{n+1}} \ldots \beta_l^{m_{n+l}}$$

$$\times e^{t\left(\sum_{j=1}^n \kappa_j m_j + \sum_{j=1}^l \bar{\kappa}_j m_{n+j}\right)}. \quad (1.4.21)$$

Setting $\beta_j = c_j \alpha_j$, $j = 1,\ldots,l$, and

$$\gamma_j = \alpha_j e^{\kappa_j t}, \quad j = 1,\ldots,n, \quad (1.4.22)$$

where $\kappa_j = -\lambda_j + i\bar{\lambda}_j$, $j = 1,\ldots,n$, and $\bar{\kappa}_j = -\mu_j + i\bar{\mu}_j$, $j = 1,\ldots,l$, we solve equalities (1.4.20) for $\gamma_1,\ldots,\gamma_n$. Then we get that the γ_j are series in $x_1,\ldots,x_n$, whose coefficients are polynomials in t. The coefficients of these polynomials are, in their turn, trigonometric polynomials with frequencies $\bar{\lambda}_j - \bar{\mu}_j$, $j = 1,\ldots,l$. This follows from the fact that the coefficients of series (1.4.20) and (1.4.21) will be the same after eliminating $\alpha_1,\ldots,\alpha_n$ with the use of equalities (1.4.22). Eliminating the γ_j from (1.4.21), we obtain the series (1.4.19) with the coefficients having the above structure. ∎

All of the above considerations apply only to the case where $\text{Re}\,\kappa_j = -\lambda_j < 0$, $j = 1,\ldots,l$, and $\text{Re}\,\bar{\kappa}_j = -\mu_j < 0$, $j = 1,\ldots,l$, here $\text{Re}\,\bar{\kappa}_j = \text{Re}\,\kappa_j$, $j = 1,\ldots,l$. In what follows, we will provide a description of the case where all of the forms $z_j^{(p)}$ in series (1.4.19) have the coefficients that are constant in t.

We now assume that in system (1.4.1) the functions of time are all periodic in t with the same period ω. Let $\rho_1,\ldots,\rho_n$ be characteristic numbers of the system

$$\frac{dx_s}{dt} = \sum_{h=1}^n p_{sh}(t) x_h$$

and $\rho_{n+1}, \ldots, \rho_{n+k}$ characteristic numbers of the system

$$\frac{dz_j}{dt} = \sum_{h=1}^{k} q_{jh}(t) z_h.$$

Denote $\frac{\ln \rho_j}{\omega}$, $j = 1, \ldots, n$, by $-\lambda_j + i\bar{\lambda}_j$, and $\frac{\ln \rho_{n+j}}{\omega}$, $j = 1, \ldots, k$, by $-\mu_j + i\bar{\mu}_j$. Then with $\lambda_j > 0$, $j = 1, \ldots, n$, and $\lambda_j = \mu_j, j = 1, \ldots, l$, for system (1.4.1) there exists the set of solutions (1.4.19) dependent on l arbitrary constants. In this case, the forms $z_j^{(p)}$ have the coefficients which can be represented as finite polynomials in t, whose coefficients are trigonometric polynomials with frequencies $\bar{\lambda}_j - \bar{\mu}_j$; these trigonometric polynomials have the coefficients that are the t-periodic functions of period ω. If $\bar{\mu}_j = \bar{\lambda}_j, j = 1, \ldots, l$, then the coefficients of $z_j^{(p)}$ can be represented as polynomials in t with the coefficients that are ω-periodic functions in t. Below is the case where the coefficients of $z_j^{(p)}$ are generally periodic functions.

If it is assumed that the coefficients of series X_s and Z_j are the trigonometric polynomials having the frequencies $\nu_1, \ldots, \nu_\sigma$ with the t-periodic coefficients of period ω, then the forms $z_j^{(p)}$ will have the coefficients that are polynomials in t, whose coefficients have the same structure as the coefficients of series X_s, Z_j, except that the new frequencies may be added to trigonometric polynomials:

$$\bar{\lambda}_j - \bar{\mu}_j, \ j = 1, \ldots, l.$$

Finally, we will discuss the case where the coefficients of series X_s and Z_j are arbitrary almost periodic functions of t. Then the forms $z_j^{(p)}$ have the coefficients that are polynomials in t. The coefficients of these polynomials will be the almost periodic functions of t if the following condition is satisfied:

$$\sum_{j=1}^{n} m_j \lambda_j + \sum_{j=1}^{l} \mu_j n_j \neq \lambda, \ \sum_{j=1}^{l} m_j + \sum_{j=1}^{l} n_j \geq 2,$$

here λ is one of the numbers $\lambda_1, \ldots, \lambda_n, \mu_1, \ldots, \mu_k$. Below we give the conditions under which the coefficients of $z_j^{(p)}$ are purely almost periodic.

We now give an example which characterizes the radius of series convergence (1.4.19). Consider the equation

$$\frac{\partial z}{\partial t} - \frac{\partial z}{\partial x} x - \frac{\partial z}{\partial y} y = -z + (z-1)^2 (x+y). \tag{1.4.23}$$

By Theorem 8, the unique holomorphic set of solutions satisfying this equation and condition $z = 0$ with $x = y = 0$ is:

$$z = \frac{ct(x+y)}{1+ct(x+y)}.$$

If $|x| < \frac{1}{2}|ct|$ and $|y| < \frac{1}{2}|ct|$, the z can be represented as the series (1.4.19). This suggests that in the general case the radius of series convergence (1.4.19) unboundedly decreases to zero as $t \to +\infty$.

Remark. Since the proof of Theorem 8 is based on Theorem 6, which, by Remark 3, can be extended to the case where the linear system of the first approximation is irregular (see Theorem 7), Theorem 8 can also be extended to a similar case. In doing so however, the result seems to be identical to Theorem 8 (with explicit modifications taking into account irregularity of the above system). Hence this case is left without any further discussion.

3. Replace the independent variable in system (1.3.1) by the formula $e^{-t} = z$. In order to define the functions $x_1, \ldots, x_n$, we then obtain a system of ordinary differential equations of the form

$$z\frac{dx_s}{dz} = -\sum_{j=1}^{n} p_{sj}(-\ln z)x_j + X_s(-\ln z, x_1, \ldots, x_n), \quad s = 1, \ldots, n. \quad (1.4.24)$$

System (1.4.24) can also be written as

$$z\frac{dx_s}{dz} = \sum_{j=1}^{n} \bar{p}_{sj}(z)x_j + \bar{X}_s(z, x_1, \ldots, x_n), \quad s = 1, \ldots, n. \quad (1.4.25)$$

The functions $\bar{p}_{sj}(z)$ given for $z \in (0,1]$, are real, continuous, and bounded there; the functions $\bar{X}_s$ given for $z \in (0,1], \|x\| \leq r$, are real, continuous and satisfy the Lipschitz condition

$$|\bar{X}_s(z, \bar{x}_1, \ldots, \bar{x}_n) - \bar{X}_s(z, \bar{\bar{x}}_1, \ldots, \bar{\bar{x}}_n)|$$
$$\leq L\sum_{j=1}^{n} |\bar{x}_j - \bar{\bar{x}}_j| \max\{\|\bar{x}^m\|, \|\bar{\bar{x}}^m\|\}$$

and, additionally, $\bar{X}_s(z, 0, \ldots, 0) = 0$.

Evidently, transition from system (1.3.1) to system (1.4.25) allows the entire Lyapunov theory of characteristic numbers to be transferred from the functions given for $t \in [0, +\infty)$ to the functions given for $z \in (0, 1]$.

For the purposes of uniformity, the characteristic number of the function $x(e^{-t}), t \in [0, +\infty)$ will be referred to as the characteristic number of the function $x(z), z \in (0, 1]$ so that the characteristic numbers of the system

$$z\frac{dx_s}{dz} = \sum_{j=1}^{n} p_{sj}(z)x_j, \quad s = 1, \ldots, n,$$

will be referred to as the characteristic numbers of the linear system

$$\frac{dx_s}{dt} = -\sum_{j=1}^{n} p_{sj}(e^{-t})x_j, \quad s = 1, \ldots, n.$$

This implies another form of Theorem 8. Consider the system of partial differential equations

$$z\frac{\partial z_j}{\partial z} + \sum_{s=1}^{n} \frac{\partial z_j}{\partial z}\left(\sum_{h=1}^{n} p_{sh}(z)x_h + X_s(z, x_1, \ldots, x_n, z_1, \ldots, z_k)\right)$$

Behavior of Integral Curves in the Neighborhood

$$= \sum_{h=1}^{k} q_{jh}(z)z_j + \sum_{h=1}^{n} r_{jh}(z)x_h + Z_j(z, x_1, \ldots, x_n, z_1, \ldots, z_k), \ j = 1, \ldots, k. \quad (1.4.26)$$

Assume that the functions, as in (1.4.26), satisfy the conditions:

a) the functions p_{sh}, q_{jh}, r_{jh} are real, continuous, bounded for $z \in (0,1]$, while X_s and Z_j are given for $z \in (0,1], \|x\| \le r, \left(\sum_{j=1}^{k}|z_j|^2\right)^{1/2} \le r$, and are real and continuous there;

b) the functions of X_s and Z_j satisfy condition b) of the present section (see point one) just as the functions of variables $x_1, \ldots, x_n, z_1, \ldots, z_k$. Denote by $\lambda_1, \ldots, \lambda_n$ the characteristic numbers of the system

$$z\frac{dx_s}{dz} = \sum_{h=1}^{n} p_{sh}(z)x_h, \ s = 1, \ldots, n, \quad (1.4.27)$$

by $\mu_1, \ldots, \mu_k$ the characteristic numbers of the system

$$z\frac{dz_j}{dz} = \sum_{h=1}^{k} q_{jh}(z)z_h, \ j = 1, \ldots, k. \quad (1.4.28)$$

Then the following statement holds.

If the following conditions are satisfied: 1) the functions appearing in system (1.4.26) satisfy conditions a) and b);

2) $\lambda_j > 0, \ j = 1, \ldots, n$, and l equalities $\lambda_j = \mu_j, \ j = 1, \ldots, l$, hold;

3) systems (1.4.27) and (1.4.28) are regular, then the system of equations (1.4.26) has the set of solutions

$$z_j = f_j(z, x_1, \ldots, x_n, c_1, \ldots, c_l), \ j = 1, \ldots, k, \quad (1.4.29)$$

which is dependent on l arbitrary constants $c_1, \ldots, c_l$ having the property $z_j \equiv 0$ for $x_1 = \ldots = x_n = 0$.

All the statements (formulated in Corollary to Theorem 8) concerning the construction of functions (1.4.29), are still valid for the case of explicit changes.

This will be discussed in some detail. Assume that the functions X_s and Z_j are holomorphic with respect to $x_1, \ldots, x_n, z_1, \ldots, z_k$ at sufficiently small values of $|x_s|$ and $|z_j|$ equally for all $z \in (0,1]$. Then, proceeding from the above assumptions, as in the case of the independent variable t, it is possible to conclude that the set of solutions (1.4.29) can be represented as the series

$$z_j = \sum_{p=1}^{\infty} z_j^{(p)}, \quad (1.4.30)$$

where $z_j^{(p)}$ are the homogeneous forms of the p-th power of $x_1, \ldots, x_n$ with the coefficients that are polynomials in $c_1, \ldots, c_l$ and, simultaneously, are continuous functions in z given for $z \in (0,1]$. Series (1.4.30) converge with sufficiently small $|x_j|$ for any fixed $z > 0$. However, the radius of convergence of these series unboundedly decreases to zero as $z \to +0$. Now consider the structure of coefficients of the forms $z_j^{(p)}$ in series (1.4.30) determined by various properties of system (1.4.26). If all the functions of the independent variable z appearing in system (1.4.26), and the coefficients of series X_s and Z_j are constants in integral non-negative powers of $x_1, \ldots, x_n, z_1, \ldots, z_k$, then the coefficients of $z_j^{(p)}$ are polynomials in $\ln z$ whose coefficients, in turn, are finite linear combinations of sines and cosines with the arguments that are linear combinations of

$$(\bar{\lambda}_j - \bar{\mu}_j) \ln z, \quad j = 1, \ldots, l, \tag{1.4.31}$$

with integral coefficients, where $\lambda_j + i\bar{\lambda}_j$, $j = 1, \ldots, n$; $\mu_j + i\bar{\mu}_j$, $j = 1, \ldots, k$, are eigenvalues of the matrices

$$\mathbf{P} = \{p_{sh}\}, \ s, h = 1, \ldots, n; \ \mathbf{Q} = \{q_{jh}\}, \ j, h = 1, \ldots, k.$$

In this case, series (1.4.30) converge and constitute a solution to (1.4.26) if $\lambda_j > 0$, $j = 1, \ldots, n$; $\lambda_j = \mu_j$, $j = 1, \ldots, l$.

Now turn to the general case. Assume that the functions p_{sh}, q_{jh}, r_{jh} and the coefficients of series X_s and Z_j are the functions which are holomorphic with respect to the variables $z, \varphi_1, \ldots, \varphi_N$ for sufficiently small $|z|, |\varphi_j|$, $j = 1, \ldots, N$. Suppose that with $z = \varphi_1 = \ldots = \varphi_N = 0$ the matrices $\mathbf{P}$ and $\mathbf{Q}$ have the eigenvalues with the above properties and the quantities $\varphi_1, \ldots, \varphi_N$ are the functions z satisfying the system of equations.

$$z \frac{d\varphi_s}{dz} = \sum_{j=1}^{N} \bar{p}_{sj} \varphi_j + \Phi_s, \quad s = 1, \ldots, N. \tag{1.4.32}$$

In system (1.4.32), the functions Φ_s are assumed to be developed in integral non-negative powers of variables $\varphi_1, \ldots, \varphi_N$ as the series starting at least with the second-order terms that converge for sufficiently small $|\varphi_j|$ and have constant coefficients in z. Also, assume that the $\bar{p}_{sj}$ are constants and the matrix $\bar{\mathbf{P}}$ has all eigenvalues with positive real parts. Under such assumptions of system (1.4.26) it is stated that this system has the set of solutions to (1.4.30), which is dependent on l arbitrary constants. In this case, the coefficients of $z_j^{(p)}$ are polynomials in $c_1, \ldots, c_l$ and, simultaneously, are the series arranged in integral non-negative powers of variables $z, \varphi_1, \ldots, \varphi_N$ whose coefficients have the same form as in the special case above.

To prove this statement, denote z by φ_0 in the coefficients of system (1.4.26) and find a solution to system (1.4.26) as a function of $z, \varphi_0, \ldots, \varphi_N, x_1, \ldots, x_n$. Then, to define the functions $z_1, \ldots, z_k$, we obtain a new system of partial differential equations

$$z \frac{\partial z_j}{\partial z} + \sum_{s=1}^{n} \frac{\partial z_j}{\partial x_s} \left(\sum_{h=1}^{n} p_{sh}(\varphi_0, \varphi_1, \ldots, \varphi_N) x_h \right)$$

$$+ X_s(\varphi_0, \varphi_1, \ldots, \varphi_N, x_1, \ldots, x_n, z_1, \ldots, z_k)$$

$$+ \sum_{s=0}^{N} \frac{\partial z_j}{\partial \varphi_s} \left(\sum_{h=0}^{N} \bar{p}_{sh} \varphi_h + \Phi_s(\varphi_1, \ldots, \varphi_N) \right)$$

$$= \sum_{h=1}^{k} q_{jh}(\varphi_0, \varphi_1, \ldots, \varphi_N) z_h + \sum_{h=1}^{n} r_{jh}(\varphi_0, \varphi_1, \ldots, \varphi_N) x_h$$
$$+ Z_j(\varphi_0, \varphi_1, \ldots, \varphi_N, x_1, \ldots, x_n, z_1, \ldots, z_k), \quad j = 1, \ldots, k. \quad (1.4.33)$$

System (1.4.33) sets $\bar{p}_{s0} = 0$ for $s > 0, \bar{p}_{00} = 1, \bar{p}_{0h} = 0$ for $h > 0, \Phi_0(\varphi_1, \ldots, \varphi_N) = 0$. In the stationary case (where the coefficients of system (1.4.26) are z-independent), this system differs from system (1.4.26) in the number of independent variables. Hence, system (1.4.33) has the set of solutions

$$z_j = \sum_{p=1}^{\infty} \bar{z}_j^{(p)}, \quad (1.4.34)$$

where $\bar{z}_j^{(p)}$ are the homogeneous forms of $x_1, \ldots, x_n, \varphi_0, \ldots, \varphi_N$ in p power with the coefficients that are polynomials in $c_1, \ldots, c_l$ and, simultaneously, are polynomials in $\ln z$ whose coefficients are trigonometric polynomials of the above type. Collecting in series (1.4.34) the forms that are homogeneous only in $x_1, \ldots, x_n$, we find that the statement about the structure of series (1.4.30) is true.

4. Now consider the case where the coefficients of series as in (1.4.19), in other words, as in (1.4.30) are constants in the independent variable. Consider the series (1.4.19). To this end, we will discuss the system of partial differential equations

$$\frac{\partial z_j}{\partial t} + \sum_{s=1}^{n} \frac{\partial z_j}{\partial x_s} \left(\sum_{h=1}^{n} p_{sh} x_h + X_s \right) = \sum_{h=1}^{k} q_{jh} z_h + \sum_{h=1}^{n} r_{jh} x_h + Z_j, \quad j = 1, \ldots, k, \quad (1.4.35)$$

where the functions of X_s and Z_j are taken to be the series with constant coefficients which do not contain free and linear terms in $x_1, \ldots, x_n, z_1, \ldots, z_k$. Also, we assume that p_{sh}, q_{jh}, r_{jh} are real constants. Denote by $\lambda_1, \ldots, \lambda_n$ the eigenvalues of the matrix $\mathbf{P} = \{p_{sh}\}$, and by $\kappa_1, \ldots, \kappa_k$ the eigenvalues of the matrix $\mathbf{Q} = \{q_{jh}\}$. By employing the nonsingular linear transformation of the required functions and independent variables, we reduce system (1.4.35) to the form

$$\frac{\partial \bar{z}_j}{\partial t} + \sum_{s=1}^{n} \frac{\partial \bar{z}_j}{\partial \bar{x}_s}(\epsilon_{s-1} \bar{x}_{s-1} + \lambda_s \bar{x}_s + \bar{X}_s) = \delta_{j-1} \bar{z}_{j-1} + \kappa_j \bar{z}_j + \sum_{h=1}^{n} \bar{r}_{jh} \bar{x}_h + \bar{Z}_j,$$

$$\epsilon_0 = \delta_0 = 0, \quad j = 1, \ldots, k. \quad (1.4.36)$$

Here we denote by ϵ_{s-1} and δ_{j-1} the numbers 0 and 1, so that the matrices of linear terms constitute a canonical Jordan form.

Theorem 9 [Zubov (1964)] *If 1)* Re $(\lambda_j) < 0, 1 \leq j \leq n$;
2) $\lambda_j = \kappa_j, 1 \leq j \leq l$;
3) $\epsilon_j = \delta_j, 1 \leq j \leq l-1$;
4) $\epsilon_l = 0, \bar{r}_{jh} = 0, 1 \leq j \leq l, 1 \leq h \leq l$;
5) there is no relationship of the form $\sum_{j=1}^{n} m_j \lambda_j = \kappa_h, 1 \leq h \leq k$ *for any integral non-negative* $m_j, \sum_{j=1}^{n} m_j \geq 1$, *except where* $\lambda_j = \kappa_j, 1 \leq j \leq l$, *then systems (1.4.35)*

and (1.4.36) have the set of holomorphic solutions that depend on l arbitrary constants and can be represented as the series

$$\bar{z}_j = \sum_{p=1}^{\infty} \bar{z}_j^{(p)}(c_1,\ldots,c_l,\bar{x}_1,\ldots,\bar{x}_n), \ j=1,\ldots,k, \qquad (1.4.37)$$

where $\bar{z}_j^{(p)}$ are the homogeneous forms of power p in $\bar{x}_1,\ldots,\bar{x}_n$ whose coefficients are polynomials in $c_1,\ldots,c_l$. Series (1.4.37) converges for $|c_s| \leq c_0, |\bar{x}_s| \leq r, r > 0, c_0 > 0$.

Proof: Substitute in (1.4.36) the series

$$\bar{z}_j = \sum_{p=1}^{\infty} \bar{z}_j^{(p)}(\bar{x}_1,\ldots,\bar{x}_n), \ j=1,\ldots,k, \qquad (1.4.38)$$

where $\bar{z}_j^{(p)}$ are the homogeneous forms of power p with respect to $x_1,\ldots,x_n$ with the coefficients to be defined. By comparing the forms of the same power, we obtain the system of equations to determine $\bar{z}_j^{(p)}$

$$\sum_{s=1}^{n} \frac{\partial \bar{z}_j^{(p)}}{\partial \bar{x}_s}(\epsilon_{s-1}\bar{x}_{s-1} + \lambda_s \bar{x}_s) = \delta_{j-1} \bar{z}_{j-1}^{(p)} + \kappa_j \bar{z}_j^{(p)} + \bar{R}_j^{(p)}. \qquad (1.4.39)$$

The functions $\bar{R}_j^{(p)}$ are the homogeneous forms of power p with respect to $\bar{x}_1,\ldots,\bar{x}_n$ that are dependent on expansion of functions $\bar{X}_s, \bar{Z}_j$ and on the forms $\bar{z}_j^{(\mu)}, \mu = 1,\ldots,p-1$. If the function $\bar{R}_j^{(p)}$ for $p>1$ is defined, then from condition 5) of Theorem 9 it follows that system (1.4.39) has a unique solution as the forms $\bar{z}_j^{(p)}(\bar{x}_1,\ldots,\bar{x}_n), j=1,\ldots,k$.

It will be shown that system (1.4.39) with $p=1$ has a solution as a system of linear forms $\bar{z}_j^{(1)}$ dependent on l arbitrary constants. Set

$$\bar{z}_j^{(1)} = y_j^{(1)} + y_j^{(2)}, \ j=1,\ldots,k, \qquad (1.4.40)$$

where $y_j^{(1)}$ is a linear form in $\bar{x}_1,\ldots,\bar{x}_l$, and $y_j^{(2)}$ is a linear form in $\bar{x}_{l+1},\ldots,\bar{x}_n$. Let $y_j^{(1)} = 0, \ j = l+1,\ldots,k, \ y_j^{(2)} = 0, \ j=1,\ldots,l$, then to define other forms, we obtain equations

$$\sum_{s=1}^{l} \frac{\partial y_j^{(1)}}{\partial \bar{x}_s}(\epsilon_{s-1}\bar{x}_{s-1} + \lambda_s \bar{x}_s) = \delta_{j-1} y_{j-1}^{(1)} + \kappa_j y_j^{(1)}, \ j=1,\ldots,l, \qquad (1.4.41)$$

$$\sum_{s=l+1}^{n} \frac{\partial y_j^{(2)}}{\partial \bar{x}_s}(\epsilon_{s-1}\bar{x}_{s-1} + \lambda_s \bar{x}_s) = \delta_{j-1} y_{j-1}^{(2)} + \kappa_j y_j^{(2)} + \sum_{h=l+1}^{n} \bar{r}_{jh}\bar{x}_h,$$

$$j = l+1,\ldots,k. \qquad (1.4.42)$$

Under conditions 4) and 5) in Theorem 9, system (1.4.42) has a unique solution as a system of linear forms $y_j^{(2)}$. The functions

$$x_{k_1+\ldots+k_{j-1}+l_j+1} = \frac{t^{l_j} e^{\lambda_j t}}{l_j!}, \quad 0 \le l_j \le k_j - 1, \; j = 1,\ldots,m, \; k_1 + \ldots + k_m = l,$$

constitute a solution to the system

$$\frac{d\bar{x}_i}{dt} = \epsilon_{i-1}\bar{x}_{i-1} + \lambda_i \bar{x}_i, \; i = 1,\ldots,l. \tag{1.4.43}$$

Here k_j is the size of the corresponding Jordan box. The matrix of the fundamental solution system $\mathbf{X}(t)$ for (1.4.43) is

$$\begin{pmatrix}
x_1 & 0 & 0 & 0 & \ldots & \ldots & \ldots & \ldots & \ldots & \ldots & \ldots \\
x_2 & x_1 & 0 & 0 & \ldots & \ldots & \ldots & \ldots & \ldots & \ldots & \ldots \\
x_3 & x_2 & x_1 & 0 & \ldots & \ldots & \ldots & \ldots & \ldots & \ldots & \ldots \\
\ldots & \ldots & \ldots & \ldots & \ldots & \ldots & \ldots & \ldots & \ldots & \ldots & \ldots \\
\ldots & \ldots & \ldots & \ldots & \ldots & \ldots & \ldots & \ldots & \ldots & \ldots & \ldots \\
x_{k_1} & x_{k_1-1} & x_{k_1-2} & \ldots & x_1 & 0 & 0 & 0 & 0 & \ldots & \ldots \\
0 & 0 & 0 & \ldots & 0 & x_{k_1+1} & 0 & 0 & 0 & \ldots & \ldots \\
0 & 0 & 0 & \ldots & 0 & x_{k_1+2} & x_{k_1+1} & 0 & 0 & \ldots & \ldots \\
0 & 0 & 0 & \ldots & 0 & x_{k_1+3} & x_{k_1+2} & x_{k_1+1} & 0 & \ldots & \ldots \\
\ldots & \ldots & \ldots & \ldots & \ldots & \ldots & \ldots & \ldots & \ldots & \ldots & \ldots \\
0 & 0 & 0 & \ldots & 0 & x_{k_2} & x_{k_2-1} & x_{k_2-2} & \ldots & x_{k_1+1} & 0 & \ldots \\
0 & 0 & 0 & \ldots & 0 & 0 & 0 & 0 & \ldots & 0 & x_{k_2+1} & \ldots \\
0 & 0 & 0 & \ldots & 0 & 0 & 0 & 0 & \ldots & 0 & x_{k_2+2} & \ldots \\
\ldots & \ldots & \ldots & \ldots & \ldots & \ldots & \ldots & \ldots & \ldots & \ldots & \ldots \\
\ldots & \ldots & \ldots & \ldots & \ldots & \ldots & \ldots & \ldots & \ldots & \ldots & \ldots
\end{pmatrix}.$$

Clearly,

$$Y(t) = \mathbf{X}(t)c, \tag{1.4.44}$$

where

$$c = \begin{pmatrix} c_1 \\ \vdots \\ c_l \end{pmatrix}, \quad Y = \begin{pmatrix} y_1^{(1)} \\ \vdots \\ y_l^{(1)} \end{pmatrix}$$

yields a general solution to system (1.4.43), while the functions $y_j^{(1)}, j = 1,\ldots,l$, regarded as linear forms of the variables $\bar{x}_i, i = 1,\ldots,l$, constitute a solution to system (1.4.41). We have thus defined the linear forms $\bar{z}_j^{(p)}$ with $j = 1,\ldots,l$. The forms $\bar{z}_j^{(1)}$ for $l+1 \le j \le k$ are uniquely determined from system (1.4.39) with $p = 1$.

The forms $\bar{z}_j^{(p)}$ for $p > 1$ are uniquely determined from system (1.4.39). This system of formal series (1.4.38) is completely defined. Convergence of series (1.4.38) follows from the Corollary to Theorem 8. ∎

We shall now consider the conditions under which the forms $z_j^{(p)}$, described in Corollary to Theorem 8, will have periodic and almost periodic coefficients. Return to system (1.4.1) and assume that the functions $p_{sh}(t)$, $q_{jh}(t)$ and $r_{jh}(t)$ are continuous, real, and ω-periodic. The functions X_s, Z_j are holomorphic with respect to $x_1, \ldots, x_n, z_1, \ldots, z_k$ in a sufficiently small neighborhood of the point $x_1 = \ldots = x_n = z_1 = \ldots = z_k = 0$ equally for all $t \in (-\infty, +\infty)$. As before, it is assumed that expansions of the functions X_s and Z_j do not contain free and linear terms in $x_1, \ldots, x_n, z_1, \ldots, z_k$, and their coefficients are almost periodic functions. By employing the nonsingular linear transformations of the required functions and independent variables with ω-periodic continuously differentiable coefficients, it is possible to obtain the following form for system (1.4.1)

$$\frac{\partial \bar{z}_j}{\partial t} + \sum_{s=1}^{n} \frac{\partial \bar{z}_j}{\partial \bar{x}_s}(\epsilon_{s-1}\bar{x}_{s-1} + \lambda_s \bar{x}_s + \bar{X}_s) = \delta_{j-1}\bar{z}_{j-1} + \kappa_j \bar{z}_j + \sum_{h=1}^{n} \bar{r}_{jh}\bar{x}_h + \bar{Z}_j,$$

$$\epsilon_0 = \delta_0 = 0, \ j = 1, \ldots, k, \qquad (1.4.36')$$

where $\bar{X}_s$ and $\bar{Z}_j$ are functions of the same type as X_s and Z_j, so that their expansions continue to have almost periodic coefficients.

The numbers $\lambda_1, \ldots, \lambda_n$ and $\kappa_1, \ldots, \kappa_k$ are characteristic indices respectively for system (1.4.5) and (1.4.6), and the numbers ϵ_{s-1} and δ_{j-1} are selected so that the matrices of the coefficients that are naturally determined by them have a canonical Jordan form. The variables $\bar{r}_{jh}$ are the continuous ω-periodic functions of t.

Theorem 9' *If 1) $Re\lambda_j < 0$, $1 \leq j \leq n$;*
2) $\lambda_j = \kappa_j$, $1 \leq j \leq l$;
3) $\epsilon_j = \delta_j$, $1 \leq j \leq l-1$;
4) $\epsilon_l = 0$, $\bar{r}_{jh} = 0$, $1 \leq j \leq l$, $1 \leq h \leq l$;
5) there is no relationship of the form $\sum\limits_{j=1}^{n} m_j Re\lambda_j = Re\kappa_h$, $1 \leq h \leq k$ for any non-negative integer m_j : $\sum\limits_{j=1}^{n} m_j \geq 2$ and $Re\lambda_j = Re\kappa_s$, $l+1 \leq j \leq n$, $l+1 \leq s \leq k$, then the system (1.4.36') has the set of holomorphic solutions that are dependent on l arbitrary constants and can be represented as the series

$$\bar{z}_j = \sum_{p=1}^{\infty} \bar{z}_j^{(p)}(t, c_1, \ldots, c_l, \bar{x}_1, \ldots, \bar{x}_n), \ j = 1, \ldots, k, \qquad (1.4.37')$$

where $\bar{z}_j^{(p)}$ are the homogeneous forms of power p in $\bar{x}_1, \ldots, \bar{x}_n$ whose coefficients are the polynomials in $c_1, \ldots, c_l$ with the coefficients almost periodic in t.

The proof of this theorem, just as that of Theorem 9, is based on finding solutions to a nonhomogeneous linear system with almost periodic nonhomogeneity. The conditions of Theorem 9' suggest that there is the non-resonance case whose presence makes it possible to determine uniquely the coefficients of the forms $\bar{z}_j^{(p)}$ with $p = 2$. For $p = 1$, the proof is carried out by analogy with Theorem 9. ∎

Remark. If the expansion coefficients of the functions X_s and Z_j are continuous ω-periodic, then under conditions of Theorem 9' the forms $\bar{z}_j^{(p)}$ in (1.4.37') will have

Behavior of Integral Curves in the Neighborhood

the t-periodic coefficients with period ω. In this case, condition 5) can be relaxed by imposing on the system the requirement under which there is no relationship of the form

$$\sum_{j=1}^n m_j \lambda_j = \kappa_h + \frac{i2\tilde{k}\pi}{\omega}, \ 1 \leq h \leq k, \ \tilde{k} = 0, \pm 1, \ldots,$$

for any integer $m_j > 0$ such that

$$\sum_{j=1}^n m_j \geq 2 \text{ and } \lambda_j \neq \kappa_s + \frac{i2\tilde{k}\pi}{\omega}, \ l+1 \leq j \leq n, \ l+1 \leq s \leq k, \ \tilde{k} = 0, \pm 1, \ldots.$$

5. We now discuss representations of a set of integral curves in a system of ordinary differential equations in the neighborhood of a singular point. Consider a system of n differential equations of the general form

$$z\frac{dx_s}{dz} = \sum_{h=1}^n p_{sh}(z)x_h + p_s(z)z + X_s(z, x_1, \ldots, x_n), \ s = 1, \ldots n. \quad (1.4.45)$$

Assume that the right-hand sides of system (1.4.45) satisfy the following conditions:

1) the functions $p_{sh}(z)$ and $p_s(z)$ given for $z \in (0, 1]$, are real, continuous and bounded, while the functions $X_s(z, x_1, \ldots, x_n)$ given for $z \in (0, 1], \|x\| \leq r, r > 0$, are real and continuous;

2) the functions $X_s(z, x_1, \ldots, x_n)$ can be represented as

$$X_s(z, x_1, \ldots, x_n) = \bar{X}_s(z, z_1, x_1, \ldots, x_n) \text{ with } z_1 = z; \quad (1.4.46)$$

here $\bar{X}_s(z, z_1, x_1, \ldots, x_n)$ are the functions that are given for $z \in (0, 1], |z_1| + \|x\| \leq r$, satisfy Lipschitz condition

$$|\bar{X}_s(z, \bar{z}_1, \bar{x}_1, \ldots, \bar{x}_n) - \bar{X}_s(z, \bar{\bar{z}}_1, \bar{\bar{x}}_1, \ldots, \bar{\bar{x}}_n)|$$

$$\leq L\left(|\bar{z}_1 - \bar{\bar{z}}_1| + \|\bar{x} - \bar{\bar{x}}\|\right) \max\left((|\bar{z}_1| + \|\bar{x}\|)^m, (|\bar{\bar{z}}_1| + \|\bar{\bar{x}}\|)^m\right), \ m > 0, \quad (1.4.47)$$

go identically to zero for $z_1 = x_1 = \ldots = x_n = 0$, and are continuously differentiable with respect to $z_1, x_1, \ldots, x_n$. Consider the linear system

$$\frac{dx_s}{dt} = -\sum_{h=1}^n p_{sh}(e^{-t})x_h, \ s = 1, \ldots, n. \quad (1.4.48)$$

Denote its characteristic numbers by $\lambda_1, \ldots, \lambda_n$.

Theorem 10 [Zubov (1964)] *If 1) among the characteristic numbers of system (1.4.48) there are l positive $\lambda_1 > 0, \ldots, \lambda_l > 0$;*

2) system (1.4.48) is regular;

3) the right-hand sides of system (1.4.45) satisfy the above conditions, then system (1.4.45) has a family of integral curves that is dependent on l arbitrary constants

$$x_s = x_s(z, c_1, \ldots, c_l), \ s = 1, \ldots, n, \quad (1.4.49)$$

each of which satisfies the condition $x_s \to 0$ as $z \to +0$, $s = 1, \ldots, n$.

Proof: Consider the system of partial differential equations

$$z\frac{\partial x_s}{\partial z} + \frac{\partial x_s}{\partial z_1}z_1 + \sum_{j=1}^{l} \frac{\partial x_s}{\partial y_j}\lambda_j y_j = \sum_{h=1}^{n} p_{sh}(z)x_h + p_s(z)z_1$$

$$+\bar{X}_s(z, z_1, x_1, \ldots, x_n), \quad s = 1, \ldots, n. \tag{1.4.50}$$

Point 3 in this section implies that the system of equations (1.4.50) has the solution set

$$x_s = f_s(z, z_1, y_1, \ldots, y_l, c_1, \ldots, c_l), \tag{1.4.51}$$

possessing the property $f_s(z, 0, 0, \ldots, 0, c_1, \ldots, c_l) \equiv 0$. The left-hand side of equation (1.4.50) is the total derivative of the function x_s with respect to z, that is multiplied by z, if x_s is taken to be dependent on z via the functions $z_1 = z, y_j = z^{\lambda_j}$ and independent of z.

From this it follows that the functions of (1.4.51) constitute the desired set of solutions to system (1.4.45). ∎

Remark. If system (1.4.48) is the irregular one with the coefficient of irregularity σ, then the statement of Theorem 10 remains valid if its characteristic numbers include l numbers greater than $\sigma, \lambda_j > \sigma, j = 1, \ldots, l$, here $m > \sigma/\min_{j=1,\ldots,l} \lambda_j$. This statement is similarly proved in terms of the Remark for Theorem 8.

1.5 Lyapunov Stability of Nonperturbed Motions

1. Suppose we have the system of ordinary differential equations

$$\frac{dx}{dt} = F(x, t), \tag{1.5.1}$$

where x is the vector and $F(x, t)$ is the n-dimensional vector function. Also, suppose we know the solution to this differential equation

$$x = \bar{x}(t), \tag{1.5.2}$$

defined for $t \geq 0$. By convention, the solution will be called the nonperturbed (program) motion, and any other solution will be called a perturbed motion, as distinct from the former.

Definition 3. The nonperturbed motion $x = \bar{x}(t)$ is called Lyapunov stable if for any $t_0 \geq 0$ and any number $\epsilon > 0$ there is a number $\delta(\epsilon, t_0) > 0$ such that with $\|x_0 - \bar{x}_0\| < \delta(\epsilon, t_0)$ there is $\|x(t, x_0, t_0) - \bar{x}(t)\| < \epsilon$ for $t \geq t_0$, where $x(t, x_0, t_0)$ is the perturbed motion passing through the point x_0 at the time $t = t_0$, i.e.,

$$x(t, x_0, t_0) = x_0, \quad \bar{x}(t_0) = \bar{x}_0.$$

Lyapunov Stability of Nonperturbed Motions

Definition 4. The nonperturbed motion is called asymptotically stable in the sense of Lyapunov if it is Lyapunov stable and the value of $\delta(\epsilon, t_0)$ can be chosen so that
$$\|x(t, x_0, t_0) - \bar{x}(t)\| \xrightarrow[t \to +\infty]{} 0.$$

Here $\|x\|$ means $\sqrt{\sum_{i=1}^{n} x_i^2}$.

Transform the coordinates in the original system by the formula
$$x(t) = y(t) + \bar{x}(t), \qquad (1.5.3)$$

where $y(t)$ is the desired new vector function. As a result, we get
$$\frac{dy}{dt} = F(y + \bar{x}(t), t) - F(\bar{x}(t), t). \qquad (1.5.4)$$

System (1.5.4) has a zero solution (an equilibrium)
$$y = 0. \qquad (1.5.5)$$

If this solution is taken to be a nonperturbed motion, then Definition 3 will be as follows.

Definition 5. The zero solution $y = 0$ is called Lyapunov stable if for any $t_0 \geq 0$ and any $\epsilon > 0$ there is $\delta(\epsilon, t_0) > 0$ such that for $\|y_0\| < \delta$ there is $\|y(t, y_0, t_0)\| < \epsilon$ at $t \geq t_0$, where $y(t, y_0, t_0)$ is a perturbed motion such that $y(t_0, y_0, t_0) = y_0$.

Definition 6. The real continuous function $V(x,t)$ of the $(n + 1)$ argument specified in the domain $\|x\| \leq h, t \geq 0$, where $h > 0$ and $V(0,t) = 0$ for any $t \geq 0$, is called positive-definite if there exists a real continuous function $W(x)$ specified for $\|x\| \leq h$ such that 1) $W(0) = 0$,
2) $W(x) > 0$ for $\|x\| \neq 0$,
3) $V(x,t) \geq W(x)$.

If properties 2) and 3) of the function $W(x)$ are replaced by properties 2') $W(x) < 0$ for $\|x\| \neq 0$ and
3') $V(x,t) \leq W(x)$, then the function $V(x,t)$ will be negative-definite.

Examples: $V(x, y, t) = x^2 + \frac{y^2}{1+t}$ is not a positive-definite function, while
$$V(x, y, t) = x^2 + y^2 + \frac{1}{2} xy \sin t$$

is precisely such a function.

2. Now consider the system of differential equations in vector form
$$\frac{dx}{dt} = f(x, t), \qquad (1.5.6)$$

where $f(x,t)$ is the continuous real vector function that is given for $\|x\| \leq h_0$ and $t \geq 0$, and satisfies Lipschitz condition with respect to x in this domain. Suppose that system (1.5.6) has an equilibrium $x = 0$, i.e. $f(0,t) = 0$ for all $t \geq 0$. Evidently, the unique solution

$$x = x(t, x_0, t_0) \qquad (1.5.7)$$

passes through each point $\|x_0\| < h_0$ for $t \geq t_0$.

If there exists a t-derivative of the function $V_1(t, x_0, t_0) = V(x(t, x_0, t_0), t)$, then the function $V(x,t)$ is said to be differentiable in terms of (1.5.6) along the chosen integral curve and

$$\frac{dV}{dt} = \frac{dV_1}{dt} = \sum_{s=1}^{n} \frac{\partial V}{\partial x_s} f_s(x,t) + \frac{\partial V}{\partial t}. \qquad (1.5.8)$$

Theorem 11 (theorem of stability) *In order for the zero solution to the system of differential equations (1.5.6) to be Lyapunov stable, it is necessary and sufficient that there be the function $V(x,t)$ satisfying the following conditions: 1) the function $V(x,t)$ is given for $\|x\| \leq h_0, t \geq 0$, where h_0 is a positive constant;*

2) the function $V(0,t) = 0$ for all $t \geq 0$ and with the fixed value of t is continuous in the variable x at the point $x = 0$;

3) the function $V(x,t)$ is positive-definite;

4) the function $V_1(t, x_0, t_0) = V(x(t, x_0, t_0), t)$ does not increase at $t \geq t_0 \geq 0$ for all x_0 such that $\|x_0\| \leq h_0$.

All the functions intended to resolve stability issues are called Lyapunov functions.

Proof: *Sufficiency.* Suppose that there exists the function $V(x,t)$ which satisfies the conditions of this theorem. It is essential to show that in this case the zero solution of system (1.5.6) is stable in the sense of Lyapunov. To do this, take $\epsilon > 0$ ($\epsilon < h_0$) and fix a number $t_0 > 0$. Consider the sphere $\|x\| = \epsilon$ and find the least value of $W(x)$ on this sphere. The function $W(x)$ is definite and continuous, since $V(x,t)$ is positive-definite by the condition of the theorem.

Let

$$\inf_{\|x\|=\epsilon} W(x) = \lambda > 0, \qquad (1.5.9)$$

the exact lower boundary being reached. Because of the continuity of $V(x,t)$, at the point $x = 0$ there is a number $\delta(\epsilon, t_0)$ such that

$$V(x, t_0) < \lambda \text{ for } \|x\| < \delta. \qquad (1.5.10)$$

Show that the obtained number $\delta(\epsilon, t_0)$ corresponds to the assumed number ϵ by the definition of stability. Take any point that satisfies the inequality

$$\|x_0\| < \delta(\epsilon, t_0). \qquad (1.5.11)$$

Then by the property of (1.5.10).

$$V(x_0, t_0) < \lambda. \qquad (1.5.12)$$

Next, by condition 4) of the theorem, the function $V_1(t, x_0, t_0)$ is nonincreasing for $t \geq t_0$, therefore

$$V(t, x_0, t_0) \leq V(x_0, t_0) < \lambda \text{ for } t \geq t_0. \qquad (1.5.13)$$

Lyapunov Stability of Nonperturbed Motions

Hence, $\|x(t, x_0, t_0)\| < \epsilon$ for all $t \geq t_0$; otherwise there exists an instant of time $T > t_0$ such that $\|x(T, x_0, t_0)\| = \epsilon$ and then

$$V_1(T, x_0, t_0) = V(x(T, x_0, t_0), T) \geq W(x(T, x_0, t_0)) \geq \lambda,$$

which contradicts inequality (1.5.13) for all $t \geq t_0$. The obtained contradiction proves our statement.

Necessity. Let the equilibrium be stable in the sense of Lyapunov. We will show that in such an event there exists the Lyapunov function that satisfies Theorem 11. Consider the solution (1.5.7) subject to $\|x_0\| < \delta$ and $t_0 \geq 0$. The function V at the point (x_0, t_0) will be defined as follows:

$$V(x_0, t_0) = \sup_{t \geq t_0} \|x(t, x_0, t_0)\|. \tag{1.5.14}$$

If the right-hand side of relationship (1.5.14) proves to be greater than one on some integral curve of system (1.5.6), then we set at this point $V(x_0, t_0) = 1$. Relationship (1.5.14) defines uniquely the single-valued function $V(x, t)$ given (in accordance with the remark) in the domain $\|x_0\| \leq h_0, t \geq 0$. Thus, condition 1) of Theorem 11 holds.

Since the point $x = 0$ is the equilibrium of system (1.5.6), then $V(0, t) = 0$ for $t \geq t_0$, which follows from (1.5.14). It will be shown that the function $V(x, t)$ is continuous for the fixed $t = t_0$ at the point $x = 0$. Take $\epsilon_1 > 0$. Because of the available stability there exists $\delta(\epsilon_1, t_0) > 0$ such that $\|x(t, x_0, t_0)\| < \epsilon_1$ for all $t \geq t_0$ if $\|x_0\| < \delta(\epsilon_1, t_0)$. By (1.5.14), $V(x_0, t_0) < \epsilon_1$, i.e. for $\epsilon_1 > 0$ there exists $\delta > 0$ such that as soon as $\|x_0\| < \delta$, then there is $V(x_0, t_0) < \epsilon_1$, which proves the fulfilment of condition 2) in the theorem. It can be readily seen that $V(x_0, t_0) \geq \|x(t_0, x_0, t_0)\| = \|x_0\| = W(x_0) > 0$ (positive definiteness of the function $V(x, t)$).

In order to demonstrate the validity of condition 4), it suffices to establish that the value of the function $V_1(t_1, x_0, t_0)$ is not less than $V_1(t_2, x_0, t_0)$, only if $t_2 \geq t_1 \geq t_0$. To do this, on the integral curve (1.5.7) we choose two points corresponding to the instants t_1 and t_2 and calculate at these points the value of V by formula (1.5.14). In the former case the exact upper boundary of $\|x(t, x_0, t_0)\|$ is calculated for $t \geq t_1$, while in the latter for $t \geq t_2$. Since $t_2 \geq t_1$,

$$\sup_{t \geq t_1} \|x(t, x_0, t_0)\| \geq \sup_{t \geq t_2} \|x(t, x_0, t_0)\|.$$

Therefore, the function defined by (1.5.14) along the integral curve (1.5.7) does not increase. This completely proves the necessity of conditions of Theorem 11. ∎

Corollary 1. The requirements imposed by Theorem 11 upon the function V, exclusive of condition 4), can be easily verified. On the face of it, condition 4) seems to be boundless: the integral curves of system (1.5.7) are unknown, hence verification of condition 4) seems to be impossible. However, by condition 4), the function V is monotone along any integral curve at $t \geq t_0$ and therefore, is differentiable almost for all t values at $t \geq t_0$ (by the Lebesgue theorem, the monotone function derivative exists almost everywhere). In this case the relationship $\frac{dV}{dt} \leq 0$ holds almost everywhere. Thus, condition 4) can be immediately verified by finding a total derivative of the function V in terms of the system.

Corollary 2. If the function $V(x,t)$ is positive-definite and continuous in the variable x at the point $x = 0$ uniformly in t for $t \geq 0$ and $\frac{dV}{dt} \leq 0$, then the zero solution of system (1.5.6) will be stable uniformly in $t_0 \geq 0$, i.e., in the definition of stability it is possible to choose $\delta(\epsilon, t_0) = \delta(\epsilon)$ to be independent of t_0.

Theorem 11' (theorem of asymptotic stability) [1] *Assume that there is a positive-definite function $V(x,t)$ that is continuous in x for $x = 0$ uniformly in t for $t \geq 0$ and is such that*

$$\frac{dV}{dt} = -W(x,t) \qquad (1.5.15)$$

in terms of system (1.5.6), where $W(x,t)$ is a positive-definite function. Then the equilibrium $x = 0$ is asymptotically Lyapunov stable uniformly in $t_0 \geq 0$.

Proof: Suppose there is such a function $V(x,t)$. We show that the zero solution is stable uniformly in t_0. In the proof of Theorem 11, a number $\delta(\epsilon, t_0)$ was selected from the condition $V(x,t_0) < \lambda$ for $\|x\| < \delta(\epsilon, t_0)$. Since the function $V(x,t)$ is continuous in x at the point $x = 0$ uniformly in t for $t \geq 0$, then there exists a number $\delta(\epsilon)$ such that $V(x, t_0) < \lambda$ for $\|x\| < \delta(\epsilon)$ at all $t_0 \geq 0$. Then it becomes evident that the number $\delta(\epsilon)$ corresponds to the selected number $\epsilon > 0$ from the definition of stability. This completes the proof of Corollary 2.

Show that $\|x(t, x_0, t_0)\| \to 0$ as $t \to +\infty$, i.e., for any $\epsilon' > 0$ there is $T(\epsilon')$ such that $\|x(t, x_0, t_0)\| < \epsilon'$ for $t > T(\epsilon')$. For the assumed $\epsilon' > 0$ we can find $\delta(\epsilon') = \delta' > 0$ satisfying the stability definition. Then the following two cases are possible: 1) there exists T such that $\|x(T, x_0, t_0)\| < \delta'$, $\|x_0\| < \epsilon'$ and therefore $\|x(t, x_0, t_0)\| < \epsilon'$ for all $t > T$;

2) there does not exist such T, i.e., for any $t \geq t_0$ there is always $\|x(t, x_0, t_0)\| \geq \delta'$. Since $W(x,t)$ is a positive-definite function, in this case $W(x,t) > \alpha > 0$ for $x = x(t, x_0, t_0)$. We have

$$\frac{dV(x,t)}{dt} \leq -\alpha; \qquad (1.5.15')$$

by integrating (1.5.15'), we obtain

$$V(x(t, x_0, t_0), t) \leq -\alpha(t - t_0) + V(x_0, t_0). \qquad (1.5.16)$$

Since $\|x(t, x_0, t_0)\| > \delta'$ for all $t \geq t_0$ and the function $V(x,t)$ is positive-definite, then $V(x(t, x_0, t_0), t) > \beta > 0$. The right-hand side of inequality (1.5.16) approaches $-\infty$ as t increases. The obtained contradiction shows that case 2) is not possible. This proves that only case 1) is legitimate, i.e. $\|x(t, x_0, t_0)\| \to \infty$ as $t \to +\infty$. ∎

By the instability of the solution $x \equiv 0$ to system (1.5.6) will be meant the property that is opposite to the stability uniform in $t_0 \geq 0$.

Theorem 12 (theorem of instability) *For the zero solution of system (1.5.6) to be Lyapunov unstable, it is necessary and sufficient that for $\|x\| \leq h_0, t \geq 0$ there be given the functions $V(x,t)$ and $W(x,t)$ possessing the following properties: 1) the*

---
[1] Fundamental investigations in this field are contained in the works [Lyapunov (1950)], [Krasovskii (1952)], [Barbashin (1951)], [Letov (1962)], [Rumjancev (1963)], [Chetaev (1965)], [Zubov (1964)].

Lyapunov Stability of Nonperturbed Motions

function $V(x,t)$ is bounded in some neighborhood of the point $x = 0$ for all $t \geq 0$ right away, i.e., $V(x,t) < l$ for $\|x\| \leq h_0, t \geq 0$;

2) in the neighborhood of the point $x = 0$, which is arbitrarily small, there is at least one point $\bar{x}$ such that $V(\bar{x}, t_0) > 0$ for some $t_0 \geq 0$;

3) by system (1.5.6), the complete derivative function $V(x,t)$ satisfies the relationship $\frac{dV}{dt} = \lambda V + W(x,t)$, where $\lambda > 0$, and $W(x,t)$ is a non-negative function for $\|x\| < h_0$ and is real and continuous.

Proof: *Necessity.* Suppose there is instability, which means that there exists such $\epsilon \in (0, h_0)$ that, whatever $\delta \in (0, \epsilon)$ may be assumed, there are necessarily a point (x_0, t_0) ($\|x_0\| < \delta, t_0 \geq 0$) and an instant of time $\tau \geq t_0$ such that $\|x(\tau, x_0, t_0)\| \geq \epsilon$. Take any point (x_0, t_0) satisfying the inequalities $\|x_0\| < \delta, t_0 \geq 0, \delta < \epsilon < h_0$. Then two cases are possible: either 1) $\|x(t, x_0, t_0)\| < \epsilon$ for $t \geq t_0$, or 2) there exists an instant $\tau \geq t_0$ such that $\|x(\tau, x_0, t_0)\| = \epsilon$ and $\|x(t, x_0, t_0)\| < \epsilon$ for all $t_0 \leq t < \tau$. Let $V(x_0, t_0) = 0$ in the first case and $V(x_0, t_0) = e^{-(\tau - t_0)} = e^{t_0 - \tau}$ in the second.

Thus, the function $V(x,t)$ is defined at any point of the set $\{(x,t) : \|x\| < \epsilon, t \geq 0\}$ and is bounded therein: $V(x,t) < 1$. As noted above, the second type points x_0 exist in an arbitrarily small neighborhood of the point $x = 0$. Therefore, the constructed function $V(x,t)$ satisfies condition 2). We shall now show that the function $V(x,t)$ also satisfies condition 3). Indeed, if the point $(\tilde{x}, \tilde{t})$ refers to the first type, then $V_1(t, \tilde{x}, \tilde{t}) = 0$ for $t \geq \tilde{t}$, since $\|x(t, \tilde{x}, \tilde{t})\| < \epsilon$ for all $t \geq \tilde{t}$ and, therefore, $\frac{dV}{dt} = V$ along any such a motion. If the point $(\tilde{x}, \tilde{t})$ refers to the second type, then $V(\tilde{x}, \tilde{t}) = e^{\tilde{t} - \tau}$, whence it follows that $\frac{dV}{dt} = V$ by system (1.5.6), since $V_1(t, \tilde{x}, \tilde{t}) = e^{\tilde{t} - \tau}$ for all $\tilde{t} \leq t < \tau$.

Thus, condition 3) holds along any motion $x(t, \tilde{x}, \tilde{t}), t > \tilde{t}$ for $\lambda = 1$ and $W = 0$.

Sufficiency. Suppose there exist the functions $V(x,t), W(x,t)$ satisfying the theorem. Show that $x = 0$ is unstable. Conversely, let an equilibrium $x = 0$ be stable uniformly in $t_0 \geq 0$. Then for any $\epsilon \in (0, h_0)$ there is $\delta > 0$ such that with $\|x_0\| < \delta$ for all $t_0 \geq 0$ there will be at once $\|x(t, x_0, t_0)\| < \epsilon < h_0$ for all $t \geq t_0$. Then for any such motion $x(t, x_0, t_0)$ the function $V_1(t, x_0, t_0)$ is bounded for all $t > t_0$ (property 1). By property 2), there exists the point (x_0, t_0) such that $V(x_0, t_0) > 0$. By property 3), the function $V(x(t, x_0, t_0), t)$ satisfies at $t \geq t_0$ the linear differential equation

$$\frac{dV(x(t, x_0, t_0), t)}{dt} = \lambda V(x(t, x_0, t_0), t) + W(x(t, x_0, t_0), t)$$

with initial condition $V(x(t_0, x_0, t_0), t_0) = V(x_0, t_0) > 0$. Integrating this equation and passing to the inequality, we obtain $V(x(t, x_0, t_0), t) \geq V(x_0, t_0) e^{\lambda(t - t_0)}$ for $t \geq t_0$, which contradicts the boundedness of the function $V(x(t, x_0, t_0), t)$. The obtained contradiction shows that the equilibrium $x = 0$ is unstable. ∎

3. We shall now consider the linear system of differential equations

$$\frac{dx}{dt} = \mathbf{P}(t)x, \qquad (1.5.17)$$

where the functions $p_{si}(t)$ are real, continuous, bounded, and given for $t \geq 0$. System (1.5.17) may be regarded as a linear approximation of system (1.5.6), therefore the behavior of solutions to system (1.5.17) is essentially related to the behavior of solutions to system (1.5.6). To make it more precise, we may say that if the zero solution to system (1.5.17) is asymptotically stable, then the zero solution to system (1.5.6) will also be asymptotically stable on rather broad assumptions about nonlinear terms. This implies that the asymptotic stability criterion for linear systems is important in the solution of the general stability problem of the zero solution to system (1.5.6).

If the coefficients in system (1.5.17) are all constant, then the solution of the asymptotic stability problem amounts to investigating the roots of equation $|\mathbf{P} - \lambda \mathbf{E}| = 0$. In the general case, the issue remains open.

Theorem 13 *For any solution $x(t)$ passing through the point $x = x_0$ at $t = t_0$ in system (1.5.15) to satisfy inequalities of the form*

$$a_1 \|x_0\| e^{-b_1(t-t_0)} \leq \|x(t)\| \leq a_2 \|x_0\| e^{-b_2(t-t_0)} \text{ for } t > t_0, \qquad (1.5.18)$$

it is necessary and sufficient that there be two quadratic forms

$$V(x,t) = x^* \mathbf{A}(t) x, \quad W(x,t) = x^* \mathbf{B}(t) x,$$

satisfying the conditions: 1) the function $V(x,t)$ is positive-definite and $W(x,t)$ is negative-definite; these functions are such that

$$\alpha_1 \|x\|^2 \leq V(x,t) \leq \alpha_2 \|x\|^2, \quad -\beta_1 \|x\|^2 \leq W(x,t) \leq -\beta_2 \|x\|^2,$$

where $a_i, b_i, \alpha_i, \beta_i, i = 1, 2$ are positive constants;
2) by system (1.5.17),

$$\frac{dV}{dt} = W. \qquad (1.5.19)$$

Proof: *Necessity.* Let $\mathbf{Y}(t)$ be a fundamental matrix of system (1.5.17). Then

$$x(t) = \mathbf{Y}(t) \mathbf{Y}^{-1}(t_0) x_0. \qquad (1.5.20)$$

Take the quadratic form $W(x,t)$ as $W(x) = -x^* x$. Set

$$V(x_0, t_0) = -\int_{t_0}^{+\infty} W(x(\tau)) d\tau. \qquad (1.5.21)$$

Clearly, $W(x,t)$ satisfies the inequalities for $\beta_1 = \beta_2 = 1$ and $\frac{dV}{dt} = W$. Show that $V(x_0, t_0)$ is the quadratic form. Indeed,

$$V(x_0, t_0) = \int_{t_0}^{+\infty} x^*(\tau) x(\tau) d\tau = x_0^* \int_{t_0}^{+\infty} \mathbf{Y}^{-1*}(t_0) \mathbf{Y}^*(\tau) \mathbf{Y}(\tau) \mathbf{Y}^{-1}(t_0) d\tau \, x_0;$$

by inequality (1.5.18), we obtain

$$\int_{t_0}^{+\infty} a_1^2 \|x_0\|^2 e^{-2b_1(\tau - t_0)} d\tau \leq V(x_0, t_0)$$

$$= \int_{t_0}^{+\infty} \|x(\tau)\|^2 d\tau \leq \int_{t_0}^{+\infty} a_2^2 \|x_0\|^2 e^{-2b_2(\tau - t_0)} d\tau,$$

or
$$\frac{1}{2b_1}a_1^2\|x_0\|^2 \leq V(x_0,t_0) \leq \frac{1}{2b_2}a_2^2\|x_0\|^2.$$

It is clear that the function $V(x,t)$ is positive-definite and there are positive numbers α_1, α_2 such that
$$\alpha_1\|x\|^2 \leq V(x,t) \leq \alpha_2\|x\|^2.$$

Sufficiency. From Theorem 11' we have asymptotic stability of an equilibrium if there are the functions V and W satisfying the theorem. Show that any integral curve of system (1.5.17) satisfies inequalities (1.5.18). Indeed, divide both sides of equality (1.5.19) by V and integrate between the limits t_0 and t, then

$$\int_{t_0}^{t}\frac{dV}{V}d\tau = \int_{t_0}^{t}\frac{W}{V}d\tau,$$

whence

$$V_1(t,x_0,t_0) = V(x(t,x_0,t_0),t) = V(x_0,t_0)\exp\left(\int_{t_0}^{t}\frac{W}{V}d\tau\right). \quad (1.5.22)$$

The function V_1 on the left-hand side of equality (1.5.22) is the value of the function V on the selected integral curve of system (1.5.17). Evaluate the integral on the right-hand side of (1.5.22). To do this, consider the expression under the integral sign

$$\frac{W}{V} = \frac{x^*\mathbf{B}(t)x}{x^*\mathbf{A}(t)x}.$$

If the numerator and denominator of the ratio are replaced by their maximum and minimum values from the conditions of Theorem 13, then we obtain the estimates

$$-\frac{\beta_1}{\alpha_1} \leq \frac{W}{V} \leq -\frac{\beta_2}{\alpha_2}.$$

Apply these to the relationship (1.5.22):

$$V(x_0,t_0)e^{-\frac{\beta_1}{\alpha_1}(t-t_0)} \leq V_1(t,x_0,t_0) \leq V(x_0,t_0)e^{-\frac{\beta_2}{\alpha_2}(t-t_0)}.$$

Replace the function $V(x_0,t_0)$ with its least value (condition 1 in Theorem 13), and the function $V_1(t,x_0,t_0)$ with its greatest value from the same inequality, then

$$\alpha_1\|x_0\|^2 e^{-\frac{\beta_1}{\alpha_1}(t-t_0)} \leq \alpha_2\|x(t)\|^2;$$

similarly,
$$\alpha_1\|x(t)\|^2 \leq \alpha_2\|x_0\|^2 e^{-\frac{\beta_2}{\alpha_2}(t-t_0)}.$$

Extracting the square root on both sides of the latter inequalities and grouping them, we obtain inequality (1.5.18). ∎

Remark. Write relationship (1.5.19) in expanded form:

$$\frac{dV}{dt} = \frac{dx^*}{dt}\mathbf{A}(t)x + x^*\frac{d\mathbf{A}(t)}{dt}x + x^*\mathbf{A}(t)\frac{dx}{dt}.$$

Replacing $\frac{dx}{dt}$ and $\frac{dx^*}{dt}$ from system (1.5.17), we obtain the equality

$$\frac{dV}{dt} = x^*\left(\mathbf{P}^*(t)\mathbf{A}(t) + \frac{d\mathbf{A}(t)}{dt} + \mathbf{A}(t)\mathbf{P}(t)\right)x = x^*\mathbf{B}(t)x,$$

which holds along any trajectory of the system, i.e.

$$\frac{d\mathbf{A}(t)}{dt} + \mathbf{P}^*(t)\mathbf{A}(t) + \mathbf{A}(t)\mathbf{P}(t) = \mathbf{B}(t). \tag{1.5.23}$$

Equation (1.5.23) is called Lyapunov's matrix equation.

If $\mathbf{P} = \text{const}$, then the matrix $\mathbf{A} = \text{const}$, and (1.5.23) is an algebraic matrix equation of the form

$$\mathbf{P}^*\mathbf{A} + \mathbf{A}\mathbf{P} = \mathbf{B}. \tag{1.5.24}$$

To solve the problem of asymptotic stability $x = 0$ in system (1.5.17), it is necessary and sufficient that equation (1.5.23) or (1.5.24) would have a solution as the positive-definite matrix $\mathbf{A}$ provided that the matrix $\mathbf{B}$ is negative-definite.

Chapter 2

On Behavior of Trajectories in the Neighborhood of a Periodic Orbit

2.1 Autonomous Dynamical Systems

1. Consider a real n-dimensional Euclidean space E_n of points $(x_1, \ldots, x_n)$ and a system of n differential equations of the form

$$\frac{dX}{dt} = F(X), \tag{2.1.1}$$

where X is the vector from E_n and F is the vector function with the values from E_n. Assume that the right-hand side of system (2.1.1) is given throughout E_n, is continuous therein, and satisfies the Lipschitz condition in each bounded region $G \subset E_n$ with a finite constant L_G, i.e., for $X, Y \in G$ we have

$$\|F(X) - F(Y)\| < L_G \|X - Y\|. \tag{2.1.2}$$

Here $\|Z\|$ is the Euclidean length of the vector Z:

$$\|Z\| = \sqrt{\sum_{j=1}^{n} z_j^2} \quad \text{and} \quad Z = (z_1, \ldots, z_n).$$

Also, assume that any solution to system (2.1.1) is defined for all the values of $t \in (-\infty, +\infty)$, which does not impose any additional restrictions on the derivation of theorems for geometry of integral curves [Nemyckij and Stepanov (1949)]. These assumptions suggest that

a) for any point $X_0 \in E_n$ there is a unique integral curve $X = X(t, X_0)$ of system (2.1.1) that satisfies the condition $X(0, X_0) = X_0$ and is given for $t \in (-\infty, +\infty)$;

b) the vector function $X(t, X_0)$ is continuous in all its independent variables, i.e., given $t_1 \in (-\infty, +\infty)$ and $X_1 \in E_n$, then for every $\epsilon > 0$ there is a $\delta > 0$ such that for

$$\|X_2 - X_1\| + |t_2 - t_1| < \delta$$

we get
$$\|X(t_2, X_2) - X(t_1, X_1)\| < \epsilon;$$

c) if X_1 is the point on the integral curve $X(t, X_0)$ corresponding to t_1, and X_2 is the point on the integral curve $X(t, X_1)$ corresponding to t_2, then X_2 is the point of the integral curve $X(t, X_0)$ corresponding to $t_1 + t_2$, i.e.

$$X_2 = X(t_1 + t_2, X_0) = X(t_2, X(t_1, X_0))$$

for any finite real t_1, t_2. This property is called the group property, since the vector function $X(t, X_0)$ may be regarded as a group of transformations of E_n onto itself that is dependent on one real parameter of t.

The above three properties stem from the general theorems in section 1.1. They contain all the features peculiar to the families of integral curves that are determined in E_n by a system of equations as in (2.1.1). Therefore, we have to study the families of curves located in E_n that satisfy the above properties and do not involve any particular system of differential equations as in (2.1.1). This problem leads us to the notion of a dynamical system in E_n [Nemyckij and Stepanov (1949)].

Definition 7. The vector function $X = X(t, X_0)$ is called a dynamical system in E_n, if it satisfies the following conditions:

1) for any fixed value of the second variable $X_0 \in E_n$, the vector function $X = X(t, X_0)$ is defined for all $t \in (-\infty, +\infty)$ and $X(0, X_0) = X_0$;

2) the vector function $X = X(t, X_0)$ is continuous in all its independent variables by property b);

3) for any finite real t_1, t_2 we have

$$X(t_2, X(t_1, X_0)) = X(t_1 + t_2, X_0).$$

The point $X = X(t, X_0)$ with fixed X_0 is called the representative point of the phase space E_n; the vector function $X = X(t, X_0)$ with fixed X_0 is called the motion of dynamical system, the collection of representative points for all $t \in (-\infty, +\infty)$ for some motion is termed the trajectory of dynamical system; the collection of all representative points for $t \in [0, +\infty)$ (respectively $t \in (-\infty, 0]$) is called the positive semitrajectory (respectively the negative semitrajectory) of the dynamical system. Trajectories are designated as $X((-\infty, +\infty), X_0)$, and semitrajectories as $X([0, +\infty), X_0)$ (respectively $X((-\infty, 0], X_0)$).

Definition 8. The set $A \subset E_n$ is called invariant for a dynamical system $X(t, X_0)$ if $X(t, X_0) \in A$ for all $t \in (-\infty, +\infty)$ provided that $X_0 \in A$.

The invariance property is that each invariant set is a combination of the dynamical system trajectories and, conversely, any set-theoretic sum of dynamical system trajectories is an invariant set.

Theorem 14 *If $X(t, X_0)$ is a dynamical system in E_n, then for any sufficiently small $\epsilon > 0$ and some arbitrarily large $0 < T < +\infty$ there is $\delta > 0$, $\delta = \delta(\epsilon, T, X_0)$ so small that for $\|X_0 - Y_0\| < \delta$ there is $\|X(t, X_0) - X(t, Y_0)\| < \epsilon$ for $|t| \leq T$.*

Geometrically, the problem is that the motions starting at sufficiently near points X_0, Y_0 stay arbitrarily close to one another for any initially fixed period $[-T, +T]$.

Autonomous Dynamical Systems

Proof: Fix the point X_0. Assume that arbitrary finite numbers $\epsilon > 0$ and $T > 0$ are given and Theorem 14 is incorrect. Then there exist the sequence of points Y_0^k, the number $\epsilon_1 > 0$, and the sequence of real numbers t_k satisfying the conditions $\|X_0 - Y_0^k\| \xrightarrow[k \to \infty]{} 0$, while $\|X(t_k, Y_0^k) - X(t_k, X_0)\| \geq \epsilon_1$, where $t_k \in [-T, +T]$ with $k = 1, 2, \ldots$.

Choose from the sequence t_k the convergent subsequence $t_{n_k} \to t_0$. Then from the preceding inequality we have $\|X(t_{n_k}, Y_0^{n_k}) - X(t_{n_k}, X_0)\| \geq \epsilon_1$, which is impossible, since each vector under the norm has the same limit as $k \to +\infty$ by property 2) of the dynamical system, that is, $X(t_0, X_0)$. The obtained contradiction shows that Theorem 14 is valid. ∎

This theorem is often called the theorem of integral continuity and provides the main tool to study the properties of dynamical system.

Theorem 15 *The closure $\bar{A}$ of the invariant set A is also an invariant set.*

Proof: If $X_0 \in \bar{A}$ and $X_0 \in A$, then by Definition 8, $X(t, X_0) \in A \subset \bar{A}$ for all $t \in (-\infty, +\infty)$. Consequently, of interest is only the case where $X_0 \in \bar{A}$ and $X_0 \bar{\in} A$. Then, by the definition of the closure A, the set contains the sequence of points $X_n \in A$ such that $X_n \xrightarrow[n \to \infty]{} X_0$ and $X(t, X_n) \xrightarrow[n \to \infty]{} X(t, X_0)$ for any finite t by property 2) of dynamical system. Hence, $X(t, X_0) \in \bar{A}$. At the same time, the relationship $X(t, X_0) \bar{\in} A$ must hold, or else $X_0 = X(-t, X(t, X_0)) \in A$ due to invariance of A, which is impossible. Thus, both the closure of the invariant set A and the boundary of the invariant set constitute an invariant set (if the emply set is always thought of as invariant). ∎

Corollary. If A is an invariant set, $\bar{A}$ is its closure, and $\bar{A}\setminus A$ is the set of boundary points and is nonempty, then it is a nonempty set-theoretic sum of the trajectories of dynamical system.

Definition 9. The point X_0 is called the rest point of dynamical system if it constitutes an invariant set, i.e., $X(t, X_0) = X_0$ for $t \in (-\infty, +\infty)$.

Theorem 16 *The set of all rest points is invariant and closed.*

Proof: Let A be the set of all rest points. Then $X(t, X_0) \in A$ for $t \in (-\infty, +\infty)$ follows from $X_0 \in A$, hence A is invariant. Show that A is closed, i.e. $A = \bar{A}$; clearly, $A \subset \bar{A}$. Show that the inverse inclusion is also valid: $\bar{A} \subset A$. If $X_0 \in \bar{A}$, then there exists a sequence $X_n \in A$ such that $\|X_0 - X_n\| \xrightarrow[n \to \infty]{} 0$ and $\|X(t, X_n) - X(t, X_0)\| \xrightarrow[n \to \infty]{} 0$. Consequently, $\|X_n - X(t, X_0)\| \xrightarrow[n \to \infty]{} 0$, whence $X(t, X_0) = X_0$ for $t \in (-\infty, +\infty)$. Thus, X_0 is the point of rest and, therefore, $\bar{A} \subset A$. ∎

Theorem 17 *The motions of dynamical system, as distinct from the rest points, cannot be adjacent to the rest points in a finite time.*

Proof: Let X_0 be a rest point, and $X(t, Y_0)$ the motion adjoining it in a finite time $X(\bar{t}, Y_0) = X_0$, then $Y_0 = X(-\bar{t}, X_0) = X_0$. Therefore, the starting point of such a motion necessarily coincides with the rest point. This means that there is no motion which is distinct from the rest point and adjoints it in a finite time. ∎

Theorem 18 *If an arbitrarily small neighborhood of the point X_0 contains a semitrajectory of a dynamical system, then X_0 is its rest point.*

Proof: Suppose the reverse is true: X_0 is not the rest point. Then there exists a finite number t_0 such that

$$X(t_0, X_0) \neq X_0 \quad \text{and} \quad \|X_0 - Y_0\| = a > 0, \quad Y_0 = X(t_0, X_0).$$

From Theorem 14 for each $\epsilon > 0$ and $T = |t_0|$ it is possible to select such a $\delta > 0$ that with $\|X_0 - X_1\| < \delta$ there will be $\|X(t, X_0) - X(t, X_1)\| < \epsilon$ for $|t| \leq T$. With $\epsilon = \frac{1}{2a}$, all the motions starting in the δ-neighborhood of the point X_0 turn out to be clear of the $\frac{1}{2a}$-neighborhood of this point, which is impossible, since on the assumption of the theorem there must exist a semitrajectory entirely confined to any sufficiently small neighborhood of the point X_0. The obtained contradiction shows that X_0 is the rest point. ∎

Corollary. If the motion $X(t, X_0) \xrightarrow[\text{or } t \to -\infty]{t \to +\infty} Y_0$, then Y_0 is the rest point.

Definition 10. The rest point X_0 of dynamical system is referred to as isolated, if there is a neighborhood that does not contain any other rest points. All of the other rest points that do not have this property are called nonisolated.

2. All the rest points of dynamical system can be classified into two categories: Lyapunov stable and unstable. The isolated Lyapunov stable points of rest, in turn, may be subdivided into two classes: asymptotically stable in the sense of Lyapunov and stable, but not asymtotically.

Definition 11. The rest point X_0 is called Lyapunov stable, if for every $\epsilon > 0$ there is $\delta > 0$ such that for $\|X_1 - X_0\| < \delta$ there is $\|X(t, X_1) - X_0\| < \epsilon$ for $t \geq 0$. The rest points that do not possess such property are called Lyapunov unstable.

By the theory of functions, the Lyapunov stable points of rest in a dynamical system are the points where the vector function $X(t, X_1)$, as the second argument function, is uniformly continuous in $t \in [0, +\infty)$.

Definition 12. The rest point X_0 of dynamical system is called asymptotically Lyapunov stable if for each $\epsilon > 0$ there is $\delta > 0$ such that for $\|X_1 - X_0\| < \delta$ there is $\|X(t, X_1) - X_0\| < \epsilon$ with $t \geq 0$, and $\|X(t, X_1) - X_0\| \xrightarrow[t \to +\infty]{} 0$.

Definition 13. The point X_0 is called ω-limit (α-limit) for the motion $X(t, X_1)$, if there exists a sequence of real numbers $t_1, t_2, \ldots, t_n, \ldots$, such that $t_n \to +\infty$ ($t_n \to -\infty$) and $X(t_n, X_1) \to X_0$ as $n \to \infty$.

Theorem 19 *The set of all ω-limit points for any motion is invariant and closed (likewise, the set of all α-limit points for any motion is invariant and closed).*

Proof: Take some motion $X(t, X_0)$ and assume that it has ω-limit points. Denote by Ω the set of all ω-limit points of the motion $X(t, X_0)$. Show that Ω is invariant. Let $Y_0 \in \Omega$, then there exists a sequence $t_n \xrightarrow[n \to \infty]{} +\infty$ such that $X_n = X(t_n, X_0) \xrightarrow[n \to \infty]{} Y_0$. It is apparent that $X(t, X_n) \xrightarrow[n \to \infty]{} X(t, Y_0)$ because of the continuity of $X(t, X_0)$ in variables t, X_0. Then the point $X(t, Y_0) \in \Omega$ for any $t \in (-\infty, +\infty)$, since there exists a sequence $t_n + t$ such that $X(t_n + t, X_0) \xrightarrow[n \to \infty]{} X(t, Y_0)$. Thus, the set Ω is invariant. Show that it is closed. Indeed, let Y be a limit point for the set Ω. Then there exists a sequence $Y_n \in \Omega$, $Y_n \xrightarrow[n \to \infty]{} Y$. Using the property of Euclidean distance, we find

$$\|X(t, X_0) - Y\| \leq \|Y - Y_n\| + \|X(t, X_0) - Y_n\|.$$

Autonomous Dynamical Systems

For any $\epsilon > 0$ there is a variable n_ϵ such that $\|Y - Y_n\| < \epsilon/2$ for $n \geq n_\epsilon$. Since $Y_n \in \Omega$, it is possible to select such t_ϵ that $\|X(t_\epsilon, X_0) - Y_n\| < \epsilon/2$. Choosing the sequence $\epsilon_i \xrightarrow[i \to \infty]{} +0$, it is possible to construct a sequence τ_i such that $\tau_i \to +\infty$ and $X(\tau_i, X_0) \to Y$ as $i \to \infty$. This means that $Y \in \Omega$ and Ω is closed. ∎

Corollary. If the motion $X(t, X_0)$ for $t \geq 0$ stays in a bounded set $G \subset E_n$, then this motion has ω-limit points and their set $\Omega \subset \bar{G}$.

If X_0 is a rest point, then the α- and ω-limit points of the motion $X(t, X_0)$ coincide with X_0. The motion $X(t, X_0)$ is called periodic, if there is a number $T > 0$ such that $X(t + T, X_0) = X(t, X_0)$ for $t \in (-\infty, +\infty)$ and $X(t, X_0) \not\equiv X_0$, the α- and ω-limit sets of the periodic motion are coincident with each other and span the entire trajectory of the periodic motion.

Partition the collection of all invariant sets of dynamical system into two. One of these contains all the stable invariant sets, while the other contains the unstable sets. From the category of stable invariant sets it is possible to select the class of asymptotically stable invariant sets as in the case of rest points.

Denote $\rho(X, M) = \inf_{Y \in M} \|X - Y\|$, $M \subset E_n$, $S(M, \epsilon) = \{X : X \in E_n, \rho(X, M) < \epsilon\}$, $\epsilon > 0$, where $\rho(X, M)$ is a distance between the point X and the set M, and $S(M, \epsilon)$ is the ϵ-neighborhood of the set M.

Definition 14. The invariant set M is called Lyapunov stable if for every $\epsilon > 0$ there is $\delta > 0$ such that for $X_0 \in S(M, \delta)$ there is $X(t, X_0) \in S(M, \epsilon)$ for all $t \geq 0$. The invariant sets, which do not have such a property, are called unstable.

Definition 15. The invariant set M is called asymptotically Lyapunov stable if for every $\epsilon > 0$ there is $\delta > 0$ such that for $X_0 \in S(M, \delta)$ there is $X(t, X_0) \in S(M, \epsilon)$ for all $t \geq 0$ and $\rho(X(t, X_0), M) \xrightarrow[t \to +\infty]{} 0$.

Discussion of the behavior of integral curves of the ordinary differential equations in the neighborhood of the oscillatory mode involves analysis of the bounded invariant sets, since the coordinates of the oscillating object vary in the bounded region of the phase space. With this in mind, we further consider the bounded invariant sets only.

Definitions 14 and 15 contain the notions of orbital stability and asymptotic orbital stability. Next, section 2.2 deals with construction of a periodic orbit neighborhood from the standpoint of both the orbital stability and the coordinate stability.

3. We shall now characterize the classes of motion.

Definition 16. Let the dynamical system $X(t, X_0)$ be given in the n-dimensional Euclidean space E_n. The point X_0 is called Poisson-stable as $t \to +\infty$ (P_+-stable) if the motion $X(t, X_0)$ has the point X_0 as its ω-limit point. The point X_0 is called Poisson-stable as $t \to -\infty$ (P_--stable), if X_0 is α-limit for the motion $X(t, X_0)$. If the point X_0 is simultaneously the α- and ω-limit point for the motion $X(t, X_0)$, then X_0 is called Poisson-stable (P-stable).

The Poisson stability means that for any two numbers $\epsilon > 0$ and $T > 0$ it is possible to find at least two numbers $t_1 > T$ and $t_2 < -T$ such that the representative points $X(t_i, X_0)$ fall within an ϵ-neighborhood of the point X_0.

Theorem 20 *The set of all Poisson-stable points X_0 is invariant.*

Proof: Denote by B the collection of all points X_0 that are P-stable. Show that

$X_0 \in B$ implies $X(t, X_0) \in B$ for $t \in (-\infty, +\infty)$. By the definition of P stability, there exist two sequences t_i and τ_i, $t_i \xrightarrow[i \to \infty]{} +\infty$, $\tau_i \xrightarrow[i \to \infty]{} -\infty$, such that

$$X(t_i, X_0) = X_i \xrightarrow[i \to \infty]{} X_0, \quad X(\tau_i, X_0) = Y_i \xrightarrow[i \to \infty]{} X_0.$$

By the continuity of the vector function $X(t, X_0)$ for any $\bar{t} \in (-\infty, +\infty)$ we obtain

$$X(\bar{t}, X_i) \xrightarrow[i \to \infty]{} X(\bar{t}, X_0), \quad X(\bar{t}, Y_i) \xrightarrow[i \to \infty]{} X(\bar{t}, X_0).$$

Consequently,

$$X(\bar{t} + t_i, X_0) \to X(\bar{t}, X_0), \quad X(\bar{t} + \tau_i, X_0) \to X(\bar{t}, X_0)$$

as $i \to \infty$, which means that the point $X(\bar{t}, X_0)$ is both the α-limit and the ω-limit point for the motion $X(t, X(\bar{t}, X_0))$, therefore $X(\bar{t}, X_0) \in B$. ∎

Definition 17. The motion $X(t, X_0)$ is called central if $X_0 \in \bar{B}$, where $\bar{B}$ is a closure of the invariant set B of all P-stable points.

This definition suggests that the central motion either describes the trajectory entirely composed of its α- and ω-limit points or the trajectory of this motion is at the boundary of the set B.

To study the structure of the set of central motions, we introduce the notion of the wandering point.

Definition 18. The point X_0 is called wandering, if there exists a neighborhood $S(X_0, \epsilon)$ such that it is possible to select the number $T > 0$ possessing the following property: any motion starting in this neighborhood does not return thereto for $|t| > T$. Denote by W_1 the set of all wandering points.

Theorem 21 *The set W_1 is open and invariant.*

Proof: Let $X_0 \in W_1$. This means that there exists the neighborhood $S(X_0, \epsilon)$ and the number $T > 0$ satisfying the above Definition 18. Let the point $X_1 \in S(X_0, \epsilon)$ be nonwandering. Then there exists a neighborhood of this point $S(X_1, \epsilon_1) \subset S(X_0, \epsilon)$ such that at least one motion starts there, returning at a particular t, $|t| > T$ to $S(X_1, \epsilon_1)$. But this contradicts the definiton of a wandering point X_0. Thus, W_1 is open.

Show that W_1 is invariant. Indeed, let $X_0 \in W_1$ and $X_1 = X(t, X_0)$, where $t \in (-\infty, +\infty)$ is a fixed real number. Let X_1 be a nonwandering point. Then there exists the neighborhood sequence $S(X_1, \epsilon_k)$, $\epsilon_k \xrightarrow[k \to \infty]{} +0$, and the sequence $T_k \xrightarrow[k \to \infty]{} \infty$ such that $S(X_1, \epsilon_k) \cap X(T_k, S(X_1, \epsilon_k))$ is nonempty and is denoted by $X(t, S(X_1, \epsilon_k))$ that is the image of $S(X_1, \epsilon_k)$ in transforming $X(t, X_0)$. In this case, because of the continuity of the vector-valued function $X(t, X_0)$, there exists the sequence of neighborhoods $S(X_0, \delta_k)$, $\delta_k \xrightarrow[k \to \infty]{} +0$ such that all the motions starting in $S(X_1, \epsilon_k)$ fall within $S(X_0, \delta_k)$ at time $-t$. But then $X(T_k, S(X_0, \delta_k)) \cap S(X_0, \delta_k)$ is not empty. Hence, X_0 is the nonwandering point. The obtained contradiction shows that W_1 is invariant. ∎

If $W_1 \neq E_n$, then the set $E_n \setminus W_1 = \mu_1$ is invariant and closed; hence, μ_1 can be regarded as a new space where the same dynamical system is defined. Denoting

Autonomous Dynamical Systems

by W_2 the collection of all points wandering with respect to μ_1, we have that W_2 is invariant and open with respect to μ_1. Therefore, it is possible to construct the closed invariant set $\mu_2 = \mu_1 \setminus W_2$. Continuing this process unboundedly, we obtain a sequence of embedded invariant sets $\mu_1 \supset \mu_2 \supset \mu_3 \supset \ldots$. Set $\mu = \bigcap_{i=1}^{\infty} \mu_i$, where μ is the closed invariant set.

Theorem 22 *The set μ is the collection of all trajectories of central motions.*

Proof: The theorem states that $\mu = \bar{B}$. Show that $\mu \subset \bar{B}, \bar{B} \subset \mu$. Since the sets μ_i, $i = 1, 2, \ldots$, are invariant, the wandering points for μ_i are not P-stable. Hence $B \subset \mu_i$, $i = 1, 2, \ldots$, and therefore $B \subset \mu$. Since μ is closed, $\bar{B} \subset \mu$.

We now show that $\mu \subset \bar{B}$, i.e., the P-stable points are contained in any neighborhood $S(X_0, \epsilon)$ of an arbitrary point $X_0 \in \mu$. Choose a sequence of the increasing positive numbers T_i such that $T_i \xrightarrow[i \to \infty]{} +\infty$. For any point $X \in \mu$, set $\Sigma(X, \epsilon) = S(X, \epsilon) \cap \mu$. The set μ is invariant and is composed of the μ-nonwandering points. There exists $t_1 > T_1$ such that $\Sigma \cap X(t_1, \Sigma)$ is not empty, where $\Sigma = \Sigma(X_0, \epsilon)$, $\epsilon > 0$. Since the sets $\Sigma, X(t_1, \Sigma)$ are open with respect to μ, there exist $X_1 \in \mu, \epsilon_1 > 0$ such that $\Sigma(X_1, \epsilon_1) \subset \Sigma \cap X(t_1, \Sigma)$.

Let $\Sigma_1 = \Sigma(X_1, \epsilon_1/2)$, then there is $t_2 > T_2$ such that $\Sigma_1 \cap X(-t_2, \Sigma_1)$ is not empty and such $X_2 \in \mu, \epsilon_2 > 0$ are found, then $\Sigma(X_2, \epsilon_2) \subset \Sigma_1 \cap X(-t_2, \Sigma_1)$; in this case $\epsilon_2 \leq \epsilon_1/2$. Introduce the notation $\Sigma_2 = \Sigma(X_2, \epsilon_2/2)$. Next, there exist $X_3 \in \mu, \epsilon_3 > 0$ such that $\Sigma(X_3, \epsilon_3) \subset \Sigma_2 \cap X(t_3, \Sigma_2)$, where $t_3 > T_3, \epsilon_3 \leq \epsilon_2/2$. Also, set $\Sigma(X_3, \epsilon_3/2) = \Sigma_3$. Next, we define $X_4 \in \mu, \epsilon_4 > 0$, for which $\Sigma(X_4, \epsilon_4) \subset \Sigma_3 \cap X(-t_4, \Sigma_3)$, where $t_4 > T_4, \epsilon_4 \leq \epsilon_3/2$, etc. Proceeding with this process unboundedly and noting that $\Sigma_i \subset \Sigma_{i-1}, i = 2, 3, \ldots$, and the diameter of the set Σ_i does not exceed ϵ_i, where $\epsilon_i \leq \frac{\epsilon}{2^{i-1}}, i = 1, 2, \ldots$, at the intersection of the sets Σ_i we obtain the point Y_0,

$$Y_0 = \bigcap_{i=1}^{\infty} \Sigma_i.$$

Show that the point Y_0 is P_--stable. Let there be given arbitrary numbers $T > 0, \delta > 0$. Select a natural i such that $T_{2i+1} > T, \epsilon_{2i} < \delta$. By construction,

$$Y_0 \in \Sigma(X_{2i+1}, \epsilon_{2i+1}) \quad \text{and} \quad \Sigma_{2i} = \Sigma(X_{2i}, \epsilon_{2i}/2) \subset \Sigma(Y_0, \delta),$$

since

$$\delta > \epsilon_{2i}, \quad \|Y_0 - X_{2i}\| < \epsilon_{2i}/2.$$

Consequently,

$$Y_0 \in \Sigma(X_{2i+1}, \epsilon_{2i+1}) \subset \Sigma_{2i} \cap X(t_{2i+1}, \Sigma_{2i}).$$

Hence, using the transformation of the group with the parameter $-t_{2i+1}$, obtain

$$X(-t_{2i+1}, Y_0) \in \Sigma_{2i} \cap X(-t_{2i+1}, \Sigma_{2i}) \subset \Sigma(Y_0, \delta),$$

where $-t_{2i+1} < -T_{2i+1} < -T$. Thus, the point Y_0 is P_--stable. P_+-stability can be proved in much the same way. ■

From the way the set of central motions is constructed it follows that the central motions constitute the principal framework for a dynamical system.

Thus, to describe stationary oscillations, we may use only those functions which, in some regular manner, return to a neighborhood of their values assumed by these functions. A wide class of such functions is generated by recurrent [Birkhoff (1927)] motions.

Definition 19. The closed invariant nonempty set M is called minimal if it does not have a proper subset with the same properties.

Definition 20. The motion $X(t, X_0)$ of dynamical system is called recurrent if for every $\epsilon > 0$ there is a number $T_\epsilon > 0$, such that any arc of temporal length T_ϵ, taken on the trajectory $X((-\infty, +\infty), X_0)$, approximates the entire trajectory of the motion $X(t, X_0)$ to the nearest ϵ.

Consider the arc $X(t, X_0)$, $t \in (\alpha, \alpha + T_\epsilon)$. Construct the spheres of radius ϵ centered at each point of this arc. The motion is recurrent if it stays in the set just constructed for all $t \in (-\infty, +\infty)$ and any $\alpha \in (-\infty, +\infty)$.

Theorem 23 (Birkhoff theorem) *Every motion starting in the bounded minimal set is recurrent.*

Proof: Suppose $X(t, X_0)$, $X_0 \in M$ is not recurrent and M is a bounded minimal set. Then there are at least one number $\epsilon > 0$ and two sequences $T_k \xrightarrow[k \to \infty]{} +\infty$ and $\alpha_k \in (-\infty, +\infty)$ such that it is possible to select the sequence t_k satisfying the condition $\|X(t_k, X_0) - X(t, X_0)\| \geq \epsilon$ for all $t \in (\alpha_k, \alpha_k + T_k)$. Choose from the sequence α_k a subsequence such that the sequence $X(\alpha_{n_k}, X_0) \xrightarrow[k \to \infty]{} X_1, X_1 \in M$, and choose from the sequence t_{n_k} a subsequence τ_k so that $X(\tau_k, X_0) \xrightarrow[k \to \infty]{} Y_1$. We have that the point Y_1 is exterior to the ϵ-neighborhood of the semitrajectory $X([0, +\infty), X_1)$. Indeed, if there exists $t > 0$ such that $\|Y_1 - X(t, X_1)\| < \epsilon$, then we have that $\|X(\tau_k, X_0) - X(\alpha_{n_k} + t, X_0)\| < \epsilon$ with a sufficiently large k, where α_{n_k} is the value of α_k that corresponds to τ_k. The last inequality contradicts the above construction. The set of ω-limit points of the motion $X(t, X_1)$ is invariant, closed, nonempty, and is contained in M. However, there exists at least one point Y_1 from the set M, which does not appear in this ω-limit set. This contradicts the minimality of M, i.e., any motion starting in M is recurrent. ∎

Definition 21. The motion $X(t, X_0)$ is called recurrent if for every $\epsilon > 0$ there is a number $L_\epsilon > 0$ such that for every fixed t in any interval $(\alpha, \alpha + L_\epsilon)$ there is a number τ satisfying the inequality $\|X(t + \tau, X_0) - X(t, X_0)\| < \epsilon$. This definition is equivalent to that given above. In general, the number τ appearing in that definition is dependent on t and ϵ, and is called the t, ϵ-almost period of the motion $X(t, X_0)$.

The recurrent motion $X(t, X_0)$ is called almost periodic if for every $\epsilon > 0$ the almost periods τ can be selected irrespective of t.

The foregoing suggests that oscillations in a dynamical system (if they are of a regular nature) are described by recurrent motions which, in a special case, can be almost periodic and, in a more special case, periodic motions.

The results of this chapter are related to studies on the behavior of integral curves of a stationary system of differential equations in the neighborhood of a periodic orbit.

2.2 Preliminary Studies on Properties of the Neighborhood of a Periodic Orbit

1. Consider a system of differential equations

$$\frac{dy_s}{dt} = f_s(y_1, y_2, \ldots, y_n, y_{n+1}), \quad s = 1, \ldots, n+1. \tag{2.2.1}$$

Assume that the right-hand sides of system (2.2.1) are given in some region of the Euclidean $(n+1)$-dimensional space and are twice continuously differentiable therein. Also, suppose system (2.2.1) has a periodic solution

$$y_s = \varphi_s(t), \quad s = 1, \ldots, n+1, \tag{2.2.2}$$

with period 2π whose orbit M is lying entirely inside the above region.

Denote by Y the vector $Y = (y_1, \ldots, y_{n+1})$, and by $F(Y)$ the vector function $F(Y) = (f_1, \ldots, f_{n+1})$. Then system (2.2.1) can be written as

$$\frac{dY}{dt} = F(Y).$$

The integral curve of this system, whose plot passes through the point Y_0 at $t = 0$, is denoted by

$$Y = Y(t, Y_0). \tag{2.2.3}$$

Denote by $\rho(Y, M)$ the distance between the point Y and the set M. Also, denote the collection of all points Y such that $\rho(Y, M) < \delta$ by $S(M, \delta)$, where δ is a positive number. Here

$$\rho(Y, M) = \inf_{\bar{Y} \in M} \rho(Y, \bar{Y}), \quad \rho(Y, \bar{Y}) = \|Y - \bar{Y}\| = \sqrt{\sum_{i=1}^{n+1}(Y_i - \bar{Y}_i)^2}.$$

Next, let the vector $\Phi(t)$ denote the periodic solution $\Phi(t) = (\varphi_1(t), \ldots, \varphi_{n+1}(t))$.

Definition 22. Periodic solution (2.2.2) of system (2.2.1) is called orbitally stable [Zubov (1962b)] if for each $\epsilon > 0$ it is possible to select $\delta(\epsilon) > 0$ such that for $Y_0 \in S(M, \delta)$ we obtain $Y(t, Y_0) \in S(M, \epsilon)$ at all $t \geq 0$. In addition, if a number $\delta(\epsilon)$ can be chosen so that $\rho(Y(t, Y_0), M) \to 0$ as $t \to +\infty$, then the periodic solution (2.2.2) to system (2.2.1) is called orbit-asymptotically stable or a periodic self-oscillation of system (2.2.1).

Since the properties of the integral curves of system (2.2.1) mentioned in Definition 22 can also be considered for $t \leq 0$, $t \to -\infty$, we will distinguish such stability in positive and negative directions.

Definition 23. Periodic solution (2.2.2) to system (2.2.1) is called Lyapunov stable [Lyapunov (1950)] if, regardless of what number $t_0 \in (-\infty, +\infty)$ is fixed, for each $\epsilon > 0$ it is possible to select $\delta(\epsilon) > 0$ such that with $\rho(Y_0, \Phi(t_0)) < \delta$ we obtain $\rho(Y(t - t_0, Y_0), \Phi(t)) < \epsilon$ for all $t \geq t_0$.

In compliance with the remark to Definition 22 the stability of periodic solution (2.2.2) may also be considered in the negative direction.

Henceforth, this chapter describes the behavior of integral curves (2.2.3) of system (2.2.1) in a sufficiently small neighborhood of the orbit M of periodic solutions as in (2.2.2) and defines the cases where the periodic solution (2.2.2) to system (2.2.1) is orbitally stable, orbit-asymptotically stable, Lyapunov stable, and the cases where periodic and almost periodic motions appear in the above neighborhood of the closed curve M. Again, existence conditions are established for the family of integral curves of system (2.2.1) unboundedly approximating M, and methods are found for representation of such families.

In what follows, we shall study behavior of integral curves of system (2.2.1) in a sufficiently small, but fixed neighborhood $S(M, \delta)$ of the set M without special stipulations made each time for this circumstance. Note that the framework of section 2.1 makes it possible to establish basic properties of the orbit M neighborhood from the standpoint of behavior of the integral curves of system (2.2.1) in this neighborhood.

Let $Y_0 \in S(M, \delta)$. Then the integral curve $Y = Y(t, Y_0)$ has one of the following two properties: either $Y(t, Y_0) \in S(M, \delta)$ for $t \in (-\infty, +\infty)$, or $Y(t, Y_0)$ leaves $S(M, \delta)$ at time $t \neq 0$. In the latter case we have integral curves of two types. The curves leaving $S(M, \delta)$, both with an increase and decrease in time, refer to the first type, while those staying in $S(M, \delta)$ for $t \geq 0$ or $t \leq 0$ refer to the second type. If the semitrajectory of the integral curve $Y(t, Y_0)$ stays in $S(M, \delta)$, then it has the limit set $\overline{M} \subset S(M, \delta)$. Two cases are possible here: either $\overline{M} = M$ or $\overline{M}$ does not coincide with M. In the first case, the semitrajectory of the integral curve $Y(t, Y_0)$ approaches arbitrarily close the orbit M of the periodic solution (2.2.2) to system (2.2.1) — the asymptotic curve. In the second case, in the neighborhood $S(M, \delta)$ there may be motions of every type mentioned in section 2.1. Below are given conditions under which $\overline{M}$ is composed, specifically, of periodic orbits or almost periodic motions. It is interesting to note that the second semitrajectory of the integral curve $Y(t, Y_0)$ can also stay in $S(M, \delta)$; therefore, its limit set can either coincide with M, then $Y(t, Y_0)$ is called doubly asymptotic, or not coincide with M. In the latter this limit set can be filled with every possible motion.

2. We shall introduce a special coordinate system in the neighborhood of orbit M. The hyperplane P_t, which passes through a point $\Phi(t)$ with the fixed t and is defined by the equation

$$(Y - \Phi(t), F(t)) = 0, \qquad (2.2.4)$$

where $F(t) = F(\Phi(t))$, is a normal hyperplane with respect to the closed curve M drawn through the point $\Phi(t)$. Here (a, b) is the scalar product of vectors a and b.

Let $Q_t = P_t \cap S(M, \delta)$. Assume that a number δ is such that the sets Q_t, $t \in [0, 2\pi]$ do not meet. Take on Q_0 the point Y_0 and the number $t \in [0, 2\pi]$. If the number $\rho(Y_0, \Phi_0)$ is sufficiently small ($\Phi_0 = \Phi(0)$), then it can be stated that the integral curve $Y = Y(t, Y_0)$ will initially intersect Q_t at the time

$$\tau = \tau(t, Y_0). \qquad (2.2.5)$$

Preliminary Studies on Properties

From the property of function (2.2.5) it follows that the vector

$$Z(t) = Y(\tau, Y_0) - \Phi(t) \qquad (2.2.6)$$

lies in Q_t. Hence,

$$(Z(t), F(t)) = 0. \qquad (2.2.7)$$

Construct a system of differential equations which is satisfied by functions (2.2.6) and (2.2.5). Differentiating (2.2.6) and (2.2.7) with respect to t and excluding $\dfrac{dY(\tau, Y_0)}{d\tau}$ with the help of system (2.2.1), we find

$$\frac{d\tau}{dt} = \frac{F^2(t) - (Z, \dot{F}(t))}{(F(Z + \Phi(t)), F(t))}, \qquad (2.2.8)$$

$$\frac{dZ}{dt} = \frac{F^2(t) - (Z, \dot{F}(t))}{(F(Z + \Phi(t)), F(t))} F(Z + \Phi(t)) - F(t), \qquad (2.2.9)$$

where $F^2(t) = (F(t), F(t))$. To define the functions τ and Z, we consider (2.2.8) and (2.2.9) as an independent system of differential equations.

Introduce on the hyperplane P_t a coordinate system, choosing its origin to be the point $\Phi(t)$, and the unit vectors of axes to be the vectors $\bar{B}_1(t), \ldots, \bar{B}_n(t)$ that are pairwise orthogonal and continuously differentiable. We take the vector $\bar{B}_{n+1}(t)$ to be the unit vector of a tangent to the curve at the point $\Phi(t)$

$$\bar{B}_{n+1}(t) = \frac{F(t)}{\sqrt{F^2(t)}}.$$

Denote by $x_1, \ldots, x_{n+1}$ the coordinates of the point Y in the new system. Then we get

$$y_s = \varphi_s(t) + \sum_{j=1}^{n+1} b_{sj}(t) x_j, \quad s = 1, \ldots, n+1. \qquad (2.2.10)$$

Converting relationships (2.2.10), we find

$$x_s = \sum_{j=1}^{n+1} b_{js}(t)(y_j - \varphi_j(t)), \quad s = 1, \ldots, n+1, \qquad (2.2.11)$$

where $b_{sj}(t)$ is the s-th component of $\bar{B}_j(t)$, $j = 1, \ldots, n+1$. Note that the transformation matrix for (2.2.11) is transposed with respect to the transformation matrix for (2.2.10), since the latter is orthogonal. Setting in (2.2.11) $z_s = y_s - \varphi_s(t)$, we replace in (2.2.9) and (2.2.8) the desired functions. Then, to define the new functions $x_1, \ldots, x_{n+1}$ and τ, we obtain the system

$$\frac{d\tau}{dt} = \frac{F^2(t) - (\mathbf{B}(t)X, \dot{F}(t))}{(F(\mathbf{B}(t)X + \Phi(t)), F(t))}, \qquad (2.2.12)$$

$$\frac{dX}{dt} = \frac{d\mathbf{B}^*(t)}{dt} \cdot \mathbf{B}(t)X + \mathbf{B}^*(t) \left(\frac{F^2(t) - (\mathbf{B}(t)X, \dot{F}(t))}{F(\mathbf{B}(t)X + \Phi(t)), F(t))} \right.$$

$$\times F(\mathbf{B}(t)X + \Phi(t)) - F(t)\Big). \qquad (2.2.13)$$

Lemma 4 *The function $H(Z,t) = (Z, F(t))$ is the integral of system (2.2.9) which goes to zero on all integral curves $Z = Z(t, Z_0, t_0)$, $Z = Z_0$ at $t = t_0$, since the point $Y_0 = Z_0 + \Phi(t_0)$ lies in Q_{t_0}.*

Proof: The total derivative with respect to t that is calculated from the function $H(Z,t)$ by system (2.2.9), is identically zero. Therefore, the function H retains its initial value for all $t \geq t_0$, if it is calculated on the integral curves of system (2.2.9). If $Z_0 = Y_0 - \Phi(t_0)$, where Y_0 is a point on Q_{t_0}, then

$$(Z_0, F(t_0)) = ((Y_0 - \Phi(t_0)), F(t_0)) = 0.$$

Hence, on the integral curves defined by such an initial data there will be $H(Z,t) = 0$. ∎

We shall use Lemma 4 as the basis for further transformations of system (2.2.13) and first derive the equations for x_{n+1}. It follows from formula (2.2.11) that

$$x_{n+1} = \frac{(Z, F(t))}{\sqrt{F^2(t)}} = \frac{H(Z,t)}{\sqrt{F^2(t)}},$$

hence

$$\frac{dx_{n+1}}{dt} = \frac{d}{dt}\left(\frac{1}{\sqrt{F^2(t)}}\right)\sqrt{F^2(t)}\,x_{n+1},$$

therefore

$$x_{n+1} = x_{n+1}(0)\sqrt{\frac{F^2(0)}{F^2(t)}}.$$

Since, to solve the problem stated in this section, we may restrict the discussion to the behavior of the integral curves of system (2.2.1) passing at $t = 0$ through Q_0, we assume $x_{n+1}(0) = 0$, and hence $x_{n+1} \equiv 0$.

After replacing the function τ by the formula $\tau = \theta + t$ we come to the system to be discussed in more detail

$$\frac{d\theta}{dt} = g(x_1, \ldots, x_n, t), \qquad (2.2.14)$$

$$\frac{dx_s}{dt} = g_s(x_1, \ldots, x_n, t), \quad s = 1, \ldots, n, \qquad (2.2.15)$$

where

$$g(x_1, \ldots, x_n, t) = \frac{F^2(t) - (\mathbf{B}(t)X, \dot{F}(t))}{(F(\mathbf{B}(t)X + \Phi(t)), \dot{F}(t))} - 1,$$

and the functions $g_s(x_1, \ldots, x_n, t)$ are the first n components of the vector

$$\frac{d\mathbf{B}^*(t)}{dt}\mathbf{B}(t)X + \mathbf{B}^*(t)\left(\frac{F^2(t) - (\mathbf{B}(t)X, \dot{F}(t))}{(F(\mathbf{B}(t)X + \Phi(t)), F(t))}\right.$$

Preliminary Studies on Properties 81

$$\times F(\mathbf{B}(t)X + \Phi(t)) - F(t)\bigg).$$

Here in both cases $x_{n+1} = 0$ for the vector X.

Now establish a relation between the solutions of system (2.2.1) and the solutions of systems (2.2.14) and (2.2.15). Let $Y_0 \in Q_0$. Suppose $Y = Y(t, Y_0)$ is a solution to system (2.2.1), and $\tau = \tau(t, Y_0)$ is the time at which this solution initially intersects Q_t. Then the functions

$$x_s = x_s(t, x_1^0, \ldots, x_n^0, 0) = \sum_{j=1}^{n+1} b_{js}(t)(y_j(\tau, Y_0) - \varphi_j(t)), \quad s = 1, \ldots, n, \quad (2.2.16)$$

satisfy system (2.2.15) and initial conditions

$$x_s = x_s^0 = \sum_{j=1}^{n+1} b_{js}(0)(y_j^0 - \varphi_j^0) \text{ for } t = 0, \quad s = 1, \ldots, n.$$

The function $\theta = \tau(t, Y_0) - t$ is the solution of equation (2.2.14) determined by the condition $\theta = 0$ for $t = 0$. We now assume that the functions

$$\theta = \theta(t, x_1^0, \ldots, x_n^0, 0), \ x_s = x_s(t, x_1^0, \ldots, x_n^0, 0), \quad s = 1, \ldots, n,$$

constitute a solution to system (2.2.14) — (2.2.15) with initial conditions $\theta = 0$, $x_s = x_s^0$ at $t = 0$. Set $\tau = \theta + t$; then

$$y_s(\tau, Y_0) = \varphi_s(t) + \sum_{j=1}^{n} b_{sj}(t)x_s(t, x_1^0, \ldots, x_n^0, 0), \quad s = 1, \ldots, n+1 \quad (2.2.17)$$

is a solution to system (2.2.1), if y_s are taken to be the functions of τ. In this case, $y_s = y_s^0 = \sum_{j=1}^{n} b_{sj}(0)x_j^0 + \varphi_s^0$, $s = 1, \ldots, n+1$ for $t = 0$, where Y_0 is a point on Q_0.

3. Relationships (2.2.16) and (2.2.17) form the basis for studies on behavior of the integral curves of system (2.2.1) in the neighborhood of the periodic solution $\Phi(t)$ with the help of system (2.2.14) and (2.2.15).

Lemma 5 *For any two finite numbers $T > 0$ and $\delta > 0$ there is a number $\delta_1(T, \delta) > 0$ such that $\rho(Y(t - t_0, Y_0), \Phi(t)) < \delta$ for $|t - t_0| \leq T$ if $\rho(Y_0, \Phi(t_0)) < \delta_1$, $t_0 \in [0, 2\pi]$.*

Proof: Suppose the statement of Lemma 5 is not valid. Then three sequences Y_0^k, t_0^k and t_k are found to be such that $\rho(Y_0^k, \Phi(t_0^k)) \to 0$ as $k \to \infty$, while

$$\rho(Y(t_k - t_0^k, Y_0^k), \Phi(t_k)) \geq \delta. \quad (2.2.18)$$

From these sequences we choose convergent subsequences. To save notation, let $Y_0^k \to \bar{Y}_0$, $t_0^k \to \bar{t}_0$, $t_k \to \bar{t}$ as $k \to \infty$. Then

$$Y(t_k - t_0^k, Y_0^k) \xrightarrow[k \to \infty]{} Y(\bar{t} - \bar{t}_0, \bar{Y}_0) = \Phi(\bar{t}),$$

since
$$\bar{Y}_0 = \Phi(\bar{t}_0), \quad \Phi(t_k) \xrightarrow[k\to\infty]{} \Phi(\bar{t}).$$
At the same time, inequality (2.2.18) holds, which contradicts the relationships. This contradiction shows that the statement is true. ∎

Lemma 6 *For any finite number $T > 2\pi$ there is a number $\delta_2 = \delta_2(T) > 0$ such that for $\rho(Y_0, \Phi(t_0)) < \delta_2$, $t_0 \in [0, 2\pi]$ there exists an instant $\bar{t}_1 = t(Y_0, t_0)$ such that $0 \leq \bar{t}_1 - t_0 \leq T$ and $Y(\bar{t}_1 - t_0, Y_0) \in Q_0$.*

Proof: Choose a positive number t_2 such that $2\pi + t_2 < T$, $t_2 < 2\pi$ and

$$((\Phi(t_2) - \Phi(0)), F(0))((\Phi(-t_2) - \Phi(0)), F(0)) < 0. \qquad (2.2.19)$$

Inequality (2.2.19) implies that the points $\Phi(t_2)$ and $\Phi(-t_2)$ lie on different sides of the plane P_0. Next, choose a number $\epsilon > 0$ so small that the closed spheres of radius ϵ with their centers at the points $\Phi(t_2)$ and $\Phi(-t_2)$ are entirely on the same side of the plane P_0 as their centers. According to Lemma 5, for a number T chosen by ϵ there is a number $\delta_2(T, \epsilon)$ such that $\rho(Y(t - t_0, Y_0), \Phi(t)) < \epsilon$ for $|t - t_0| \leq T$, if $\rho(Y_0, \Phi(t_0)) < \delta_2$, $t_0 \in [0, 2\pi]$. For $t = 2\pi - t_2$ the function $\Phi(t) = \Phi(-t_2)$. For $t = 2\pi + t_2$ the function $\Phi(t) = \Phi(t_2)$, therefore

$$\rho(Y(2\pi - t_2 - t_0, Y_0), \Phi(-t_2)) < \epsilon, \quad \rho(Y(2\pi + t_2 - t_0, Y_0), \Phi(t_2)) < \epsilon.$$

The left-hand side of inequality (2.2.19) made up of these points remains negative. Hence, the variable $((Y(t - t_0, Y_0) - \Phi(0)), F(0))$ changes its sign on the interval $[2\pi - t_2, 2\pi + t_2]$, so there exists $\bar{t}_1 = t(Y_0, t_0) \in [t_0 + 2\pi - t_2, t_0 + 2\pi + t_2] \subset [t_0, t_0 + T]$ such that $((Y(\bar{t}_1 - t_0, Y_0) - \Phi(0)), F(0)) = 0$. ∎

Lemma 7 *If for every $\epsilon > 0$ there is $\delta > 0$ such that for $\rho(Y_0, \Phi(0)) < \delta$, $Y_0 \in Q_0$, we obtain $\rho(Y(t, Y_0), \Phi(t)) < \epsilon$ for all $t \geq 0$, then the periodic solution $\Phi(t)$ of system (2.2.1) is stable in the sense of Lyapunov.*

Proof: Suppose for every $\epsilon > 0$ there is $\delta(\epsilon) > 0$ such that the conditions of Lemma 7 are satisfied. For the numbers $\delta(\epsilon)$ and $T > 2\pi$ it is possible to select a number $\delta_1 = \delta_1(T, \epsilon)$ such that $\rho(Y(t - t_0, Y_0), \Phi(t)) < \delta$ for $|t - t_0| \leq T$, $t_0 \in [0, 2\pi]$, if $\rho(Y_0, \Phi(t_0)) < \delta_1$. Based on Lemma 6, the number δ_1 can also be chosen so small that the integral curve $Y(t - t_0, Y_0)$ will intersect Q_0 at $t = \bar{t}_1 = t(Y_0, t_0)$, where $0 \leq \bar{t}_1 - t_0 \leq T$. Denote by $\bar{Y}_0$ the point of intersection of that integral curve with Q_0. Then $\rho(\bar{Y}_0, \Phi(0)) < \delta$ and

$$\rho(Y(t, \bar{Y}_0), \Phi(t)) < \epsilon \quad \text{for all} \quad t \geq 0. \qquad (2.2.20)$$

Thus, before intersecting Q_0 the integral curve $Y(t - t_0, Y_0)$ satisfies the inequality $\rho(Y(t - t_0, Y_0), \Phi(t)) < \delta \leq \epsilon$, and after intersecting Q_0 the relationship (2.2.20). Strictly speaking, the proof of stability in the sense of Lyapunov has not been obtained in full, since $t_0 \in [0, 2\pi]$. The proof may be completed with the following remark. If $t_0 \in (-\infty, +\infty)$, then $t_0 = \tilde{t}_0 + 2m\pi$, $\tilde{t}_0 \in [0, 2\pi]$, where m is an integer. The integral curve $Y(t - t_0, Y_0)$ identically coincides with $Y(\tilde{t} - \tilde{t}_0, Y_0)$, where $\tilde{t} + 2\pi m = t$. ∎

Preliminary Studies on Properties

Lemma 8 *If for every $\epsilon > 0$ there is $\delta(\epsilon) > 0$ such that for $\rho(Y_0, \Phi(0)) < \delta$, $Y_0 \in Q_0$ we get $\rho(Y(t, Y_0), M) < \epsilon$ for all $t \geq 0$, then the periodic solution $\Phi(t)$ of system (2.2.1) is orbitally stable. Moreover, if the number $\delta(\epsilon)$ can be chosen so that $\rho(Y(t, Y_0), M) \to 0$ as $t \to +\infty$, then the periodic solution $\Phi(t)$ is a periodic self-oscillation.*

Proof: For $\epsilon > 0$ we find $\delta > 0$ satisfying the conditions of Lemma 8. Let $Y_0 \in S(M, \delta_3)$, where $\delta_3 = \min(\delta_1(\delta, T), \delta_2(T))$, $T > 2\pi$. Then it is possible to select a number $t_0 \in [0, 2\pi]$ such that Y_0 would lie in Q_{t_0}. Hence, there exists $\bar{t}_1 = t(Y_0, t_0)$ that is the time of interection of $Y(t - t_0, Y_0)$ with the part Q_0 of the hyperplane P_0, here $t_0 \leq t_1 \leq t_0 + T$. Prior to the time of intersection we obtain $\rho(Y(t - t_0, Y_0), \Phi(t)) < \delta$, and hence, $\rho(Y(t - t_0, Y_0), M) < \epsilon$.

Denote by $\bar{Y}_0$ the point of intersection of the integral curve with Q_0. Then after the time of intersection the integral curve under consideration coincides with $Y(t, \bar{Y}_0)$ and hence the following inequalities hold for all $t \geq 0$

$$\rho(Y(t, \bar{Y}_0), M) < \epsilon, \quad \rho(Y(t, Y_0), M) < \epsilon.$$

Note that $\rho(Y(t, Y_0), M) \to 0$ as $t \to +\infty$ if an appropriate condition holds for the integral curves starting on Q_0 at the points Y_0 which satisfy the inequality $\rho(Y_0, M) < \delta$. ∎

4. We now focus our attention on stability of the periodic solution involved.

Theorem 24 [Andronov (1956)] *The periodic solution $\Phi(t)$ of system (2.2.1) cannot be asymptotically Lyapunov stable.*

Proof: Assume that the periodic solution $\Phi(t)$ is asymptotically stable. Then, by Definition 4, for every number $\epsilon > 0$ and for $t_0 = 0$, there is an appropriate δ. The number $h \in (0, 2\pi)$ can be chosen so small that $\rho(\Phi(h), \Phi(0)) < \delta$ and then $\rho(\Phi(t+h), \Phi(t)) \to 0$ as $t \to +\infty$. Since $\Phi(t)$ is a non-vanishing continuous 2π-periodic function which, by uniqueness, has no points of self-intersection,

$$\inf_{t \in [0, 2\pi]} \rho(\Phi(t+h), \Phi(t)) = \delta_h > 0.$$

The obtained contradiction shows that Theorem 24 is valid. ∎

Theorem 25 *For the periodic solution $\Phi(t)$ of system (2.2.1) to be orbitally stable (orbit-asymptotically stable), it is necessary and sufficient that the zero solution of system (2.2.15) be stable (asymptotically stable) in the sense of Lyapunov.*

Proof: *Necessity.* Let the periodic solution $\Phi(t)$ of system (2.2.1) be orbitally stable (orbit-asymptotically stable). Show that the zero solution to system (2.2.15) is stable in the sense of Lyapunov (asymptotically Lyapunov stable), in other words, for any $t_0 \geq 0$ and $\epsilon > 0$ there is $\delta'(\epsilon, t_0)$ such that for $\|X_0\| < \delta'$ we get $\|X(t, X_0, t_0)\| < \epsilon$ for all $t \geq t_0$ (in the case of asymptotic stability $\|X(t, X_0, t_0)\| \to 0$ as $t \to +\infty$). Indeed, since $\Phi(t)$ is orbitally stable, then for $\epsilon > 0$ it is possible to select $\delta > 0$ such that for $Y_0 \in S(M, \delta)$ we get $Y(t, Y_0) \in S(M, \epsilon/\sqrt{n})$ at all $t \geq 0$.

Fix the number $t_0 \geq 0$ and choose the point $Y_0 \in Q_{t_0}$ determined by

$$y_s^0 - \varphi_s(t_0) = \sum_{j=1}^{n} b_{sj}(t_0) x_j^0, \quad s = 1, \ldots, n+1, \quad \text{where} \quad \|X_0\| < \delta/\sqrt{n+1}.$$

Denote by $\tau = \tau(t, Y_0, t_0)$ the time of the first intersection of the integral curve $Y(t - t_0, Y_0)$ with Q_t. Then the integral curve $X = X(t, X_0, t_0)$ of system (2.2.15) for $t \geq t_0$ is determined by the formula

$$x_s(t, X_0, t_0) = \sum_{j=1}^{n+1} b_{js}(t)(y_j(\tau - t_0, Y_0) - \varphi_j(t)), \quad s = 1, \ldots, n.$$

Hence, for $\|X_0\| < \delta' = \delta/\sqrt{n+1}$ we obtain $\|X(t, X_0, t_0)\| < \epsilon$ for all $t \geq t_0$ since

$$\rho(Y(\tau - t_0, Y_0), \Phi(t)) = \rho(Y(\tau - t_0, Y_0), M) < \epsilon/\sqrt{n}.$$

Thus, the zero solution of system (2.2.15) is Lyapunov stable uniformly in $t_0 \geq 0$. If the periodic solution $\Phi(t)$ to system (2.2.1) is a self-oscillation of system (2.2.1), then the last relationship implies that

$$\|X(t, X_0, t_0)\| \to 0 \quad \text{as} \quad t \to +\infty.$$

Sufficiency. Let the zero solution of system (2.2.15) be stable (asymptotically stable) in the sense of Lyapunov. Fix $t_0 = 0$. Then for every $\epsilon > 0$ there is $\delta > 0$ such that for $\|X_0\| < \delta$ we obtain $\|X(t, X_0, 0)\| \leq \epsilon/\sqrt{n+1}$ for all $t \geq 0$ ($\|X(t, X_0, t_0)\| \to 0$ as $t \to +\infty$).

Denote by $\tau = \tau(t, X_0)$ the function $\theta(t, X_0) + t$, where θ is a solution to equation (2.2.14) satisfying the initial condition $\theta = 0$ for $t = 0$. Take an arbitrary point $Y_0 \in Q_0$ such that $\rho(Y_0, \Phi(0)) < \delta/\sqrt{n}$, and determine X_0 from the formulas

$$x_s^0 = \sum_{j=1}^{n+1} b_{js}(0)(y_j^0 - \varphi_j^0), \quad s = 1, \ldots, n.$$

Then $\|X_0\| < \delta$ and, therefore, $\|X(t, X_0, 0)\| < \epsilon/\sqrt{n+1}$ for all $t \geq 0$.

From formula (2.2.17) it follows that $\rho(Y(\tau, Y_0), \Phi(t)) < \epsilon$ for $t \geq 0$. Hence, $\rho(Y(t, Y_0), M) < \epsilon$ for all $t \geq 0$. In the case of the zero asymptotically Lyapunov stable solution to system (2.2.15) we have $\rho(Y(t, Y_0), M) \to 0$ as $t \to +\infty$. By applying Lemma 8, we obtain the required statement. ∎

Theorem 26 *For the periodic solution $\Phi(t)$ of system (2.2.1) to be Lyapunov stable, it is necessary and sufficient that the zero solution $\theta = 0$, $X = 0$ to system (2.2.14) — (2.2.15) be Lyapunov stable.*

Proof: *Necessity.* Let the periodic solution $\Phi(t)$ of system (2.2.1) be Lyapunov stable. Show that the zero solution to system (2.2.14) — (2.2.15) has the same property. Let L be the length of the closed curve M. Break this curve by the points $A_0, \ldots, A_{m-1}$ into m arcs, each of which has length L/m. On the arc $[A_j, A_{j+1}]$ we choose a point B_j so that the sphere with its center at this point would contain on its boundary the points A_j and A_{j+1}, $j = 0, \ldots, m-1$, $A_m = A_0$. Set $\rho_1 = \sup\limits_{\substack{0 \leq j \leq m-1 \\ A_0 \in M}} \rho(A_j, B_j)$. Then it turns out that $\rho_1 > 0$ for any m and $\rho_1 \to 0$ as $m \to \infty$.

Next, choose a number $\rho_2 > 0$ such that the closed spheres $\bar{S}(B_j, \rho_2)$ would not

Preliminary Studies on Properties

intersect, when $0 \leq j \leq m-1$, whatever the position may be occupied by the point A_0 on M. The number ρ_2 can always be chosen in this manner, here $\rho_2 \to 0$, then $m \to \infty$.

Let $A_1 = \Phi(t)$, where $t \in [0, 2\pi]$ is any fixed number. Then the number m can be chosen so that the points B_0 and B_1 lie on different sides of the plane P_t. The number ρ_2 may also be chosen so that the spheres $\bar{S}(B_0, \rho_2)$ and $\bar{S}(B_1, \rho_2)$ would not have the points common to P_t. Fix the number t_0. From the stability of the periodic solution $\Phi(t)$ to system (2.2.1) it follows that for $\epsilon > 0$ there is its associated $\delta(\epsilon) > 0$ (see Definition 23). Take the number ρ_2 to be ϵ. Construct a solution to system (2.2.14) — (2.2.15) as

$$X = X(t, X_0, t_0), \quad \theta = \theta(t, X_0, t_0), \quad X = X_0, \quad \theta = 0 \quad \text{with} \quad t = t_0.$$

Construct a solution to system (2.2.1), $Y = Y(t - t_0, Y_0)$, where Y_0 is related to X_0 by

$$y_s^0 - \varphi_s(t_0) = \sum_{j=1}^{n} b_{sj}(t_0) x_j^0, \quad s = 1, \ldots, n+1.$$

Then the function θ is defined by $\theta = \tau - t$, where $\tau = \tau(t, Y_0, t_0)$ is the first time of intersection of the integral curve $Y(t - t_0, Y_0)$ with Q_t. Estimate the function θ. Let $\|X_0\| < \delta/\sqrt{n+1}$, then $\rho(Y_0, \Phi(t_0)) < \delta$. Hence, $\rho(Y(t - t_0, Y_0), \Phi(t)) < \rho_2$ for all $t \geq t_0$. Set $A_1 = \Phi(t), B_0 = \Phi(t'), B_1 = \Phi(t'')$, where $t_0 \leq t' < t < t''$, and t', t'' are sufficiently close to t. Then we have $\rho(Y(t - t_0, Y_0), \Phi(t)) < \rho_2$ for $t = t'$ and $t = t''$. So, the points $Y(t' - t_0, Y_0)$ and $Y(t'' - t_0, Y_0)$ lie on different sides of the plane P_t. This implies that in $[t', t'']$ there is a point τ such that $Y(\tau - t_0, Y_0)$ lies on Q_t. From this it follows that $|\tau - t|$ does not exceed $t'' - t'$. The last number is not greater than the sum of the time lengths of arcs $[A_0, A_1], [A_1, A_2]$. Since the time length of each arc $[A_j, A_{j+1}]$ does not exceed $\frac{L}{mv}$, $v = \inf_{t \in [0, 2\pi]} \sqrt{F^2(t)}$, then we have $|\tau - t| \leq \frac{2L}{mv}$.

Thus, if $\|X_0\| < \delta/\sqrt{n+1}$, then $|\theta| \leq \frac{2L}{mv}$.

Now estimate $\|X(t, X_0, t_0)\|$. The vector-valued function $\Phi(t)$ is uniformly continuous. Therefore, for any $\epsilon > 0$ there is a number $\epsilon_1 \in (0, \epsilon]$ such that for $|\tau - t| < \epsilon_1$ there is $\rho(\Phi(\tau), \Phi(t)) < \frac{\epsilon}{2\sqrt{n}}$. Choose a number m_0 such that $\frac{2L}{m_0 v} < \epsilon_1$. For the number $\rho_2 > 0$ defined in terms of m_0, there is δ such that $\rho(Y(t - t_0, Y_0), \Phi(t)) < \min(\rho_2, \frac{\epsilon}{2\sqrt{n}})$ for all $t \geq t_0$ if $\rho(Y_0, \Phi(t_0)) < \delta$. Show that the number $\delta/\sqrt{n+1}$ is the desired number. Indeed, from the relationships

$$x_s(t, X_0, t_0) = \sum_{j=1}^{n+1} b_{js}(t)(y_j(\tau - t_0, Y_0) - \varphi_j(\tau))$$

$$+ \sum_{j=1}^{n+1} b_{js}(t)(\varphi_j(\tau) - \varphi_j(t)), \quad s = 1, \ldots, n,$$

we obtain $\|X(t, X_0, t_0)\| < \epsilon$, $|\theta| < \epsilon_1 \leq \epsilon$ for all $t \geq t_0$ when $\|X_0\| < \delta/\sqrt{n+1}$. Since δ is taken to be t_0-independent, the zero solution of system (2.2.14) — (2.2.15) is Lyapunov stable uniformly in $t_0 \geq 0$.

Sufficiency. Let the zero solution of system (2.2.14) and (2.2.15) be Lyapunov stable. Show that the periodic solution $\Phi(t)$ to system (2.2.1) is Lyapunov stable. Let $\epsilon > 0$ and $t_0 = 0$. Find a number $\delta(\epsilon) > 0$ such that for $\|X_0\| < \delta$ there is $\|X(t, X_0, 0)\| < \epsilon, |\theta| < \epsilon$ for all $t \geq 0$. Set

$$y_s^0 = \varphi_s^0 + \sum_{j=1}^n b_{sj}(0) x_j^0, \quad s = 1, \ldots, n+1,$$

then we have

$$y_s(t+\theta, Y_0) - \varphi_s(t+\theta) = \varphi_s(t) - \varphi_s(t+\theta) + \sum_{j=1}^n b_{sj}(t) x_j(t, X_0, 0),$$

$$s = 1, \ldots, n+1.$$

From this it follows that $\rho(Y(t, Y_0), \Phi(t)) < \epsilon_1$ for all $t \geq 0$ as soon as $\rho(Y_0, \Phi(0)) < \delta_1, Y_0 \in Q_0$, where $\epsilon_1 > 0$ is an arbitrary number. The number $\delta_1 > 0$ is selected so that there is $\delta_1 \leq \delta(\epsilon)/\sqrt{n}$, where $\epsilon \in (0, \frac{\epsilon_1}{2\sqrt{n+1}})$ is such that for all $|\theta| < \epsilon$ we obtain

$$\rho(\Phi(t+\theta), \Phi(t)) < \frac{\epsilon_1}{2\sqrt{n+1}} \quad \text{for all} \quad t \geq 0.$$

By applying Lemma 7, we see that the periodic solution $\Phi(t)$ of system (2.2.1) is Lyapunov stable. ∎

Theorem 27 *In order for system (2.2.1) to have a periodic solution whose orbit is located in a sufficiently small neighborhood of M, it is necessary and sufficient that system (2.2.15) has a periodic solution of $2\pi n_1$ period lying in a sufficiently small neighborhood of the point $X = 0$, where $n_1 \geq 1$ is an integer.*

Proof: *Necessity.* Let system (2.2.1) have a periodic solution $Y = \Psi(t)$ whose graph is located in a sufficiently small neighborhood of M. Suppose the periodic solution has a period T and intersects Q_0 at the point Y_0 for $t = 0$. Denote by $\tau = \tau(t)$ the time of the first intersection of the solution $\Psi(t)$ with Q_t.

Show that the functions

$$x_s(t) = \sum_{j=1}^{n+1} b_{js}(t)(\psi_j(\tau(t)) - \varphi_j(t)), \quad s = 1, \ldots, n \quad (2.2.21)$$

are t-periodic solutions of system (2.2.15) with period $2\pi n_1$, where n_1 is the number of intersections of $\Psi(t)$ with Q_0 for $t \in (0, T]$. Functions (2.2.21) constitute a solution to system (2.2.15), which follows from (2.2.16). It now remains to show that they are t-periodic with period $2\pi n_1$. For this purpose, we will consider the equation, which is satisfied by the function τ,

$$\frac{d\tau}{dt} = \frac{F^2(t) - ((\Psi(\tau) - \Phi(t)), \dot{F}(t))}{(F(t), F(\Psi(\tau)))}. \quad (2.2.22)$$

The right-hand side of equation (2.2.22) is a 2π-periodic function in t and, simultaneously, a T-periodic function in τ. Denote by $\tau = \tau(t)$ the solution to equation (2.2.22) determined by the condition $\tau = 0$ for $t = 0$. The function $\tau(t+2\pi n_1)$ also provides a solution to equation (2.2.22). Here $\tau(2\pi n_1) = T$. The function $\tau(t+2\pi n_1) - T$ is the solution to system (2.2.22) vanishing at $t = 0$. Hence, by the theorem of uniqueness, the following identity holds

$$\tau(t + 2\pi n_1) = \tau(t) + T. \quad (2.2.23)$$

Next, using (2.2.23), it can be readily established that the functions (2.2.21) are $2\pi n_1$-periodic in t.

Sufficiency. Let $X(t, X_0, 0)$ be a $2\pi n_1$-periodic solution to system (2.2.15). Show that system (2.2.1) has a periodic solution of period $T = \tau(2\pi n_1) = \theta(2\pi n_1) + 2\pi n_1$, where $\theta(t)$ is a solution to equation (2.2.14) with initial condition $\theta = 0$ at $t = 0$. In fact, after substituting the periodic vector function $X(t, X_0, 0)$, the right-hand side of (2.2.14) becomes the $2\pi n_1$-periodic function in t. Therefore, the relationship $\tau(t + 2\pi n_1) = \tau(t) + T$ holds identically, here $\tau(t) = \theta(t) + t$. The functions defined by relationships (2.2.17) constitute a solution to system (2.2.1) if they are regarded as the functions of an independent variable τ. Show that they are T-periodic in τ. Replace in (2.2.17) the independent variable t with $t+2\pi n_1$. Then the right-hand side remains unchanged, and the argument on the left-hand side acquires an increment in T, hence the following identity holds

$$Y(\tau, Y_0) = Y(T + \tau, Y_0). \quad \blacksquare$$

A few general remarks on the background of the problem are in order. The approach applied here to construction of the neighborhood of a periodic orbit has led us to a system of differential equations which allows us to solve the problem completely by investigating the behavior of solutions to this system in the neighborhood of its zero solution. In this case, the solutions to system (2.2.1) and (2.2.14) — (2.2.15) are related by formulas (2.2.16) and (2.2.17) that are geometrically meaningful.

The problem of constructing the neighborhood of a periodic solution to system (2.2.1) has a broad background of its own [Nemyckij and Stepanov (1949)], [Birkhoff (1932)], [Hadamard (1901)]. The traditional approach to the problem is to transform the original system to some other system which finally allows one to judge about the behavior of projections of the integral curves of system (2.2.1) onto the normal hyperplane P_t changing its position in space with time. In this respect, the main apparatus is transformation which is carried out by the integral curves of the phase space into themselves via a period. Departure from this approach to the above problem makes it possible to apply the basic ideas of Academician A.M.Lyapunov. The subsequent sections employ Lyapunov's ideas and their recent developments in full.

2.3 Unity Roots of Characteristic Equation

1. In what follows we assume that the right-hand sides of system (2.2.1) are the single-valued analytic functions of variables $y_1, \ldots, y_{n+1}$ in a sufficiently small neighborhood of the closed curve M. Then the right-hand sides of (2.2.14) and (2.2.15) are also the single-valued analytic functions of variables $x_1, \ldots, x_n$, whose expansions as series in integral non-negative powers of variables $x_1, \ldots, x_n$ have continuous real coefficients that are 2π-periodic in t:

$$\frac{d\vartheta}{dt} = g(x_1, \ldots, x_n, t) = \sum_{m=1}^{\infty} X^{(m)}(x_1, \ldots, x_n, t), \qquad (2.3.1)$$

$$\frac{dx_s}{dt} = g_s(x_1, \ldots, x_n, t) = \sum_{j=1}^{n} \rho_{sj}(t) x_j + \sum_{m=2}^{\infty} X_s^{(m)}(x_1, \ldots, x_n, t),$$

$$s = 1, \ldots, n, \qquad (2.3.2)$$

where $X^{(m)}, X_s^{(m)}$ are homogeneous forms in m powers of variables $x_1, \ldots, x_n$ with real coefficients that are 2π-periodic in t. The functions $p_{sj}(t)$ are also continuous, real, and 2π-periodic.

Next, consider the linear system

$$\frac{dx_s}{dt} = \sum_{j=1}^{n} \rho_{sj}(t) x_j, \quad s = 1, \ldots, n. \qquad (2.3.3)$$

A fundamental system of solutions to (2.3.3) will be denoted by $\mathbf{X}(t), \mathbf{X}(0) = \mathbf{E}$, where $\mathbf{X}(t)$ is the nonsingular square matrix of order n, and $\mathbf{E}$ is the unit matrix. It is well known that the fundamental system of solutions $\mathbf{X}(t)$ is determined from the formula

$$\mathbf{X}(t) = \mathbf{\Psi}(t) \mathbf{X}(2\pi)^{\frac{t}{2\pi}}, \qquad (2.3.4)$$

where $\mathbf{\Psi}(t)$ is a 2π-periodic matrix. It follows from (2.3.4) that the behavior of solutions to system (2.3.3) is determined by the roots of the characteristic equation $\det(\mathbf{X}(2\pi) - \rho \mathbf{E}) = 0$ and by the canonical structure of the matrix $\mathbf{X}(2\pi)$.

System (2.3.3) is reducible. By employing the nonsingular linear transformation with periodic real coefficients above the desired functions, it can be reduced to a system with constant coefficients. The linear transformation becomes

$$z_s = \sum_{j=1}^{\infty} g_{sj}(t) x_j, \quad s = 1, \ldots, n. \qquad (2.3.5)$$

After such a replacement of variables the equations (2.3.1), (2.3.2) become

$$\frac{d\vartheta}{dt} = \sum_{m=1}^{\infty} \bar{X}^{(m)}(z_1, \ldots, z_n, t), \qquad (2.3.6)$$

$$\frac{dz_s}{dt} = \sum_{j=1}^{n} a_{sj} z_j + \sum_{m=2}^{\infty} \bar{X}_s^{(m)}(z_1, \ldots, z_n, t), \quad s = 1, \ldots, n, \qquad (2.3.7)$$

where a_{sj} are the real constants, and $\bar{X}^{(m)}, \bar{X}_s^{(m)}$ are the forms to power of m for $z_1, \ldots, z_n$ with the coefficients that are 2π-periodic in t.

Unity Roots of Characteristic Equation

Denote by $\mathbf{A}$ the matrix of linear approximation coefficients for system (2.3.7), by $\lambda_1, \ldots, \lambda_n$ the roots of the equation

$$|\mathbf{A} - \lambda \mathbf{E}| = 0, \qquad (2.3.8)$$

referred to as the characteristic equation of the first approximation system for the equations (2.3.7), and by $\rho_1, \ldots, \rho_n$ the roots of the equation $\det(\mathbf{X}(2\pi) - \rho \mathbf{E}) = 0$.

Suppose the following inequalities hold

$$|\rho_j| < 1, \quad j = 1, \ldots, n. \qquad (2.3.9)$$

Then we have that $\operatorname{Re}(\lambda_j) < 0$, $j = 1, \ldots, n$, since

$$\rho_j = e^{\lambda_j 2\pi}. \qquad (2.3.10)$$

This case occupies a special place and is the subject of investigation in Andronov's theorem which states that, with the conditions (2.3.9) satisfied, the periodic solution (2.2.2) of system (2.2.1) is a periodic self-oscillation and, in addition, is Lyapunov stable. If the notions adopted in the stability theory of motion are employed, then the case (2.3.9) can be characterized as nonsingular, the zero solution of system (2.3.2) being asymptotically Lyapunov stable. Then, by Theorem 25, the periodic solution (2.2.2) of system (2.2.1) is a periodic self-oscillation.

In order to define Lyapunov stability of the periodic solution (2.2.2) to system (2.2.1), it is sufficient to indicate that any solution to system (2.3.2) satisfies the inequality

$$\|X(t, X_0, t_0)\| \le a \|X_0\| e^{-c(t-t_0)}, \qquad (2.3.11)$$

where a, c are positive constants that are independent of X_0 for all sufficiently small $\|X_0\|$.

Integrating (2.3.1), we find

$$\vartheta = \vartheta_0 + \int_{t_0}^{t} \sum_{m=1}^{\infty} X^{(m)}(x_1, \ldots, x_n, t) dt.$$

Hence, using (2.3.11), we obtain $|\vartheta| \le |\vartheta_0| + b \|X_0\|$ for all sufficiently small $\|X_0\|$, where b is a positive constant independent of X_0. The last inequality shows that the zero solution of systems (2.3.1) — (2.3.2) in the case of (2.3.9) is Lyapunov stable. But the periodic solution (2.2.2) to system (2.2.1) is also Lyapunov stable as it follows from Theorem 26.

2. If inequalities (2.3.9) fail to hold for all the roots of the characteristic equation, then there is a variety of integral curve behaviors in the neighborhood of periodic solution (2.2.2). The nature of such cases is more complicated than what we have in fulfilling (2.3.9). Concurrent with the complication of qualitative construction of the periodic solution neighborhood, the analytic difficulties experienced in studies also become more complicated.

Assume that among the roots of the characteristic equation there are $k < n$ roots equal to unity $\rho_j = 1$, $j = 1, \ldots, k$, and, additionally, $|\rho_j| < 1$, $j = k+1, \ldots, n$. In other words, the roots of equation (2.3.8) satisfy the relationships

$$\lambda_j = 0, \quad j = 1, \ldots, k, \qquad \operatorname{Re}(\lambda_j) < 0, \quad j = k+1, \ldots, n.$$

Further it will be assumed that the elementary divisors of the matrix $\mathbf{A}$ corresponding to the zero roots of equation (2.3.8) are prime. Set $N = n - k$. From the assumptions about the roots of the characteristic equation and the structure of matrix $\mathbf{A}$ it is inferred that, in particular, the transform can be chosen such that equation (2.3.6) and system (2.3.7) become

$$\frac{d\vartheta}{dt} = \sum_{m=1}^{\infty} \bar{X}^{(m)}, \tag{2.3.12}$$

$$\frac{d\bar{x}_s}{dt} = \sum_{m=2}^{\infty} \bar{X}_s^{(m)}, \quad \bar{X}_s^{(m)} \bigg|_{\bar{x}_1 = \ldots = \bar{x}_k = 0} = 0, \quad s = 1, \ldots, k.$$

$$\frac{d\bar{y}_j}{dt} = \sum_{l=1}^{N} b_{jl} \bar{y}_l + \sum_{m=2}^{\infty} \bar{Y}_j^{(m)}, \quad j = 1, \ldots, N, \tag{2.3.13}$$

where $\bar{X}^{(m)}$, $\bar{X}_s^{(m)}$, $\bar{Y}_j^{(m)}$ are homogeneous forms to the power of m for variables $\bar{x}_1, \ldots, \bar{x}_k, \bar{y}_1, \ldots, \bar{y}_N$ with real coefficients that are 2π-periodic in t. The variables b_{jl} are real numbers whose matrix has eigenvalues with negative real parts. We shall seek a solution to system (2.3.13) in the form of the series

$$\bar{x}_s = c_s + \sum_{m=2}^{\infty} u_s^{(m)}, \quad s = 1, \ldots, k, \tag{2.3.14}$$

$$\bar{y}_j = \sum_{m=1}^{\infty} v_j^{(m)}, \quad j = 1, \ldots, N, \tag{2.3.15}$$

where $u_s^{(m)}$ and $v_j^{(m)}$ are homogeneous forms to the powers of m for arbitrary constants $c_1, \ldots, c_k$, whose coefficients are 2π-periodic functions of t to be defined. Substituting series (2.3.14) and (2.3.15) in system (2.3.13) and equating the homogeneous forms to the same powers on the right and left, we obtain the sequence of equation systems

$$\frac{du_s^{(m)}}{dt} = U_s^{(m)}, \quad s = 1, \ldots, k, \quad m = 2, 3 \ldots, \tag{2.3.16}$$

$$\frac{dv_j^{(m)}}{dt} = \sum_{l=1}^{N} b_{jl} v_l^{(m)} + V_j^{(m)}, \quad j = 1, \ldots, N, \quad m = 1, 2, \ldots. \tag{2.3.17}$$

The functions $U_s^{(m)}$, $V_j^{(m)}$ are homogeneous forms to the powers, $c_1, \ldots, c_k$. The coefficients of these forms depend on the coefficients of the forms, $\bar{X}_s^{(m_1)}, \bar{Y}_j^{(m_1)}, m_1 \leq m$, and on the coefficients of the forms, $U_s^{(m_1)}$ and $V_j^{(m_1)}, m_1 < m$; or to make it more precise, the form coefficients $U_s^{(m)}, V_j^{(m)}$ are polynomials in the above variables. This

Unity Roots of Characteristic Equation

suggests that if $U_s^{(m_1)}$ and $V_j^{(m_1)}$ are derived as the periodic coefficient forms for $m_1 < m$, then $U_s^{(m)}, V_j^{(m)}$ are also the forms with periodic coefficients.

We shall now show that the first nonperiodic coefficient will be found in the definition of the forms $u_s^{(m)}, s = 1, \ldots, k$. The presence of the nonperiodic coefficient in the definition of the forms $u_s^{(m)}$ is the subject of the general case. The case, where all coefficients of the forms $u_s^{(m)}$ can be defined as nonperiodic, is called a special case as distinct from the general case.

3. Consider the general case. First show that the statement about the first nonperiodic coefficient is true and the forms $v_j^{(m)}$ are defined uniquely as the forms with periodic coefficients if the coefficients of forms $V_j^{(m)}, j = 1, \ldots, N$, are periodic. Indeed, consider the linear system of differential equations

$$\frac{dX}{dt} = \mathbf{B}X + P(t),$$

where $P(t)$ is a continuous 2π-periodic n-dimensional vector function, and $\mathbf{B}$ is a constant real matrix whose eigenvalues have non-negative real parts or at least are such that the matrix $\mathbf{E} - e^{2\pi \mathbf{B}}$ is nonsingular. The general solution of this system becomes

$$X = e^{\mathbf{B}t}X_0 + \int_0^t e^{\mathbf{B}(t-\tau)} P(\tau) d\tau.$$

In order for X to be a 2π-periodic vector function, it is necessary and, by the periodicity of $P(t)$, sufficient that $X(0) = X(2\pi)$. The developed form of the last equality is

$$X_0 = e^{2\pi \mathbf{B}} X_0 + \int_0^{2\pi} e^{\mathbf{B}(2\pi-\tau)} P(\tau) d\tau,$$

whence we get that

$$X_0 = (\mathbf{E} - e^{2\pi \mathbf{B}})^{-1} \int_0^{2\pi} e^{\mathbf{B}(2\pi-\tau)} P(\tau) d\tau.$$

From this it follows that the periodic solution of such a linear system is unique and takes the form

$$X = e^{\mathbf{B}t} (\mathbf{E} - e^{2\pi \mathbf{B}})^{-1} \int_0^{2\pi} e^{\mathbf{B}(2\pi-\tau)} P(\tau) d\tau + \int_0^t e^{\mathbf{B}(t-\tau)} P(\tau) d\tau.$$

Thus, if we denote by $v^{(m)}$ the N-dimensional vector whose components are the desired forms $v_j^{(m)}$, and by $V^{(m)}$ the vector whose components are the given forms $V_j^{(m)}$ with 2π-periodic coefficients, then the $v^{(m)}$ taken to be a vector, 2π-periodic with respect to t, is determined uniquely from the formula

$$v^{(m)} = e^{\mathbf{B}t} (\mathbf{E} - e^{2\pi \mathbf{B}})^{-1} \int_0^{2\pi} e^{\mathbf{B}(2\pi-\tau)} V^{(m)} d\tau + \int_0^t e^{\mathbf{B}(t-\tau)} V^{(m)} d\tau. \quad (2.3.18)$$

Hence, the first nonperiodic coefficient occurs in the series (2.3.14).

Note that the necessary and sufficient condition for the existence of a periodic solution to system (2.3.16) is to satisfy the equalities

$$\int_0^{2\pi} U_s^{(m)} dt = 0, \quad s = 1, \ldots, k. \tag{2.3.19}$$

Denote by μ the value of m such that the equality (2.3.19) fails to hold for the first time for at least one $s \leq k$. Denote by $\bar{u}_s^{(\mu)}$ the solution of the system (2.3.16) the right-hand side of which is replaced by the forms

$$U_s^{(\mu)} - \frac{1}{2\pi}\int_0^{2\pi} U_s^{(\mu)} dt, \quad s = 1, \ldots, k.$$

In equation (2.3.12) and system (2.3.13) we now make such a replacement of the desired functions:

$$\bar{x}_s = \xi_s + \sum_{m=1}^{\mu-1} u_s^{(m)} + \bar{u}_s^{(\mu)}, \quad s = 1, \ldots, k,$$

$$\bar{y}_j = \sum_{m=1}^{\mu} v_j^{(m)} + \eta_j, \quad j = 1, \ldots, N. \tag{2.3.20}$$

In formulas (2.3.20), $u_s^{(m)}, v_j^{(m)}$ are the forms with respect to $\xi_1, \ldots, \xi_k$, obtained from the above forms by replacing c_s with ξ_s. Then we get the following system

$$\frac{d\vartheta}{dt} = \Theta,$$

$$\frac{d\xi_s}{dt} = \psi_s + \Xi_s, \quad s = 1, \ldots, k, \tag{2.3.21}$$

$$\frac{d\eta_j}{dt} = \sum_{l=1}^{N} b_{jl}\eta_l + H_j, \quad j = 1, \ldots, N.$$

The functions Ξ_s can be represented as

$$\Xi_s = \sum_{m=0}^{\infty} \Xi_s^{(m)},$$

where $\Xi_s^{(m)}$ are the homogeneous forms of $\xi_1, \ldots, \xi_k$ with coefficients that are analytic in $\eta_1, \ldots, \eta_N$ and 2π-periodic in t. In this case, $\Xi_s^{(m)} \equiv 0$ for $m < \mu$ if $\eta_1 = \ldots = \eta_N = 0$.

Denote by $Z_s^{(\mu)}$ the values of the form $\Xi_s^{(\mu)}$ obtained for $\eta_1 = \ldots = \eta_N = 0$. Consider the system of equations

$$\frac{d\xi_s}{dt} = Z_s^{(\mu)}, \quad s = 1, \ldots, k. \tag{2.3.22}$$

The right-hand sides of this system are the homogeneous forms of order μ with respect to $\xi_1, \ldots, \xi_k$ with real coefficients. The necessary and sufficient conditions for asymptotic Lyapunov stability of the zero solution to system (2.3.22) can be expressed in terms of such coefficients. These conditions can be expressed in terms of the function v that is a solution to the system of equations

$$\sum_{j=1}^{k} \frac{\partial v}{\partial \xi_j} Z_j^{(\mu)} = W,$$

$$\sum_{j=1}^{k} \frac{\partial v}{\partial \xi_j} \xi_j = (m' + 1 - \mu)v, \qquad (2.3.23)$$

where W is the homogeneous function of order m' that is positive-definite. A necessary and sufficient condition for asymptotic Lyapunov stability of the zero solution to system (2.3.22) is the existence of a solution to system (2.3.23) as the negative-definite homogeneous function of order $m' + 1 - \mu$. Moreover, system (2.3.23) admits the solution v in the closed form expressed in terms of the functions $Z_j^{(\mu)}$ and W.

Next, we seek a solution to the equation $\frac{d\vartheta}{dt} = \Theta$ as the series

$$\vartheta = \sum_{m=0}^{\infty} \vartheta^{(m)}, \qquad (2.3.24)$$

where $\vartheta^{(m)}$ is the form to the power of m for $\xi_1, \ldots, \xi_k$, whose coefficients are to be determined and are analytic in $\eta_1, \ldots, \eta_N$ and periodic in t. Substituting (2.3.24) in the first equation of system (2.3.21) and equating the forms of the same dimension for $\xi_1, \ldots, \xi_k$, we obtain for definition of the forms $\vartheta^{(m)}$ the system of equations

$$\frac{\partial \vartheta^{(m)}}{\partial t} + \sum_{j=1}^{N} \frac{\partial \vartheta^{(m)}}{\partial \eta_j} \left(\sum_{l=1}^{N} b_{jl}\eta_l + H_j^{(0)} \right) = \Theta^{(m)}, \qquad (2.3.25)$$

where $H_j^{(0)}$ is the value of the function H_j when $\xi_1 = \ldots = \xi_k = 0$, and $\Theta^{(m)}$ are homogeneous functions to the power of m for $\xi_1, \ldots, \xi_k$ with coefficients that are analytic in $\eta_1, \ldots, \eta_N$ and periodic in t if $\Theta^{(m_1)}, m_1 < m$ are defined to be the forms of the same nature.

In this manner the forms $\Theta^{(m)}$ can always be defined if the following relationship holds

$$\int_0^{2\pi} \Theta^{(m)} dt \equiv 0 \quad \text{with} \quad \eta_1 = \ldots = \eta_N = 0. \qquad (2.3.26)$$

Omitting the proof of this proposition for now, we denote by ν the value of m for which the identities (2.3.26) fail to hold for the first time.

Theorem 28 *If the zero solution of system (2.3.22) is asymptotically stable, then the periodic solution (2.2.2) of system (2.2.1) is a periodic self-oscillation. In addition, if $\nu + 1 > \mu$, then the periodic solution (2.2.2) of system (2.2.1) is Lyapunov stable.*

Proof: Our immediate task is to transform system (2.3.21) in order to annihilate the forms $\Xi_s^{(m)}, m < \mu$ and make it possible for the forms $\Xi_s^{(m)}$ to have real constant coefficients. For this purpose, we consider the system of equations

$$\frac{\partial x_s}{\partial t} + \sum_{j=1}^{N} \frac{\partial x_s}{\partial \eta_j} \left(\sum_{l=1}^{N} b_{jl}\eta_l + H_j \right) = \Xi_s, \quad s = 1, \ldots, k. \qquad (2.3.27)$$

Assume that in system (2.3.27) the functions ξ_s in H_j are replaced by the desired functions $x_1, \ldots, x_k$. We seek a solution to system (2.3.27) as the series

$$x_s = \xi_s + \sum_{m=1}^{\infty} \xi_s^{(m)}, \quad s = 1, \ldots, k, \qquad (2.3.28)$$

where the functions $\xi_s^{(m)}$ are the homogeneous forms of $\xi_1, \ldots, \xi_k$ with coefficients that are analytic in $\eta_1, \ldots, \eta_N$, periodic in t and are to be defined. Substituting (2.3.28) in (2.3.27) and equating the forms of the same dimension on the right and left for $\xi_1, \ldots, \xi_k$, we obtain a sequence of equation systems to define the functions $\xi_s^{(m)}$:

$$\frac{\partial \xi_s^{(m)}}{\partial t} + \sum_{j=1}^{N} \frac{\partial \xi_s^{(m)}}{\partial \eta_j} \left(\sum_{l=1}^{N} b_{jl} \eta_l + H_j^{(0)} \right) = \Xi_s^{(m)}, \quad s = 1, \ldots, k, \quad m = 0, 1, \ldots . \quad (2.3.29)$$

Such a selection of the forms $\xi_s^{(m)}$ from system (2.3.29) becomes possible whenever

$$\int_0^{2\pi} \Xi_s^{(m)} dt \equiv 0 \quad \text{with} \quad \eta_1 = \ldots = \eta_N = 0. \qquad (2.3.30)$$

Condition (2.3.30) holds for $m < \mu$ but not for $m = \mu$. In the last case the right-hand sides of system (2.3.29) are substituted by

$$\Xi_s^{(m)} - \frac{1}{2\pi} \int_0^{2\pi} \hat{\Xi}_s^{(m)} dt,$$

where the sign $\wedge$ stands for substitution of $\eta_1, \ldots, \eta_N$ by zeros. The solution to the modified system (2.3.29) with $m = \mu$ will be denoted by $\hat{\xi}_s^{(m)}$. Lying aside the problem of solution existence in the above form for conditions (2.3.30), we make a substitution in system (2.3.21):

$$\xi_s = \bar{\xi}_s + \sum_{m=1}^{\mu-1} \xi_s^{(m)} + \hat{\xi}_s^{(\mu)}, \quad s = 1, \ldots, k. \qquad (2.3.31)$$

The forms appearing in equalities (2.3.31) are calculated for $\xi_s = \bar{\xi}_s$. After transformation, system (2.3.21) becomes

$$\frac{d\vartheta}{dt} = \bar{\Theta}, \qquad (2.3.32)$$

$$\frac{d\bar{\xi}_s}{dt} = \bar{\Xi}_s, \quad s = 1, \ldots, k$$

$$\frac{d\eta_j}{dt} = \sum_{l=1}^{N} b_{jl} \eta_l + \bar{H}_j, \quad j = 1, \ldots, N. \qquad (2.3.33)$$

In system (2.3.33), the functions $\bar{\Xi}_s$ become

$$\bar{\Xi}_s = \sum_{m=\mu}^{\infty} Z_s^{(m)}, \quad s = 1, \ldots, k, \qquad (2.3.34)$$

Unity Roots of Characteristic Equation

where $Z_s^{(m)}$ are the forms to the m-th power for variables $\xi_1, \ldots, \xi_k$, here the forms $Z_s^{(\mu)}$ either have real constant coefficients or will have them after repeated transformation of the same character. The functions $\bar{H}_j$ have the property that the collection of their lower dimension terms independent of $\eta_1, \ldots, \eta_N$ constitutes the forms for $\bar{\xi}_1, \ldots, \bar{\xi}_k$ to the power $\mu_1 > \mu$. Furthermore, we show that the zero solution to system (2.3.33) is asymptotically Lyapunov stable. For this purpose, we find a negative-definite homogeneous function v satisfying system (2.3.23), and a negative-definite quadratic form v_1 such that

$$\sum_{j=1}^{N} \frac{\partial v_1}{\partial \eta_j} \left(\sum_{l=1}^{N} b_{jl} \eta_l \right) = \sum_{l=1}^{N} \eta_l^2.$$

Compute the total derivative of the function $V = v + v_1$ in terms of system (2.3.33). Then we get a positive-definite function W_1 of variables $\bar{\xi}_1, \ldots, \bar{\xi}_k, \eta_1, \ldots, \eta_N$:

$$\frac{dV}{dt} = \sum_{s=1}^{k} \frac{\partial v}{\partial \bar{\xi}_s} \bar{\Xi}_s + \sum_{j=1}^{N} \frac{\partial v_1}{\partial \eta_j} \left(\sum_{l=1}^{N} b_{jl} \eta_l + \bar{H}_j \right) = W_1;$$

it follows that

$$W_1 = W + \sum_{l=1}^{N} \eta_l^2 + \sum_{s=1}^{k} \frac{\partial v}{\partial \bar{\xi}_s} \sum_{m=\mu+1}^{\infty} Z_s^{(m)} + \sum_{j=1}^{N} \frac{\partial v_1}{\partial \eta_j} \bar{H}_j. \quad (2.3.35)$$

The third summand in (2.3.35) has order with respect to $\bar{\xi}_1, \ldots, \bar{\xi}_k$ no less than $m'+1$ and, therefore, does not affect the sign of the form W_1 for sufficiently small $|\bar{\xi}_s|$, $|\eta_j|$. Denote by $S_j^{(\mu_1)}$ the forms to the least power for $\bar{\xi}_1, \ldots, \bar{\xi}_k$ which appear in the expansion of the functions $\bar{H}_j$ and are independent of $\eta_1, \ldots, \eta_N$. Then $\sum_{j=1}^{N} \frac{\partial v_1}{\partial \eta_j} S_j^{(\mu_1)}$ can be estimated in terms of the quadratic form of the quantities $\eta_1, \ldots, \eta_N$ and the form to the power $2\mu_1$ for $\bar{\xi}_1, \ldots, \bar{\xi}_k$; indeed,

$$\left| \sum_{j=1}^{N} \frac{\partial v_1}{\partial \eta_j} S_j^{(\mu_1)} \right| \leq \frac{\epsilon^2}{2} \sum_{j=1}^{N} \left(\frac{\partial v_1}{\partial \eta_j} \right)^2 + \frac{1}{2\epsilon^2} \sum_{j=1}^{N} \left(S_j^{(\mu_1)} \right)^2. \quad (2.3.36)$$

If the number m', the order of the function W, is selected so that the equality $m' = \mu + 1$ holds, then by applying the estimate (2.3.36), we arrive at a conclusion about validity of the above statement. Thus, the zero solution to system (2.3.33), and, hence to (2.3.2), proves to be asymptotically Lyapunov stable, because all the transformations employed in our reasoning do not disturb stability. From this, and from Theorem 25, it is inferred that the periodic solution (2.2.2) to system (2.2.1) is a periodic self-oscillation.

We now show that for $\nu + 1 > \mu$ the periodic solution (2.2.2) to system (2.2.1) will also be Lyapunov stable. To do this, we first estimate $\|\Xi\|$, where Ξ is the vector with components $\bar{\xi}_s$. Calculate the total derivative of the function v with respect to t in terms of system (2.3.33):

$$\frac{dv}{dt} = \sum_{s=0}^{k} \frac{\partial v}{\partial \bar{\xi}_s} \sum_{m=\mu}^{\infty} Z_s^{(m)} = W + \sum_{s=1}^{k} \frac{\partial v}{\partial \bar{\xi}_s} \sum_{m=\mu+1}^{\infty} Z_s^{(m)}.$$

For sufficiently small values $|\bar{\xi}_s|, |\eta_j|$ the following two inequalities hold:

$$-a_1 \|\Xi\|^{m'-\mu+1} \leq v \leq -a_2 \|\Xi\|^{m'-\mu+1}, \qquad (2.3.37)$$

$$b_1 \|\Xi\|^{m'} \leq \frac{dv}{dt} \leq b_2 \|\Xi\|^{m'}, \qquad (2.3.38)$$

where a_i and b_i, $i = 1, 2$ are positive constants. Integrating the inequalities (2.3.38) along the integral curves of system (2.3.33) and considering the estimate (2.3.37), we get that the following inequalities hold for all $t \geq t_0$

$$c_1 \|\Xi_0\|(1 + c_2\|\Xi_0\|^{\mu-1}(t-t_0))^{-\frac{1}{\mu-1}} \leq \|\Xi\|$$

$$\leq d_1 \|\Xi_0\|(1 + d_2\|\Xi_0\|^{\mu-1}(t-t_0))^{-\frac{1}{\mu-1}}. \qquad (2.3.39)$$

Estimate the solution of equation (2.3.32). For this purpose we perform the transformation:

$$\vartheta = \varphi + \sum_{m=0}^{\nu-1} \vartheta^{(m)},$$

where $\vartheta^{(m)}$ is the form of variables $\xi_1, \ldots, \xi_k$, satisfying the equation (2.3.25). Then the function φ satisfies the equation

$$\frac{d\varphi}{dt} = \sum_{m=0}^{\infty} \Phi^{(m)}, \qquad (2.3.40)$$

where $\Phi^{(m)}$ is the homogeneous form to the power m for $\xi_1, \ldots, \xi_k$. Performing transformation (2.3.31) in equation (2.3.40), we find that its right-hand side for sufficiently small $|\xi_s|, |\eta_j|$, $s = 1, \ldots, k$, $j = 1, \ldots, N$ satisfies the inequalities

$$-a_3 \|\Xi\|^{\nu} \leq \frac{d\varphi}{dt} \leq a_4 \|\Xi\|^{\nu}, \qquad (2.3.41)$$

where a_3, a_4 are positive constants. Integrating (2.3.41) and using (2.3.39), we obtain

$$|\varphi| \leq a_5 \|\Xi\|^{\nu+1-\mu} \quad \text{with} \quad t \geq t_0,$$

where a_5 is the positive constant independent of Ξ_0 with a sufficiently small $\|\Xi_0\|$. In the last inequality the φ is a solution to equation (2.3.40) that is determined by the initial condition $\varphi = 0$ for $t = t_0$.

Since the preliminary transformations of system (2.3.1) and (2.3.2) are not concerned with the problem of stability, it is possible to state that for $\nu + 1 > \mu$ the zero solution to system (2.3.1) and (2.3.2) is Lyapunov stable. Therefore, the periodic solution (2.2.2) to system (2.2.1) is not only orbit-asymptotically stable, but also Lyapunov stable.

Remark. The proof of the possibility to define the forms participating in the above transformations was omitted from equations (2.3.25), (2.3.29) in order to purge the proof for Theorem 28 of excess details.

Unity Roots of Characteristic Equation

We now fill up the gap. For this purpose we will consider not equations of the forms as such, but equations for their coefficients which can be reduced in the general case to the form

$$\frac{\partial x}{\partial t} + \sum_{j=1}^{N} \frac{\partial x}{\partial \eta_j} \left(\sum_{l=1}^{N} b_{jl}\eta_l + H_j^{(0)} \right) = y, \quad (2.3.42)$$

where the right-hand side of equation (2.3.42) is the known analytic function of variables $\eta_1, \ldots, \eta_N$ and, simultaneously, the 2π-periodic function in t:

$$y = a_0(t) + \sum_{m_1+\ldots+m_N=1}^{\infty} a^{(m_1,\ldots,m_N)}(t) \eta_1^{m_1} \ldots \eta_N^{m_N} = a_0(t) + \sum_{m=1}^{\infty} \bar{y}^{(m)}.$$

We will seek a solution to equation (2.3.42) in the form

$$x = b_0(t) + \sum_{m=1}^{\infty} x^{(m)}, \quad (2.3.43)$$

where $\bar{y}^{(m)}, x^{(m)}$ are homogeneous forms to the power m for $\eta_1, \ldots, \eta_N$ with the coefficients that are 2π-periodic in t. Substituting (2.3.43) in (2.3.42) and equating the forms of the same dimension on both sides, we find

$$\frac{db_0}{dt} = a_0,$$

$$\frac{\partial x^{(m)}}{\partial t} + \sum_{j=1}^{N} \frac{\partial x^{(m)}}{\partial \eta_j} \sum_{l=1}^{N} b_{jl}\eta_l = \bar{y}^{(m)}, \quad m = 1, 2, \ldots. \quad (2.3.44)$$

From system (2.3.44) it follows that the necessary condition for the existence of a solution to (2.3.42), which is analytic in $\eta_1, \ldots, \eta_N$ and 2π-periodic in t, is the fulfilment of the equation

$$\int_0^{2\pi} a_0 dt = 0.$$

The forms $\bar{y}^{(m)}$ have periodic coefficients if $x^{(m_1)}, m_1 < m$, are defined as the forms with periodic coefficients. Show that the form $x^{(m)}$ is uniquely defined to be the form with periodic coefficients if the coefficients of the form $\bar{y}^{(m)}$ are periodic. Towards this end, substituting in (2.3.44) the desired forms $x^{(m)}$ in expanded form, and equating the coefficients on the right and left with the terms of the same dimension for $\eta_1, \ldots, \eta_N$, we obtain the linear system

$$\frac{dZ}{dt} = \mathbf{B}_m Z + P_m(t), \quad (2.3.45)$$

where the vector Z is made up of the coefficients of the form $x^{(m)}$, and the components of the vector P_m are the corresponding coefficients of the form $\bar{y}^{(m)}$.

The familiar Lyapunov's theorem suggests that the eigenvalues of the matrix $\mathbf{B}_m$ are linear combinations with integral non-negative coefficients of the eigenvalues of the matrix $\mathbf{B}$ and, therefore, have negative real parts. From the above reasoning it follows that system (2.3.45) has a unique 2π-periodic solution in t. So it has been shown that the forms $x^{(m)}$ are uniquely defined as the forms with periodic coefficients.

Now, the series $\sum_{m=1}^{\infty} x^{(m)}$ is determined uniquely. Show that it is convergent. To this end, consider the system

$$\frac{dx}{dt} = y - a_0,$$

$$\frac{d\eta_j}{dt} = \sum_{l=1}^{N} b_{jl}\eta_l + H_j^0, \quad j = 1, \ldots, N. \tag{2.3.46}$$

Lyapunov's first method implies that system (2.3.46) has a set of solutions which can be represented as series

$$x = \sum_{m_1 + \ldots + m_N = 1}^{\infty} L^{(m_1, \ldots, m_N)}(t) \alpha_1^{m_1} \ldots \alpha_N^{m_N} e^{t \sum_{j=1}^{N} m_j \kappa_j}, \tag{2.3.47}$$

$$\eta_s = \sum_{m_1 + \ldots + m_N = 1}^{\infty} L_s^{(m_1, \ldots, m_N)}(t) \alpha_1^{m_1} \ldots \alpha_N^{m_N} e^{t \sum_{j=1}^{N} m_j \kappa_j}. \tag{2.3.48}$$

Series (2.3.47), (2.3.48) converge for $t \geq 0$ if $|\alpha_1|, \ldots, |\alpha_N|$ are sufficiently small. The Jacobian of the functions $\eta_1, \ldots, \eta_N$ for quantities $\beta_1 = \alpha_1 e^{\kappa_1 t}, \ldots, \beta_N = \alpha_N e^{\kappa_N t}$ (κ_1, $\ldots$, κ_N are eigenvalues of matrix **B**) is nonzero for sufficiently small $|\beta_1|, \ldots, |\beta_N|$. By converting series (2.3.48) and eliminating with the help of these conversions the quantities $\beta_1, \ldots, \beta_N$ from series (2.3.47), we obtain a holomorphic function of the quantities $\eta_1, \ldots, \eta_N$, satisfying equation (2.3.42), where y is replaced by $y - a_0$.

We have thus found the holomorphic solution which coincides with the formally constructed series, since the coefficients of expansion of the holomorphic series into integral non-negative powers of $\eta_1, \ldots, \eta_N$ have 2π period in t. This follows from the fact that the coefficients of series (2.3.47), (2.3.48) are polynomials with 2π-periodic coefficients in t. So is the case with the coefficients of the above-mentioned holomorphic series. It remains to show that after substituting y by $y - a_0$ the holomorphic series can satisfy an equation of the form (2.3.42), with periodic coefficients of the holomorphic series only. In fact, equation (2.3.45) with $m = 1$ has the periodic nonhomogeneity of $P_1(t)$. Therefore, a unique solution, which does not contain exponential terms, is periodic, $P_2(t)$ also being a periodic vector function. Further, taking $P_m(t)$ as a 2π-periodic function, it may be stated that the solution of system (2.3.45) represented as a polynomial to the powers of t with periodic coefficients, is defined as periodic. Thus, series (2.3.43) formally satisfying equation (2.3.42) turns out to be convergent. So the gap in the preceding reasoning is completely filled up. ∎

We now discuss a special case where all the roots of the characteristic equation are equal to unity. Then $\lambda_j = 0$, $j = 1, \ldots, n$. As above, assume that elementary divisors of the matrix **A** are prime. Then it is possible to select the transform (2.3.5) such that system (2.3.2) becomes

$$\frac{dz_s}{dt} = \sum_{m=2}^{\infty} \bar{X}_s^{(m)}, \quad s = 1, \ldots, n, \tag{2.3.49}$$

where $\bar{X}_s^{(m)}$ are homogeneous forms to the power of m for $z_1,\ldots,z_n$ with 2π-periodic coefficients in t. We seek a solution to system (2.3.49) as the series

$$z_s = c_s + \sum_{m=2}^{\infty} u_s^{(m)}, \qquad (2.3.50)$$

where $u_s^{(m)}$ are homogeneous forms to the power of m for arbitrary constants $c_1,\ldots,c_n$ with 2π-periodic coefficients in t to be defined. Substituting series (2.3.50) in (2.3.49) and equating on the left and right the forms of the same dimension with respect to $c_1,\ldots,c_n$, we obtain the system

$$\frac{du_s^{(m)}}{dt} = U_s^{(m)}, \quad s=1,\ldots,n, \quad m=2,3,\ldots. \qquad (2.3.51)$$

Assume that in the solution of (2.3.51) there is the general case where there exists an integer μ such that the following inequality holds for at least one $s \le n$

$$\int_0^{2\pi} U_s^{(\mu)} dt \ne 0.$$

Denote by $\bar{u}_s^{(\mu)}$ a solution to the system

$$\frac{du_s^{(\mu)}}{dt} = U_s^{(\mu)} - \frac{1}{2\pi}\int_0^{2\pi} U_s^{(\mu)} dt.$$

Replace in system (2.3.49) the desired functions by the formulas

$$z_s = \Psi_s + \sum_{m=1}^{\mu-1} \bar{u}_s^{(m)} + \bar{u}_s^{(\mu)}, \quad s=1,\ldots,n. \qquad (2.3.52)$$

The homogeneous forms appearing in (2.3.52) are calculated for $c_s = \Psi_s$. After such a replacement, to define the functions $\Psi_1,\ldots,\Psi_n$, we obtain

$$\frac{d\Psi_s}{dt} = \sum_{m=\mu}^{\infty} \Psi_s^{(m)}, \quad s=1,\ldots,n. \qquad (2.3.53)$$

In this case, the forms $\Psi_s^{(\mu)}, s=1,\ldots,n$ have constant real coefficients. By repeating the transformation of system (2.3.53), it is possible to make such adjustments that the terms constituting the $(\mu+1)$-th order form would also have real coefficients. In general, using a finite sequence of such transformations, we can have that in the forms of the system transformed up to order $(\mu+l)$ the coefficients are real and constant. Further, in the case of necessity the system (2.3.53) is said to be obtained from application of the sequence of such transformations. If the zero solution of the system

$$\frac{d\Psi_s}{dt} = \Psi_s^{(\mu)}, \quad s=1,\ldots,n, \qquad (2.3.54)$$

is asymptotically Lyapunov stable, then, as it follows from Theorem 28, the periodic solution (2.2.2) to system (2.2.1) is orbit-asymptotically stable. Furthermore, if $\nu +$

$1 > \mu$, then it is Lyapunov stable. Here ν is the power of the form $\vartheta^{(m)}$ for which a nonperiodic coefficient appears for the first time in the solution of equation (2.3.6) as a series

$$\vartheta = \sum_{m=1}^{\infty} \vartheta^{(m)},$$

where $\vartheta^{(m)}$ is a homogeneous form to the power of m for variables $z_1, \ldots, z_n$ with 2π-periodic coefficients in t that are to be defined.

2.4 The Case of Several Complex Roots with Unity Moduli

1. Assume that the characteristic equation has $2k \leq n$ complex roots with unity moduli. The remaining roots have moduli that are less than unity. In the case where the complex roots with unity moduli include the equal ones, we assume that their corresponding elementary divisors of the matrix $\mathbf{A}$ are prime. Under such assumptions it is possible to select transforms (2.3.5) such that the system (2.3.1) — (2.3.2) becomes

$$\frac{d\vartheta}{dt} = \Theta, \tag{2.4.1}$$

$$\frac{dx_s}{dt} = -\lambda_s y_s + X_s,$$

$$\frac{dy_s}{dt} = \lambda_s x_s + Y_s, \quad s = 1, \ldots, k, \tag{2.4.2}$$

$$\frac{dz_j}{dt} = \sum_{l=1}^{N} b_{jl} z_l + Z_j, \quad j = 1, \ldots, N \quad (N + 2k = n).$$

From the above assumptions it follows that the eigenvalues of the matrix $\mathbf{B}$, whose elements are real constants b_{jl}, have negative real parts. Without loss of generality, the numbers $\lambda_1, \ldots, \lambda_k$ can be taken to be positive. The functions Θ, X_s, Y_s, Z_j are holomorphic in x_s, y_s, z_j in the neighborhood of the point $x_1 = \ldots = x_k = y_1 = \ldots = y_k = z_1 = \ldots = z_N = 0$ and are periodic in t. Expansions of the functions X_s, Y_s, Z_j do not contain the terms below the second dimension with respect to $x_1, \ldots, x_k, y_1, \ldots, y_k, z_1, \ldots, z_N$. The function Θ vanishes at $x_1 = \ldots = x_k = y_1 = \ldots = y_k = z_1 = \ldots = z_N = 0$.

We bring system (2.4.2) to a form such that the functions X_s, Y_s contain no terms that are independent of $x_1, \ldots, x_k, y_1, \ldots, y_k$. To this end, consider the system of partial differential equations

$$\frac{\partial x_s}{\partial t} + \sum_{j=1}^{N} \frac{\partial x_s}{\partial z_j} \left(\sum_{l=1}^{N} b_{jl} z_l + Z_j \right) = -\lambda_s y_s + X_s;$$

$$\frac{\partial y_s}{\partial t} + \sum_{j=1}^{N} \frac{\partial y_s}{\partial z_j} \left(\sum_{l=1}^{N} b_{jl} z_l + Z_j \right) = \lambda_s x_s + Y_s, \quad s = 1, \ldots, k. \tag{2.4.3}$$

The Case of Several Complex Roots with Unity Moduli

We seek a solution to this system as series

$$x_s = \sum_{m=2}^{\infty} x_s^{(m)}, \quad y_s = \sum_{m=2}^{\infty} y_s^{(m)}, \qquad (2.4.4)$$

where $x_s^{(m)}, y_s^{(m)}$ are homogeneous forms to the power of m for the quantities $z_1, \ldots, z_N$ with coefficients that are periodic in t and are to be defined. Substituting series (2.4.4) in system (2.4.3) and equating the forms to the same power on the left and right, we obtain the sequence of systems

$$\frac{\partial x_s^{(m)}}{\partial t} + \sum_{j=1}^{N} \frac{\partial x_s^{(m)}}{\partial z_j} \sum_{l=1}^{N} b_{jl} z_l = -\lambda_s y_s^{(m)} + X_s^{(m)},$$

$$\frac{\partial y_s^{(m)}}{\partial t} + \sum_{j=1}^{N} \frac{\partial y_s^{(m)}}{\partial z_j} \sum_{l=1}^{N} b_{jl} z_l = \lambda_s x_s^{(m)} + Y_s^{(m)}, \quad s = 1, \ldots, k, \quad m = 2, 3 \ldots. \qquad (2.4.5)$$

Evidently, the forms $X_s^{(m)}$ and $Y_s^{(m)}$ have periodic coefficients if $x_s^{(m_1)}$ and $y_s^{(m_1)}$ are defined as periodic coefficients for $m_1 < m$. Show that the forms $x_s^{(m)}, y_s^{(m)}$ are uniquely defined to be the forms with periodic coefficients if $X_s^{(m)}, Y_s^{(m)}$ have periodic coefficients. To prove this statement, we make in system (2.4.5) a linear nonsingular replacement in the desired functions such that system (2.4.5) becomes

$$\frac{\partial \bar{x}_s^{(m)}}{\partial t} = \sum_{j=1}^{N} \frac{\partial \bar{x}_s^{(m)}}{\partial z_j} \sum_{l=1}^{N} b_{jl} z_l = i\lambda_s \bar{x}_s^{(m)} + \bar{X}_s^{(m)},$$

$$\frac{\partial \bar{y}_s^{(m)}}{\partial t} = \sum_{j=1}^{N} \frac{\partial \bar{y}_s^{(m)}}{\partial z_j} \sum_{l=1}^{N} b_{jl} z_l = -i\lambda_s \bar{y}_s^{(m)} + \bar{Y}_s^{(m)}.$$

By the Euler formula, $\bar{x}_s^{(m)} = \frac{1}{m} \sum_{j=1}^{N} \frac{\partial \bar{x}_s^{(m)}}{\partial z_j} z_j$, therefore the functions $\bar{x}_s^{(m)}, \bar{y}_s^{(m)}$ satisfy the system

$$\frac{\partial \bar{x}_s^{(m)}}{\partial t} + \sum_{j=1}^{N} \frac{\partial \bar{x}_s^{(m)}}{\partial z_j} \left(\sum_{l=1}^{N} \left(b_{jl} - \frac{i\lambda_s \delta_{jl}}{m} \right) z_l \right) = \bar{X}_s^{(m)},$$

$$\frac{\partial \bar{y}_s^{(m)}}{\partial t} + \sum_{j=1}^{N} \frac{\partial \bar{y}_s^{(m)}}{\partial z_j} \left(\sum_{l=1}^{N} \left(b_{jl} + \frac{i\lambda_s \delta_{jl}}{m} \right) z_l \right) = \bar{Y}_s^{(m)}. \qquad (2.4.6)$$

Each of these equations falls in the category which has already been discussed.

In the preceding section it was proved that similar equations have a unique solution as the form with periodic coefficients. Consequently, each system in (2.4.5) for $m = 2, 3, \ldots$ has a unique solution as the form to the power m with periodic coefficients. Now it remains to show that the formally constructed series (2.4.4) converge for sufficiently small $|z_1|, \ldots, |z_N|$. Convergence of series (2.4.4) is proved by Lyapunov's first method. We now contemplate the method of attack. For system (2.4.2) we construct the set of solutions depending on N arbitrary constants as a special type series based on the eigenvalues of the matrix **B**. Elimination of these arbitrary constants gives rise to $2k$ holomorphic functions that are integrals in system (2.4.3). Expansion of the functions in terms of integral positive powers of $z_1, \ldots, z_N$ has the coefficients that are periodic coefficient polynomials with respect to t. Next, direct calculations make it possible to show that such type series can satisfy system (2.4.3) only where their coefficients are periodic functions. We have thus completed the proof of convergence of series (2.4.4).

Denote series (2.4.4) by u_s, v_s and replace the desired functions in system (2.4.1) — (2.4.2) by

$$x_s = \xi_s + u_s, \quad y_s = \eta_s + v_s. \tag{2.4.7}$$

Then system (2.4.1) — (2.4.2) becomes

$$\frac{d\vartheta}{dt} = \bar{\Theta}, \tag{2.4.8}$$

$$\frac{d\xi_s}{dt} = -\lambda_s \eta_s + \Xi_s,$$

$$\frac{d\eta_s}{dt} = \lambda_s \xi_s + H_s, \tag{2.4.9}$$

$$\frac{dz_j}{dt} = \sum_{l=1}^{N} b_{jl} z_l + \bar{Z}_j, \quad s = 1, \ldots, k, \quad j = 1, \ldots, N.$$

The functions $\bar{\Theta}, \bar{Z}_j$ are of the same character as the functions Θ and Z_j in system (2.4.1) — (2.4.2). Compared to the functions X_s, Y_s, the functions Ξ_s, H_s have the additional property

$$\Xi_s = H_s \equiv 0, \quad s = 1, \ldots, k \quad \text{for} \quad \xi_1 = \ldots = \xi_k = \eta_1 = \ldots = \eta_k = 0.$$

Replace the desired functions in system (2.4.8) and (2.4.9) by the formulas

$$\xi_s = r_s \cos \theta_s, \quad \eta_s = r_s \sin \theta_s, \quad s = 1, \ldots, k, \tag{2.4.10}$$

where r_s and θ_s are the new desired functions. To define them together with the functions $\theta, z_1, \ldots, z_N$, obtain the system of differential equation

$$\frac{d\theta}{dt} = \tilde{\Theta}, \tag{2.4.11}$$

$$\frac{d\theta_s}{dt} = \lambda_s + \Theta_s,$$

$$\frac{dr_s}{dt} = R_s, \quad s = 1, \ldots, k, \tag{2.4.12}$$

$$\frac{dz_j}{dt} = \sum_{l=1}^{N} b_{jl} z_l + \tilde{Z}_j, \quad j = 1, \ldots, N. \tag{2.4.13}$$

The Case of Several Complex Roots with Unity Moduli 103

The functions $R_s = \Xi_s \cos\theta_s + H_s \sin\theta_s, \tilde{\Theta}, \tilde{Z}_j$ are analytic in $r_1,\ldots,r_k, z_1,\ldots,z_N$ and are expanded in integral positive power series of these variables, converging in a sufficiently small neighborhood of the point $r_1 = \ldots = r_k = z_1 = \ldots = z_N = 0$. In this case, expansions of the functions R_s, Z_j start with terms of at least second dimension in the quantities $r_1,\ldots,r_k, z_1,\ldots,z_N$. The function $\tilde{\Theta}$ vanishes at $r_1 = \ldots = r_k = z_1 = \ldots = z_N = 0$.

Assume that the functions $\Theta_s = (H_s \cos\theta_s - \Xi_s \sin\theta_s) r_s^{-1}$ are also analytic functions in $r_1,\ldots,r_k, z_1,\ldots,z_N$ possessing the same property as the function $\tilde{\Theta}$. This assumption is a constraint. Its significance will be discussed below. Note that the coefficients in the expansions of the right-hand sides of system (2.4.11) — (2.4.13) into non-negative integral power series of variables $r_1,\ldots,r_k, z_1,\ldots,z_N$ are the trigonometric polynomials of arguments $\theta_1,\ldots,\theta_k$ with the coefficients periodic in t. By the existence theorem, system (2.4.11) — (2.4.13) has a solution determined by initial data $\theta = \theta_0$, $\theta_s = \theta_s^0$, $r_s = r_s^0$, $z_j = z_j^0$ for $t=0$. Suppose the functions $\theta_1,\ldots,\theta_k, r_1,\ldots,r_k, z_1,\ldots,z_N$ are given by the equalities

$$\theta_s = \theta_s(t, \theta_1^0,\ldots,\theta_k^0, r_1^0,\ldots,r_k^0, z_1^0,\ldots,z_N^0), \quad s=1,\ldots,k, \qquad (2.4.14)$$

$$r_s = r_s(t, \theta_1^0,\ldots,\theta_k^0, r_1^0,\ldots,r_k^0, z_1^0,\ldots,z_N^0), \quad s=1,\ldots,k,$$

$$z_j = z_j(t, \theta_1^0,\ldots,\theta_k^0, r_1^0,\ldots,r_k^0, z_1^0,\ldots,z_N^0), \quad j=1,\ldots,N. \qquad (2.4.15)$$

If $\theta_1^0,\ldots,\theta_k^0$ from equalities (2.4.14) are expressed and then eliminated from (2.4.15), we get the functions r_s and z_j which should satisfy the system of equations

$$\frac{\partial r_s}{\partial t} + \sum_{\sigma=1}^{k} \frac{\partial r_s}{\partial \vartheta_\sigma}(\lambda_\sigma + \Theta_\sigma) = R_s, \quad s=1,\ldots,k,$$

$$\frac{\partial z_j}{\partial t} + \sum_{\sigma=1}^{k} \frac{\partial z_j}{\partial \vartheta_\sigma}(\lambda_\sigma + \Theta_\sigma) = \sum_{l=1}^{N} b_{jl} z_l + \tilde{Z}_j, \; j=1,\ldots,N. \qquad (2.4.16)$$

We seek a solution to system (2.4.16) in the form of series

$$r_s = c_s + \sum_{m=2}^{\infty} r_s^{(m)}, \quad s=1,\ldots,k, \qquad (2.4.17)$$

$$z_j = \sum_{m=1}^{\infty} z_j^{(m)}, \quad j=1,\ldots,N. \qquad (2.4.18)$$

The functions $r_s^{(m)}, z_j^{(m)}$ are homogeneous forms to the power of m for arbitrary constants $c_1,\ldots,c_k$ with coefficients that are trigonometric polynomials of the arguments $\theta_1,\ldots,\theta_k$ whose coefficients, in turn, are periodic in t functions. To save notation, such type of functions will be called the "asterisk" (*) type functions. So the forms $r_s^{(m)}, z_j^{(m)}$ have the (*) type coefficients to be defined. Substituting series (2.4.17) and (2.4.18) in system (2.4.16) and equating on the left and right the functions to the same power with respect to $c_1,\ldots,c_k$, we get a sequence of equation systems

$$\frac{\partial r_s^{(m)}}{\partial t} + \sum_{\sigma=1}^{k} \frac{\partial r_s^{(m)}}{\partial \theta_\sigma} \lambda_\sigma = R_s^{(m)}, \quad s=1,\ldots,k, \qquad (2.4.19)$$

$$\frac{\partial z_j^{(m)}}{\partial t} + \sum_{\sigma=1}^{k} \frac{\partial z_j^{(m)}}{\partial \theta_\sigma} \lambda_\sigma = \sum_{l=1}^{N} b_{jl} z_l^{(m)} + \tilde{Z}_j^{(m)}, \quad j = 1, \ldots, N. \tag{2.4.20}$$

The functions $R_s^{(m)}, \tilde{Z}_j^{(m)}$ are homogeneous forms to the power of m for $c_1, \ldots, c_k$ with the (*) type coefficients if the forms $r_s^{(m_1)}, z_j^{(m_1)}, m_1 < m$ are defined as the forms with the same type coefficients. Also, suppose the numbers $1, \lambda_1, \ldots, \lambda_k$ are rationally independent, i.e. there is no such set of simultaneous nonzero integers $m, m_1, \ldots, m_k$ such that

$$\sum_{s=1}^{k} m_s \lambda_s + m = 0.$$

In the definition of the forms $r_s^{(m)}, z_j^{(m)}$, the type (*) coefficients may be disturbed for the first time when series (2.4.17) is constructed. To show the validity of this statement, we have to establish that system (2.4.20) always has a solution $z_j^{(m)}$ as the form with the type (*) coefficients if the coefficients of the forms $Z_j^{(m)}$ are of the (*) type. Indeed, denote by φ_j the coefficient with some term fixed in the form $z_j^{(m)}$, which is the same for all $j = 1, \ldots, N$, and by Φ_j the coefficient with a similar term in $Z_j^{(m)}$. Then, to define the functions $\varphi_1, \ldots, \varphi_N$, we obtain a system of differential equations

$$\frac{\partial \varphi_j}{\partial t} + \sum_{\sigma=1}^{k} \frac{\partial \varphi_j}{\partial \theta_\sigma} \lambda_\sigma = \sum_{l=1}^{N} b_{jl} \varphi_l + \Phi_j, \quad j = 1, \ldots, N. \tag{2.4.21}$$

From the above assumption it follows that the functions Φ_j take the form

$$\Phi_j = \sum_{p=-q}^{q} \sum_{\nu_1 + \ldots + \nu_k = p} q_j^{(\nu_1, \ldots, \nu_k)}(t) e^{i \sum_{\sigma=1}^{k} \nu_\sigma \theta_\sigma}. \tag{2.4.22}$$

We seek a solution to system (2.4.21) as

$$\varphi_j = \sum_{p=-q}^{q} \sum_{\nu_1 + \ldots + \nu_k = p} f_j^{(\nu_1, \ldots, \nu_k)}(t) e^{i \sum_{\sigma=1}^{k} \nu_\sigma \theta_\sigma}, \tag{2.4.23}$$

where $f_j^{(\nu_1, \ldots, \nu_k)}(t)$ are 2π-periodic functions of t to be defined. Substituting relationship (2.4.23) in system (2.4.21) and considering equality (2.4.22), we equate on both sides the coefficients to the same power of $e^{i\theta_\sigma}, \sigma = 1, \ldots, k$ and, to define the required functions $f_j^{(\nu_1, \ldots, \nu_k)}$, we obtain the system of equations

$$\frac{df_j^{(\nu_1, \ldots, \nu_k)}}{dt} = \sum_{l=1}^{N} \left(b_{jl} - i\sigma_{jl} \sum_{\sigma=1}^{k} \lambda_\sigma \nu_\sigma \right) f_j^{(\nu_1, \ldots, \nu_k)} + q_j^{(\nu_1, \ldots, \nu_k)}, \quad j = 1, \ldots, N. \tag{2.4.24}$$

Since the eigenvalues of the matrix

$$\mathbf{B} - i\mathbf{E} \sum_{\sigma=1}^{k} \lambda_\sigma \nu_\sigma$$

The Case of Several Complex Roots with Unity Moduli   105

have non-negative real parts, the system (2.4.24) has a unique 2π-periodic solution in t.

We now discuss calculation of the forms $r_s^{(m)}$ satisfying system (2.4.19). First show that in fulfilling the condition

$$\int_0^{2\pi} \bar{R}_s^{(m)} dt \equiv 0, \quad s = 1, \ldots, k, \qquad (2.4.25)$$

where $\bar{R}_s^{(m)}$ is a collection of terms in the form $R_s^{(m)}$ that are independent of $e^{i\theta_\sigma}$, $\sigma = 1, \ldots, k$, the forms $r_s^{(m)}$ are defined, their coefficients being of the (*) type. Denote by u_s the coefficient for some term in the form $r_s^{(m)}$, and by U_s the coefficient for a similar term in the form $R_s^{(m)}$. Now, if the coefficients of the forms $R_s^{(m)}$ are of the (*) type, then

$$U_s = \sum_{p=-q}^{q} \sum_{\nu_1+\ldots+\nu_k=p} G_s^{(\nu_1,\ldots,\nu_k)}(t) e^{i\sum_{\sigma=1}^{k} \nu_\sigma \theta_\sigma}. \qquad (2.4.26)$$

Condition (2.4.25) is fulfilled if the following relationship holds for all coefficients U_s of the form $R_s^{(m)}$

$$\int_0^{2\pi} G_s^{(0,\ldots,0)}(t) dt = 0, \quad s = 1, \ldots, k.$$

With the assumption that the condition (2.4.25) has been fulfilled, we seek u_s as

$$u_s = \sum_{p=-q}^{q} \sum_{\nu_1+\ldots+\nu_k=p} F_s^{(\nu_1,\ldots,\nu_k)}(t) e^{i\sum_{\sigma=1}^{k} \nu_\sigma \theta_\sigma}. \qquad (2.4.27)$$

Substituting (2.4.26) and (2.4.27) in (2.4.19) and equating on both sides the terms to the same powers of $e^{i\theta_\sigma}$, $\sigma = 1, \ldots, k$, we find for the required functions $F_s^{(\nu_1,\ldots,\nu_k)}$ the system of equations

$$\frac{dF_s^{(\nu_1,\ldots,\nu_k)}}{dt} + i\left(\sum_{\sigma=1}^{k} \nu_\sigma \lambda_\sigma\right) F_s^{(\nu_1,\ldots,\nu_k)} = G_s^{(\nu_1,\ldots,\nu_k)}, \quad s = 1, \ldots, k. \qquad (2.4.28)$$

We first assume that in equation (2.4.28) the numbers $(\nu_1, \ldots, \nu_k)$ are not simultaneously zero, then from the above assumption of rational independence of the numbers $1, \lambda_1, \ldots, \lambda_k$ it follows that a linear combination $\sum_{\sigma=1}^{k} \nu_\sigma \lambda_\sigma$ never reduces to an integer. Hence, equation (2.4.28) has a unique 2π-periodic solution in t. If $\sum_{\sigma=1}^{k} \nu_\sigma^2 = 0$, then equation (2.4.28) becomes

$$\frac{dF_s^{(0,\ldots,0)}}{dt} = G_s^{(0,\ldots,0)}.$$

This equation has a periodic solution only where condition (2.4.25) is satisfied. When condition (2.4.25) is satisfied, the functions $F_s^{(0,\ldots,0)}$ are defined to be accurate to an arbitrary contant which, in particular, can be chosen such that the forms $r_s^{(m)}$ satisfy the equalities $r_s^{(m)} \equiv 0$ for $t = 0, \theta_1 = \theta_1^0, \ldots, \theta_k = \theta_k^0$.

Thus, in finding series (2.4.17) and (2.4.18), two basically different cases may be encountered. In the first case, there is a number μ such that with at least one $s \leq k$ condition (2.4.25) does not hold for $m = \mu$. This case is referred to as the general case. The second case is encountered where condition (2.4.25) holds for all m. This is a special case.

2. We now discuss the general case in some detail. Define the forms $\bar{r}_s^{(m)}$ as a solution to system (2.4.19) if their right-hand sides are replaced by the forms

$$R_s^{(m)} - \frac{1}{2\pi} \int_0^{2\pi} R_s^{(m)} dt$$

for $m = \mu$.

Replace in system (2.4.11) — (2.4.13) the required functions $r_1, \ldots, r_k, z_1, \ldots, z_N$ by the formulas

$$r_s = \rho_s + \sum_{m=1}^{\mu-1} r_s^{(m)} + \bar{r}_s^{(\mu)}, \quad s = 1, \ldots, k, \quad z_j = \sum_{m=1}^{\mu} z_j^{(m)} + \xi_j, \quad j = 1, \ldots, N, \quad (2.4.29)$$

where $\rho_1, \ldots, \rho_k, \xi_1, \ldots, \xi_k$ are the new required functions. Then we get the system of equations

$$\frac{d\theta}{dt} = \Phi, \tag{2.4.30}$$

$$\frac{d\theta_s}{dt} = \lambda_s + \bar{\Phi}_s, \quad s = 1, \ldots, k, \tag{2.4.31}$$

$$\frac{d\rho_s}{dt} = P_s, \quad s = 1, \ldots, k, \tag{2.4.32}$$

$$\frac{d\xi_j}{dt} = \sum_{l=1}^{N} b_{jl} \xi_l + Y_j, \quad j = 1, \ldots, N.$$

Denote by $P_s^{(0)}, Y_j^{(0)}$ the values of functions P_s and Y_j when $\xi_1 = \ldots = \xi_N = 0$. The collection of lower dimension terms with respect to $\rho_1, \ldots, \rho_k$ in the expansions of the functions $P_s^{(0)}$ constitutes the forms $P_s^{(\mu)}$ to the power μ with real coefficients, and the collection of lower dimension terms in the functions $Y_j^{(0)}$ expanded in powers of $\rho_1, \ldots, \rho_k$ constitutes the forms $Y_j^{(\mu_1)}$. In this case $\mu_1 > \mu$. Introduce the following equation system necessary for further discussion

$$\frac{d\rho_s}{dt} = P_s^{(\mu)}, s = 1, \ldots, k. \tag{2.4.33}$$

Transform equation (2.4.30). We seek its solution as a series

$$\theta = \sum_{m=0}^{\infty} \bar{\theta}_m, \tag{2.4.34}$$

where $\bar{\theta}_m$ is the form to the power m with respect to $\rho_1, \ldots, \rho_k$ with coefficients that are analytic in $\xi_1, \ldots, \xi_N$ whose expansion coefficients are, in turn, of the (*)

type. Substituting (2.4.34) in (2.4.30) and equating the same power forms on both sides with respect to $\rho_1, \ldots, \rho_k$, for definition of the forms $\bar{\theta}_m$ we get the sequence of equations

$$\frac{\partial \bar{\theta}_m}{\partial t} + \sum_{\sigma=1}^{k} \frac{\partial \bar{\theta}_m}{\partial \theta_\sigma}(\lambda_\sigma + \tilde{\Phi}_\sigma) + \sum_{j=1}^{N} \frac{\partial \bar{\theta}_m}{\partial \xi_j}\left(\sum_{l=1}^{N} b_{jl}\xi_l + \tilde{Y}_j\right) = T_m, \qquad (2.4.35)$$

where $\tilde{\Phi}_\sigma$ and $\tilde{Y}_j$ are the values of $\bar{\Phi}_\sigma$ and Y_j for $\rho_1 = \ldots = \rho_k = 0$. In defining the forms $\bar{\theta}_m$ from equations (2.4.35), two possibilities are considered: either

$$\int_0^{2\pi} \bar{T}_m dt \equiv 0 \qquad (2.4.36)$$

for all $m \geq 0$ or there is a number $\nu > 0$ such that identity (2.4.36) does not hold for $m = \nu$. Here $\bar{T}_m$ stands for a collection of terms in the form T_m that are independent of the quantities $\xi_1, \ldots, \xi_N, e^{i\theta_1}, \ldots, e^{i\theta_k}$.

We now show that it is possible to define the form $\bar{\theta}_m$ as required when condition (2.4.36) is satisfied. Equation (2.4.35) is said to be set up for some coefficient of the term fixed in the form $\bar{\theta}_m$. For short, denote by $\bar{\theta}_m$ this coefficient, and by $\bar{T}_m$ the corresponding coefficient in the form T_m. We seek a solution to equation (2.4.35) as the series

$$\bar{\theta}_m = \sum_{p=0}^{\infty} \theta_m^{(p)},$$

where $\theta_m^{(p)}$ is the form to the power p with respect to variables $\xi_1, \ldots, \xi_N$ with the coefficients to be defined. In order to define these forms, we obtain the system of equations

$$\frac{\partial \theta_m^{(p)}}{\partial t} + \sum_{\sigma=1}^{k} \frac{\partial \theta_m^{(p)}}{\partial \theta_\sigma}\lambda_\sigma + \sum_{j=1}^{N} \frac{\partial \theta_m^{(p)}}{\partial \xi_j}\left(\sum_{l=1}^{N} b_{jl}\xi_l\right) = T_m^{(p)}. \qquad (2.4.37)$$

Denote by Z the vector composed of the coefficients of the form $\theta_m^{(p)}$, and by Q_p the vector whose components are the coefficients of the form $T_m^{(p)}$. Then, to determine Z, we find the equation

$$\frac{\partial Z}{\partial t} + \sum_{\sigma=1}^{k} \frac{\partial Z}{\partial \theta_\sigma}\lambda_\sigma + \mathbf{B}^{(p)} Z = Q_p. \qquad (2.4.38)$$

The matrix $\mathbf{B}^{(p)}$ has eigenvalues with negative real parts, since its eigenvalues are linear combinations of the eigenvalues of the matrix $\mathbf{B}$. Therefore, for $p > 0$ we will show, as above, that equation (2.4.37) has a unique solution such that the components of the vector Z are the type (*) functions. For $p = 0$ the vector Z satisfies the equation

$$\frac{\partial Z}{\partial t} + \sum_{\sigma=1}^{k} \frac{\partial Z}{\partial \theta_\sigma}\lambda_\sigma = Q_0. \qquad (2.4.39)$$

Equation (2.4.39) will have a solution of (*) type only if

$$\int_0^{2\pi} \bar{Q}_0 dt = 0,$$

where $\bar{Q}_0$ is the collection of the terms independent of $e^{i\theta_1},\ldots,e^{i\theta_k}$. The last statement is proved in the same manner as in constructing the function $r_s^{(m)}$. Thus, condition (2.4.36) is necessary for definition of the forms $\bar{\theta}_m$ in a specified form. Note that to complete the proof it remains to establish series convergence

$$\sum_{p=0}^{\infty} Q_m^{(p)}.$$

The proof of this convergence will be laid aside for the time being.

Theorem 29 *If among the roots of the characteristic equation there are k pairs of complex roots with unity moduli, i.e. $e^{\pm i\lambda_j}, j = 1,\ldots, k$, and the other roots have moduli less than unity, then the periodic solution (2.2.2) to system (2.2.1) is asymptotic-orbitally stable when the following conditions are satisfied:*

1) the numbers $1, \lambda_1, \ldots, \lambda_k$ are rationally independent;

2) the zero solution to system (2.4.33) is asymptotically Lyapunov stable. In addition, if $\nu + 1 > \mu$, then the periodic solution (2.2.2) to system (2.2.1) is Lyapunov stable.

Proof: We base our proof on the construction of Lyapunov's function, which resolves the raised problem, and the estimates obtained with the use of this function for some coordinates of a moving point. We first transform system (2.4.32) in order to annihilate in the expansions of the functions P_s the terms which generate the forms to the power not exceeding μ with respect to $\rho_1, \ldots, \rho_k$ and vanish at $\xi_1, \ldots, \xi_N = 0$. To do this, consider the following system of partial differential equations

$$\frac{\partial \rho_s}{\partial t} + \sum_{\sigma=1}^{k} \frac{\partial \rho_s}{\partial \theta_\sigma}(\lambda_\sigma + \bar{\theta}_\sigma) + \sum_{j=1}^{N} \frac{\partial \rho_s}{\partial \xi_j}\left(\sum_{l=1}^{N} b_{jl}\xi_l + Y_j\right) = P_s, \quad s = 1,\ldots, k. \quad (2.4.40)$$

We seek a solution to system (2.4.40) as series

$$\rho_s = c_s + \sum_{m=1}^{\infty} \rho_s^{(m)}, \quad (2.4.41)$$

where $\rho_s^{(m)}$ are the forms to the power m with respect to arbitrary constants $c_1, \ldots, c_k$. Substituting (2.4.41) in (2.4.40) and equating on both sides the forms to the same powers of $c_1, \ldots, c_k$, we obtain for their determination the sequence of equation systems

$$\frac{\partial \rho_s^{(m)}}{\partial t} + \sum_{\sigma=1}^{k} \frac{\partial \rho_s^{(m)}}{\partial \theta_\sigma}(\lambda_\sigma + \bar{\Phi}_\sigma^{(0)})$$

$$+ \sum_{j=1}^{N} \frac{\partial \rho_s^{(m)}}{\partial \xi_j}\left(\sum_{l=1}^{N} b_{jl}\xi_l + Y_j^{(0)}\right) = P_s^{(m)}, \quad s = 1, \ldots, k. \quad (2.4.42)$$

The functions $\bar{\Phi}_\sigma^{(0)}$ and $Y_j^{(0)}$ are obtained from $\bar{\Phi}_\sigma$ and Y_j for $\rho_1 = \ldots = \rho_k = 0$. The functions $P_s^{(m)}$ are homogeneous forms to the power m with respect to $c_1, \ldots, c_k$ whose coefficients are the holomorphic functions $\xi_1, \ldots, \xi_N$ expanded to have the type (*) coefficients if the functions $\rho_s^{(m_1)}, m_1 < m$, are the same.

The Case of Several Complex Roots with Unity Moduli

Show that for $m < \mu$ the functions $\rho_s^{(m)}$ are defined as holomorphic functions $\xi_1, \ldots, \xi_N$ whose expansions have the type (*) coefficients. For this purpose we seek $\rho_s^{(m)}$ as series

$$\rho_s^{(m)} = \sum_{p=0}^{\infty} \rho_s^{(m,p)}, \qquad (2.4.43)$$

where $\rho_s^{(m,p)}$ are homogeneous forms to the power p with respect to $\xi_1, \ldots, \xi_N$ with the type (*) coefficients to be determined. Substituting (2.4.43) in (2.4.42) and setting the terms equal on the left and right with the same powers of $\xi_1, \ldots, \xi_N$, we get the equation system

$$\frac{\partial \rho_s^{(m,p)}}{\partial t} + \sum_{\sigma=1}^{k} \frac{\partial \rho_s^{(m,p)}}{\partial \theta_\sigma} \lambda_\sigma + \sum_{j=1}^{N} \frac{\partial \rho_s^{(m,p)}}{\partial \xi_j} \sum_{l=1}^{N} b_{jl} \xi_l = P_s^{(m,p)}. \qquad (2.4.44)$$

For the coefficients of the forms $\rho_s^{(m,p)}$ we get systems of the form (2.4.38) and (2.4.39); it follows that in order to determine forms $\rho_s^{(m,p)}$ as required it is necessary and sufficient that there be

$$\int_0^{2\pi} \bar{P}_s^{(m,0)} dt = 0, \qquad (2.4.45)$$

where $\bar{P}_s^{(m,0)}$ are the terms in the form $P_s^{(m,0)}$, containing no $e^{i\theta_\sigma}$, $\sigma = 1, \ldots, k$. In any case, condition (2.4.45) holds for $m < \mu$. When $m = \mu$, we determine the forms $\rho_s^{(m,p)}$ from the system, as in (2.4.42), with the right-hand side as

$$P_s^{(\mu,0)} - \frac{1}{2\pi} \int_0^{2\pi} \bar{P}_s^{(\mu,0)} dt.$$

Strictly speaking, we have to prove convergence of the series obtained here. We lay aside the proof of this convergence for the time being. In system (2.4.32), we make a replacement

$$\rho_s = q_s + \sum_{m=1}^{\mu} \rho_s^{(m)}, \qquad (2.4.46)$$

where the forms $\rho_s^{(m)}$ are assumed to be computed for $c_s = q_s$. As a result of replacement (2.4.46), where the $q_1, \ldots, q_k$ are considered to be the new required functions, we come to a system of equations as in (2.4.32) with the only additional condition that the expansions of functions Q_s in integral positive powers of $q_1, \ldots, q_s$ contain no terms of order below μ. Here Q_s is the right-hand side of (2.4.32) corresponding to $\frac{dq_s}{dt}$.

For short, system (2.4.32) itself is said to possess the obtained property. This assumption fails to impose additional restrictions on the solution of the stability problem, since replacement (2.4.46) does not affect this problem. Moreover, if the forms $\rho_s^{(1)}$ are not identically zeros, then, by the same procedure it is possible to make a second transformation of system (2.4.32), as (2.4.46), so that the expansions of functions P_s start with the μ-th forms having constant real coefficients. We assume that such a transformation in system (2.4.32) can be made if and when necessary.

Denote by v the positive-definite function of order $m'+1-\mu$ satisfying the differential equation

$$\sum_{s=1}^{k} \frac{\partial v}{\partial \rho_s} P_j^{(\mu)} = w,$$

where w is a negative-definite homogeneous form to the power m'. Denote by v_1 the quadratic form of variables $\xi_1, \ldots, \xi_k$ satisfying the equation

$$\sum_{j=1}^{k} \frac{\partial v_1}{\partial \xi_j} \sum_{l=1}^{N} b_{jl}\xi_l = -\sum_{j=1}^{N} \xi_j^2,$$

and set $V = v + v_1$. The function V is a positive-definite function of the quantities $\rho_1, \ldots, \rho_k, \xi_1, \ldots, \xi_N$. Find the total derivative of this function in terms of system (2.4.32):

$$\frac{dV}{dt} = \sum_{s=1}^{k} \frac{\partial V}{\partial \rho_s} P_s + \sum_{j=1}^{N} \frac{\partial V}{\partial \xi_j} \left(\sum_{l=1}^{N} b_{jl}\xi_l + Y_j \right). \qquad (2.4.47)$$

Show that the right-hand side of (2.4.47) is a negative-definite function of variables $\rho_1, \ldots, \rho_k, \xi_1, \ldots, \xi_N$. For this purpose, set $m' = \mu + 1$, then

$$\frac{dV}{dt} = W + \sum_{s=1}^{k} \frac{\partial V}{\partial \rho_s}(P_s - P_s^{(\mu)}) + \sum_{j=1}^{N} \frac{\partial V}{\partial \xi_j} Y_j,$$

where $W = w - \sum_{j=1}^{N} \xi_j^2$. From this notation of the function $\frac{dV}{dt}$ it follows that it is possible to select a number $r_0 > 0$ such that for $\sum_{s=1}^{k} \rho_s^2 < r_0$ and $\sum_{j=1}^{N} \xi_j^2 < r_0$ we get

$$\frac{dV}{dt} < \frac{1}{2} W. \qquad (2.4.48)$$

The quantities $\theta_1, \ldots, \theta_k$ are now said to be arbitrary real continuously differentiable functions of an independent variable t given for $t \geq 0$. Inequality (2.4.48) implies that a zero solution to system (2.4.32) is asymptotically stable uniformly in the above-mentioned functions $\theta_1, \ldots, \theta_k$. Since the transformations, which resulted in obtaining system (2.4.32) from system (2.3.2), do not affect the problem of stability, we have that a zero solution to system (2.3.2) is asymptotically Lyapunov stable. Consequently, by Theorem 25, the periodic solution (2.2.2) to system (2.2.1) is orbit-asymptotically stable.

We now show that if the inequality $\nu + 1 > \mu$ holds, then a periodic solution (2.2.2) to system (2.2.1) is Lyapunov stable. This statement is said to be proved if Lyapunov's stability of a zero solution to system (2.3.1) — (2.3.2) is established, which, in turn, may be found if Lyapunov's stability of a zero solution to system (2.4.30) — (2.4.32) is proved, considering $\theta_1, \ldots, \theta_k$ to be the above-mentioned functions. Replace in equation (2.4.30) the required function θ by

$$\theta = \varphi + \sum_{m=0}^{\nu+1} \bar{\theta}_m.$$

Then, to define φ, we obtain the equation $\frac{d\varphi}{dt} = R$. In this case, if the inequalities $\sum_{s=1}^{k} \rho_s^2 < \rho_0$ and $\sum_{j=1}^{N} \xi_j^2 < \rho_0$ are satisfied, the function R satisfies the inequality

$$|R| < a \left(\sum_{s=1}^{k} \rho_s^2\right)^{\frac{\nu}{2}}, \qquad (2.4.49)$$

where $\rho_0 > 0$ is a sufficiently small constant, and $a > 0$ is independent of the functions $\theta_1, \ldots, \theta_k$.

Differentiating the function V, by the first group of equations (2.4.32), we have that its total derivative, in terms of this group of equations, satisfies the inequality

$$\frac{dV}{dt} < \frac{1}{2}w \quad \text{for} \quad \sum_{s=1}^{k} \rho_s^2 < \rho_0.$$

As in the preceding section, from the latter inequality it is possible to obtain an estimate for the magnitude $\left(\sum_{s=1}^{k} \rho_s^2\right)^{\frac{1}{2}} = \|\rho\|$. This estimate is

$$a_1\|\rho_0\| \left(1 + a_2\|\rho_0\|^{\mu-1}(t-t_0)^{-\frac{1}{\mu-1}}\right) \leq \|\rho\|$$

$$\leq b_1\|\rho_0\| \left(1 + b_2\|\rho_0\|^{\mu-1}(t-t_0)^{-\frac{1}{\mu-1}}\right), \quad t \geq t_0. \qquad (2.4.50)$$

From inequalities (2.4.49) and (2.4.50) and condition $\nu + 1 > \mu$ we find $|\varphi| \leq |\varphi_0| + d\|\rho_0\|^{\nu+1-\mu}$ for $t \geq t_0$, whence it follows that a zero solution to system (2.4.30) — (2.4.32) is Lyapunov stable uniformly in the functions $\theta_1, \ldots, \theta_k$. Hence, the periodic solution (2.2.2) to system (2.2.1) is also Lyapunov stable. ∎

Theorem 29 also holds where $n = 2k$. If the zero solution to system (2.4.33) is not asymptotically stable, then without considering the terms of dimension above μ it is possible, in certain conditions, to guarantee availability of a family or several families of integral curves approaching arbitrarily close the orbit M of the periodic solution (2.2.2) to system (2.2.1).

2.5 Perturbation Theory of Periodic Orbits

1. Consider the system of differential equations of order $n + 1$

$$\frac{dy_s}{dt} = f_s(y_1, \ldots, y_{n+1}), \quad s = 1, \ldots, n+1, \qquad (2.5.1)$$

or, in vector form,

$$\frac{dY}{dt} = F(Y).$$

Referring to system (2.5.1), assume that it has a 2π-periodic solution

$$Y = \Phi(t). \qquad (2.5.2)$$

Next, assume that the right-hand sides of system (2.5.1) are real single-valued analytic functions in a sufficiently small neighborhood $S(M,\delta)$ of the orbit M for the periodic solution (2.5.2). In parallel with system (2.5.1), we also consider the system

$$\frac{dY}{dt} = F(Y) + R(t,Y), \quad (2.5.3)$$

where $R(t,Y)$ is the vector function given for $Y \in S(M,\delta)$, $t \in (-\infty, +\infty)$ (or for $t \in [0,+\infty)$), that is real and continuous throughout its domain.

Examination of system (2.5.3) raises the question as to: how the integral curves of system (2.5.3) behave in a sufficiently small neighborhood of the orbit M of the periodic solution (2.5.2) to system (2.5.1) dependent on the properties of the function F and the perturbation influence R. Before resolving this problem, it should be noted that if the numbers $\delta_2 = \sup_{\substack{Y \in S(M,\delta) \\ t \geq 0}} \|R(t,Y)\|$ and $\delta_1 = \rho(Y_0, M)$ are sufficiently small, then the integral curve $Y = Y(t, Y_0, t_0)$ of system (2.5.3) passing through the point $Y_0 \in S(M,\delta)$ at $t = t_0 \geq 0$ stays in $S(M,\delta)$ as long as required. Indeed, let $\bar{Y}_0 \in M$ be such that $\|Y_0 - \bar{Y}_0\| < \delta_1$. Denote by $\bar{Y} = \bar{Y}(t, \bar{Y}_0, t_0)$ the integral curve of system (2.5.1) passing through the point $\bar{Y}_0$ at $t = t_0 \geq 0$. This integral curve will be a periodic solution to system (2.5.1), since it is possible to select a constant h such that $\bar{Y}(t, \bar{Y}_0, t_0) = \Phi(t+h)$. Denote by λ a Lipschitz constant for the functions $f_s(Y)$ in the domain $S(M,\delta)$, then

$$\|Y(t,Y_0,t_0) - \bar{Y}(t,\bar{Y}_0,t_0)\| \leq \delta_1 e^{\lambda n(t-t_0)} + \frac{\delta_2}{\lambda n}(e^{\lambda n(t-t_0)} - 1) \quad (2.5.4)$$

for all $t \geq t_0$ (see section 1.1). Inequality (2.5.4) holds as long as the integral curve of system (2.5.3) stays in the domain $S(M,\delta)$. Inequality (2.5.4) implies that the integral curve $Y = Y(t, Y_0, T_0)$ will be in the domain $S(M,\delta)$ for all t's satisfying the inequality

$$t_0 \leq t \leq t_0 + \frac{1}{\lambda n}\ln\frac{\lambda n \delta}{\lambda n \delta_1 + \delta_2}.$$

Relationship (2.5.4) and its consequence make it possible to discuss the behavior of the integral curves of system (2.5.3) starting in a sufficiently small neighborhood of the periodic orbit of system (2.5.1) only in the finite time interval. Of prime importance is the dependence of the behavior of these integral curves on the properties of the functions R and F on the infinite time interval $t \in (-\infty, +\infty)$ and semi-infinite interval $t \in [0, +\infty)$ or $t \in [t_0, +\infty)$.

We shall now consider the behavior of the integral curves of system (2.5.3) governed by properties of F and R in the unbounded time interval. From (2.5.4) it follows that the integral curves of system (2.5.3) starting in any normal hyperplane to the orbit M in its δ_1-neighborhood with time increase, intersect this normal hyperplane any finite number of times if the numbers δ_1 and δ_2 are sufficiently small. Because of this, to study the behavior of the integral curves of system (2.5.3), we may use a coordinate system as in section 2.2. Consider two normal hyperplanes to the orbit M passing through the points $\Phi(t_0)$ and $\Phi(t), t \in [t_0, t_0 + 2\pi)$. From the point Y_0 located on the first hyperplane we release the integral curve $Y = Y(t, Y_0, t_0)$ of system (2.5.3). Denote by $\tau = \tau(t, Y_0, t_0)$ the time at which this integral curve initially intersects the normal hyperplane passing through the point $\Phi(t)$ so that the point $Y(\tau, Y_0, t_0)$ lies in the same normal hyperplane. This implies the relationship

$$((Y(\tau, Y_0, t_0) - \Phi(t)), F(\Phi(t))) = 0.$$

In the normal hyperplane passing through the point $\Phi(t)$ we construct the Cartesian system of coordinates taking axes as the vectors $B_1, \ldots, B_n$ whose components are real continuously differentiable 2π-periodic functions of t. We take B_{n+1} to be the vector

$$B_{n+1} = \frac{F(t)}{\|F(t)\|},$$

where $F(t) = F(\Phi(t))$. Any vector joining the points $\Phi(t)$ and Y can be decomposed in the basis $B_1, \ldots, B_{n+1}$ so that the following relationship holds

$$Y - \Phi(t) = \sum_{j=1}^{n+1} B_j(t) x_j, \qquad (2.5.5)$$

where $x_1, \ldots, x_{n+1}$ are Cartesian coordinates of the vector $Y - \Phi(t)$ in the basis $B_1, \ldots, B_{n+1}$. Denote by $Z(t)$ the vector function $Y(\tau(t), Y_0, t_0) - \Phi(t)$. Then, from the above relationship, we have

$$(Z(t), B_{n+1}) \equiv 0,$$

for the vector $Z(t)$ lies in the normal hyperplane passing through the point $\Phi(t)$, and therefore

$$(Z(t), F(t)) \equiv 0. \qquad (2.5.6)$$

Construct a system of differential equations which serve to define the functions τ and Z. Further, using relationships (2.5.5) and (2.5.6), we pass in this system to the new required functions $x_1, \ldots, x_{n+1}$. The vector function $Z(t)$ satisfies the relationship

$$\frac{dZ(t)}{dt} = \frac{d\tau}{dt} \frac{dY(\tau, Y_0, t_0)}{d\tau} - \frac{d\Phi}{dt} = \frac{d\tau}{dt}(F(Z + \Phi) + R(\tau, Z + \Phi)) - F(t). \qquad (2.5.7)$$

If the integral curve of system (2.5.3) starting in the normal hyperplane, which passes through the point $\Phi(t_0)$, stays in $S(M, \delta)$, then relationship (2.5.6) holds identically, therefore

$$\left(\frac{dZ}{dt}, F(t)\right) + \left(Z, \frac{dF}{dt}\right) = 0.$$

Hence, using relationship (2.5.7), we get

$$\frac{d\tau}{dt} = \frac{F^2(t) - (Z, \dot{F})}{(F(Z + \Phi) + R(\tau, Z + \Phi), F(t))}. \qquad (2.5.8)$$

By (2.5.8), from (2.5.7) we find

$$\frac{dZ}{dt} = \frac{F^2(t) - (Z, \dot{F})}{(F(Z+\Phi) + R(\tau, Z+\Phi), F(t))}(F(Z+\Phi) + R(\tau, Z+\Phi)) - F(t). \quad (2.5.9)$$

Next, our task is to study the behavior of the integral curves of system (2.5.3) on a semi-infinite time interval. These integral curves intersect the normal hyperplane $\Phi(t_0)$ as many times as desired, if they stay in a sufficiently small neighborhood M. Therefore, it suffices to consider only the behavior of those integral curves of system (2.5.3) which start at $t = t_0$ on the normal hyperplane passing through the point $\Phi(t_0)$, so as to obtain the overall picture of the behavior of the integral curves of system (2.5.3). System (2.5.8) — (2.5.9) may be regarded as a system of differential equations intended for definition of the functions Z and τ. Any integral curve of system (2.5.3) which starts in the normal hyperplane passing through the point $\Phi(t_0)$, is uniquely defined via a solution to system (2.5.8) — (2.5.9). In this case, initial data are taken to be $Z = Z_0 = Y_0 - \Phi(t_0)$, $\tau = t_0$ for $t = t_0$ where Y_0 is such that $(Y_0 - \Phi(t_0), F(t_0)) = 0$. Then we get

$$Y(\tau, Y_0, t_0) = Z(t) + \Phi(t) \quad (2.5.10)$$

if the function Y is considered as a function of the parameter τ, i.e. if it satisfies the system

$$\frac{dY}{d\tau} = F(Y) + R(\tau, Y).$$

This statement is verified by simple differentiation of both parts of (2.5.10) with respect to τ.

In system (2.5.8) — (2.5.9) we now consider the new required functions using relationship (2.5.5) and setting $Z = \mathbf{B}(t)X$, where $\mathbf{B}$ is the matrix whose j-th column is composed of the vector B_j components. By the orthogonality of the matrix $\mathbf{B}$, we have the equality $\mathbf{B}^{-1} = \mathbf{B}^*$, hence $X = \mathbf{B}^*Z$. Since it will be further assumed that $(Z_0, F(t_0)) = 0$, the equality $x_{n+1} \equiv 0$ holds for all $t \geq t_0$ because

$$x_{n+1} = (B_{n+1}, Z(t)) = \frac{(F(t), Z(t))}{\|F(t)\|}. \quad (2.5.11)$$

The left-hand side of relationship (2.5.6) is the integral of system (2.5.8) — (2.5.9), therefore (2.5.11) implies the identity $x_{n+1} \equiv 0$. To define the new desired vector function X and functions τ, we get the system of equations

$$\frac{d\tau}{dt} = \frac{F^2(t) - (\mathbf{B}X, \dot{F})}{(F(\mathbf{B}X + \Phi) + R(\tau, \mathbf{B}X + \Phi), F(t))}, \quad (2.5.12)$$

$$\frac{dX}{dt} = \frac{d\mathbf{B}^*}{dt}\mathbf{B}X + \mathbf{B}^*\bigg(\frac{F^2(t) - (\mathbf{B}X, \dot{F})}{(F(\mathbf{B}X + \Phi) + R(\tau, \mathbf{B}X + \Phi), F(t))}$$

$$\times (F(\mathbf{B}X + \Phi) + R(\tau, \mathbf{B}X + \Phi)) - F(t)\bigg). \quad (2.5.13)$$

Denote by $f(t, X)$ the function obtained from the right-hand side of equation (2.5.12) for $R \equiv 0$, and by $\bar{G}(t, X)$ the vector function obtained from the right-hand side of system (2.5.13) for $R \equiv 0$. Then system (2.5.12) — (2.5.13) can be rewritten as

$$\frac{d\tau}{dt} = f(t, X) + r_1(\tau, t, X), \qquad (2.5.14)$$

$$\frac{dX}{dt} = \bar{G}(t, X) + R_2(\tau, t, X), \qquad (2.5.15)$$

where

$$r_1 = \frac{F^2(t) - (\mathbf{B}X, \dot{F})}{(F(\mathbf{B}X + \Phi) + R(\tau, \mathbf{B}X + \Phi), F(t))} - \frac{F^2(t) - (\mathbf{B}X, \dot{F})}{(F(\mathbf{B}X + \Phi), F(t))},$$

$$R_2 = \mathbf{B}^*\left(\frac{F^2(t) - (\mathbf{B}X, \dot{F})}{(F(\mathbf{B}X + \Phi) + R(\tau, \mathbf{B}X + \Phi), F(t))}(F(\mathbf{B}X + \Phi)\right. \qquad (2.5.16)$$

$$\left. + R(\tau, \mathbf{B}X + \Phi)) - \frac{F^2(t) - (\mathbf{B}X, \dot{F})}{(F(\mathbf{B}X + \Phi), F(t))}F(\mathbf{B}X + \Phi)\right).$$

Formulas (2.5.16) show that the function r_1 and the vector function R_2 are uniquely defined via R (if the function F and the matrix $\mathbf{B}$ are said to be given once and for all) and that the numbers

$$\delta_4 = \sup_{\substack{-\infty < t < +\infty, \\ \tau \geq 0, \|X\| < \delta}} |r_1(\tau, t, X)|; \qquad \delta_5 = \sup_{\substack{-\infty < t < +\infty, \\ \tau \geq 0, \|X\| < \delta}} \|R_2(\tau, t, X)\|$$

are as small as desired for a sufficiently small δ_2. Further, the vector function $R(t, Y)$ can be uniquely defined via the function r_1 and the vector function R_2:

$$R(\tau, Y) = -F(Y)$$

$$+ \frac{F(t) + \mathbf{B}(\bar{G}(t, \mathbf{B}^*(Y - \Phi(t))) + R_2(\tau, t, \mathbf{B}^*(Y - \Phi(t)))) + \dot{\mathbf{B}}\mathbf{B}^*(Y - \Phi(t))}{f(t, \mathbf{B}^*(Y - \Phi(t))) + r_1(\tau, t, \mathbf{B}^*(Y - \Phi(t)))}.$$

(2.5.17)

From (2.5.17) it follows that the number δ_2 is arbitrarily small if the numbers δ_4 and δ_5 are sufficiently small. In view of this, the functions r_1 and the vector function R_2 in system (2.5.14) — (2.5.15) can be regarded as initial perturbations.

2. Consider the stability of the solution $\Phi(t)$ to system (2.5.1) under perturbations $R(t, Y)$ on the right-hand sides of this system.

Definition 24 [Malkin (1966)]. The periodic solution $\Phi(t)$ to system (2.5.1) is called stable under persistent perturbations if for each $\epsilon > 0$ it is possible to select two numbers $\epsilon_1 > 0$ and $\epsilon_2 > 0$ such that any integral curve $Y = Y(t, Y_0, t_0)$ of system (2.5.3) starting at the point $Y_0 \in S(M, \epsilon_1)$, for $t = t_0$ continues to stay for all $t \geq t_0$ in the domain $S(M, \epsilon)$ under any choice of the vector function $R(t, Y)$ possessing the above-mentioned properties and satisfying the inequality $\|R(t, Y)\| < \epsilon_2$ for any $t \geq 0, Y \in S(M, \delta)$.

Theorem 30 *If the periodic solution (2.5.2) to system (2.5.1) is a periodic self-oscillation, then it is stable under persistent perturbations.*

Proof: In section 1.5 it is shown that if a periodic solution to system (2.5.1) is a periodic self-oscillation, then there are two functions V and W which possess the following properties: the function $V(Y)$ given for $Y \in S(M,\delta)$ is real, continuously differentiable, and satisfies the inequality $V(Y) > 0$ for $Y \in S(M,\delta)\setminus M$ and the identity $V(Y) \equiv 0, Y \in M$; the function W given in $S(M,\delta)$ is real, continuous, and satisfies the inequality $W < 0$ for $Y \in S(M,\delta)\setminus M$ and the identity $W \equiv 0, Y \in M$. In this case, the functions V and W are related to one another by

$$\sum_{j=1}^{n+1} \frac{\partial V}{\partial y_j} f_j(Y) = W(Y). \tag{2.5.18}$$

Choose a number $\delta > \epsilon > 0$ and find a number

$$\lambda = \inf_{\rho(Y,M)=\epsilon} V(Y).$$

Because of the continuity of the function V, the number λ is positive for any fixed ϵ. For the number λ we find a number $\epsilon_1 < \epsilon$ such that $\epsilon_1 > 0$ and $V(Y) < \lambda$ for $\rho(Y, M) < \epsilon_1$. Furthermore, let

$$\mu = \sup_{\epsilon_1 \le \rho(Y,M) \le \epsilon} W(Y).$$

Because of the continuity of the function W, the number $\mu < 0$. Next, choose the number $\epsilon_2 > 0$ so small that the inequality $\mu + \epsilon_2 \nu < 0$ holds, here

$$\nu = \sup_{\epsilon_1 \le \rho(Y,M) \le \epsilon} \sqrt{\sum_{j=1}^{n+1} \left(\frac{\partial V}{\partial y_j}\right)^2}.$$

We now show that all the integral curves of system (2.5.3) starting in the domain $S(M, \epsilon_1)$ for $t = t_0 \ge 0$, stay in the domain $S(M, \epsilon)$ for all $t \ge t_0$ under any choice of the function $R(t, Y)$ which is given for $t \ge 0, Y \in S(M,\delta)$, is real, continuous, and satisfies the inequality $\|R\| \le \epsilon_2$ all over the domain of definition. Moreover, it suffices to assume that $\|R\| \le \epsilon_2$ for all $\epsilon_1 \le \rho(Y, M) \le \epsilon, t \ge 0$. Compute the total derivative of the function V along the integral curve of system (2.5.3). Then from (2.5.18) we get

$$\frac{dV}{dt} = W + \sum_{j=1}^{n+1} \frac{\partial V}{\partial y_j} R_j(t, Y), \tag{2.5.19}$$

where R_j are components of the vector R. The right-hand side of equality (2.5.19) assumes negative values for all $t \ge 0, Y$, satisfying the relationship $\epsilon_1 \le \rho(Y, M) \le \epsilon$, and under such conditions we get

$$\frac{dV}{dt} \le \mu + \nu\epsilon_2 < 0.$$

Suppose the above statement is not true, i.e., there exists the integral curve $Y = Y(t)$ of system (2.5.3) such that $Y(t_0) \in S(M, \epsilon_1)$, and there is a number $\bar{t} > t_0$ satisfying the equality $\rho(Y(\bar{t}), M) = \epsilon$. Further, $\bar{t}$ is said to be the least number of this kind. Denote by $\bar{\bar{t}}$ the time which satisfies the conditions $\bar{\bar{t}} > t_0$, $\rho(Y(\bar{\bar{t}}), M) = \epsilon_1$, such that $\bar{\bar{t}}$ is the greatest number of this kind. Then the function $V(t) = V(Y(t))$ on the interval $(\bar{\bar{t}}, \bar{t})$ is found to have the property $V(\bar{t}) \geq \lambda$, $V(\bar{\bar{t}}) < \lambda$ and $\frac{dV}{dt} < 0$ for $t \in (\bar{\bar{t}}, \bar{t})$. The latter inequality holds because the arc of the integral curve $Y(t)$ joining the points $Y(\bar{t})$ and $Y(\bar{\bar{t}})$ lies entirely in the closed domain $\epsilon_1 \leq \rho(Y, M) \leq \epsilon$. So we have that $V(\bar{t}) > V(\bar{\bar{t}})$ and $\frac{dV}{dt} < 0$ for $t \in (\bar{\bar{t}}, \bar{t})$. Consequently, the above assumption is not true, therefore all the integral curves of system (2.5.3) starting in the domain $S(M, \epsilon_1)$ stay in the domain $S(M, \epsilon)$ for $t \geq t_0$. ∎

Remark. Theorem 30 does not provide means for comprehensive studies on the qualitative construction of the neighborhood of the periodic orbit M as regards the behavior of the integral curves of system (2.5.3), because even though the vector function R is time-independent, all the motions discussed in section 2.1 may be encountered in the domain $S(M, \epsilon)$. Therefore, to obtain some specific information on the behavior of the integral curves of system (2.5.3) in the neighborhood of $S(M, \epsilon)$, it is essential to make further assumptions about the vector function $R(t, Y)$.

3. Expand the function $\bar{G}(t, X)$ as a series in integral non-negative powers of variables $x_1, \ldots, x_n$ and denote by $\mathbf{P}(t)$ the matrix of linear term coefficients in $x_1, \ldots, x_n$. In parallel with system (2.5.14) and (2.5.15), we will consider the system of linear equations

$$\dot{X} = \mathbf{P}(t)X. \tag{2.5.20}$$

Characteristic indices of system (2.5.20) will be denoted by $\lambda_1, \ldots, \lambda_n$.

Theorem 31 *Suppose the following conditions are satisfied:*

1) the vector function $R(t, Y)$ is given for $Y \in S(M, \delta)$ and is independent of t, $R(t, Y) = R(Y)$, being real and continuous;

2) the vector function $R_2(t, X)$ and the function $r_1(t, X)$ satisfy the inequalities

$$\|R_2(t, X)\| \leq a(M + L\|X\|), \tag{2.5.21}$$

$$\|R_2(t, \bar{X}) - R_2(t, \bar{\bar{X}})\| \leq aL\|\bar{X} - \bar{\bar{X}}\|, \|\bar{X}\| < \delta, \|\bar{\bar{X}}\| < \delta, \quad t \in (-\infty, +\infty), \tag{2.5.22}$$

$$|r_1(t, X)| \leq a(M + L\|X\|), \tag{2.5.21'}$$

$$|r_1(t, \bar{X}) - r_1(t, \bar{\bar{X}})| \leq aL\|\bar{X} - \bar{\bar{X}}\|, \|\bar{X}\| < \delta, \|\bar{\bar{X}}\| < \delta, t \in (-\infty, +\infty), \tag{2.5.22'}$$

where $a \geq 0$ is the parameter;

3) characteristic indices of system (2.5.20) have nonzero real parts, including k indices with negative real parts $\lambda_1, \ldots, \lambda_k$ and $n - k$ with positive real parts $\lambda_{k+1}, \ldots, \lambda_n$. Then the following statements are true:

1. It is possible to select so small positive numbers $a_0 > 0$ and $\delta_0 > 0$ that for all $a \geq 0, a \leq a_0$ system (2.5.15) has a unique 2π-periodic solution $X = \bar{X}(t)$ located in the domain $\|X\| \leq \delta_0$. In a sufficiently small neighborhood of the 2π-periodic solution $X = \bar{X}(t)$ there is a family of integral curves $X = X(t, c_1, \ldots, c_k)$ of system (2.5.15) which is dependent on k arbitrary constants located in this neighborhood for all $t \geq t_0$ and satisfies the inequality

$$\|\bar{X}(t) - X(t, c_1, \ldots, c_k)\| \leq b\|c\|e^{-\bar{\lambda}(t-t_0)}, \qquad (2.5.23)$$

where $\bar{\lambda}$ and b are positive constants, $\|c\| = \sqrt{\sum_{j=1}^{k} c_j^2}$. Inequality (2.5.23) is true for all $t \geq t_0$ and for all real $c_1, \ldots, c_k$ satisfying the relationship $\|c\| \leq c_0$. Any integral curve of system (2.5.15) starting in the mentioned neighborhood of the solution $X = \bar{X}(t)$ is either a member of the family of integral curves or leaves this neighborhood in a finite time. If $k = n$, then the 2π-periodic solution $X = \bar{X}(t)$ is asymptotically stable; if $k = 0$, then all the integral curves starting in a sufficiently small neighborhood of the 2π-periodic solution $X = \bar{X}(t)$ leave it in a finite time.

2. For any positive $a \leq a_0$, system (2.5.3) has a unique integral manifold located in the domain $S(M, \delta_0)$. This integral manifold can be represented in parametric form

$$Y = \bar{Y}(\theta), \qquad (2.5.24)$$

where $\bar{Y}(\theta)$ is 2π-periodic vector function θ. Any motion

$$Y = \bar{Y}(t, t_0), \qquad (2.5.25)$$

defined by the formula $\bar{Y}(t, t_0) = \bar{Y}(\tau(t, t_0))$, where $\tau(t, t_0)$ is the solution of the equation

$$\dot{\tau} = h(\tau), \qquad (2.5.26)$$

satisfying the condition $\tau = t_0$ for $t = t_0$,

$$h(\tau) = \frac{(F(\mathbf{B}(\tau)\bar{X}(\tau) + \Phi(\tau)) + R(\mathbf{B}(\tau)\bar{X}(\tau) + \Phi(\tau)), F(\tau))}{F^2(\tau) - \left(\mathbf{B}(\tau)\bar{X}(\tau), \frac{dF(\tau)}{d\tau}\right)} \qquad (2.5.27)$$

and is a periodic solution to system (2.5.3) given by the condition $Y = \mathbf{B}(t_0)\bar{X}(t_0) + \Phi(t_0)$ with $t = t_0$.

The integral manifold (2.5.24) has the property that there exists a sufficiently small neighborhood such that for each $t_0 > 0$ it is possible to select the family of integral curves $Y = Y(t, c_1, \ldots, c_k)$ of system (2.5.3) satisfying the inequality

$$\|\bar{Y}(t, t_0) - Y(t, c_1, \ldots, c_k)\| \leq b\|c\|e^{-\bar{\lambda}(t-t_0)} \qquad (2.5.28)$$

for $t \geq t_0$, $\|c\| \leq c_0$. The integral curves of system (2.5.3) starting in the mentioned neighborhood and not entering the family, leave this neighborhood in a finite time.

Proof: Instead of system (2.5.15), consider the system of differential equations of a more general form

$$\dot{X} = \mathbf{P}(t)X + \chi(t, X) + Q(t, X), \qquad (2.5.29)$$

where $\mathbf{P}(t)$ is the matrix with real continuous 2π-periodic elements which is in fact the matrix of system (2.5.20). The vector functions $\chi(t, X), Q(t, X)$ given for $\|X\| < \delta, t \in$

Perturbation Theory of Periodic Orbits

$(-\infty, +\infty)$ are real and continuous, their components being 2π-periodic functions in t. The vector function $\chi(t, X)$ is such that

$$\|\chi(t, X)\| \le a(M + L\|X\|), \quad \|X\| < \delta, \tag{2.5.30}$$

$$\|\chi(t, \bar{X}) - \chi(t, \bar{\bar{X}})\| \le aL\|\bar{X} - \bar{\bar{X}}\|, \quad \|\bar{X}\| < \delta, \quad \|\bar{\bar{X}}\| < \delta. \tag{2.5.31}$$

The vector function $Q(t, X)$ has partial derivatives of the first order with respect to $x_1, \ldots, x_n$ that are uniformly continuous in $t \in (-\infty, +\infty)$ for $\|X\| \le \delta$ and go to zero along with $Q(t, X)$ when $X = 0$.

For the purposes of further discussion we will employ some results obtained in chapter 1, especially those described in sections 1.3 and 1.4. In parallel with system (2.5.29) we consider the system of integral equations

$$X(t) = \int_{-\infty}^{+\infty} \mathbf{G}(t, \tau)(\chi(\tau, X) + Q(\tau, X))d\tau. \tag{2.5.32}$$

In order to solve system (2.5.32), we employ a step-by-step method. Setting $X^{(0)} \equiv 0$, for $p = 0, 1, \ldots$, we get

$$X^{(p+1)}(t) = \int_{-\infty}^{+\infty} \mathbf{G}(t, \tau)(\chi(\tau, X^{(p)}) + Q(\tau, X^{(p)}))d\tau. \tag{2.5.33}$$

Assume that the vector function $X^{(p)}(t)$ satisfies the relationship

$$\|X^{(p)}(t)\| \le h \quad \text{for all} \quad p = 0, 1, \ldots, \quad t \in (-\infty, +\infty). \tag{2.5.34}$$

Then, denoting by $\gamma > 0$ the constant which satisfies the inequality

$$\int_{-\infty}^{+\infty} \|\mathbf{G}(t, \tau)\|d\tau \le \gamma,$$

from (2.5.33) we get

$$\|X^{(p+1)}(t)\| \le \int_{-\infty}^{+\infty} \|\mathbf{G}(t, \tau)\| \left(\|\chi(\tau, X^{(p)})\| + \|Q(\tau, X^{(p)})\| \right) d\tau. \tag{2.5.35}$$

Since the partial derivatives with respect to $x_1, \ldots, x_n$ of the function $Q(t, X)$ vanishes at $X = 0$ and are uniformly continuous in t when $\|X\| \le \delta$, we have

$$\|Q(t, \bar{X}) - Q(t, \bar{\bar{X}})\| \le \lambda(h)\|\bar{X} - \bar{\bar{X}}\| \tag{2.5.36}$$

for $\|\bar{X}\| \le h$, $\|\bar{\bar{X}}\| \le h$, where $\lambda(h)$ is the function strictly decreasing to zero as $h \to +0$ and is given for $h \in [0, \delta]$. In fact, if $q(t, X)$ is one of the components of the vector function $Q(t, X)$, then from the finite increment formula we obtain

$$|q(t, \bar{X}) - q(t, \bar{\bar{X}})| = \left| \sum_{j=1}^{n} \frac{\partial q(t, \bar{X} + \theta(\bar{\bar{X}} - \bar{X}))}{\partial x_j}(\bar{x}_j - \bar{\bar{x}}_j) \right|,$$

where $0 \leq \theta \leq 1$. Since the functions $\frac{\partial q}{\partial x_j}$ are uniformly continuous in $t \in (-\infty, +\infty)$ and vanish at $X = 0$, then there is a function $\lambda_1(h)$ such that $\left|\frac{\partial q(t, \bar{X} + \theta(\bar{\bar{X}} - \bar{X}))}{\partial x_j}\right| \leq \lambda_1(h)$ for all $\|\bar{X}\| \leq h$, $\|\bar{\bar{X}}\| \leq h$, because the latter inequalities imply $\|\bar{X} + \theta(\bar{\bar{X}} - \bar{X})\| \leq h$, $h \in [0, \delta]$. In this case $\lambda_1(h) \to +0$ as $h \to +0$. This implies validity of (2.5.36). Using (2.5.30) and (2.5.36), and (2.5.34), estimate the right-hand side of (2.5.35). We get

$$\|X^{(p+1)}(t)\| \leq \int_{-\infty}^{+\infty} \|\mathbf{G}(t, \tau)\|(aM + aLh + h\lambda(h)) dt$$

$$\leq \gamma(aM + aLh + h\lambda(h)). \tag{2.5.37}$$

Estimate the function $X^{(p+1)} - X^{(p)}$. From (2.5.33) we have

$$\|X^{(p+1)}(t) - X^{(p)}(t)\| \leq \int_{-\infty}^{+\infty} \|\mathbf{G}(t, \tau)\| \left(\left\| \chi(\tau, X^{(p)}) - \chi(\tau, X^{(p-1)}) \right\| + \left\| Q(\tau, X^{(p)}) - Q(\tau, X^{(p-1)}) \right\| \right) d\tau$$

$$\leq \int_{-\infty}^{+\infty} \|\mathbf{G}(t, \tau)\|(aL + \lambda(h)) \left\| X^{(p)} - X^{(p-1)} \right\| d\tau. \tag{2.5.38}$$

Of course, inequality (2.5.38) holds for condition (2.5.34) $\|X^{(p)}\| \leq h, \|X^{(p-1)}\| \leq h, h \leq \delta$. Consider inequality (2.5.38) for $p = 1$:

$$\|X^{(2)}(t) - X^{(1)}(t)\| \leq \int_{-\infty}^{+\infty} \|\mathbf{G}(t, \tau)\|(aL + \lambda(h))\|X^{(1)}\| d\tau \leq \gamma(aL + \lambda(h))h, \tag{2.5.39}$$

where it is assumed that $\|X^{(1)}(t)\| \leq h$. From (2.5.38) and (2.5.39) under condition (2.5.34) we have

$$\|X^{(p+1)} - X^{(p)}\| \leq \gamma^p (aL + \lambda(h))^p h. \tag{2.5.40}$$

We now estimate $X^{(1)}$:

$$\|X^{(1)}\| \leq \gamma aM. \tag{2.5.41}$$

From (2.5.41) it follows that $\|X^{(1)}\| \leq h$ for

$$a \leq \frac{h}{\gamma M} = a_1(h). \tag{2.5.42}$$

Further, the condition $\|X^{(p+1)}\| \leq h$ implies the inequality $\gamma(aM + aLh + h\lambda(h)) \leq h$, whence

$$a \leq h \frac{\frac{1}{\gamma} - \lambda(h)}{M + Lh} = a_2(h). \tag{2.5.43}$$

Set $a_0(h) = \min(a_1(h), a_2(h))$. Choose the number h_0 so small that for $h \leq h_0$ the following relationship holds

$$\gamma(a_0(h)L + \lambda(h)) < 1. \tag{2.5.44}$$

The series
$$\gamma aM + \gamma(aL + \lambda(h))h + \ldots + \gamma^p(aL + \lambda(h))^p h + \ldots \qquad (2.5.45)$$
is majorant for the series
$$\sum_{p=1}^{\infty} \|X^{(p)}(t) - X^{(p-1)}(t)\|. \qquad (2.5.46)$$
When the condition (2.5.44) is satisfied, the series (2.5.45) is convergent; hence, the series (2.5.46) and the series
$$\sum_{k=1}^{\infty} \left(X^{(p)}(t) - X^{(p-1)}(t)\right) \qquad (2.5.47)$$
uniformly converge in $t \in (-\infty, +\infty)$. Denote by $\bar{X}(t)$ the sum of the series (2.5.47). From the estimate obtained above we get $\|\bar{X}(t)\| \leq h$ for $t \in (-\infty, +\infty)$, if $h \leq h_0$, and
$$\|\bar{X}(t)\| \leq \gamma aM + \frac{\gamma h(aL + \lambda(h))}{1 - \gamma(aL + \lambda(h))} \quad \text{for} \quad h \leq h_0, \ a \leq a_0(h).$$

We now show that the vector function $X = \bar{X}(t)$ is a 2π-periodic solution to system (2.5.29). Indeed, if the vector function $X^{(p)}$, $\|X^{(p)}\| \leq h \leq h_0$ is a 2π-periodic function, then the vector function $X^{(p+1)}$ defined by (2.5.33) is also a 2π-periodic function. In fact, differentiating both sides of (2.5.33), we get that $X^{(p+1)}(t)$ satisfies the system of linear differential equations
$$\dot{X} = \mathbf{P}(t)X + H(t), \qquad (2.5.48)$$
where $H(t)$ is a 2π-periodic function
$$H(t) = \chi(t, X^{(p)}(t)) + Q(t, X^{(p)}(t)).$$

In section 1.2 it is shown that such a system has a unique 2π-periodic solution. It remains to demonstrate that it can be written as in (2.5.33). To this end, we make a nonsingular linear transformation of the system (2.5.48) with periodic continuously differentiable coefficients and bring it to the form
$$\dot{X}_1 = \mathbf{A}_1 X_1 + H_1(t),$$
$$\dot{X}_2 = \mathbf{A}_2 X_2 + H_2(t), \qquad (2.5.49)$$
where $\mathbf{A}_1$ and $\mathbf{A}_2$ are the constant element matrices of orders k and $n-k$ having the eigenvalues $\lambda_1, \ldots, \lambda_k$ and $\lambda_{k+1}, \ldots, \lambda_n$, respectively, and $H_1(t)$ and $H_2(t)$ are the 2π-periodic vector functions. From section 1.2 it follows that the 2π-periodic solution of the system (2.5.49) is given by
$$X_1(t) = \int_{-\infty}^{t} e^{\mathbf{A}_1(t-\tau)} H_1(\tau) d\tau, \quad X_2(t) = -\int_{t}^{+\infty} e^{\mathbf{A}_2(t-\tau)} H_2(\tau) d\tau. \qquad (2.5.50)$$

Switching in (2.5.50) to the old desired function and writing the obtained expressions uniformly in terms of the matrix $\mathbf{G}(t,\tau)$, we get what is required. Since the vector function $X^{(1)}(t)$ is 2π-periodic, then each of the vector functions $X^{(p)}$ for $p > 1$ is also 2π-periodic. As a result of uniform convergence of (2.5.47), the vector function $X = \bar{X}(t)$ is a 2π-periodic solution to system (2.5.32). Thus, substituting this solution, we get the identity

$$\bar{X}(t) = \int_{-\infty}^{+\infty} \mathbf{G}(t,\tau)(\chi(\tau,\bar{X}) + Q(\tau,\bar{X}))d\tau. \qquad (2.5.51)$$

Differentiating both sides with respect to t, we have that the vector function $X = \bar{X}(t)$ is also a 2π-periodic solution to system (2.5.29).

We now demonstrate that this solution is a unique 2π-periodic solution to system (2.5.29) satisfying the condition $\|X\| \leq h$. Conversely, let the system (2.5.29) additionally have at least one solution of such type $X = \tilde{X}(t)$ which is different from $\bar{X}(t)$ and lies in the domain $\|\tilde{X}\| \leq h \leq h_0$. Substituting this solution in system (2.5.29), we get the identity

$$\dot{\tilde{X}}(t) = \mathbf{P}(t)\tilde{X} + \chi(t,\tilde{X}) + Q(t,\tilde{X}). \qquad (2.5.52)$$

Multiplying both sides of (2.5.52) by $\mathbf{G}(t,\tau)$ and then integrating with respect to τ from $-\infty$ to t and from t to $+\infty$, we obtain

$$\mathbf{G}(t,t-0)\tilde{X}(t) + \int_{-\infty}^{t} \mathbf{G}(t,\tau)\mathbf{P}(\tau)\tilde{X}(\tau)d\tau$$

$$= \int_{-\infty}^{t} \mathbf{G}(t,\tau)(\mathbf{P}(\tau)\tilde{X}(t) + \chi(\tau,\tilde{X}) + Q(\tau,\tilde{X}))d\tau,$$

$$-\mathbf{G}(t,t+0)\tilde{X}(t) + \int_{t}^{+\infty} \mathbf{G}(t,\tau)\mathbf{P}(\tau)\tilde{X}(\tau)d\tau$$

$$= \int_{t}^{+\infty} \mathbf{G}(t,\tau)(\mathbf{P}(\tau)\tilde{X}(\tau) + \chi(\tau,\tilde{X}) + Q(\tau,\tilde{X}))d\tau.$$

Adding these relationships termwise, we find that $\tilde{X}(t)$ is also a solution to system (2.5.32). From this, and from (2.5.51), we have

$$\|\tilde{X}(t) - \bar{X}(t)\| = \left\|\int_{-\infty}^{+\infty} \mathbf{G}(t,\tau)\left(\chi(\tau,\tilde{X}) - \chi(\tau,\bar{X}) + Q(\tau,\tilde{X}) - Q(\tau,\bar{X})\right)d\tau\right\|,$$

whence

$$\|\tilde{X} - \bar{X}\| \leq \gamma(aL + \lambda(h))c,$$

where $c = \sup_{-\infty < t < +\infty} \|\tilde{X}(t) - \bar{X}(t)\|$. From the latter inequality we also find that $c \leq \gamma(aL + \lambda(h))c$ for $a \leq a_0$, $h \leq h_0$. For $c > 0$ the latter inequality is not possible, since $\gamma(aL + \lambda(h)) < 1$, hence, $c = 0$, and $\tilde{X} \equiv \bar{X}$.

In the system (2.5.29) we now replace the required vector function X by $X = \bar{X}(t) + u$. Then, to define u, we obtain the system of equations

$$\dot{u} = \mathbf{P}(t)u + \tilde{\chi}(t,u) + \tilde{Q}(t,u), \qquad (2.5.53)$$

where

$$\tilde{\chi}(t,u) = \chi(t,\bar{X}+u) - \chi(t,\bar{X}), \quad \tilde{Q}(t,u) = Q(t,\bar{X}+u) - Q(t,\bar{X}). \qquad (2.5.54)$$

Perturbation Theory of Periodic Orbits

From (2.5.54) it follows that

$$\|\tilde{\chi}(t,\bar{u}) - \tilde{\chi}(t,\bar{\bar{u}})\| \le aL\|\bar{u} - \bar{\bar{u}}\|, \quad \tilde{\chi} = \tilde{Q} = 0 \quad \text{with} \quad u = 0. \tag{2.5.55}$$

The inequality (2.5.55) holds for $\|\bar{u}\| < \delta_1$, $\|\bar{\bar{u}}\| \le \delta_1$, where $\delta_1 = \delta - \bar{h}, \bar{h} = \sup_{-\infty < t < +\infty} \|\bar{X}(t)\|$. The function $\tilde{Q}(t,u)$ is continuously differentiable with respect to $u_1, \ldots, u_n$ and its partial derivatives are uniformly continuous in $t \in (-\infty, +\infty)$ for $\|u\| \le \delta_1$, $\|\tilde{Q}(t,\bar{u}) - \tilde{Q}(t,\bar{\bar{u}})\| \le \lambda(h_1)\|\bar{u} - \bar{\bar{u}}\|$ for $\|\bar{u}\| \le h_1 - \bar{h}$, $h_1 \in (\bar{h}, \delta)$. Next, set $h_0 \le \frac{1}{2}\delta$ and $h_1 = 2\bar{h}$, then

$$h_1 \le 2aM\gamma + \frac{2\gamma h(aL + \lambda(h))}{1 - \gamma(aL + \lambda(h))} = k(a),$$

where $k(a) \to +0$ as $a \to +0$. Here h is considered to be a fixed constant $h \le \bar{h}$. This choice of h_1 shows that the function $\tilde{\chi}(t,u) + \tilde{Q}(t,u) = \bar{\chi}(t,u)$ satisfies the inequality

$$\|\bar{\chi}(t,\bar{u}) - \bar{\chi}(t,\bar{\bar{u}})\| \le \alpha\|\bar{\bar{u}} - \bar{u}\|, \tag{2.5.56}$$

where α is a positive parameter, $\alpha = aL + \lambda(k(a))$, $\|\bar{u}\| \le \bar{h}, \|\bar{\bar{u}}\| \le \bar{h}$. Hence, the system (2.5.53) becomes

$$\dot{u} = \mathbf{P}(t)u + \bar{\chi}(t,u). \tag{2.5.57}$$

In section 1.3 it is shown that such a system has a solution set that is dependent on k arbitrary constants $u = u(t, c_1, \ldots, c_k)$ and satisfies the inequality

$$\|u\| \le b\|c\|e^{-\hat{\lambda}(t-t_0)}, \quad t \ge t_0, \quad \|c\| \le c_0, \tag{2.5.58}$$

b and $\hat{\lambda}$ being positive constants, $c = \sqrt{\sum_{j=1}^{k} c_j^2}$. The inequality (2.5.58) holds, if $\alpha \le \alpha_0$, where α_0 is a sufficiently small positive constant. The constants b and $\hat{\lambda}$ are independent of $c_1, \ldots, c_k$, but they are dependent, to a certain extent, on α_0, so if α_0 is sufficiently small, then $\hat{\lambda}$ may be as close to

$$\bar{\lambda} = \inf_{j=1,\ldots,k} (-\text{Re}(\lambda_j))$$

as desired. It remains to show that any integral curve of system (2.5.57) located in a sufficiently small neighborhood of $u = 0$, for all $t \ge t_0$ is necessarily contained in the indicated family of integral curves. That is, any integral curve of system (2.5.57), which is not a member of this family and starts in the above-mentioned sufficiently small neighborhood of the point $u = 0$, leaves this neighborhood in a finite time $\bar{t}$, where $\bar{t} > t_0$. We first show that any integral curve staying in a sufficiently small neighborhood of the point $u = 0$ for all $t \ge t_0$ approaches this point unboundedly and no more slowly than by the exponential law. This property of the integral curve and the theorem of uniqueness imply that the integral curve will belong to the above-constructed family. Assume that for all $t \ge t_0$ in the region $\|u\| \le \gamma_0$ there is the integral curve $u = u(t)$ of system (2.5.57). Then, substituting the function $u(t)$ in system (2.5.57), we obtain the identity

$$\dot{u}(t) = \mathbf{P}(t)u(t) + \bar{\chi}(t, u(t)).$$

Similarly, from this we obtain another identity

$$u(t) = \int_{t_0}^{+\infty} \mathbf{G}(t,\tau)\bar{\chi}(\tau, u(\tau))d\tau + \mathbf{G}(t,t_0)u(t_0). \tag{2.5.59}$$

The matrix $\mathbf{G}(t,\tau)$ satisfies the inequality

$$\|\mathbf{G}(t,\tau)\| \leq \gamma_1 e^{-\lambda|t-\tau|}, \quad \lambda > 0. \tag{2.5.60}$$

By applying the inequality (2.5.60), we obtain

$$\|u(t)\| \leq \int_{t_0}^{+\infty} \alpha\gamma_1 e^{-\lambda|t-\tau|} \|u(\tau)\| d\tau + \gamma_1 e^{-\lambda(t-t_0)} \|u(t_0)\|$$

$$\leq \frac{2\gamma_0\gamma_1\alpha}{\lambda} + \gamma_1\gamma_0 e^{-\lambda(t-t_0)} = \gamma_0 \left(\frac{2\gamma_1\alpha}{\lambda} + \gamma_1 e^{-\lambda(t-t_0)} \right). \tag{2.5.61}$$

It is assumed that α_0 is so small that $2\gamma_1\alpha/\lambda < \frac{1}{2}\beta, \alpha \leq \alpha_0$, where $\beta < 1$. Selecting now from the condition $\gamma_1 e^{-\lambda T} \leq \frac{1}{2}\beta$ the number T, from (2.5.61) we have that $\|u(t)\| \leq \gamma_0\beta$ for $t \geq T + t_0$. Replacing in the identity (2.5.59) t_0 by $t_0 + T$, from (2.5.61) we obtain analogously the inequality

$$\|u(t)\| \leq \gamma_0\beta^2 \quad \text{with} \quad t \geq 2T + t_0,$$

whence

$$\|u(t)\| \leq \gamma_0 e^{-\mu(t-t_0)}, \quad t \geq t_0, \quad \mu > 0. \tag{2.5.62}$$

So the vector function $u(t)$ decreases unboundedly to zero as $t \to +\infty$ and satisfies inequality (2.5.62). The vector function $u(t)$ satisfies the system of integral equations

$$u(t) = \int_{t_0}^{+\infty} \mathbf{G}(t,\tau)\bar{\chi}(\tau, u(\tau))d\tau + \mathbf{G}(t,t_0)c_1,$$

where $c_1 = u(t_0)$. Hence, because of the available uniqueness (see section 1.3), $u(t)$ is coincident with one of the integral curves entering the above family $u = u(t, c_1, \ldots, c_k)$.

We now return to system (2.5.15) which is a special case of system (2.5.29), therefore all the results obtained for system (2.5.29) are applied to the solution of system (2.5.15). Thus, it is possible to select two numbers a_0 and δ_0 such that the system (2.5.15) for any $a \leq a_0, a \geq 0$ has a 2π-periodic solution $X = \bar{X}(t)$ located in the region $\|X\| \leq \delta_0$, and $\|\bar{X}(t)\| \leq \delta(a)$, where $\delta(a)$ is a function, $\delta(a) \to +0$ as $a \to +0$. Moreover, in a sufficiently small neighborhood of the 2π-periodic solution there is the family of integral curves $X = X(t, c_1, \ldots, c_k) = \bar{X}(t) + u(t, c_1, \ldots, c_k)$ given for all $t \geq t_0$ and satisfying the inequality

$$\|X(t, c_1, \ldots, c_k) - \bar{X}(t)\| \leq b\|c\|e^{-\bar{\lambda}(t-t_0)}$$

for any $t \geq t_0$, $\|c\| \leq c_0$.

In this case, every integral curve of system (2.5.15) that is not a member of this family and starts in a sufficiently small neighborhood of the integral curve $X = \bar{X}(t)$ leaves it in a finite time. If $k = n$, then the solution $X = \bar{X}(t)$ is asymptotically Lyapunov stable; and if $k = 0$, then all the integral curves, except for $X = \bar{X}(t)$, starting in its arbitrarily small neighborhood, leave this neighborhood in a finite time $\bar{t} > t_0$.

We shall now study the properties of system (2.5.3). Construct a solution to system (2.5.14) — (2.5.15)

$$\tau = \tau(t, X^{(0)}, t_0),$$

$$X = X(t, X^{(0)}, t_0),$$

such that $\tau = \tau_0 = t_0$, $X = X^{(0)}$ for $t = t_0$. The vector function

$$Y(\tau, Y^{(0)}, \tau_0) = \Phi(t) + \mathbf{B}(t)X(t, \bar{X}^{(0)}, t_0), \tag{2.5.63}$$

where

$$Y^{(0)} = \Phi(t_0) + \mathbf{B}(t_0)X^{(0)} \tag{2.5.64}$$

is a solution of system (2.5.3) when t is replaced by τ, that is to say, it satisfies the system

$$\frac{dY}{d\tau} = F(Y) + R(Y). \tag{2.5.65}$$

In this case, τ is the above-constructed function of t and, therefore, satisfies the differential equation

$$\frac{d\tau}{dt} = f(t, X(t, X^{(0)}, t_0)) + r_1(t, X(t, X^{(0)}, t_0)) \tag{2.5.66}$$

and the initial condition $\tau = \tau_0 = t_0$ for $t = t_0$. As a solution to system (2.5.15) we take its 2π-periodic solution $X = \bar{X}(t)$ and introduce the following notation

$$\bar{Y}(\tau) = \Phi(\tau) + \mathbf{B}(\tau)\bar{X}(\tau). \tag{2.5.67}$$

The independent variable t may be considered as a function of the parameter τ, then it will satisfy the equation

$$\frac{dt}{d\tau} = \frac{1}{f(t, \bar{X}(t)) + r_1(t, \bar{X}(t))}. \tag{2.5.68}$$

In the system (2.5.65) and (2.5.68) we replace t by τ and obtain

$$Y(t, Y^{(0)}, t_0) = \bar{Y}(\tau), \tag{2.5.69}$$

where τ is a solution to the equation

$$\dot{\tau} = \frac{1}{f(\tau, \bar{X}(\tau)) + r_1(\tau, \bar{X}(\tau))} = \frac{((F(\bar{Y}(\tau)) + R(\bar{Y}(\tau))), F(\tau))}{F^2(\tau) - ((\bar{Y}(\tau) - \Phi(\tau)), \dot{F}(\tau))} \quad (2.5.70)$$

with initial condition $\tau = \tau_0 = t_0$ for $t = t_0$. So the system (2.5.3) has an integral manifold which can be represented as

$$Y = \bar{Y}(\tau) = \Phi(\tau) + \mathbf{B}(\tau)\bar{X}(\tau). \quad (2.5.71)$$

The integral manifold (2.5.71) located in the domain $S(M, \delta)$ exists for any $a \leq a_0$ and is unique in its kind. Also, $\bar{Y}(\tau)$ is a 2π-periodic vector function. If any solution to system (2.5.3) is written as

$$Y = Y(t, Y^{(0)}, t_0) = \Phi(\tau) + \mathbf{B}(\tau)(\bar{X}(\tau) + U(\tau, U^{(0)}, \tau_0))$$

$$= \bar{Y}(\tau) + \mathbf{B}(\tau)U(\tau, U^{(0)}, \tau_0), \quad (2.5.72)$$

then investigation of properties of the function U shows that there exists a neighborhood of the integral manifold $Y = \bar{Y}(\tau)$ such that it is possible to indicate the family of integral curves $Y = Y(t, c_1, \ldots, c_k)$ of system (2.5.3) which is dependent on k arbitrary constants and satisfies the inequality

$$\|Y(t, c_1, \ldots, c_k) - \bar{Y}(\tau)\| \leq b_1 \|c\| e^{-\hat{\lambda}(t-t_0)} \quad (2.5.73)$$

for all $t \geq t_0, \|c\| \leq c_0, a \leq a_0$. This family lies entirely in the above-mentioned neighborhood, and the other integral curves starting in this neighborhood leave it in a finite time. In fact, the variable τ is a function of the independent variable t which is determined by the condition $\tau = \tau_0 = t_0$ for $t = t_0$ and satisfies the equation

$$\frac{d\tau}{dt} = \frac{((F(\bar{Y}(\tau) + \mathbf{B}(\tau)U(\tau, U^{(0)}, \tau_0)) + R(\bar{Y}(\tau) + \mathbf{B}(\tau)U)), F(\tau))}{F^2(\tau) - ((\bar{Y}(\tau) - \Phi(\tau) + \mathbf{B}(\tau)U), \dot{F}(\tau))}. \quad (2.5.74)$$

If the vector function $U(\tau, U^{(0)}, \tau_0)$ is the integral curve which is a member of the family $U = U(\tau, c_1, \ldots, c_k)$, then with a sufficiently small c_0 the τ is defined for $t \geq t_0$ as a strictly increasing function of t. If a sufficiently small neighborhood is involved, where the solution U stays, then the τ also tends strictly monotonously to $+\infty$ as $t \to +\infty$. Therefore, all the integral curves not entering into the family $u(\tau, c_1, \ldots, c_k)$ leave a sufficiently small neighborhood of the point $u = 0$, and the corresponding integral curves of system (2.5.3) abandon the neighborhood of the integral manifold $Y = \bar{Y}(\tau)$. If $k = n$, then the integral manifold is asymptotically stable. It may be assumed that we have thereby completed the proof of Theorem 31, since the periodicity of any solution to system (2.5.3) located on the manifold $Y = \bar{Y}(\tau)$ will be proved below. ∎

We shall now consider the case where the periodic solution (2.5.2) of system (2.5.1) is the periodic self-oscillation of this system and the characteristic indices $\lambda_1, \ldots, \lambda_n$ of system (2.5.20) may involve the variables with zero real parts.

Perturbation Theory of Periodic Orbits

Theorem 32 *If 1) the system (2.5.1) has a periodic self-oscillation (2.5.2), 2) the vector function $R(t, Y) = R(Y)$ is explicitly independent of t and satisfies the Lipschitz condition in the domain $S(M, \delta)$ and the inequality $\|R(Y)\| \leq am_1$, where $a \geq 0$ is the parameter, then there exists a positive number $a_0 > 0$ such that for any $a \leq a_0$, $a \geq 0$ the system (2.5.3) has at least one periodic solution with period $T(a)$. This periodic solution lies in the domain $S(M, \delta(a))$, where $\delta(a)$ is a continuous positive function given for $a > 0$ and is such that $\delta(a) \to +0$ as $a \to +0$.*

Proof: Since the solution (2.5.2) to system (2.5.1) is a periodic self-oscillation, the zero solution to the system obtained from (2.5.15) for $R_2 \equiv 0$, is asymptotically Lyapunov stable (see section 2.2). Then there exist two functions $V(t, X)$ and $W(t, X)$ (see section 1.5). The function V is positive-definite, continuously differentiable with respect to its arguments, and admits the infinitesimal higher limit, while the function W is negative-definite and is related to the function V by

$$W = \frac{\partial V}{\partial t} + \sum_{s=1}^{n} \frac{\partial V}{\partial x_s} g_s(t, X), \quad (2.5.75)$$

where $g_s(t, X)$ are components of the vector $\bar{G}(t, X)$ (see system (2.5.15)). From condition 2) of Theorem 32 it follows that the vector function $R_2(t, X)$ in system (2.5.15) satisfies the inequality

$$\|R_2(t, X)\| \leq am_1 \quad \text{for all} \quad t \in (-\infty, +\infty), \quad \|X\| \leq \delta. \quad (2.5.76)$$

We say that the functions V and W are given for any $t \in (-\infty, +\infty)$, $\|X\| \leq \bar{\delta} \leq \delta$. The number $\bar{\delta}$ is thought to be chosen so that the following relations hold

$$V(t, X) \geq V_1(x), \quad W(t, X) \leq -V_2(X) \quad \text{for all} \quad \|X\| \leq \bar{\delta}, \ t \in (-\infty, +\infty),$$

where $V_i(X)$ are continuous functions that are given for $\|X\| \leq \bar{\delta}$ and take positive values at $X \neq 0$, and $V_i(0) = 0, i = 1, 2$. Set $\lambda = \inf\limits_{\substack{t\in(-\infty,+\infty),\\ \|X\|=\bar{\delta}}} V(t, X)$ and select $\delta_0 < \bar{\delta}$ so that for $\|X\| \leq \delta_0, t \in (-\infty, +\infty)$ there is $V(t, X) < \lambda$. Denote by $-\beta(\epsilon)$ the magnitude $\sup\limits_{\substack{t\in(-\infty,+\infty),\\ \|X\|\in[\epsilon,\bar{\delta}]}} W(t, X)$, where $0 < \epsilon < \bar{\delta}$. Choose the number a_0 so that

$$a_0 \tilde{m} - \beta(\delta_0) < 0, \quad (2.5.77)$$

where $\tilde{m} = m_1 \bar{m}$,

$$\bar{m} = \sup_{\substack{t\in(-\infty,+\infty),\\ \|X\|\leq\bar{\delta}}} \sqrt{\sum_{s=1}^{n}\left(\frac{\partial V}{\partial x_s}\right)^2}. \quad (2.5.78)$$

In the equality (2.5.78) $\bar{m} < +\infty$, it is always possible to choose V to be the function with uniformly constrained partial derivatives [Malkin (1954)].

Differentiate now the function V in terms of system (2.5.15):

$$\dot{V} = \frac{\partial V}{\partial t} + \sum_{s=1}^{n} \frac{\partial V}{\partial x_s}(g_s + r_s^{(2)}) = W + \sum_{s=1}^{n} \frac{\partial V}{\partial x_s} r_s^{(2)}, \quad (2.5.79)$$

here $r_s^{(2)}$ are components of the vector R_2. From (2.5.78), (2.5.79) we have

$$\dot{V} \leq W + \|R_2\| \sqrt{\sum_{s=1}^{n}\left(\frac{\partial V}{\partial x_s}\right)^2} \leq W + a\tilde{m}.$$

Finally we get
$$\dot{V} = W + a\tilde{m}. \qquad (2.5.80)$$

Show that all the integral curves of system (2.5.15) stay in the region $\|X\| \leq \bar{\delta}$ for all $t \geq t_0$ if $\|X_0\| \leq \delta_0$. This folows from the fact that for $\|X_0\| \in [\delta_0, \bar{\delta}], t \in (-\infty, +\infty)$ by (2.5.77) and (2.5.80) we get $\dot{V} < \beta(\delta_0) + a_0\tilde{m} < 0$ for all $0 \leq a \leq a_0$. Also, show that there exists a number $\bar{T}$ such that all the integral curves of system (2.5.15) starting at $t = t_0$ in the region $\|X\| \leq \delta_0$ stay there for $t \geq \bar{T} + t_0$. Take $\epsilon > 0, \epsilon < \delta_0$ so close to δ_0 as to have $-\beta(\epsilon) + a_0\tilde{m} < 0$. Suppose this statement is not true. Then there are two series of points $X_k^{(0)}$ and numbers T_k such that $T_k \to +\infty$ as $k \to \infty$ and $\|X_k^{(0)}\| \leq \delta_0, k = 1, 2, \ldots$. In addition, there exists an integral curve $X(t, X_k^{(0)}, t_0)$ such that
$$\|X(t_0 + T_k, X_k^{(0)}, t_0)\| > \delta_0, \quad k = 1, 2, \ldots.$$

It is assumed that a_0' is selected such that $-\beta(\bar{\delta}_0) + a_0'\tilde{m} < 0$, where $\bar{\delta}_0$ is selected for ϵ just as δ_0 has been selected for $\bar{\delta}$. All the integral curves starting at $\|X_0\| \leq \bar{\delta}_0$ stay in the region $\|X\| \leq \epsilon$ for all $t \geq t_0$, and hence in the region $\|X\| \leq \delta_0$. Next, all of the integral curves starting in the region $\|X\| \in [\bar{\delta}_0, \delta_0]$ stay in the region $\|X\| \leq \bar{\delta}$, therefore the relationship $\dot{V} = -\beta(\delta_0) + a_0'\tilde{m} < 0$ holds along the arcs of such integral curves as long as the arcs of these integral curves stay in the region $\|X\| \in [\bar{\delta}_0, \bar{\delta}]$. Then all these integral curves enter into the region $\|X\| \leq \bar{\delta}_0$ in a finite time $\bar{T}$; otherwise the function V would have to assume negative values, because on the integral curves, as long as they stay in the region $\|X\| \in [\bar{\delta}_0, \delta_0]$, there will be $V \leq V_0 - \alpha(t - t_0)$, where $V_0 = V(t_0), \alpha = \beta(\bar{\delta}_0) - a_0'\tilde{m} > 0$. Hence for all $t \geq \bar{T} + t_0$ the integral curve $X = X(t, X^{(0)}, t_0)$ stays in the region $\|X\| \leq \epsilon < \delta_0$, whatever $X^{(0)}, \|X^{(0)}\| \leq \delta_0$. Let N be a natural number such that $2\pi N > T$. Consider the mapping of the closed sphere $\|X^{(0)}\| \leq \delta_0$ onto itself: $X = X(t_0 + 2\pi N, X^{(0)}, t_0)$. This mapping satisfies all the conditions of the Brauer theorem [Kantorovič and Akilov (1977)]. Hence in the region $\|X\| < \delta_0$ there is a point $\bar{X}^{(0)}$ such that

$$\bar{X}^{(0)} = X(t_0 + 2\pi N, \bar{X}^{(0)}, t_0). \qquad (2.5.81)$$

The right-hand side of (2.5.15) is $2\pi N$-periodic, and hence the solution to this system $X = \bar{X}(t) = X(t, \bar{X}^{(0)}, t_0)$ is $2\pi N$-periodic. Note that the obtained solution exists for all $a \leq a_0$ and is located in the region $\|X\| \leq \delta_0 = \delta(a_0)$. Evidently, for any $a \leq a_0$ the system (2.5.15) has the periodic solution $X = X(t)$ located in the region $\|X\| \leq \delta(a)$, where $\delta(a)$ is the continuous function assuming positive values for $a \neq 0, \delta(0) = 0$.

It remains to show that system (2.5.3) also has a periodic solution of the period $T(a)$ contained in the domain $S(M, \delta(a))$ for $0 \leq a \leq a_0$. Consider the vector function

$$\bar{Y}(\vartheta) = \Phi(\vartheta) + \mathbf{B}(\vartheta)\bar{X}(\vartheta). \qquad (2.5.82)$$

Perturbation Theory of Periodic Orbits

The proof of Theorem 31 implies that $Y = \bar{Y}(\vartheta)$ is the integral manifold of system (2.5.3). Now show that it is a periodic orbit of this system with period $T(a)$. Find a solution to system (2.5.14) — (2.5.15) with initial data $\tau = \tau_0 = t_0, X = X^{(0)} = \bar{X}^{(0)} = \bar{X}(t_0)$ when $t = t_0$. Then the vector function

$$Y = Y(\tau, \bar{Y}_0, \tau_0) = Y(t(\tau), t_0, \bar{X}^{(0)}) \tag{2.5.83}$$

is a solution to system (2.5.3) if it is viewed as a function of the parameter τ. In fact,

$$\frac{dY}{d\tau} = \frac{d\bar{Y}}{dt}\frac{dt}{d\tau} = \left(\frac{d\Phi}{dt} + \frac{d(\mathbf{B}\bar{X})}{dt}\right)\frac{dt}{d\tau} = F(Y) + R(Y),$$

where t is the function of τ satisfying the initial condition $t = t_0$ for $\tau = \tau_0$ and equation (2.5.68). If $t = t(\tau)$ is a solution to this equation, then the function $t(\tau) + 2\pi N$ is also a solution to equation (2.5.68). The function $t = t(\tau)$ monotonously increases for sufficiently small a's. Hence there is a number $T(a)$ such that $t(\tau_0 + T(a)) = 2\pi N + t_0$. Then the function $t(\tau + T(a))$ is also the solution of equation (2.5.68) which coincides with the solution $t(\tau) + 2\pi N$ by uniqueness, since for $\tau = \tau_0$ we have $t(\tau_0 + T(a)) = t(\tau_0) + 2\pi N$. It follows that $t = \tau \frac{2\pi N}{T(a)} + p(\tau)$, where $p(\tau)$ is a $T(a)$-periodic function in τ. Eliminating t from (2.5.83), we get

$$Y = Y(\tau, \bar{Y}_0, \tau_0) = \Phi\left(\tau\frac{2\pi N}{T(a)} + p(\tau)\right)$$

$$+ \mathbf{B}\left(\tau\frac{2\pi N}{T(a)} + p(\tau)\right)\bar{X}\left(\tau\frac{2\pi N}{T(a)} + p(\tau)\right). \tag{2.5.84}$$

From the form of the right-hand side of (2.5.84) it follows that $Y(\tau, \bar{Y}_0, \tau_0)$ is a periodic solution to the system $\frac{dy}{d\tau} = F(y) + R(y)$ with period $T(a)$. ∎

It should be noted that after establishing the existence of a periodic orbit in system (2.5.3), it is essential to study the behavior of integral curves of this system in a sufficiently small neighborhood of this periodic orbit and, in particular, to tackle the problem of its uniqueness.

Chapter 3

Natural and Forced Oscillations in Systems with Many Degrees of Freedom

3.1 Basic Properties of Periodic Dynamical Systems

1. A broad class of physical systems can be described by the system of differential equations

$$\frac{dX}{dt} = F(t, X), \qquad (3.1.1)$$

so that the development of oscillations in these physical systems is determined by the properties of solutions to system (3.1.1). The right-hand side of this system satisfies the following conditions:
   1) the vector-valued function $F(t, X)$ given for all $t \in (-\infty, +\infty)$, $X \in E_n$, is real and continuous in all its independent variables;
   2) in each finite region $G \subset E_n$, the function $F(t, X)$ satisfies the Lipschitz condition

$$\|F(t, X) - F(t, Y)\| \le L_G \|X - Y\|, \quad X, Y \in G, \quad t \in (-\infty, +\infty),$$

where L_G is a positive constant which is generally dependent on the region G;
   3) the function $F(t, X)$ is 2π-periodic in t for each $X \in E_n$, i.e. the following identity is true

$$F(t + 2\pi, X) \equiv F(t, X).$$

Under these conditions, for every point $X^{(0)} \in E_n$ and every number $t_0 \in (-\infty, +\infty)$ there is a vector function $X = X(t, X^{(0)}, t_0)$ satisfying system (3.1.1) and initial condition $X = X^{(0)}$ at $t = t_0$. This vector-valued function is continuous in all its independent variables. It is assumed to be defined for all $t \in (-\infty, +\infty)$. Introduce new required function $x_{n+1} = t$, a new required vector-valued function $\bar{X} = \{x_1, \ldots, x_n, x_{n+1}\}$, where $\{x_1, \ldots, x_n\} = X$, and a new independent variable

τ by the formula $dt/d\tau = 1$. Then we get the system of differential equations of order $(n+1)$

$$\frac{d\bar{X}}{d\tau} = \bar{F}(\bar{X}), \tag{3.1.2}$$

where $\bar{F} = \{f_1, \ldots, f_n, 1\}, \{f_1, \ldots, f_n\} = F$.

Now, under the above transformation, system (3.1.1) goes over to system (3.1.2). Since x_{n+1} tends to $+\infty$ as $\tau \to +\infty$, the dynamical system introduced has no ω-limit points. This excludes the possibility of investigation of studies on such dynamical systems by the methods described in section 2.1, while reduction of system (3.1.1) to form (3.1.2) creates only an illusion that nonautonomous systems can be investigated by using dynamical theory. This leads to introduction of the concept of a periodic dynamical system. If this concept is not to be introduced formally, we first consider basic properties of the integral curves defined by system (3.1.1).

To this end, we replace in system (3.1.1) the independent variable by $t = t_0 + \tau$ and denote by

$$X = X(\tau, X^{(0)}, t_0) \tag{3.1.3}$$

the solution of the obtained system

$$\frac{dX}{d\tau} = F(t_0 + \tau, X), \tag{3.1.4}$$

satisfying the initial condition $X = X^{(0)}$ with $\tau = 0$. The vector function (3.1.3) is periodic in t_0 with 2π period for any values of $\tau \in (-\infty, +\infty)$ and under any initial conditions $X^{(0)} \in E_n$. Indeed, by designation, the function $X(\tau, X^{(0)}, t_0 + 2\pi)$ is a solution to the system

$$\frac{dX}{d\tau} = F(\tau + t_0 + 2\pi, X) \tag{3.1.5}$$

and satisfies the initial condition $X = X^{(0)}$ for $\tau = 0$. But, by periodicity, the right-hand side of (3.1.5) coincides with that of (3.1.4), and then, by uniqueness, the following identity holds

$$X(\tau, X^{(0)}, t_0) \equiv X(\tau, X^{(0)}, t_0 + 2\pi).$$

Explicitly, we consider the $(n+1)$-dimensional space to be the product of two spaces: E_n and a real axis space. Draw the integral curve of system (3.1.4) through the point $(X^{(0)}, t_0)$ and mark on it the point corresponding to the time $\tau = \tau_1, X^{(1)} = X(\tau_1, X^{(0)}, t_0), t_1 = t_0 + \tau_1$. Next, draw the integral curve through the point $(X^{(1)}, t_1)$ and mark on it the point corresponding to the time $\tau = \tau_2$,

$$X^{(2)} = X(\tau_2, X^{(1)}, t_1), \quad t_2 = \tau_2 + t_1.$$

Hence

$$X^{(2)} = X(\tau_2, X(\tau_1, X^{(0)}, t_0), t_0 + \tau_1), \quad t_2 = t_0 + \tau_1 + \tau_2.$$

Basic Properties of Periodic Dynamical Systems

The point $(X^{(2)}, t_2)$ coincides with that lying on the integral curve, passing through $(X^{(0)}, t_0)$, and corresponding to the time $\tau = \tau_1 + \tau_2$, hence the following equality holds
$$X(\tau_1 + \tau_2, X^{(0)}, t_0) = X(\tau_2, X(\tau_1, X^{(0)}, t_0), t_0 + \tau_1).$$

Definition 25. A periodic dynamical system D_p is said to be given in E_n if there exists a vector-valued function $X(\tau, X^{(0)}, t_0)$ that is given for all $t_0 \in (-\infty, +\infty), \tau \in (-\infty, +\infty), X^{(0)} \in E_n$ and has the properties:

1) for any $X^{(0)} \in E_n, t_0 \in (-\infty, +\infty)$ the unique curve $X = X(\tau, X^{(0)}, t_0)$ given for all $\tau \in (-\infty, +\infty)$ is defined and is such that $X = X^{(0)}$ for $\tau = 0$,

2) the function $X = X(\tau, X^{(0)}, t_0)$ is continuous in all its independent variables,

3) the following relationship is satisfied identically

$$X(\tau_1 + \tau_2, X^{(0)}, t_0) \equiv X(\tau_2, X(\tau_1, X^{(0)}, t_0), t_0 + \tau_1), \qquad (3.1.6)$$

4) the following identity holds

$$X(\tau, X^{(0)}, t_0) \equiv X(\tau, X^{(0)}, t_0 + 2\hat{\pi}).$$

In order to study the basic properties of the integral curves of system (3.1.1), we need to examine a periodic dynamical system, since it has the same basic properties as the integral curves of system (3.1.4), except that it has the advantage of not being related to any specific form of the right-hand sides of system (3.1.4).

Theorem 33 *Given $X^{(0)} \in E_n, t_0 \in (-\infty, +\infty)$ for any $\epsilon > 0, T > 0$ it is possible to select $\delta = \delta(T, \epsilon) > 0$ such that for $\|X^{(0)} - Y^{(0)}\| + |s_0 - t_0| < \delta$ we get $\|X(\tau, X^{(0)}, t_0) - X(\tau, Y^{(0)}, s_0)\| < \epsilon$ for $|\tau| \leq T$.*

Proof: Suppose the reverse is true. Then there exist sequences $X_k^{(0)} \to X^{(0)}$ and $t_0^{(k)} \to t_0$ as $k \to \infty, \tau_k \in [-T, T]$ and a number $\epsilon_1 > 0$ such that

$$\|X(\tau_k, X^{(0)}, t_0) - X(\tau_k, X_k^{(0)}, t_0^{(k)})\| \geq \epsilon_1. \qquad (3.1.7)$$

From the sequence τ_k we choose a convergent subsequence which, for convenience of notation, will also be denoted by τ_k so that $\tau_k \to \tau$ as $k \to \infty, \tau \in [-T, T]$. Then by property 2) of system D_p we get

$$X(\tau_k, X^{(0)}, t_0) \to X(\tau, X^{(0)}, t_0) \quad \text{as} \quad k \to \infty.$$

From property 2) of D_p it also follows that

$$X(\tau_k, X_k^{(0)}, t_0^{(k)}) \to X(\tau, X^{(0)}, t_0) \quad \text{as} \quad k \to \infty,$$

but this contradicts the inequality (3.1.7). Hence, the assumption is not true and Theorem 33 is valid. ∎

In what follows we will use the following designations and terms. The vector-valued function $X(\tau, X^{(0)}, t_0)$ with fixed $X^{(0)}$ and t_0 will be called the motion. The point of this motion with fixed τ will be referred to as the representative point.

The collection of representative points for all $\tau \in (-\infty, +\infty)$ of some particular motion will be identified as the trajectory of this motion and will be designated as $X((-\infty, +\infty), X^{(0)}, t_0)$. The set of all representative points for $\tau \geq 0$ and $\tau \leq 0$ will be called the positive (or respectively, negative) semitrajectory of motion and will be denoted by $X([0, +\infty), X^{(0)}, t_0)$, or $X((-\infty, 0], X^{(0)}, t_0)$, respectively.

Definition 26. The set $A \subset E_n$ is called an invariant set of D_p, if for every $X^{(0)} \in A$ there is $t_0 \in (-\infty, +\infty)$ such that the trajectory $X((-\infty, +\infty), X^{(0)}, t_0)$ of the motion $X(\tau, X^{(0)}, t_0)$ is contained in A.

This definition suggests that the set $A \subset E_n$ is invariant if and only if it represents a set-theoretical sum of trajectories D_p. A simple example of the invariant set is provided by the rest point of D_p. The point $X^{(0)}$ is called the rest point of D_p if there is t_0 such that

$$X(\tau, X^{(0)}, t_0) \equiv X^{(0)} \quad \text{for all} \quad \tau \in (-\infty, +\infty).$$

From this it follows that for any t_0 we get

$$X(\tau, X^{(0)}, t_0) \equiv X^{(0)} \quad \text{for all} \quad \tau \in (-\infty, +\infty).$$

Consequently, $X^{(0)} = X((-\infty, +\infty), X^{(0)}, t_0)$.

Another simple example of the invariant set is provided by the trajectory of a periodic motion. The motion $X(\tau, X^{(0)}, t_0)$ is called periodic if there is $T > 0$ such that

$$X(\tau + T, X^{(0)}, t_0) \equiv X(\tau, X^{(0)}, t_0) \quad \text{for all} \quad \tau \in (-\infty, +\infty).$$

Note that periodic motions of two different types are to be distinguished in D_p. One type is available when the number T is commensurable with 2π, and the other when T is incommensurable with 2π. In order for the first type periodic motion to exist in D_p, it is necessary and sufficient that there exist a point $X^{(0)} \in E_n$ and a number $t_0 \in (-\infty, +\infty)$ such that the following equation holds for some natural N

$$X(2\pi N, X^{(0)}, t_0) = X^{(0)}. \tag{3.1.8}$$

In fact, if such a solution exists, then its period T is expressed as $T = 2\pi N/M$, where M and N are some natural numbers. Then this periodic motion also has a period MT, hence

$$X(MT, X^{(0)}, t_0) = X(2\pi N, X^{(0)}, t_0) = X^{(0)}.$$

If there exist such $X^{(0)}$, t_0 for which the relation (3.1.8) holds, then from properties 3) and 4) of the system D_p we obtain

$$X(\tau + 2\pi N, X^{(0)}, t_0) = X(\tau, X(2\pi N, X^{(0)}, t_0), t_0 + 2\pi N)$$

$$= X(\tau, X^{(0)}, t_0 + 2\pi N) = X(\tau, X^{(0)}, t_0) \quad \text{for all} \quad \tau \in (-\infty, +\infty).$$

It follows that if D_p has no periodic motion of the first type, then the system D_p has no self-intersecting motions with the time length of loops commensurable with 2π.

Basic Properties of Periodic Dynamical Systems

In order for the system D_p to have the second type periodic motion $X(\tau, X^{(0)}, t_0)$, it is necessary and sufficient that there be a point $X^{(0)} \in E_n$ and numbers $t_0 \in (-\infty, +\infty), T > 0$ (T being incommensurable with 2π) such that

$$X(T, X^{(0)}, t_0) = X^{(0)} \quad \text{and} \quad X(\tau, X^{(0)}, t_0) \equiv X(\tau, X^{(0)}, t_0 + T)$$

for all $\tau \in (-\infty, +\infty)$.

Let A be an invariant set of D_p. Then for any $t_0 \in (-\infty, +\infty)$ there is a point $X^{(0)} \in A$ such that the motion $X(\tau, X^{(0)}, t_0)$ for all $\tau \in (-\infty, +\infty)$ is contained in A. Denote by A_{t_0} the set of all points $X^{(0)}$ possessing the above property. Then we have that $A = \bigcup_{t_0 \in (-\infty, +\infty)} A_{t_0}$. The set A_{t_0} will be called the section of the set A corresponding to the value of the parameter t_0. If the set of representative points $X(\tau, X^{(0)}, t_0)$ for all $X^{(0)} \in B \subset E_n$ is designated as $X(\tau, B, t_0)$, then the invariant set A can be represented via some section of A_{t_0} as

$$A = \bigcup_{\tau \in (-\infty, +\infty)} X(\tau, A_{t_0}, t_0).$$

It follows that any two sections of the invariant set A are homeomorphic with respect to each other, i.e. it is possible to establish a one-to-one and bicontinuous correspondence between them.

Theorem 34 *If the set $A \subset E_n$ is an invariant set of D_p, then its closure $\bar{A}$ is also an invariant set of D_p.*

Proof: Let $X^{(0)} \in \bar{A}$. Show that there is t_0 such that

$$X(\tau, X^{(0)}, t_0) \in \bar{A} \quad \text{for all} \quad \tau \in (-\infty, +\infty).$$

Two cases are possible here: either $X^{(0)} \in A$ and then such a t_0 exists because of the invariance of A, or $X^{(0)} \in \bar{A}$ and then there is a sequence

$$X_k^{(0)} \in A, \quad X_k^{(0)} \to X^{(0)} \quad \text{as} \quad k \to \infty.$$

Consider the latter. By Definition 26, for each $X_k^{(0)}$ there is $t_0^{(k)}$ such that $X((-\infty, +\infty), X_k^{(0)}, t_0^{(k)}) \subset A$. Set $t_0^{(k)} = 2\pi m_k + s_0^{(k)}$, where m_k is an integer such that $s_0^{(k)} \in [0, 2\pi]$. Choose from the sequence $s_0^{(k)}$ the sequence converging to $t_0 \in [0, 2\pi]$ and also denoted by $s_0^{(k)}$.

Show that the motion $X(\tau, X^{(0)}, t_0)$ has the property that

$$X((-\infty, +\infty), X^{(0)}, t_0) \subset \bar{A}.$$

By property 4) of the system D_p, we get

$$X(\tau, X_k^{(0)}, t_0^{(k)}) = X(\tau, X_k^{(0)}, s_0^{(k)}).$$

It follows from the continuity of the vector-valued function $X(\tau, X^{(0)}, t_0)$ that

$$X(\tau, X_k^{(0)}, s_0^{(k)}) \to X(\tau, X^{(0)}, t_0) \quad \text{as} \quad k \to \infty.$$

Hence, for any fixed τ we have that $X(\tau, X^{(0)}, t_0) \in \bar{A}$ and $X(\tau, X^{(0)}, t_0) \bar{\in} \bar{A}$ for all $\tau \in (-\infty, +\infty)$, since for $\tau = 0$ we have $X^{(0)} \bar{\in} A$. ∎

Remark. Incidentally, the proof of Theorem 34 has established the fact that if A is an invariant set of D_p, then the set of its boundary points $\bar{A} \setminus A$ is also an invariant set of D_p.

We discuss in some detail the properties of the rest points of D_p. The set of all rest points of D_p is invariant and closed. In fact, each rest point is an invariant set. Hence, an arbitrary combination of such points, specifically the set of all rest points, is also an invariant set. We only need to show that the collection of all rest points is a closed invariant set. Let $X_k^{(0)} \to X^{(0)}$ as $k \to \infty$ and $X_k^{(0)}, k = 1, 2, \ldots$, are the rest points of D_p. Then there exists such a $t_0 \in (-\infty, +\infty)$ that we have

$$X(\tau, X_k^{(0)}, t_0) \equiv X_k^{(0)} \quad \text{for all} \quad \tau \in (-\infty, +\infty).$$

The fact that
$$X(\tau, X_k^{(0)}, t_0) \to X(\tau, X^{(0)}, t_0) \quad \text{as} \quad k \to \infty$$

implies that the equality $X(\tau, X^{(0)}, t_0) = X^{(0)}$ holds for all $\tau \in (-\infty, +\infty)$, therefore $X^{(0)}$ is the rest point.

As with the dynamical system, the following statements hold for the system D_p. The motion $X(\tau, X^{(0)}, t_0)$, as distinct from the rest point, does not enter into some particular rest point of D_p in a finite time. If some trajectory of D_p lies in any sufficiently small neighborhood of the point $X^{(0)}$, then $X^{(0)}$ is the rest point. Show the validity of this statement. In fact, if $X^{(0)}$ is not the rest point, then taking an arbitrary t_0 we can construct the motion $X(\tau, X^{(0)}, t_0)$. Then there exist $\bar{\tau}$ such that $X^{(1)} = X(\bar{\tau}, X^{(0)}, t_0) \neq X^{(0)}$. Set $\|X^{(1)} - X^{(0)}\| = a$. According to Theorem 33, for the numbers $a/4$ and $\bar{\tau}$ it is possible to select $\delta = \delta(\bar{\tau}, \frac{a}{4})$ such that for

$$\|X^{(0)} - Y^{(0)}\| < \delta \quad \text{there is} \quad \|X(\tau, X^{(0)}, t_0) - X(\tau, Y^{(0)}, t_0)\| < a/4$$

$$\text{for all} \quad \tau \in [-|\bar{\tau}|, |\bar{\tau}|].$$

This inference contradicts the condition, for in any sufficiently small neighborhood of the point $X^{(0)}$ there is a point $Y^{(0)}$ such that for all $\tau \in (-\infty, +\infty)$ the following inequality is satisfied

$$\|X(\tau, Y^{(0)}, t_0) - X^{(0)}\| < a/4.$$

Thus, the above statement is valid.

Definition 27. The point $\bar{X}^{(0)} \in E_n$ is called an ω-limit (α-limit) point of the motion $X = X(\tau, X^{(0)}, t_0)$ if there is a sequence

$$t_m \to +\infty \quad (t_m \to -\infty) \quad \text{as} \quad m \to \infty$$

such that
$$X^{(m)} = X(t_m, X^{(0)}, t_0) \to \bar{X}^{(0)} \quad \text{as} \quad m \to \infty.$$

Theorem 35 *The set of all $\omega(\alpha)$-limit points of some particular motion $X = X(\tau, X^{(0)}, t_0)$ is invariant and closed.*

Basic Properties of Periodic Dynamical Systems

Proof: Let Ω be a nonempty set of all ω-limit points of the motion $X(\tau, X^{(0)}, t_0)$. Show that Ω is invariant. Let $\bar{X}^{(0)} \in \Omega$, then there is a sequence $t_m \to +\infty$ such that $X^{(m)} = X(t_m, X^{(0)}, t_0) \to \bar{X}^{(0)}$ as $m \to \infty$. Set $t_m = 2\pi k_m + t_0^{(m)}$, where an integer k_m is chosen such that $t_0^{(m)} \in [0, 2\pi]$. From the sequence $t_0^{(m)}$ we choose a convergent subsequence and also denote it by $t_0^{(m)}$. Let $t_0^{(m)} \to \bar{t}_0$ as $m \to \infty$. Show that $X((-\infty, +\infty), \bar{X}^{(0)}, t_0 + \bar{t}_0) \in \Omega$. Indeed, by properties 3) and 4) of the system D_p, for any $\tau \in (-\infty, +\infty)$ we get

$$X(\tau + t_m, X^{(0)}, t_0) = X(\tau, X^{(m)}, t_0 + t_m) = X(\tau, X^{(m)}, t_0 + t_0^{(m)}).$$

By the continuity of the motion,

$$X(\tau, X^{(m)}, t_0 + t_0^{(m)}) \to X(\tau, \bar{X}^{(0)}, t_0 + \bar{t}_{(0)}) \quad \text{as} \quad m \to \infty.$$

Thus, the point $X(\tau, \bar{X}^{(0)}, t_0 + \bar{t}_0)$ is ω-limit for the motion $X(\tau, X^{(0)}, t_0)$, therefore Ω is invariant.

Show that Ω is closed. Let $\bar{X}_m^{(0)} \in \Omega$ and $\bar{X}_m^{(0)} \to \bar{X}^0$ as $m \to \infty$. Then there are sequences $t_k^{(m)} \to +\infty$ such that $X(t_k^{(m)}, X^{(0)}, t_0) \to \bar{X}_m^{(0)}$ as $k \to \infty$. Take the sequence of positive numbers $\epsilon_1, \epsilon_2, \ldots, \epsilon_l \to +0$ as $l \to +\infty$. For a number ϵ_l, it is possible to choose N_l such that for all $m > N_l$ we get $\|\bar{X}^{(0)} - \bar{X}_m^{(0)}\| < \epsilon_l/2$. Fix one of such m's, then we can choose for it $k(m, l)$ such that for all $k > k(m, l)$ we get $\|\bar{X}_m^{(0)} - X(t_k^{(m)}, X^{(0)}, t_0)\| < \epsilon_l/2$. For some particular $k > k(m, l)$ and for some $m > N_l$ we set $t_l = t_k^{(m)}$, then $\|\bar{X}^{(0)} - X(t_l, X^{(0)}, t_0)\| < \epsilon_l$. The sequence t_l can be selected so that $t_l \to +\infty$ as $l \to \infty$. Since $X(t_l, X^{(0)}, t_0) \to \bar{X}^{(0)}$ as $l \to \infty$, then $\bar{X}^{(0)} \in \Omega$ and Ω is closed. For the set A — the set of α-limit points of the motion $X(\tau, X^{(0)}, t_0)$ — the proof is analogous. ∎

Corollary. Any motion $X = X(\tau, X^{(0)}, t_0)$ remaining for all $\tau \geq 0$ ($\tau \leq 0$) in the bounded region G, has the $\omega(\alpha)$-limit set located entirely in G.

2. We now discuss the behavior of motions in the neighborhood of D_p invariant sets. Put $A_{t_0+\tau} = X(\tau, A_{t_0}, t_0)$.

Definition 28. The bounded invariant set A of the system D_p is called stable if for every $\epsilon > 0$ it is possible to select $\delta > 0$ such that for any t_0 the motion $X(\tau, X^{(0)}, t_0)$ set up in the δ-neighborhood of A_{t_0} remains in the ϵ-neighborhood of the set $A_{t_0+\tau}$ for all $\tau \geq 0$. In other words, if $\rho(X, B)$ stands for the distance from the point X to the set B, then the invariant set A of the system D_p is stable if for any $\epsilon > 0$ there is $\delta > 0$ such that for each t_0 with $\rho(X^{(0)}, A_{t_0}) < \delta$ there is

$$\rho(X(\tau, X^{(0)}, t_0), A_{t_0+\tau}) = \rho(X(\tau, X^{(0)}, t_0), X(\tau, A_{t_0}, t_0)) < \epsilon \quad \text{for all} \quad \tau \geq 0.$$

In addition, if $\rho(X(\tau, X^{(0)}, t_0), A_{t_0+\tau}) \to 0$ as $\tau \to +\infty$ then the invariant set A of the system D_p is called asymptotically stable.

Theorem 36 *In order for the bounded invariant set A of the system D_p to be stable, it is necessary and sufficient that for every $\epsilon > 0$ there be $\delta(\epsilon) > 0$, where*

$$\delta(\epsilon) = \inf_{\substack{\tau \in (-\infty, 0), \\ t_0 \in (-\infty, +\infty), \\ \rho(X^{(0)}, A_{t_0}) = \epsilon}} \rho(X(\tau, X^{(0)}, t_0), A_{t_0+\tau}).$$

Proof: *Necessity.* Assume that A is stable, then for every $\epsilon > 0$, by Definition 28, there is the corresponding $\delta > 0$. Therefore, for any motion $X(\tau, X^{(0)}, t_0)$ such that $\rho(X^{(0)}, A_{t_0}) = \epsilon$ there is

$$\rho(X(\tau, X^{(0)}, t_0), A_{t_0+\tau}) \geq \delta \quad \text{for all} \quad \tau \leq 0.$$

Hence, the condition of Theorem 36 is satisfied.

Sufficiency. We show that all the motions set up in the $\delta(\epsilon)$-neighborhood of any of the sets A_{t_0}, remain in the ϵ-neighborhood of $A_{t_0+\tau}$ for all $\tau \geq 0$, where $\delta(\epsilon)$ is the quantity determined by ϵ under the conditions of the theorem. In fact, otherwise there exists a point $(X^{(0)}, t_{01})$ such that

$$\rho(X(t_{01} - t_0, X^{(0)}, t_0), A_{t_{01}}) = \epsilon, \quad t_{01} > t_0.$$

But the motion $X(\tau, X_1^{(0)}, t_{01})$, $X_1^{(0)} = X(t_{01} - t_0, X^{(0)}, t_0)$ at the time $\tau = -(t_{01} - t_0) < 0$ falls within the δ-neighborhood of the set $A_{t_{01}+\tau}$, which is impossible. ∎

To be noted is that the stability of the bounded invariant set A of the system D_p is determined in such a manner that only the distance from the representative point to the section of this set is used. Therefore, without loss of generality in the solution of the stability problem, the invariant set A can be regarded as closed.

Theorem 37 *In order for the closed, bounded invariant set A of the system D_p to be asymptotically stable, it is necessary and sufficient that the following conditions be satisfied:*

1) there is no motion $X(\tau, X^{(0)}, t_0)$ exterior to A such that its α-limit points enter into A,

2) there exists a sufficiently small neighborhood of some section of the set A such that the motions $X(\tau, X^{(0)}, t_0)$ not belonging to A and remaining in an arbitrarily small neighborhood of the set $A_{t_0+\tau}$ for all $\tau \in (-\infty, +\infty)$, do not pass through it.

Proof: *Necessity.* If A is asymptotically stable, then for every $\epsilon > 0$ there is $\delta(\epsilon) > 0$ such that for $\rho(X^{(0)}, A_{t_0}) < \delta(\epsilon)$ we have

$$\rho(X(\tau, X^{(0)}, t_0), A_{t_0+\tau}) < \epsilon \quad \text{for all} \quad \tau \geq 0, \quad t_0 \in (-\infty, +\infty).$$

If the motion $X(\tau, X^{(0)}, t_0)$ has an α-limit point $\bar{X}^{(0)} \in A$, then there is a sequence $\tau_m \to -\infty$ as $m \to \infty$ such that

$$X(\tau_m, X^{(0)}, t_0) \to \bar{X}^{(0)} \quad \text{as} \quad m \to \infty.$$

Then

$$\rho(X(\tau_m, X^{(0)}, t_0), A_{t_0+\tau_m}) \to 0 \quad \text{as} \quad m \to \infty,$$

which is not possible, since with $\rho(X^{(0)}, A_{t_0}) = \epsilon$ for all $\tau \in (-\infty, 0]$ the following inequality holds

$$\rho(X(\tau, X^{(0)}, t_0), A_{t_0+\tau}) \geq \delta(\epsilon).$$

Hence, condition 1) is satisfied.

Basic Properties of Periodic Dynamical Systems

Now assume that condition 2) is violated, then for any $\epsilon > 0$ it is possible to choose a motion $X(\tau, X^{(0)}, t_0)$ such that

$$X(\tau, X^{(0)}, t_0) \bar{\in} A \quad \text{and} \quad \rho(X(\tau, X^{(0)}, t_0), A_{t_0+\tau}) < \epsilon \quad \text{for all} \quad \tau \in (-\infty, +\infty).$$

By the boundedness of $\tau \leq 0$ (the set A is taken to be bounded and closed) this motion has α-limit points which, by condition 1), cannot lie in A. Let $\bar{X}^{(0)}$ be an α-limit point of the motion $X(\tau, X^{(0)}, t_0)$, then there is $\bar{t}_0$ such that the motion $X(\tau, \bar{X}^{(0)}, \bar{t}_0)$ is entirely composed of α-limit points of the motion $X(\tau, X^{(0)}, t_0)$. Now choose $\epsilon_1 < \epsilon$ and find δ for it by the definition of stability. Then, from the condition $\rho(X^{(0)}, A_{t_0}) < \delta$ it follows that

$$\rho(X(\tau, X^{(0)}, t_0), A_{t_0+\tau}) < \epsilon_1 < \epsilon \quad \text{for all} \quad \tau \geq 0.$$

The motion $X(\tau, \bar{X}^{(0)}, \bar{t}_0)$ is such that

$$\rho(X(\tau, \bar{X}^{(0)}, \bar{t}_0), A_{\bar{t}_0+\tau}) \to 0 \quad \text{as} \quad \tau \to +\infty.$$

Therefore, there is $\tau_1 > 0$ such that $\rho(X(\tau_1, \bar{X}^{(0)}, \bar{t}_0), A_{\bar{t}_0+\tau_1}) < \delta$. Then for all $\tau \geq 0$ we get

$$\rho(X(\tau, X_1^{(0)}, t_{01}), A_{t_{01}+\tau}) < \epsilon_1,$$

where

$$X_1^{(0)} = X(\tau_1, \bar{X}^{(0)}, \bar{t}_0), \quad t_{01} = \bar{t}_0 + \tau_1.$$

Consequently, all the motions $X(\tau, X^{(0)}, t_0)$ possessing the property $\|\bar{X}^{(0)} - X^{(0)}\| + |\bar{t}_0 - t_0| < \rho_1$ for all sufficiently small ρ_1 are governed by the condition

$$\rho(X(\tau, X^{(0)}, t_0), A_{t_0+\tau}) < \epsilon_1 < \epsilon \quad \text{for all} \quad \tau \geq \tau_1.$$

The point $\bar{X}^{(0)}$ is α-limit. Therefore, there exists a sequence $\tau_m \xrightarrow[m \to \infty]{} -\infty$ such that $X(\tau_m, X^{(0)}, t_0) \xrightarrow[m \to \infty]{} \bar{X}^{(0)}$ and, additionally, $t_0 + \tau_m = 2\pi k_m + t_{0m}$, where the number k_m is selected so that $t_{0m} \in [0, 2\pi]$ and $t_{0m} \to \bar{t}_0$ as $m \to \infty$. Thus, there are sequences $X_m \to \bar{X}^{(0)}$, $t_{0m} \to \bar{t}_0$ as $m \to \infty$, where $X_m = X(\tau_m, X^{(0)}, t_0)$, and by property 4) of D_p we get

$$X(\tau, X_m, t_{0m} + 2\pi k_m) = X(\tau, X_m, t_{0m}) = X(\tau, X_m, t_0 + \tau_m).$$

Hence, by property 3) of D_p we get

$$X(\tau, X_m, t_{0m}) = X(\tau + \tau_m, X^{(0)}, t_0),$$

therefore, for $\tau = -\tau_m$ we have $X(\tau, X_m, t_{0m}) = X^{(0)}$ and, hence,

$$\rho(X(\tau, X_m, t_{0m}), A_{t_{0m}+\tau}) = \rho(X(\tau + \tau_m, X^{(0)}, t_0), A_{t_0+\tau+\tau_m})$$
$$= \rho(X^{(0)}, A_{t_0}) \quad \text{for} \quad \tau = -\tau_m.$$

If it is assumed that $\rho(X^{(0)}, A_{t_0}) = \epsilon$, then this yields a contradiction, because, on the one hand, for sufficiently large m there must be

$$\rho(X(\tau, X_m, t_{0m}), A_{t_{0m}+\tau}) < \epsilon_1 < \epsilon \quad \text{for all} \quad \tau > \tau_1,$$

while, on the other, this inequality fails to hold for arbitrarily large values of τ, if the points (X_m, t_{0m}) are sufficiently close to the point $(\bar{X}^{(0)}, \bar{t}_0)$. This contradiction shows that condition 2) is satisfied.

Sufficiency. Assume that the conditions of Theorem 37 are satisfied and show that A is an asymptotically stable invariant set. First show that the condition of Theorem 36 is satisfied and hence the set A is stable. To this end, we take a sufficiently small $\epsilon > 0$, for which there is no motion $X(\tau, X^{(0)}, t_0)$ not belonging to A such that

$$\rho(X(\tau, X^{(0)}, t_0), A_{t_0+\tau}) < \epsilon \quad \text{for all} \quad \tau \in (-\infty, +\infty),$$

and assume that the condition of Theorem 36 does not hold for this ϵ. Considering that $\rho(X(\tau, X^{(0)}, t_0), A_{t_0+\tau})$ is a 2π-periodic function in t_0, we study only the values of $t_0 \in [0, 2\pi]$. Then there are sequences $t_{0k} \xrightarrow[k\to\infty]{} t_0, X_k \xrightarrow[k\to\infty]{} \bar{X}^{(0)}$ such that $\rho(X_k, A_{t_{0k}}) = \epsilon$ and

$$\delta_k = \inf_{\tau \leq 0} \rho(X(\tau, X_k, t_{0k}), A_{t_{0k}+\tau}) \to 0 \quad \text{as} \quad k \to \infty.$$

Choose the sequence $\epsilon_k \to 0$. If necessary, from t_{0k}, X_k it is possible to select subsequences such that for the motion $X(\tau, X_k, t_{0k})$ there is the value of the parameter $\tau = \tau_k$ satisfying the condition $\rho(X(\tau_k, X_k, t_{0k}), A_{t_0+\tau_k}) < \epsilon_k$. Denote by $\bar{X}_k$ the point corresponding to the instant τ_k, and by τ_{0k} the quantities $t_0 + \tau_k$. The motion $X(\tau, \bar{X}_k, \tau_{0k})$ cannot remain in an ϵ-neighborhood of the set $A_{t_{0k}+\tau}$ for all $\tau \leq 0$, or else there will be the α-limit point of this motion. The set of all such points is invariant, and hence there is the motion $X(\tau, X^{(0)}, t_0)$ satisfying the condition $\rho(X(\tau, X^{(0)}, t_0), A_{t_0+\tau}) \leq \epsilon$, which is not possible by the choice of ϵ. Thus any motion $X(\tau, \bar{X}_k, \tau_{0k})$ leaves the ϵ-neighborhood of $A_{t_0+\tau}$ for some particular $\tau < 0$. Let it be for $\tau = \theta_k$. Then $\theta_k = 2\pi m_k + \bar{\theta}_{0k}$, where the integer m_k is such that $\bar{\theta}_{0k} \in [0, 2\pi]$. Set $\bar{\bar{X}}_k = X(\theta_k, \bar{X}_k, \tau_{0k})$ and choose sequences such that

$$\bar{\bar{X}}_k \to \bar{\bar{X}}(0), \quad \bar{\theta}_{0k} \to \bar{\theta}_0 = \bar{t}_0 \quad \text{as} \quad k \to \infty.$$

We show that the motion $X(\tau, \bar{X}^{(0)}, \bar{t}_0)$ for all $\tau \leq 0$ remains in a closed ϵ-neighborhood of the set $A_{\bar{t}_0+\tau}$, and the motion $X(\tau, \bar{\bar{X}}^{(0)}, \bar{\theta}_0)$ remains in an ϵ-neighborhood of the set $A_{\bar{\theta}_0+\tau}$ for all $\tau \leq 0$.

We focus on the proof of the first statement (the second statement is proved in the same way). First show that $\tau_k \to -\infty$ as $k \to \infty$. If this is not the case, then there must exist a subsequence τ_{n_k} such that $\tau_{n_k} \to \theta_0$ as $k \to \infty$. Denote by $\bar{X}$ the limit point of the sequence $\bar{X}_k$. It may be said that $\bar{X}_k \to \bar{X}$ as $k \to \infty$. Note that $X(-\tau_k, \bar{X}_k, t_{0k} + \tau_k) = X_k$. In any finite interval of τ variation, the motion $X(\tau, \bar{X}_k, t_{0k})$ approaches the motion $X(\tau, \bar{X}^{(0)}, t_0)$ uniformly in τ as $k \to \infty$. This holds for $\tau = \tau_k$, therefore $X(\theta_0, \bar{X}^{(0)}, t_0) = \bar{X} \in A_{t_0+\theta_0}$, which is not possible because

Basic Properties of Periodic Dynamical Systems 141

of the invariance of the set A. So $\tau_k \to -\infty$ as $k \to \infty$. If it is assumed that the motion $X(\tau, \bar{X}^{(0)}, \bar{t}_0)$ leaves the closed ϵ-neighborhood of the set $A_{\bar{t}_0+\tau}$ in a finite time $\tau < 0$, then the same property will be exhibited by all the motions set up in a sufficiently small neighborhood of the point $(\bar{X}_0, \bar{t}_0)$. But this is not so, for the motions $X(\tau, \bar{X}_k, \theta_k)$ arising in an arbitrarily small neighborhood of this point stay long enough in an ϵ-neighborhood of $A_{\theta_{0k}+\tau}$ for $\tau < 0$. Thus, there exists a motion $X(\tau, \bar{X}^{(0)}, \bar{t}_0)$, whose negative semitrajectory remains in a closed ϵ-neighborhood of $A_{\bar{t}_0+\tau}$ for all $\tau \leq 0$. Consequently, there exists an α-limit motion $X(\tau, X^{(0)}, t_0)$ confined to a closed ϵ-neighborhood of the set $A_{t_0+\tau}, \tau \in (-\infty, +\infty)$, which contradicts the choice of ϵ. Thus the condition of Theorem 36 is satisfied, hence the set A is stable.

We shall now show that for the thus chosen ϵ and $\delta(\epsilon)$, any motion $X(\tau, X^{(0)}, t_0)$ has the property

$$\rho(X(\tau, X^{(0)}, t_0), A_{t_0+\tau}) \to 0 \quad \text{as} \quad \tau \to +\infty,$$

only if $\rho(X^{(0)}, A_{t_0}) < \delta(\epsilon)$.

Indeed, if not there is the motion $X(\tau, X^{(0)}, t_0)$ that is set up in the $\delta(\epsilon)$-neighborhood of A_{t_0} and does not tend to $A_{t_0+\tau}$ as $\tau \to +\infty$, hence there exists $\epsilon_1 > 0$ such that

$$\rho(X(\tau, X^{(0)}, t_0), A_{t_0+\tau}) > \epsilon_1 > 0 \quad \text{for all} \quad \tau \geq 0.$$

Therefore, there exists at least one ω-limit trajectory of this motion lying entirely in the ϵ-neighborhood of $A_{t_0+\tau}, \tau \in (-\infty, +\infty)$, which contradicts the choice of ϵ. From this it follows that the set A is asymptotically stable. ∎

Definition 29. *The motion $X(\tau, X^{(0)}, t_0)$ is said to tend to the bounded closed invariant set A of the system D_p as $\tau \to +\infty$ if $\rho(X(\tau, X^{(0)}, t_0), A_{t_0+\tau}) \to 0$ as $\tau \to +\infty$.*

Lemma 9 *If for any sufficiently small $\epsilon > 0$ there are no motions $X(\tau, X^{(0)}, t_0)$ such that*

$$\rho(X(\tau, X^{(0)}, t_0), A_{t_0+\tau}) < \epsilon \quad \text{for} \quad \tau \in (-\infty, +\infty),$$

then, to avoid the occurrence of the motions

$$X(\tau, X^{(0)}, t_0) \to A \quad \text{as} \quad \tau \to -\infty,$$

it is necessary and sufficient to avoid the occurrence of the motions $X(\tau, X^{(0)}, t) \bar{\in} A$ having their α-limiting points in A.

Proof: *Necessity.* Suppose there is no motion $X(\tau, X^{(0)}, t_0) \to A$ as $\tau \to -\infty$. Then we show that there is no motion $X(\tau, X^{(0)}, t_0) \bar{\in} A$ with α-limit points in A. Suppose the reverse is true and take ϵ such that in a closed neighborhood of A_{t_0} there are no points $X^{(0)}$ having the property $\rho(X(\tau, X^{(0)}, t_0), A_{t_0+\tau}) \leq \epsilon$ for $\tau \in (-\infty, +\infty)$. Then, by the reasoning employed in the proof of Theorem 37, we construct two points $(X^{(0)'}, t'_0), (\bar{X}^{(0)}, \bar{t}_0)$ such that the motion $X(\tau, \bar{X}^{(0)}, \bar{t}_0)$ for all $\tau \geq 0$ remains in the closed ϵ-neighborhood of $A_{\bar{t}_0+\tau}$, and the motion $X(\tau, X^{(0)'}, t'_0)$ remains in a closed ϵ-neighborhood of $A_{t'_0+\tau}$ for all $\tau \leq 0$. The latter motion does not tend to A as $\tau \to -\infty$, hence there exists a sufficiently small $\epsilon' > 0$ such that the inequality

$\rho(X(\tau, X^{(0)'}, t'_0), A_{t'_0+\tau}) \geq \epsilon'$ is satisfied on some particular sequence $\tau_k \to -\infty$ as $k \to \infty$. From the sequence $X(\tau_k, X^{(0)'}, t'_0)$ we choose the subsequence converging to $X^{(0)}$. Then $X^{(0)}$ is an α-limit point of the motion $X(\tau, X^{(0)'}, t'_0)$. So there is a motion $X(\tau, X^{(0)}, t_0)$ whose trajectory enters into the set of all α-limit points of the motion $X(\tau, X^{(0)'}, t'_0)$ and, therefore, $X(\tau, X^{(0)}, t_0)$ is located in a closed ϵ-neighborhood of A, which is not possible because of the choice of $\epsilon > 0$. We have thus completed the proof of necessity.

Sufficiency. Here the proof is left to the reader's discretion. ∎

Theorem 38 *In order for the bounded closed invariant set A of the system D_p to be asymptotically stable, it is necessary and sufficient that the following conditions one satisfied:*

1) there are no motions that does not belong to A and tend to A as $\tau \to -\infty$,

2) there exists a sufficiently small neighborhood of A containing no integral motions of the system D_p.

Proof: follows from Theorem 37 and Lemma 9. ∎

3. Now we focus on the basic type motions in the system D_p.

Definition 30. The motion $X(\tau, X^{(0)}, t_0)$ of the system D_p is called Poisson stable in the positive direction, i.e. P_+ stable (respectively, stable in the negative direction, i.e. P_- stable), if there is a sequence $t_m \to +\infty$ (respectively, $t_m \to -\infty$) such that

$$X(t_m, X^{(0)}, t_0) \to X^{(0)} \quad \text{and} \quad t_m \to 0 \pmod{2\pi} \quad \text{as} \quad m \to \infty.$$

The motion $X(\tau, X^{(0)}, t_0)$ is called P stable if it is simultaneously P_+ and P_- stable.

The relation $t_m \to 0 \pmod{2\pi}$ signifies that it is possible choose a sequence of integers k_m such that $t_m - 2\pi k_m \to 0$ as $m \to \infty$.

Theorem 39 *If the motion $X(\tau, X^{(0)}, t_0)$ is $P_+, (P_-, P)$ stable, then any motion $X(\tau, X^{(1)}, t_0 + \alpha)$, where $X^{(1)} = X(\alpha, X^{(0)}, t_0)$, is also $P_+(P_-, P)$ stable.*

Proof: Consider the sequence t_m satisfying Definition 30 and show that $X(t_m, X^{(1)}, t_0 + \alpha) \to X^{(1)}$ as $m \to \infty$. Indeed, by property 3) of the system D_p we have that

$$X(t_m, X^{(1)}, t_0 + \alpha) = X(\alpha + t_m, X^{(0)}, t_0) = X(\alpha, X^{(m)}, t_m + t_0),$$

where $X^{(m)} = X(t_m, X^{(0)}, t_m) \to X^{(0)}, t_m + t_0 \to t_0 \pmod{2\pi}$ as $m \to \infty$. From this it follows that

$$X(\alpha, X^{(m)}, t_m + t_0) \to X(\alpha, X^{(0)}, t_0) = X^{(1)} \quad \text{as} \quad m \to \infty. \blacksquare$$

The definition of Poisson stability for the motion $X(\tau, X^{(0)}, t_0)$ implies that in the case of $P_+(P_-)$ stability the point $X^{(0)}$ is $\omega(\alpha)$-limit for this motion. Should the motion $X(\tau, X^{(0)}, t_0)$ be P stable, $X^{(0)}$ is both α- and ω-limit point for this motion. Since Ω and A are the closed invariant sets, the entire trajectory $X((-\infty, +\infty), X^{(0)}, t_0)$ is composed of the limit points of the motion $X(\tau, X^{(0)}, t_0)$. If the motion $X(\tau, X^{(0)}, t_0)$ is P_+ stable, then

$$X((-\infty, +\infty), X^{(0)}, t_0) \subset X((0, +\infty), X^{(0)}, t_0) \quad \text{and} \quad X((-\infty, +\infty), X^{(0)}, t_0) \subset \Omega.$$

Basic Properties of Periodic Dynamical Systems

Since Ω is closed and invariant, we get $A \subset \Omega$. If A is not empty and the collection of limit points for the points of the trajectory $X((-\infty, +\infty), X^{(0)}, t_0)$ is denoted by B, then we have that $A \subset B, B \subset \Omega$ and, therefore, $A \subset \Omega$. If the motion $X(\tau, X^{(0)}, t_0)$ is P_- stable, then the inverse inclusion $A \supset \Omega$ holds. With P stability, we get $\Omega = A = X((-\infty, +\infty), X^{(0)}, t_0)$.

Remark. If the motion $X(\tau, X^{(0)}, t_0)$ is $P_+(P_-, P)$ stable, then there is such sequence of integers n_k that $X(2\pi n_k, X^{(0)}, t_0) \to X^{(0)}$ as $k \to \infty$.

The P_+ and P_- stable motions may be illustrated by referring to the periodic motion. Indeed, let T be a period of the motion $X(\tau, X^{(0)}, t_0)$, then $X(mT, X^{(0)}, t_0) = X^{(0)}$. If T is commensurable with 2π, then there are sequences of n_k, m_k, such that $Tn_k = 2\pi m_k$. If T is incommensurable with 2π, then there is the sequence of $2\pi m_k$ approximating $n_k T$ as close as desired, so that $X(2\pi m_k, X^{(0)}, t_0) \to X^{(0)}$ as $k \to \infty$.

Consider the collection of all P_+ stable motions. Their trajectories form some invariant set A. Establish the structure of section A_{t_0} for this set. Consider the collection of all points from E_n with rational coordinates and the collection of open spheres with all kinds of rational radii and centers located in every possible point with rational coordinates. Let U be a system of the thus constructed neighborhoods u. Form the sets $u^* = u \setminus \left(u \bigcap_{m=1}^{\infty} X(-2\pi m, u, t_0) \right)$. Put $v^+ = \bigcup_{u^* \in U} u^*$. Then $A_{t_0} = E_n \setminus v^+$. Show that any motion $X(\tau, X^{(0)}, t_0)$, where $X^{(0)} \in A_{t_0}$, is P_+ stable. Indeed, if $X^{(0)} \in A_{t_0}$ and u_m is its neighborhood, then $X^{(0)} \in E_n \setminus v^+$. Consequently, $X^{(0)} \bar{\in} v^+$, hence $X^{(0)} \bar{\in} u^*$ and, in particular, $X^{(0)} \bar{\in} u_m^*$. From $X^{(0)} \bar{\in} u_m^*$ follows the existence of k such that

$$X^{(0)} \in u_m \cap X(2\pi k, u_m, t_0), \quad \text{therefore} \quad X(2\pi k, X^{(0)}, t_0) \in u_m \cap X(2\pi k, X^{(0)}, t_0).$$

This suggests that the motion $X(\tau, X^{(0)}, t_0)$ is P_+ stable. The section B_{t_0} of the trajectory set B is constructed in the same manner for all P_- stable motions. The section C_{t_0} of the trajectory set for all P stable motions is given by the formula $C_{t_0} = A_{t_0} \cap B_{t_0}$. The set of all Poisson stable motions is closely related to the so-called central motions lying in the set $\bar{C}$. Further, the construction will be found for the sets of such motions.

Definition 31. We say that in the system D_p there is recurrence of regions if for any $t_0 \in (-\infty, +\infty)$ and any region G it is possible to choose a sequence $n_k \xrightarrow[k \to \infty]{} +\infty$ such that $G \cap X(2\pi n_k, G, t_0) \neq \Lambda$ (Λ is an empty set).

If, using the parameter $-2\pi n_k$, we transform the above relationship, we get $G \cap X(-2\pi n_k, G, t_0) \neq \Lambda$, i.e. a positive direction recurrence implies a negative direction recurrence.

Definition 32. The motion $X(\tau, X^{(0)}, t_0)$ is called wandering with respect to the system D_p given in E_n if there exist a number $N > 0$ and a neighborhood u of a point $X^{(0)} \in E_n$ such that $u \cap X(2\pi m, u, t_0) = \Lambda$ for all $m \geq N$.

As above, it may be shown that the positive wandering motion is the negative wandering motion. It follows from Definition 32 that any motion $X(\tau, \bar{X}^{(0)}, t_0), \bar{X}^{(0)} \in u$ is also wandering, for otherwise there exists a neighborhood u_1 of the point $\bar{X}^{(0)}, u_1 \subset u$ and a sequence $n_k \xrightarrow[k \to \infty]{} +\infty$ such that $u_1 \cap X(\tau_k, u_1, t_0) \neq \Lambda$, $\tau_k = 2\pi n_k$. Hence, $u \cap X(2\pi n_k, u, t_0) \neq \Lambda$, which is not possible.

Let τ_1 be any real number. If $X(\tau, X^{(0)}, t_0)$ is a wandering motion, then the motion $X(\tau, X_1, t_0 + \tau_1)$ is also wandering, where $X_1 = X(\tau_1, X^{(0)}, t_0)$. Indeed, since the motion $X(\tau, X^{(0)}, t_0)$ is wandering, the relationship $u \cap X(2\pi m, u, X^{(0)}) = \Lambda$, $m \geq N$, holds for some N and some neighborhood u. Take the image of this relationship corresponding to the parameter τ_1 and denote by u_1 an open neighborhood of X_1 defined by $u_1 = X(\tau_1, u, t_0)$. Then we have $u_1 \cap X(2\pi m, u_1, t_0 + \tau_1) = \Lambda$ for all $m \geq N$. Thus, the above statement holds.

Denote by B_1 the set of all points lying on every possible trajectories of the wandering motions. The set B_1 is open and invariant. Let M_1 be the closure of the invariant set $E_n \backslash B_1$. The set M_1 is an invariant set of nonwandering motions. If there exists an open invariant set G contained entirely in M_1, then in the system D_p given in G there is recurrence of regions. Indeed, suppose there is no recurrence of regions. Then there exists $t_0 \in (-\infty, +\infty)$ such that some motion $X(\tau, X^{(0)}, t_0)$, $X^{(0)} \in G$ is wandering, which is not possible.

If the system D_p has the motion restricted for all $\tau \geq 0$, then the set M_1 is not empty. Show this. Let $X(\tau, X^{(0)}, t_0)$ be the motion restricted for all $\tau \geq 0$. Then its ω-limit set Ω is invariant and closed. Take $X_1 \in \Omega$ and t_1 such that $X(\tau, X_1, t_1) \in \Omega$ for all $\tau \geq 0$. By the boundedness and closedness of Ω, there is a subsequence $n_k \xrightarrow[k \to \infty]{} +\infty$ such that $X(2\pi n_k, X_1, t_1) \to X_2 \in \Omega$ as $k \to \infty$. Show that $X(\tau, X_2, t_1)$ is a nonwandering motion. Indeed, otherwise there is a number $N > 0$ and a neighborhood u of the point X_2 such that $u \cap X(2\pi n_k, u, t_1) = \Lambda$ for all $n_k > N$. At the same time, $X_{n_k} \in u$, $X_{n_{k+1}} \in u$ for all sufficiently large n_k, but since $X_{n_{k+1}} = X(2\pi(n_{k+1} - n_k), X_{n_k}, t_1)$, the relation $u \cap X(2\pi n_k, u, t_1) = \Lambda$ is not possible. So, the motion $X(\tau, X_2, t_1)$ is nonwandering, and hence M_1 is nonempty.

If the system D_p is given in the bounded closed set, any wandering motion $X(\tau, X^{(0)}, t_0)$ can be outside $S(\epsilon, M_1)$ (the ϵ-neighborhood of the nonwandering motions set M_1) for $|\tau| \leq T < +\infty$ only. Indeed, suppose the opposite is true. Then we have that there exist such number $\epsilon > 0$, such wandering motion $X(\tau, X^{(0)}, t_0)$ and such sequence $\tau_k \xrightarrow[k \to \infty]{} +\infty$ that for all $k = 1, 2, \ldots$ the relation $X(\tau_k, X^{(0)}, t_0) \bar{\in} S(\epsilon, M_1)$ holds. Then there exists the limit set of the motion $X(\tau, X^{(0)}, t_0)$, at least one point of which lies outside the ϵ-neighborhood of the set M_1. This point is the limit for a sequence of the form $X(2\pi n_k + \bar{\tau}, X^{(0)}, t_0)$ and, as follows from earlier statement, cannot be wandering. Hence, it belongs to the ϵ-neighborhood of M, which yields a contradiction.

Selecting M_2 of all nonwandering motions for the system D_p determined in M_1, and continuing this process, we obtain the series of sets $M_1, M_2, \ldots$. Introduce the notation $\Gamma = \bigcup_{i=1}^{\infty} M_i$ (Γ is a closed invariant set). Each motion contained in Γ is called central.

Theorem 40 *In the set of central motions, the everywhere dense set C is the set of Poisson stable motions.*

Proof: Let the motion $X(\tau, X^{(0)}, t_0)$ be central. Show that for any $\epsilon > 0$ there is a point $\bar{X}^{(0)} \in S(\epsilon, X^{(0)})$ such that the motion $X(\tau, \bar{X}^{(0)}, t_0)$ is P stable. The point $X^{(0)}$ is nonwandering, therefore there is $n_1 > 1$ such that $S_\epsilon \cap X(2\pi n_1, S_\epsilon, t_0) \neq \Lambda$,

Basic Properties of Periodic Dynamical Systems

$S_\epsilon = S(\epsilon, X_0)$. Let $X_1^{(0)} \in S_\epsilon \cap X(2\pi n_1, S_\epsilon, t_0)$ together with its ϵ_1-neighborhood. The point $X_1^{(0)}$ is nonwandering, hence there is $\nu_1 < -n_1 - 1$ such that

$$S_{\epsilon_1} \cap X(2\pi\nu_1, S_{\epsilon_1}, t_0) \neq \Lambda, \quad S_{\epsilon_1} = S(\epsilon_1, X_1^{(0)}).$$

Suppose the point $X_2^{(0)}$ is such that its ϵ_2-neighborhood enters into this set. Proceeding from S_{ϵ_2} and ν_1, we construct the point $X_3^{(0)}$ and its neighborhood S_{ϵ_3}, etc. In this case, it may be said that

$$\epsilon_k \leq \epsilon/2^k, \quad k = 1, 2, \ldots, \quad \text{and} \quad X_k^{(0)} \to \bar{X}^{(0)} \quad \text{as} \quad k \to \infty.$$

We show that the motion $X(\tau, \bar{X}^{(0)}, t_0)$ is P stable. For this purpose, we first establish that it is P_+ stable. Let $\delta > 0$ be given. The number k is taken so large that $\epsilon/2^k < \delta$. By construction, there exists ν_{k+1} such that $X(-2\nu_{k+1}\pi, S(\epsilon_{k+1}, X_k^{(0)}), t_0)$ lies in $S(\epsilon_{k+1}, X_{k+1}^{(0)}) \subset S(\epsilon_k, X_k^{(0)}) \subset S(\bar{X}^{(0)}, \delta)$. ∎

Now we turn to recurrent motions.

Definition 33. The bounded closed nonempty invariant set containing no proper subset with the same properties is called minimal.

Any system D_p having a bounded closed nonempty invariant set necessarily has a minimal set. This suggests that if the system D_p has at least one motion $X(\tau, X^{(0)}, t_0)$ restricted for all $\tau \geq 0$, then it also has at least one minimal set.

Theorem 41 *Any motion of the dynamical system D_p contained in a minimal set is recurrent.*

Proof: Suppose the reverse is true. Then there exists $\epsilon > 0$ such that it is possible to select three sequences: $\tau_j^{(0)}, \tau_j, s_j$; $s_j \xrightarrow[j\to\infty]{} +\infty$, here $X(\tau_j^{(0)}, X^{(0)}, t_0)$ is outside the ϵ-neighborhood of the arc $X(\tau, X^{(0)}, t_0), \tau \in [\tau_j, \tau_j + s_j]$. From these sequences we choose those subsequences which are also denoted by $\tau_j^{(0)}, \tau_j, s_j$ and are such that $\tau_j^{(0)} \to \tau^{(0)}$ (mod 2π), $X(\tau_j^{(0)}, X^{(0)}, t_0) \to \bar{X}^{(0)}$, $X(\tau_j, X^{(0)}, t_0) \to X_1^{(0)}$, $\tau_j \to \tau_{01}$ (mod 2π) as $j \to \infty$. By closure and invariance, the motion $X(\tau, X_1^{(0)}, t_0 + \tau_{01})$ belongs to the minimal set M involved; moreover, $\bar{X}^{(0)} \in M$. Show that the distance of the point $\bar{X}^{(0)}$ to the motion $X(\tau_1, X_1^{(0)}, t_0 + \tau_{01})$ satisfies the relation

$$\rho(\bar{X}^{(0)}, X(\tau, X_1^{(0)}, t_0 + \tau_{01})) \geq \epsilon \quad \text{for all} \quad \tau \in (-\infty, +\infty).$$

Indeed, otherwise there exists a point $\bar{\tau} \in (-\infty, +\infty)$ such that we get $\rho(\bar{X}^{(0)}, X(\bar{\tau}, X_1^{(0)}, t_0 + \tau_{01})) < \epsilon$. Then, from some particular j, the property of the previously chosen sequences is violated. Thus, $\rho(\bar{X}^{(0)}, X((-\infty, +\infty), X_1^{(0)}, t_0 + \tau_{01})) \geq \epsilon$. Consequently, the ω-limit set of the motion $X(\tau, X_1^{(0)}, t_0 + \tau_{01})$ is bounded, closed, nonempty, invariant, and noncoincident with M, since $\bar{X}^{(0)} \bar{\in} \Omega$. Hence it follows that M is not minimal. The obtained contradiction suggests that Theorem 41 is valid. ∎

Theorem 41 may also be formulated as follows: if $X^{(0)}$ is a point of a minimal set, then there exists a motion $X(\tau, X^{(0)}, t_0)$ that is recurrent and contained in this minimal set.

Now we show that the statement, which in a sense is inverse to Theorem 41, is legitimate, viz. if $X(\tau, X^{(0)}, t_0)$ is a recurrent motion, then the closure of its trajectory $\overline{X((-\infty, +\infty), X^{(0)}, t_0)}$, is a minimal set. Denote this set by M. If there exists the minimal set $A \subset M$, then $X^{(0)} \bar{\in} A$ and $\rho(X^{(0)}, A) = \bar{\epsilon} > 0$. Let $X_1 \in A$; then there is a sequence τ_k such that $X(\tau_k, X^{(0)}, t_0) \xrightarrow[k \to \infty]{} X_1$. In this case the point X_1 may be chosen so that $\tau_k = 2\pi n_k$. Set $X_k = X(2\pi n_k, X^{(0)}, t_0)$ and $\epsilon = \frac{1}{2}\bar{\epsilon}$. By the definition of recurrence, L_ϵ is found and $\delta > 0$ is chosen so small that for $\tau \in [-L_\epsilon, +L_\epsilon]$ all the motions arising in the δ-neighborhood of X_1 with the parameter t_0 do not deviate more than $\frac{1}{2}\bar{\epsilon}$ from $X(\tau, X_1, t_0)$. Then these arcs of motion remain in an ϵ-neighborhood of the set A. Consequently, $X^{(0)}$ lies outside the ϵ-neighborhood of the arc $X(\tau, X_k, t_0), \tau \in [-L_\epsilon, +L_\epsilon]$ with any sufficiently large k, which is not possible by the definition of recurrence. So the bounded, closed nonempty invariant set $A \subset M$ is nonexistent. Hence, M is a minimal set. ∎

3.2 Small Parameter Method

1. Consider the system of equations

$$\dot{X} = F(t, X), \qquad (3.2.1)$$

the right-hand sides of which have the following properties:

1) given for $t \in (-\infty, +\infty)$, $X \in E_n$ they are real and continuous in collection of all their variables;

2) they satisfy the Lipschitz condition in any finite region $G \subset E_n$:

$$\|F(t, X) - F(t, Y)\| \leq L_G \|X - Y\|, \quad X \in G, \quad Y \in G, \quad t \in (-\infty, +\infty),$$

where L_G is the constant dependent on the region G;

3) the independent variable t explicitly enters into the right-hand sides of the system (3.2.1) that are 2π-periodic functions in t for any $X \in E_n$:

$$F(t + 2\pi, X) = F(t, X);$$

4) the solution of the system (3.2.1), $X = X(t, X^{(0)}, 0)$ is defined for all $t \geq 0$, $X^{(0)} \in E_n$, where $X = X^{(0)}$ for $t = 0$.

Introduce a one-parameter family of vector-valued functions depending on the integer parameter $N \geq 0$:

$$Y = Y(N, X) = X(2\pi N, X, 0), \quad X \in E_n. \qquad (3.2.2)$$

Each function of this family is given for $X \in E_n$, and is real and continuous. Also, the relation $Y = Y(N, X)$ for any $N \geq 0$ may be regarded as the one-to-one and bicontinuous map of E_n onto itself. As noted in section 3.1, the system (3.2.1) may have two types of periodic solutions. The first type involves the solutions whose period is commensurable with 2π, while the second includes those whose period is

Small Parameter Method

incommensurable with 2π. This section deals with the periodic solutions of system (3.2.1) assigned to the first type.

As is shown in section 3.1, if a periodic solution is to exist in system (3.2.1), it is necessary and sufficient that for some $N > 0$ there be a real solution to the equation

$$Y(N, X) = X. \qquad (3.2.3)$$

Indeed, denote the real solution of equation (3.2.3) by $X^{(0)}$. Then, from (3.2.2) and (3.2.3) we obtain

$$Y(N, X^{(0)}) = X(2\pi N, X^{(0)}, 0) = X^{(0)}.$$

Hence, the solution to system (3.2.1) $X = X(t, X^{(0)}, 0)$ is periodic, $2\pi N$ being one of the periods. If its smallest period is T, then $MT = 2\pi N$. Also, $T = 2\pi N/M$, and, conversely, if $X = X(t, X^{(0)}, 0)$ is a periodic solution to system (3.2.1) with period $T = 2\pi N/M$, then the relation $X(MT, X^{(0)}, 0) = X^{(0)}$ holds. Hence $X^{(0)}$ is a solution to system (3.2.3).

This statement shows that the existence of a periodic solution to system (3.2.1) is equivalent to the existence of real solutions to system (3.2.3) for some $N > 0$, and hence to the existence of fixed points in the mappings (3.2.2) of space E_n onto itself for $N > 0$. As to the existence of the $2\pi \bar{N}/M$ periodic solution of system (3.2.1), this problem is equivalent to the existence of a solution to system (3.2.3) for $N = \bar{N}$, or to the existence of fixed points of the mapping $Y(\bar{N}, X)$ of space E_n onto itself.

Now we consider the autonomous system of differential equations

$$\dot{X} = F(X), \qquad (3.2.4)$$

the right-hand sides of which satisfy conditions 1), 2), 4) given above for system (3.2.1). Set

$$Z(T, X^{(0)}) = X(T, X^{(0)}), \qquad (3.2.5)$$

where $X(T, X^{(0)})$ is a solution to system (3.2.4) which satisfies the condition $X(0, X^{(0)}) = X^{(0)}$, $X^{(0)} \in E_n$ and is a one-parameter one-to-one bicontinuous family of the mappings of E_n onto itself. It is assumed that $T > 0$. Consider the equation

$$Z(T, X) = X. \qquad (3.2.6)$$

System (3.2.4) has a periodic solution if and only if for some particular $T > 0$ the equation (3.2.6) has a real solution. Note that periodic solutions of the autonomous system (3.2.4) should not be divided into types. Since (3.2.6) is the equation for finding the fixed points of the map $Z = Z(T, X)$ of space E_n onto itself, the existence problem for periodic solutions to system (3.2.4) is equivalent to that for the fixed points of such maps. The relation of the presence of periodic solutions in the systems (3.2.1) and (3.2.4) to the existence theory for the fixed points of continuous maps makes it possible to apply general theorems of functional analysis to finding conditions for which these differential equation systems have periodic solutions.

Suppose there exists a closed nonempty set $S \subset E_n$ and a natural number $N > 0$ such that the map $Y(N, X)$ translates S into itself. In this case, the relation

$\|Y(N, X) - Y(N, \bar{X})\| \le \alpha \|X - \bar{X}\|$, where α is constant, $0 < \alpha < 1$, holds for every two points $X \in S, \bar{X} \in S$. Then system (3.2.1) has the periodic solution

$$X = X(t, X^{(0)}, 0), \quad X^{(0)} \in S,$$

where one of the periods is the number $2\pi N$. The point $X^{(0)}$ can be obtained here by the method of successive approximations. Let $X^{(1)} \in S$ be an arbitrary point. Set

$$X^{(k)} = Y(N, X^{(k-1)}), \quad k = 2, 3, \ldots.$$

Then we have that $X^{(k)} \to X^{(0)}$ as $k \to \infty$, and the estimate of the rate of convergence degree is valid here

$$\|X^{(k)} - X^{(0)}\| \le \frac{\alpha^{k-1}}{1-\alpha} \|X^{(1)} - X^{(0)}\|, \quad k = 2, 3, \ldots.$$

If $Y(N, X)$ is replaced in this reasoning by $Z(T, X)$, then the existence condition is obtained for the T-periodic solution of system (3.2.4). The validity of such assertions follows from the principle of contracting maps [Kantorovič and Akilov (1977)]. If there is $N > 0$ such that the map $Y(N, X)$ transforms the closed set $S \subset E_n$ to its bounded part, then for the existence of a periodic solution to system (3.2.1) it is sufficient that for any $X, \bar{X} \in S$ there be the following relation:

$$\|Y(N, X) - Y(N, \bar{X})\| < \|X - \bar{X}\|.$$

The most universal instrument used in finding the existence conditions for periodic solutions of systems (3.2.1) and (3.2.4) is the Schauder principle as presented in [Kantorovič and Akilov (1977)]. If there is a closed convex set $B \subset E_n$ such that for some $N > 0$ the map $Y(N, X)$ transforms it to a bounded set $\bar{B} \subset B$, then system (3.2.1) has a periodic solution with $2\pi N$ as one of its periods. Here, if $Y(N, X)$ is replaced by $Z(T, X)$, then we get the existence condition for a periodic solution to system (3.2.4). Note that the set B is called convex if, along with any two points X and $\bar{X}$, it also contains their connecting straight line segment $Y = X + t(\bar{X} - X), t \in [0, 1]$. The two above-mentioned principles may be applied only if the maps $Y(N, X), Z(T, X)$ themselves or at least their basic properties are known.

At present, to tackle the existence problem of periodic solutions and their approximate representations and to study the behavior of other motions in their neighborhood, the oscillation theory employs the so-called small parameter method. The method is as follows. Suppose the right-hand sides of system (3.2.1) are dependent on parameter μ:

$$\dot{X} = F(t, X, \mu), \tag{3.2.1'}$$

where the right-hand sides are given for all $t \in (-\infty, +\infty), X \in E_n, \mu \in [0, \mu_0]$, and are real and continuous in collection of all their arguments, and satisfy the other conditions imposed on the system (3.2.1) equally for all $\mu \in [0, \mu_0]$. When these conditions are satisfied, the map family (3.2.2) is also dependent on parameter μ:

$$Y = Y(N, X, \mu) = X(2\pi N, X, 0, \mu), \tag{3.2.2'}$$

Small Parameter Method

and again the existence problem of a periodic solution to system (3.2.1') is equivalent to that of a fixed point of the map (3.2.2') for a particular $N > 0$. Assume that the equation

$$Y(N, X, \mu) = X \qquad (3.2.3')$$

with $\mu = 0$ has the solution $X^{(0)}$. The existence and derivation of this solution are usually determined by the procedure of introducing into system (3.2.1) the small parameter μ. The solution $X^{(0)}$ of equation (3.2.3') for $\mu = 0$ corresponds to the $2\pi N$-periodic solution of system (3.2.1') for $\mu = 0$:

$$X = X_0(t, X^{(0)}, 0). \qquad (3.2.7)$$

The problem is to find a periodic solution to system (3.2.1') for $\mu \neq 0$ coming arbitrarily close to the solution (3.2.7) as $\mu \to +0$. In other words, the vector-valued function $X^{(0)}(\mu)$ is required such that $X^{(0)}(\mu) \to X^{(0)}$ as $\mu \to +0$ and the vector function $X = X(t, X^{(0)}(\mu), 0)$ is the $2\pi N$-periodic solution of system (3.2.1') for any sufficiently small $\mu > 0$, that is, we need to find the solution of equation (3.2.3'), $X = X^{(0)}(\mu)$ satisfying the condition $X \to X^{(0)}$ as $\mu \to +0$. The latter condition reduces the existence of a periodic solution of system (3.2.1') for sufficiently small $\mu > 0$ to that of the implicit function $X = X^{(0)}(\mu)$ defined by equation (3.2.3') and initial condition $X^{(0)}(\mu) \to X^{(0)}$ as $\mu \to 0$. This reduction of the existence problem of periodic solutions to that of implicit functions constitutes theoretical content of the small parameter method taken in its general form.

Note that since the solution (3.2.7) is also a $2\pi kN$-periodic solution of system (3.2.1') for $\mu = 0$ ($k > 0$ is integer), the equation $Y(kN, X, \mu) = X$ has the solution $X = X^{(0)}$ for $\mu = 0$. Therefore, it is also possible to talk about the existence of the $2k\pi N$-periodic solutions of system (3.2.1') for sufficiently small $\mu > 0$ that come arbitrarily close to the solution (3.2.7) as $\mu \to +0$ and do not amount to the $2\pi N$-periodic solutions in the general case.

Consider the autonomous case. Suppose the system (3.2.4) depends on parameter μ:

$$\dot{X} = F(X, \mu). \qquad (3.2.4')$$

It is assumed that the right-hand sides of this system given for all $X \in E_n$, $\mu \in [0, \mu_0]$, are real and continuous in collection of all their variables, and satisfy the Lipschitz condition in any finite region $G \subset E_n$ equally for all $\mu \in [0, \mu_0]$. Also, for $\mu = 0$ the system (3.2.4') has a $\bar{T}$-periodic solution

$$X = X_0(t, X^{(0)}). \qquad (3.2.8)$$

Then the equation

$$Z(T, X, \mu) = X \qquad (3.2.6')$$

with $\mu = 0$ has the solution $X = X^{(0)}, T = \bar{T}$. The problem is to find, for all sufficiently small $\mu > 0$, the periodic solution of system (3.2.4') coming arbitrarily close to the solution (3.2.8) as $\mu \to +0$. Clearly, not only the initial data, but also the period of this motion is the function of μ. So we need to find the function $T(\mu)$

and the vector-valued function $X_0(\mu)$ such that $T(\mu) \to \bar{T}, X_0(\mu) \to X^{(0)}$ as $\mu \to +0$ and the vector-valued function $X = X(t, X_0(\mu))$ is a $T(\mu)$-periodic solution of system (3.2.4′). Hence, these functions must satisfy the system (3.2.6′).

The existence problem of periodic solutions for system (3.2.4′) has thus reduced to that of the $(n+1)$ function set satisfying the system (3.2.6′). (3.2.6′) is a system of n equations. We show that the $(n+1)$-function set to be determined from this system can actually be replaced by the collection of n functions sufficient for solution of the problem involved. Indeed, at least one of the vector-valued function components (3.2.8) is not constant. Suppose such is the last component $x_n = x_n(t, X^{(0)})$. Denote by $[a, b]$ the closed interval completely filled with the values of the function x_n for $t \in [0, \bar{T}]$. Then in (a, b) it is possible to choose a number γ, and in $[0, \bar{T}]$ a number t_0 such that $x_n(t_0, X^{(0)}) = \gamma$ and $\dot{x}_n(t_0, X^{(0)}) = \gamma_1 \neq 0$. Assume that there exists the solution $X = X(t, X_0(\mu))$ of system (3.2.4′), which is periodic in t, has the period $T(\mu)$, and is such that $T(\mu) \to \bar{T}, X_0(\mu) \to X^{(0)}$ as $\mu \to +0$. The values of the function $x_n = x_n(t, X_0(\mu))$ then fill the interval $[a_\mu, b_\mu]$ located in an arbitrarily small neighborhood of $[a, b]$ if $\mu > 0$ is sufficiently small. Therefore, for all sufficiently small $\mu > 0$ we get $\gamma \in (a_\mu, b_\mu)$ and there exists a number $t_0(\mu) \in [0, T(\mu)]$ such that $x_n(t_0(\mu), X_0(\mu), \mu) = \gamma$. Replace in system (3.2.4′) the independent variable t by $t = \tau + t_0(\mu)$. With this replacement, the system (3.2.4′) remains unchanged. Its periodic solution (3.2.8) is now determined by the initial condition $X = X(t_0, X^{(0)})$ for $\tau = 0$ when $\mu = 0$, and for $\mu > 0$ by the initial condition $X = X(t_0(\mu), X_0(\mu), \mu)$ when $\tau = 0$. In this case, for any sufficiently small $\mu > 0$ the relation $x_n = \gamma$ holds for $\tau = 0$. In what follows the system (3.2.6′) is assumed to have the solution $X = X^{(0)}, T = \bar{T}$ for $\mu = 0$ and we are to find the solution of this system $X_0(\mu), T(\mu)$ for all sufficiently small $\mu > 0$, having the property $X_0(\mu) \to X^{(0)}, T(\mu) \to \bar{T}$ as $\mu \to +0$. We will seek this solution with the additional condition $x_n^{(0)}(\mu) = \gamma$ for all sufficiently small $\mu > 0$, where $x_n^{(0)}(\mu)$ is the n-th component of $X_0(\mu)$.

2. Using the small parameter method, we focus on the actual solution of the existence problem of periodic solutions for the system (3.2.1′) and (3.2.4′).

If there exists a point $X^{(0)}$ and a number $N > 0$ such that $Y(N, X^{(0)}, \mu) = X^{(0)}$ for $\mu = 0$ and

$$\frac{D(\bar{y}_1, \ldots, \bar{y}_n)}{D(x_1, \ldots, x_n)} \Big|_{\substack{X=X(0), \\ \mu=0}} \neq 0, \tag{3.2.9}$$

where $\bar{y}_j = y_j - x_j$, then the system (3.2.1′) for any sufficiently small $\mu > 0$ has a $2\pi N$-periodic solution $X = X(t, X_0(\mu), 0, \mu)$ such that $X = X_0(\mu)$ when $t = 0, X_0(\mu) \to X^{(0)}$ as $\mu \to +0$.

If there exist numbers $x_1^{(0)}, \ldots, x_{n-1}^{(0)}, \gamma, \bar{T} > 0$ such that

$$Z(\bar{T}, X^{(0)}, \mu) = X^{(0)} = \{x_1^{(0)}, \ldots, x_{n-1}^{(0)}, \gamma\} \quad \text{for} \quad \mu = 0$$

and

$$\frac{D(\bar{z}_1, \ldots, \bar{z}_n)}{D(x_1, \ldots, x_{n-1}, t)} \Big|_{\substack{t=\bar{T}, \\ X=X^{(0)}, \\ \mu=0}} \neq 0, \tag{3.2.10}$$

Small Parameter Method

where $\bar{z}_j = z_j - x_j$, then the system (3.2.4′) has the solution $X = X(t, X_0(\mu), \mu)$, which is periodic in t, and has a period $T(\mu)$ such that $T(\mu) \to \bar{T}, X_0(\mu) \to X^{(0)}$ as $\mu \to +0$, where $x_n^{(0)}(\mu) = \gamma$ for all sufficiently small $\mu > 0$.

Assume that the right-hand sides of system (3.2.1′) and (3.2.4′) are continuously differentiable with respect to variables $x_1, \ldots, x_n$. Then the functional determinants in (3.2.9) and (3.2.10) are determined, since the solutions of these systems are the continuously differentiable functions of initial data (see section 1.1). When the inequalities (3.2.9) and (3.2.10) are satisfied, the conditions of the existence theorem for implicit functions [Fichtengol'c (1969)] are satisfied, therefore the above statements are valid.

We turn to the explicit construction of the above determinants. Construct the functional determinant appearing in (3.2.9). Let $Y = Y(t, X, \mu)$ be a solution to system (3.2.1′) satisfying the initial condition $Y = X$ with $t = 0$. Substitute this solution into system (3.2.1′), then we get the identity

$$\frac{dY}{dt} = F(t, Y, \mu). \qquad (3.2.11)$$

Differentiate both sides of (3.2.11) successively with respect to $x_1, \ldots, x_n$. Then, denoting by $\mathbf{P}(t, X, \mu)$ the functional Jacobian matrix

$$\mathbf{P} = \{p_{jk}\}, \quad p_{jk} = \frac{\partial f_j}{\partial y_k}\bigg|_{Y=Y(t,X,\mu)}, \quad j,k = 1, \ldots, n,$$

we have that the functions $\frac{\partial y_j}{\partial x_k}, j, k = 1, \ldots, n$ constitute a solution to the system of differential equation

$$\frac{dY}{dt} = \mathbf{P}(t, X, \mu)Y. \qquad (3.2.12)$$

Indeed, if the identity (3.2.11) is written in scalar form, then we have the system

$$\frac{dy_j}{dt} = f_j(t, y_1, \ldots, y_n, \mu), \quad i = 1, \ldots, n.$$

Differentiating this identity with respect to x_k, we get

$$\frac{d}{dt}\frac{\partial y_j}{\partial x_k} = \sum_{i=1}^{n} \frac{\partial f_j}{\partial y_i}\frac{\partial y_i}{\partial x_k} = \sum_{i=1}^{n} p_{ji}\frac{\partial y_i}{\partial x_k}.$$

Denote by $\tilde{\mathbf{Y}}(t, X, \mu)$ the matrix of solutions obtained as follows. Set in system (3.2.12) $\mu = 0, X = X^{(0)}$. Then $Y = Y(t, X^{(0)}, 0)$ is the $2\pi N$-periodic solution to system (3.2.1′) considered for $\mu = 0$. Hence, the matrix $\mathbf{P}(t, X^{(0)}, 0)$ is also $2\pi N$-periodic. Since $Y = X$ for $t = 0$, then $\tilde{\mathbf{Y}}(t, X, \mu)$ becomes the unit matrix at $t = 0$. Thus $\tilde{\mathbf{Y}}(t, X, \mu)$ is the fundamental system of solutions for (3.2.12). Therefore, the polynomial $D(\rho) = |\tilde{\mathbf{Y}}(2\pi N, X^{(0)}, 0) - \rho \mathbf{E}|$ is characteristic for the system

$$\frac{dY}{dt} = \mathbf{P}(t, X^{(0)}, 0)Y. \qquad (3.2.13)$$

The form of the functional determinant, as in (3.2.9), implies that it is coincident with the value of the characteristic polynomial for $\rho = 1$. From this it follows that the inequality (3.2.9) holds if and only if the system (3.2.13) has no $2\pi N$-periodic solutions.

Now we focus on the case of the autonomous system of equations (3.2.4'). Its right-hand side is assumed to be continuously differentiable with respect to $x_1, \ldots, x_n$. Substitute the function $Z = Z(t, X, \mu)$ into system (3.2.4') and, differentiating the obtained identity with respect to $x_1, \ldots, x_{n-1}, t$, we have that the functions $\frac{\partial z_j}{\partial x_k} = z_j^k, j = 1, \ldots, n, k = 1, \ldots, n-1$, constitute a solution to the linear system

$$\frac{dY}{dt} = \mathbf{Q}(t, X, \mu)Y \qquad (3.2.14)$$

with the initial condition $z_j^k = \delta_{jk}$ at $t = 0$, where $\mathbf{Q}(t, X, \mu)$ is the functional Jacobian matrix

$$\mathbf{Q} = \{q_{jk}\}, \quad q_{jk} = \frac{\partial f_j}{\partial y_k}\Big|_{Y=Z(t,X,\mu)}, \quad j, k = 1, \ldots, n.$$

The equation (3.2.14) is also satisfied by the functions z_j^n, where $z_j^n = \frac{dz_j}{dt}$. As before, z_j is the j-th component of the vector $Z(t, X, \mu)$. Setting $\mu = 0, X = X_0$, we get the linear system

$$\frac{dY}{dt} = \mathbf{Q}(t, X^{(0)}, 0)Y \qquad (3.2.15)$$

which has $\bar{T}$-periodic coefficients; the functions $z_j^n, j = 1, \ldots, n$, constitute a $\bar{T}$-periodic solution to system (3.2.15). Denote by $\tilde{\mathbf{Z}}$ the matrix the j-th column of which is the solution z_j^k of system (3.2.14) for $\mu = 0, X = X^{(0)}$. Then $\tilde{\mathbf{Z}}$ is the fundamental system of solutions for system (3.2.15), since for $t = 0$ we get $\tilde{\mathbf{Z}} = \begin{pmatrix} \mathbf{E}_{n-1} & 0 \\ 0 & \gamma_1 \end{pmatrix}$, where $\mathbf{E}_{n-1}$ is the unit matrix of order $n-1$. Find the characteristic polynomial $\Delta(\rho)$ of system (3.2.15)

$$\Delta(\rho) = |\tilde{\mathbf{Z}}(\bar{T}) - \rho \mathbf{E}|. \qquad (3.2.16)$$

Since the system (3.2.15) has a periodic solution, $\Delta(1) = 0$. Denote the functional determinant in (3.2.10) by D, and the $n-1$ minor of the determinant $\Delta(1)$ located in the upper left corner, by Δ_{n-1}. Then we get

$$\Delta(1) = -\Delta_{n-1} + D = 0, \qquad (3.2.17)$$

whence

$$D = \Delta_{n-1}. \qquad (3.2.18)$$

The relation (3.2.17) is established by selecting in the lower row of the determinant $\Delta(1)$ a row of the form $(0, 0, \ldots, 0, -1)$ and expanding the determinant $\Delta(1)$ of the elements of this row. The determinant in (3.2.10) is nonzero if and only if $\Delta_{n-1} \neq 0$. This inequality holds when among the roots of the characteristic equation $\Delta(\rho) = 0$ the multiplicity of the root $\rho = 1$ does not exceed unity, or when it is multiple, but in system (3.2.15) there is only one periodic solution obtained above.

Small Parameter Method

3. Consider the cases where the functional determinants in (3.2.9), (3.2.10) become zero.

Example 1. Let the following nonautonomous quasilinear system of equations be given

$$\dot{X} = \mathbf{A}X + \Phi(t) + \mu F(t, X, \mu). \tag{3.2.19}$$

We will suppose that the vector functions Φ and F are continuous functions in their independent variables and are 2π-periodic in t. Assume that $F(t, X, \mu)$ is given for $X \in E_n, t \in (-\infty, +\infty), \mu \in [0, \mu_0]$, is real, as is $\Phi(t)$, and has continuous partial derivatives with respect to variables $x_1, \ldots, x_n$. Let the elements of the matrix $\mathbf{A}$ be real and constant, $\lambda_1, \ldots, \lambda_n$ being its eigenvalues. We will seek a $2\pi N$-periodic solution to system (3.2.19). First consider this system for $\mu = 0$:

$$\dot{X} = \mathbf{A}X + \Phi(t). \tag{3.2.19'}$$

We first note two possible, basically different cases: 1) among the quantities $\lambda_1, \ldots, \lambda_n$ there are no numbers of the form $\pm i\frac{M}{N}$; 2) among the quantities $\lambda_1, \ldots, \lambda_n$ there are numbers of the form $\pm i\frac{M}{N}$, where $M = 0, 1, \ldots$. The former case is identified as non-resonance, and the latter as resonance. In the first case, system (3.2.9') has a unique $2\pi N$-periodic solution

$$X = e^{\mathbf{A}t}C + \int_0^t e^{\mathbf{A}(t-\tau)}\Phi(\tau)d\tau, \tag{3.2.20}$$

where C is determined from the periodicity conditions

$$X(0) = X(2\pi N), \quad C = (\mathbf{E} - e^{2\pi N \mathbf{A}})^{-1}\int_0^{2\pi N} e^{\mathbf{A}(2\pi N - \tau)}\Phi(\tau)d\tau.$$

The solution (3.2.20) is a 2π-periodic solution to system (3.2.19'). In this case the system (3.2.12) becomes

$$\frac{dX}{dt} = \mathbf{A}X \tag{3.2.21}$$

(here it is assumed that in system (3.2.12) $\mu = 0, X = X^{(0)}$). In system (3.2.21) there are no $2\pi N$-periodic solutions, except for $X = 0$. Hence, it follows from the above result that for a sufficiently small $\mu > 0$, the system (3.2.19) has a $2\pi N$-periodic solution with the property that, as $\mu \to +0$, it comes arbitrarily close to the solution (3.2.20) of the linear system (3.2.19'). It should be noted that the obtained solution is actually 2π-periodic.

Now we consider the resonance case. This time the system (3.2.21) has a $2\pi N$-periodic solution. Consequently, the functional determinant in (3.2.9) for system (3.2.19) becomes zero. Solution of the existence problem for corresponding implicit functions therefore calls for additional studies. Assume that for resonance eigenvalues there are prime elementary divisors if they involve multiples. The eigenvalues $\lambda_1, \ldots, \lambda_n$ are supposed to include k zero and $2l$ nonzero resonance numbers of the form $\pm i\frac{M}{N}$, where M is a natural number. Denote these numbers by $\mu_1, \ldots, \mu_l$ and

$-\mu_1, \ldots, -\mu_l$. Proceeding from these assumptions and using nonsingular linear transformations of the desired functions $x_1, \ldots, x_n$ (the coefficients of this transformation being real and constant), the system (3.2.19) can be reduced to the form

$$\frac{du_s}{dt} = \varphi_s(t) + \mu \Phi_s(t, U, X, Y, Z, \mu),$$

$$\frac{dx_p}{dt} = -\mu_p y_p + \xi_p(t) + \mu \Xi_p(t, U, X, Y, Z, \mu),$$

$$\frac{dy_p}{dt} = \mu_p x_p + \eta_p(t) + \mu H_p(t, U, X, Y, Z, \mu), \qquad (3.2.22)$$

$$\frac{dz_r}{dt} = \sum_{i=1}^{m} p_{ri} z_i + \vartheta_r(t) + \mu \Theta_r(t, U, X, Y, Z, \mu),$$

$$s = 1, \ldots, k, \quad p = 1, \ldots, l, \quad r = 1, \ldots, m,$$

where $k + 2l + m = n$. The matrix $\mathbf{P} = \{p_{ri}\}$ has no resonance quantities among its own eigenvalues, i.e. those of the form $\pm i \frac{M}{N}$. The functions appearing in the systems (3.2.22) and (3.2.19) have the same properties. To be noted is that

$$U = (u_1, \ldots, u_k), \quad X = (x_1, \ldots, x_l), \quad Y = (y_1, \ldots, y_l), \quad Z = (z_1, \ldots, z_m).$$

Replace in system (3.2.22) the desired function x_p, y_p by the formulas

$$x_p = \bar{x}_p \cos \mu_p t + \bar{y}_p \sin \mu_p t, \quad y_p = \bar{x}_p \sin \mu_p t - \bar{y}_p \cos \mu_p t, \quad p = 1, \ldots, l. \qquad (3.2.23)$$

Then, to define the new desired functions from the system (3.2.22), we obtain the system of equations

$$\frac{dv_s}{dt} = w_s(t) + \mu W_s(t, V, Z, \mu), \qquad (3.2.24)$$

$$\frac{dz_r}{dt} = \sum_{i=1}^{m} p_{ri} z_i + \vartheta_r(t) + \mu T_r(t, V, Z, \mu), \quad s = 1, \ldots, k + 2l, \quad r = 1, \ldots, m.$$

Here, to save notation, we set

$$v_s = u_s, \quad v_{k+p} = \bar{x}_p, \quad v_{k+l+p} = \bar{y}_p, \quad s = 1, \ldots, k, \quad p = 1, \ldots, l.$$

Consider system (3.2.24) for $\mu = 0$. This system has a family of periodic solutions that is dependent on $k + 2l$ arbitrary constants of period $2\pi N$ if the following relationships hold

$$\int_0^{2\pi N} w_s(\tau) d\tau = 0, \quad s = 1, \ldots, k + 2l. \qquad (3.2.25)$$

We shall assume that the equalities (3.2.25) hold. The above-mentioned family of periodic solutions can be represented as

$$v_s = c_s + \int_0^t w_s(\tau) d\tau, \qquad (3.2.26)$$

Small Parameter Method

$$Z = e^{\mathbf{P}t}C + \int_0^t e^{\mathbf{P}(t-\tau)}T(\tau)d\tau,$$

where $T(t) = (\vartheta_1(t),\ldots,\vartheta_m(t)), C = (\mathbf{E} - e^{2\pi N \mathbf{P}})^{-1}\int_0^{2\pi N} e^{\mathbf{P}(2\pi N-\tau)}T(\tau)d\tau$. We shall seek the general solution of system (3.2.24) under condition (3.2.25) with the initial data $v_s = c_s + b_s, s = 1,\ldots, k+2l, Z = Z_0 + \Gamma$ for $t = 0$, where $Z_0 = C$. If for this solution we set up an equation of the form (3.2.3'), then we get

$$\mu Q_s = \mu \int_0^{2\pi N} W_s(\tau, V, Z, \mu)d\tau = 0, \quad s = 1,\ldots, k+2l, \tag{3.2.27}$$

$$(e^{2\pi N\mathbf{P}} - \mathbf{E})\Gamma + \mu \int_0^{2\pi N} e^{\mathbf{P}(2\pi N - \tau)}\Theta(\tau, V, Z, \mu)d\tau = 0. \tag{3.2.28}$$

These equations can be derived from (3.2.24) as follows. We set up the integral equations satisfied by the solution

$$v_s(t) = c_s + b_s + \int_0^t w_s(\tau)d\tau + \mu \int_0^t W_s(\tau, V, Z, \mu)d\tau,$$

$$Z(t) = e^{\mathbf{P}t}(C + \Gamma) + \int_0^t e^{\mathbf{P}(t-\tau)}T(\tau)d\tau + \mu \int_0^t e^{\mathbf{P}(t-\tau)}\Theta(\tau, V, Z, \mu)d\tau. \tag{3.2.29}$$

Since the functions, as in (3.2.29), are coincident at $t = 0$ and $t = 2\pi N$, we get equations (3.2.27) and (3.2.28). Since the determinant $|e^{\mathbf{P}2\pi N} - \mathbf{E}| \neq 0$ and the vector components $\Theta(t, V, Z, \mu)$ are continuously differentiable with respect to variables $v_1,\ldots,v_{k+2l}, z_1,\ldots,z_m$, then equation (3.2.28) defines Γ to be an implicit function of $\mu, b_1,\ldots, b_{k+2l}$. Note that the partial derivatives of this function with respect to variables $b_1,\ldots, b_{k+2l}$ vanish at $\mu = 0$. Since it is required to find the solutions of system (3.2.27), (3.2.28) with the property $B \to 0$, $\Gamma \to 0$ as $\mu \to +0, B = (b_1,\ldots, b_{k+2l})$, the following condition must be satisfied

$$p_s(c_1,\ldots, c_{k+2l}) = \int_0^{2\pi N} W_s(\tau, V, Z, \mu)d\tau = 0, \quad s = 1,\ldots, k+2l, \tag{3.2.30}$$

where integrands are computed on the periodic solution (3.2.26).

Thus, not any periodic solution (3.2.26) can serve as a limiting one for a periodic solution to the nonlinear system as $\mu \to +0$. Such solutions can only be the solutions (3.2.26) that satisfy (3.2.30). Fix one of them and find the functional Jacobian matrix of Q_s functions in the variables $b_1,\ldots, b_{k+2l}$, considering the vector function Γ to be the function of these variables and parameter μ which satisfies the system (3.2.28). Compute this matrix for $B = 0, \mu = 0$. Then it is found to be coincident with the functional matrix computed from the functions p_s in the variables $c_1,\ldots, c_{k+2l}$ with the values of these variables fixed as above. Consequently, if the condition (3.2.30) is satisfied and the determinant

$$\frac{D(p_1,\ldots, p_{k+2l})}{D(c_1,\ldots, c_{k+2l})} \neq 0,$$

then the original system has a $2\pi N$-periodic solution for all sufficiently small $\mu > 0$, indefinitely approaching, as $\mu \to +0$, the $2\pi N$-periodic solution of the linear system (3.2.26) determined by the condition (3.2.30). If this determinant is zero, then the investigation must be continued.

The resonance case, where for resonance eigenvalues with multiplicity more than unit there are nonprime elementary divisors, seems to be more complicated for construction of periodic solutions.

We lay aside all preliminary transformations of the above form and focus on solving the problem directly for a system of the form

$$\frac{du_s}{dt} = u_{s-1} + \varphi_s(t) + \mu \Phi_s(t, U, Z, \mu), \quad s = 1, \ldots, k, \qquad (3.2.31)$$

$$\frac{dz_r}{dt} = \sum_{i=1}^{m} p_{ri} z_i + \vartheta_r(t) + \mu \Theta_r(t, U, Z, \mu), \quad r = 1, \ldots, m,$$

where $u_0 = 0$. As a result of transformations, the right-hand sides of system (3.2.31) become $2\pi N$-periodic functions of t having the same properties as the right-hand sides of system (3.2.19). The system (3.2.31) for $\mu = 0$ has a $2\pi N$-periodic solution depending on one arbitrary constant if the following condition is satisfied

$$\int_0^{2\pi N} \varphi_1(\tau) d\tau = 0. \qquad (3.2.32)$$

Under condition (3.2.32) the above-mentioned periodic solution of system (3.2.31) for $\mu = 0$ is

$$u_1 = \gamma_1 + \int_0^t \varphi_1(\tau) d\tau,$$

$$u_2 = \gamma_2 + \int_0^t (u_1(\tau) + \varphi_2(\tau)) d\tau,$$

$$\ldots$$

$$u_k = \gamma_k + \int_0^t (u_{k-1}(\tau) + \varphi_k(\tau)) d\tau, \qquad (3.2.33)$$

$$Z = e^{\mathbf{P}t} C + \int_0^t e^{\mathbf{P}(t-\tau)} T(\tau) d\tau.$$

The constants $\gamma_1, \ldots, \gamma_{k-1}$ are chosen such that the following conditions are satisfied

$$\int_0^{2\pi N} (u_{s-1}(\tau) + \varphi_s(\tau)) d\tau = 0, \quad s = 1, \ldots, k. \qquad (3.2.34)$$

It follows from (3.2.34) that the periodic solution (3.2.33) is dependent only on one constant γ_k. We shall seek a $2\pi N$-periodic solution to the nonlinear system (3.2.31). To this end, we find the general solution with initial data $u_s = \gamma_s + b_s, s = 1, \ldots, k, Z = C + \Gamma$ and we write for it the system (3.2.3). By the above procedure we then obtain the system of equations

$$\mu Q_s = \mu \int_0^{2\pi N} \Phi_s(t, U, Z, \mu) dt = 0, \qquad (3.2.35)$$

$$(e^{2\pi N\mathbf{P}} - \mathbf{E})\Gamma + \mu \int_0^{2\pi N} e^{\mathbf{P}(2\pi N - \tau)} \Theta(\tau, U, Z, \mu) d\tau = 0. \quad (3.2.36)$$

Assume that $Q_s = p_s$ for $B = \Gamma = 0, \mu = 0, B = (b_1, \ldots, b_k)$, and the functions p_s depend on one arbitrary constant γ_k. For the existence of a solution to the equations (3.2.35) and (3.2.36) satisfying the condition $B \to 0, \Gamma \to 0$, as $\mu \to +0$, the following equalities must hold

$$p_s(\gamma_k) = 0, \quad s = 1, \ldots, k. \quad (3.2.37)$$

If there is a common root of equation (3.2.37), then we may continue the definition of the functions of interest. If the condition (3.2.37) fails to hold for all $\gamma_k \in (-\infty, +\infty)$, then the desired functions B and Γ are nonexistent and hence the original system has no $2\pi N$-periodic solutions coming arbitrarily close to the periodic solutions of the linear system as $\mu \to +0$.

Of course, the system (3.2.31) is a very special case among those of interest. However, it provides an illustrative example of the complexities due to the multiple resonance eigenvalues of the matrix $\mathbf{A}$ having nonprime elementary divisors.

Example 2. Consider the nonautonomous quasilinear system

$$\dot{X} = \mathbf{P}(t)X + \Phi(t) + \mu F(t, X, \mu), \quad (3.2.38)$$

whose right-hand sides are the functions continuous in X, t, μ and 2π-periodic in t for all $X \in E_n, \mu \in [0, \mu_1]$. We shall assume that the vector function $F(t, X, \mu)$ is continuously differentiable with respect to variables $x_1, \ldots, x_n$.

Denote by $\lambda_1, \ldots, \lambda_n$ the characteristic indices of the system

$$\dot{X} = \mathbf{P}(t)X. \quad (3.2.39)$$

In finding $2\pi N$-periodic solutions for system (3.2.38), two basically different cases are encountered: the first is the non-resonance case where the characteristic indices $\lambda_1, \ldots, \lambda_n$ do not include quantities of the form $\pm 2\pi i \frac{M}{N}, M = 0, 1, \ldots$, and the second is the resonance case where the characteristic indices include quantities of that form. In the non-resonance case, the system

$$\dot{X} = \mathbf{P}(t)X + \Phi(t) \quad (3.2.40)$$

has a unique $2\pi N$-periodic solution (which is a 2π-periodic solution). Consequently, system (3.2.38) for any sufficiently small $\mu > 0$ has a periodic solution of 2π period coming arbitrarily close to the above solution of system (3.2.40) as $\mu \to +0$.

In the resonance case, system (3.2.39) has the family of $2\pi N$-periodic solutions depending on k arbitrary constants $c_1, \ldots, c_k$. Under the known conditions on the function $\Phi(t)$ the system (3.2.40) also has the family of $2\pi N$-periodic solutions depending on the same arbitrary constants. These conditions on the function $\Phi(t)$ are supposed to be satisfied (see section 1.2). In the resonance case, the periodic solutions of system (3.2.38) coming arbitrarily close to the solutions of system (3.2.40), as $\mu \to +0$ can correspond only to those solutions of this system that are determined by the arbitrary constants $c_1, \ldots, c_k$ satisfying the system of equations

$$p_s(c_1, \ldots, c_k) = 0, \quad s = 1, \ldots, m, \quad m \geq k. \quad (3.2.41)$$

Here $k = m$ if and only if for resonance characteristic indices there are linearly independent periodic solutions of system (3.2.39), i.e. if k is the number of all characteristic indices, including their multiplicity. After this, it is possible to choose sufficient conditions under which the quantities $c_1, \ldots, c_k$ satisfying system (3.2.41) do correspond, for all sufficiently small $\mu > 0$, to the periodic solution of system (3.2.38) possessing the above-mentioned property as $\mu \to +0$.

These statements are proved by simple reduction of system (3.2.38) to the system discussed in Example 1. By the reducibility of system (3.2.39), there is a nonsingular linear transformation with periodic coefficients for the desired functions $X = \mathbf{H}(t)Y$ which reduces the matrix $\mathbf{P}(t)$ to constant matrix whose eigenvalues coincide with the characteristic indices of system (3.2.39).

Example 3. Let there be given the nonautonomous system

$$\dot{X} = F(t, X, \mu), \tag{3.2.42}$$

whose right-hand side is defined for all $X \in E_n, t \in (-\infty, +\infty), \mu \in [0, \mu_0]$, is 2π-periodic in t, and is real and twice continuously differentiable with respect to $x_1, \ldots, x_n, \mu$.

Assume that the system

$$\frac{dX}{dt} = F_0(t, X) = F(t, X, 0) \tag{3.2.43}$$

has the family of 2π-periodic solutions

$$X = X_0(t, h_1, \ldots, h_k),$$

depending on k arbitrary constants. In the system (3.2.42) we make a replacement

$$X = X_0(t, h_1, \ldots, h_k) + \mu Y,$$

where Y is the new desired function. To define it, we get the system of equations

$$\mu \frac{dY}{dt} = F(t, X_0 + \mu Y, \mu) - F_0(t, X_0) = F(t, X_0 + \mu Y, \mu) - F(t, X_0, 0)$$

$$= \mathbf{P}(t)\mu Y + \Phi(t)\mu + \mu^2 \bar{F}(t, Y, \mu).$$

Cancelling by μ, we obtain a system as in (3.2.38) with the additional condition that the system (3.2.39) has k periodic solutions with 2π period. Consequently, the general case of the nonautonomous system with the constraints on the right-hand side of the double differentiability type reduces to the system (3.2.38).

Example 4. Consider the autonomous system of differential equations

$$\frac{dX}{dt} = \mathbf{A}X + \mu F(X). \tag{3.2.44}$$

Suppose that the homogeneous system

$$\frac{dX}{dt} = \mathbf{A}X \tag{3.2.45}$$

Small Parameter Method

has a family of periodic solutions with period $\bar{T}$. This holds if and only if the eigenvalues of the matrix $\mathbf{A}$ include pure imaginary numbers of the form $\pm \frac{iM}{T}, M = 0, 1, \ldots$. Note all such eigenvalues and assume that there are k pairs of these. Then, assuming that the multiples correspond to prime elementary divisors, we can reduce system (3.2.44) to the following form by using a nonsingular linear transformation with real constant coefficients for the desired functions

$$\dot{x}_s = -\mu_s y_s + \mu \varphi_s(X, Y, Z, \mu),$$

$$\dot{y}_s = \mu_s x_s + \mu \psi_s(X, Y, Z, \mu), \quad s = 1, \ldots, k, \quad (3.2.46)$$

$$\dot{z}_j = \sum_{i=1}^{m} p_{ji} z_i + \mu g_j(X, Y, Z, \mu), \quad j = 1, \ldots, m, m + 2k = n,$$

where

$$X = (x_1, \ldots, x_k), \quad Y = (y_1, \ldots, y_k), \quad Z = (z_1, \ldots, z_m).$$

The system (3.2.46) for $\mu = 0$ has the family of $\bar{T}$-periodic solutions depending on $2k$ arbitrary constants $c_s, \gamma_s,$

$$x_s = c_s \cos \mu_s t - \gamma_s \sin \mu_s t,$$

$$y_s = c_s \sin \mu_s t + \gamma_s \cos \mu_s t, \quad s = 1, \ldots, k, \quad z_j = 0, \quad j = 1, \ldots, m. \quad (3.2.47)$$

In order to apply the theory developed above, we consider only those solutions from the family (3.2.47) for which $y_1 = 0$ at $t = 0$. Consequently, $\gamma_1 = 0$ and hence the family (3.2.47) depends on $2k - 1$ arbitrary constants. We shall seek a solution to system (3.2.46) assuming for $t = 0$ the values $x_1 = c_1 + b_1, y_1 = 0, x_s = c_s + b_s, y_s = \gamma_s + \beta_s, s = 2, \ldots, k, z_j = \delta_j, j = 1, \ldots, m$ and having the period $T = T(\mu)$. Integrating the system (3.2.46) with these initial data, we obtain the system

$$x_1 = (c_1 + b_1) \cos \mu_1 t + \int_0^t \mu \Phi_1 d\tau,$$

$$y_1 = (c_1 + b_1) \sin \mu_1 t + \int_0^t \mu \Psi_1 d\tau,$$

$$x_s = (c_s + b_s) \cos \mu_s t - (\gamma_s + \beta_s) \sin \mu_s t + \int_0^t \mu \Phi_s d\tau,$$

$$y_s = (c_s + b_s) \sin \mu_s t + (\gamma_s + \beta_s) \cos \mu_s t + \int_0^t \mu \Psi_s d\tau, \quad s = 2, \ldots, k, \quad (3.2.48)$$

$$Z = e^{\mathbf{P}t} \Delta + \mu \int_0^t e^{\mathbf{P}(t-\tau)} G d\tau,$$

where

$$\Delta = (\delta_1, \ldots, \delta_m), G = (g_1, \ldots, g_m), \mathbf{P} = \{p_{ji}\},$$

$$\Phi_s = \cos(t - \tau) \varphi_s - \sin(t - \tau) \psi_s,$$

$$\Psi_s = \sin(t - \tau) \varphi_s + \cos(t - \tau) \psi_s, \quad s = 1, \ldots, k.$$

Using now the periodicity condition, from (3.2.48) we get

$$(c_1 + b_1)\cos\mu_1 T + \int_0^T \mu\Phi_1 d\tau - (c_1 + b_1) = 0,$$

$$(c_1 + b_1)\sin\mu_1 T + \int_0^T \mu\Psi_1 d\tau = 0,$$

$$(c_s + b_s)\cos\mu_s T - (\gamma_s + \beta_s)\sin\mu_s T + \int_0^T \mu\Phi_s d\tau - (c_s + b_s) = 0, \qquad (3.2.49)$$

$$(c_s + b_s)\sin\mu_s T - (\gamma_s + \beta_s)\cos\mu_s T + \int_0^T \mu\Psi_s d\tau - (\gamma_s + \beta_s) = 0,$$

$$(e^{\mathbf{P}T} - \mathbf{E})\Delta + \mu\int_0^T e^{\mathbf{P}(T-\tau)}Gd\tau = 0.$$

For the purposes of seeking a periodic solution for which $\beta_s \to 0, b_s \to 0, \delta_j \to 0, T \to \bar{T}$ as $\mu \to +0$, we set in (3.2.49) $T = \bar{T} + \alpha(\mu)$. We shall suppose, e.g. that $c_1 \neq 0$. Then it follows from the second equation of system (3.2.49) that the quantity α is of at least the order μ. In view of this, from the remaining equations of system (3.2.49) we find

$$(c_1 + b_1)(\cos\mu_1\alpha - 1) + \int_0^T \mu\Phi_1 d\tau = 0,$$

$$(c_s + b_s)(\cos\mu_s\alpha - 1) - (\gamma_s + b_s)\sin\mu_s\alpha + \int_0^T \mu\Phi_s d\tau = 0, \qquad (3.2.50)$$

$$(c_s + b_s)\sin\mu_s\alpha + (\gamma_s + \beta_s)(\cos\mu_s\alpha - 1) + \int_0^T \mu\Psi_s d\tau = 0.$$

Cancelling μ from both sides of each equality in (3.2.50) and considering that, as follows from the second equation of system (3.2.49), $\alpha \to 0$ as $\mu \to +0$,

$$\frac{\alpha}{\mu} \xrightarrow[\mu \to +0]{} \frac{-1}{c_1\mu_1}\int_0^{\bar{T}} \Psi_1 d\tau,$$

from (3.2.50) we obtain

$$p_1 = \int_0^{\bar{T}} \Phi_1 d\tau = 0, \quad p_s = \int_0^{\bar{T}} \Phi_s d\tau + \frac{\gamma_s\mu_s}{c_1\mu_1}\int_0^{\bar{T}} \Psi_1 d\tau = 0, \qquad (3.2.51)$$

$$q_s = \int_0^{\bar{T}} \Psi_s d\tau - \frac{c_s\mu_s}{c_1\mu_1}\int_0^{\bar{T}} \Psi_1 d\tau = 0, \quad s = 2,\ldots,k,$$

where the integrands are computed on the family of periodic solutions (3.2.47) for $\gamma_1 = 0$. Thus the equalities (3.2.51) represent the necessary conditions to be satisfied by the parameters of the family (3.2.47) so that the solution of the family (3.2.47) satisfying (3.2.51) would correspond to a periodic solution of a nonlinear system. After finding solutions to system (3.2.51), under appropriate conditions it is possible to find the quantities of interest from system (3.2.48). Note that the above case of system (3.2.44) is special. In the general case, where multiple eigenvalues correspond to nonprime elementary divisors, the number of equations such as (3.2.51) becomes

Natural Oscillations of Autonomous Systems

greater than the number of parameters appearing in the family of periodic solutions for system (3.2.44) with $\mu = 0$, and solution of the problem in hand becomes much more complicated. The presence of zero eigenvalues of the matrix **A** presents no further problems.

Remark. In handling the stability problem for the periodic solutions obtained by the small parameter method, we first consider the linear system in variations. Suppose it is of the form

$$\frac{dX}{dt} = \mathbf{P}(t,\mu)X, \qquad (3.2.52)$$

where $\mathbf{P}(t,\mu)$ is the continuous matrix function which is real and 2π-periodic in t. Let $\mathbf{P}(t,0) = \mathbf{P}_0(t)$ and the zero solution of the system

$$\frac{dX}{dt} = \mathbf{P}_0(t)X \qquad (3.2.53)$$

be asymptotically stable. Then, for all sufficiently small $\mu > 0$, the zero solution of system (3.2.52) is also asymptotically stable. If among the characteristic indices of system (3.2.53) there is at least one index with positive real part, then for all sufficiently small $\mu > 0$ the zero solution of the system (3.2.52) is unstable. If among the characteristic indices there are indices whose real parts are zero and there are no characteristic indices whose real parts are positive, then the stability problem of the zero solution to system (3.2.52) for sufficiently small $\mu > 0$ is solved from the investigating basis of the terms appearing in the matrix $\mathbf{P}(t,\mu)$ and vanishing at $\mu = 0$. If, for example, $\mathbf{P}(t,\mu)$ is an analytic function of the parameter μ, then the coefficients of the characteristic equation are also analytic functions of μ and hence the characteristic indices of the system (3.2.52) are the functions of the parameter μ. Let λ be some characteristic index. Then system (3.2.52) has the solution $X = e^{\lambda t}Y$, where Y is a 2π-periodic function. Substituting this into (3.2.52) we obtain the system of differential equation to define the vector function Y. The existence conditions for the periodic solution of this system obtained as in Example 2, make it possible to find, for sufficiently small $\mu > 0$ in a fairly broad class of cases [Malkin (1956)], the signs of real parts of the characteristic indices of system (3.2.52) and hence to solve the stability problem.

3.3 Natural Oscillations of Autonomous Systems in a Neighborhood of an Equilibrium

1. Consider the system

$$\frac{dx_s}{dt} = X_s(x_1, \ldots, x_k, y_1, \ldots, y_n), \quad s = 1, \ldots, k,$$

$$\frac{dy_j}{dt} = \sum_{i=1}^{n} p_{ji} y_i + Y_j(x_1, \ldots, x_k, y_1, \ldots, y_n), \quad j = 1, \ldots, n. \qquad (3.3.1)$$

The functions X_s, Y_j are expanded as series in integral positive powers of $x_1, \ldots, x_k, y_1, \ldots, y_n$ that are convergent for sufficiently small $|x_s|, |y_j|, s = 1, \ldots, k, j = 1, \ldots, n$ and contain no terms linear in $x_1, \ldots, x_k, y_1, \ldots, y_n$. The coefficients in the expansions of X_s, Y_j and the quantities p_{ji} are real constants. Assume that the eigenvalues of the matrix $\mathbf{P}$, $\{\mathbf{P}\}_{ij} = \mathbf{p}_{ij}$, $\lambda_1, \ldots, \lambda_n$ have negative real parts. Denote by $u_j(x_1, \ldots, x_k), j = 1, \ldots, n$ the solution of the system which is holomorphic for sufficiently small $|x_s|, s = 1, \ldots, n$,

$$\sum_{i=1}^n p_{ji} y_i + Y_j(x_1, \ldots, x_k, y_1, \ldots, y_n) = 0, \quad j = 1, \ldots, n. \tag{3.3.2}$$

Theorem 42 *For the system (3.3.1) to have k holomorphic t-independent integrals of the form*

$$c_s = x_s + f_s(x_1, \ldots, x_k, y_1, \ldots, y_n),$$

where the functions f_s are expanded as convergent series in integral positive powers of $x_1, \ldots, x_k, y_1, \ldots, y_n$ which contain no terms linear in these quantities and, additionally, $f_s \equiv 0$ for $x_1 = \ldots = x_k = 0$, $f_s \equiv 0$ for $y_j = u_j(x_1, \ldots, x_k), j = 1, \ldots, n$, it is necessary and sufficient that $X_s(x_1, \ldots, x_k, u_1, \ldots, u_n) \equiv 0$.

Proof: *Necessity.* Replace in system (3.3.1) the desired functions by the formulas

$$y_j = u_j(x_1, \ldots, x_k) + \eta_j, \tag{3.3.3}$$

then the system becomes

$$\frac{dx_s}{dt} = X_s(x_1, \ldots, x_k, u_1 + \eta_1, \ldots, u_n + \eta_n) = \bar{X}_s, \tag{3.3.4}$$

$$\frac{d\eta_j}{dt} = \sum_{i=1}^n p_{ji}\eta_i + Y_j(x_1, \ldots, x_k, u_1 + \eta_1, \ldots, u_n + \eta_n)$$

$$+ \sum_{i=1}^n p_{ji} u_i - \sum_{s=1}^k \frac{\partial u_j}{\partial x_s} \bar{X}_s = \sum_{i=1}^n p_{ji}\eta_i + \bar{Y}_j(x_1, \ldots, x_k, \eta_1, \ldots, \eta_n)$$

$$- \sum_{s=1}^k \frac{\partial u_j}{\partial x_s} \bar{X}_s, \quad s = 1, \ldots, k, \quad j = 1, \ldots, n.$$

The functions $\bar{Y}_j$ in system (3.3.4) have the property that $\bar{Y}_j \equiv 0$ for $\eta_1 = \ldots = \eta_n = 0$. By the condition of Theorem 42, system (3.3.4) has k holomorphic t-independent integrals

$$c_s = x_s + f_s(x_1, \ldots, x_k, u_1 + \eta_1, \ldots, u_n + \eta_n). \tag{3.3.5}$$

Since the expansions of the functions $f_s(x_1, \ldots, x_k, y_1, \ldots, y_n)$ contain no terms that are linear in $x_1, \ldots, x_k, y_1, \ldots, y_n$ and $f_s(x_1, \ldots, x_k, u_1, \ldots, u_n) \equiv 0, s = 1, \ldots, k$, the equalities (3.3.5) are invertible. As a result of their inversion, we get the functions

$$x_s = c_s + g_s(c_1, \ldots, c_k, \eta_1, \ldots, \eta_n). \tag{3.3.6}$$

Natural Oscillations of Autonomous Systems

The functions $g_s(c_1, \ldots, c_k, \eta_1, \ldots, \eta_n)$ are expanded as convergent series in integral positive powers of quantities $c_1, \ldots, c_k, \eta_1, \ldots, \eta_n$ which contain no terms linear in these quantities. Moreover $g_s \equiv 0$ for $c_1 = \ldots = c_k = 0$, $g_s \equiv 0$ for $\eta_1 = \ldots = \eta_n = 0$. The functions (3.3.6) satisfy the system of partial differential equations

$$\sum_{j=1}^{n} \frac{\partial x_s}{\partial \eta_j} \left(\sum_{i=1}^{n} p_{ji} \eta_i + \bar{Y}_j - \sum_{i=1}^{k} \frac{\partial u_j}{\partial x_i} \bar{X}_i \right) = \bar{X}_s. \qquad (3.3.7)$$

It follows from the properties of the functions g_s that equality (3.3.6) can be rewritten as

$$x_s = c_s + \sum_{m=1}^{\infty} x_s^{(m)}, \qquad (3.3.8)$$

where $x_s^{(m)}$ are homogeneous forms to the power m with respect to the quantities $\eta_1, \ldots, \eta_n$ whose coefficients are analytic functions of the quantities $c_1, \ldots, c_k$ vanishing at $c_1 = \ldots = c_k = 0$. Substituting the series of (3.3.8) in the functions $\bar{X}_s$ and expanding these functions with respect to the powers of $\eta_1, \ldots, \eta_n$, we obtain the equality

$$\bar{X}_s = \sum_{m=0}^{\infty} \bar{X}_s^{(m)}, \qquad (3.3.9)$$

where $\bar{X}_s^{(m)}$ are the forms to the power m with respect to the quantities $\eta_1, \ldots, \eta_n$, here

$$\bar{X}_s^{(0)} = X_s(c_1, \ldots, c_k, u_1(c_1, \ldots, c_k), \ldots, u_n(c_1, \ldots, c_k)).$$

As noted above, the functions (3.3.6) satisfy system (3.3.7), therefore, substituting the series (3.3.8) into equations (3.3.7), we get the identities. By substituting the series (3.3.8) and (3.3.9) into equations (3.3.7) and equating the form coefficients with respect to $\eta_1, \ldots, \eta_n$, we obtain the algebraic systems establishing the one-to-one relationship between the coefficients of the forms $\bar{X}_s^{(m)}$ and the functions X_s and Y_j. Let

$$x_s^{(1)} = \sum_{i=1}^{n} a_{si} \eta_i \qquad (3.3.10)$$

(a_{si} in the relationships (3.3.10) are analytic functions with respect to $c_1, \ldots, c_k$, here $a_{si} = 0$ for $c_1 = \ldots = c_k = 0$). Substituting the series (3.3.8) in system (3.3.7), the forms (3.3.10) generate the terms that are independent of the quantities $\eta_1, \ldots, \eta_n$. If the series (3.3.8) are to satisfy system (3.3.7), it is essential that the free terms in system (3.3.7) be the same on the right and left. Hence we get the relationships

$$\sum_{j=1}^{n} \left(a_{sj} \sum_{l=1}^{k} \frac{\partial u_j}{\partial x_l} \bar{X}_l^{(0)} \right) = \bar{X}_s^{(0)},$$

$$\sum_{l=1}^{k} \bar{X}_l^{(0)} \left(\sum_{j=1}^{n} a_{sj} \frac{\partial u_j}{\partial x_l} - \delta_{sl} \right) = 0, \qquad (3.3.11)$$

where $\delta_{sl} = 0$ for $l \neq s$ and $\delta_{sl} = 1$ for $l = s$. The relationships (3.3.11) can be regarded as a system of linear equations intended to define the quantities $\bar{X}_s^{(0)}$, $s = 1, \ldots, k$.

The determinant in this system for sufficiently small $|c_s|, s = 1,\ldots, k$, can be made as close to the number $(-1)^k$ as desired. Thus, for sufficiently small $|c_s|$

$$\bar{X}_s^{(0)} = X_s(c_1,\ldots, c_k, u_1(c_1,\ldots, c_k),\ldots, u_n(c_1,\ldots, c_k)) \equiv 0.$$

Sufficiency. In system (3.3.1), we replace $y_j = u_j + \eta_j, j = 1,\ldots, n$, then it becomes

$$\frac{dx_s}{dt} = \bar{X}_s, \quad s = 1,\ldots, k,$$

$$\frac{d\eta_j}{dt} = \sum_{i=1}^{n} p_{ji}\eta_i + \bar{Y}_j, \quad j = 1,\ldots, n. \tag{3.3.12}$$

Consider the system of partial differential equations corresponding to this system

$$\sum_{j=1}^{n} \frac{\partial x_s}{\partial \eta_j} \left(\sum_{i=1}^{n} p_{ji}\eta_i + \bar{Y}_j \right) = \bar{X}_s, \quad s = 1,\ldots, k. \tag{3.3.13}$$

Replace the desired functions in system (3.3.13) from the formula $x_s = c_s + \xi_s$, then the new desired functions ξ_s satisfy the system

$$\sum_{j=1}^{n} \frac{\partial \xi_s}{\partial \eta_j} \left(\sum_{i=1}^{n} p'_{ji}\eta_i + \bar{Y}'_j \right) = \bar{X}'_s = \sum_{m=1}^{\infty} X_s^{(m)'}, \tag{3.3.14}$$

where p'_{ji} are analytic functions in the quantities $c_1,\ldots, c_k$ such that the eigenvalues of the matrix $\{p'_{ji}\}$ have negative real parts for sufficiently small $|c_s|$, and the functions $X_s^{(m)'}$ are the forms with respect to $\eta_1,\ldots, \eta_n$ having the coefficients that are analytic in $c_1,\ldots, c_k, \xi_1,\ldots, \xi_k$. The functions $X_s^{(1)'}$ are such that $X_s^{(1)'} = 0$ for $c_1 = \ldots = c_k = \xi_1 = \ldots = \xi_k = 0$. Hence there is a unique system of holomorphic functions $\xi_s = f_s(c_1,\ldots, c_k, \eta_1,\ldots, \eta_n)$, satisfying system (3.3.14) (see [Lyapunov (1950)], the auxiliary theorem). The functions g_s are such that $g_s \equiv 0$ for $c_1 = \ldots = c_k = 0, g_s \equiv 0$ for $\eta_1 = \ldots = \eta_n = 0$. So the system (3.3.13) has the family of solutions that is dependent on k arbitrary constants,

$$x_s = c_s + g_s(c_1,\ldots, c_k, \eta_1,\ldots, \eta_n). \tag{3.3.15}$$

Solving relationships (3.3.15) for $c_1,\ldots, c_k$, we obtain

$$c_s = x_s + \varphi_s(x_1,\ldots, x_k, \eta_1,\ldots, \eta_n).$$

Switching to the old variables, we have

$$c_s = x_s + \varphi_s(x_1,\ldots, x_k, y_1 - u_1,\ldots, y_n - u_n) = x_s + f_s(X, Y).$$

It follows from the properties of function g_s that the last relationships are the integrals of system (3.3.1), and the functions $f_s(x_1,\ldots, x_k, y_1,\ldots, y_n)$ satisfy here all the conditions of Theorem 42. ∎

Natural Oscillations of Autonomous Systems

2. Consider the case where $X_s(x_1,\ldots,x_k, u_1,\ldots,u_n) \equiv 0$, $s = 1,\ldots,k$, it is singular and amenable to treatment in full from a point of view for stability of zero solution to system (3.3.1).

Theorem 43 *If $X_s(x_1,\ldots,x_k, u_1,\ldots,u_n) \equiv 0$, then the zero solution of system (3.3.1) is stable.*

Proof: The transformation (3.3.3) of the desired functions changing system (3.3.1) to system (3.3.12) does not alter the properties of stability, hence it suffices to establish stability of the zero solution of system (3.3.12). Under the imposed conditions, system (3.3.12) has k holomorphic integrals that are representable as

$$x_s = c_s + g_s(c_1,\ldots,c_k,\eta_1,\ldots,\eta_n), \quad s = 1,\ldots,k.$$

Using these relationships, we eliminate $x_1,\ldots,x_k$ from the second group of equations (3.3.12). As a result, to define the functions $\eta_1,\ldots,\eta_n$, we obtain the system

$$\frac{d\eta_j}{dt} = \sum_{i=1}^{n} p'_{ji}\eta_i + \tilde{Y}_j. \qquad (3.3.16)$$

Expansions of the functions $\tilde{Y}_j$ in powers of $\eta_1,\ldots,\eta_n$ contain at least the second-order terms with respect to these quantities. Therefore, the zero solution of system (3.3.16) for sufficiently small $|c_s|$ is asymptotically stable, since, as noted before, matrix $\{p'_{ji}\}$ has the eigenvalue with negative real parts, and then $x_s \to c_s$ as $t \to +\infty$. ∎

We now focus on the qualitative structure of the integral curve family of system (3.3.1) located in a sufficiently small neighborhood of the origin of coordinates.

Theorem 44 *If $X_s(x_1,\ldots,x_k,u_1,\ldots,u_n) \equiv 0$, then a sufficiently small neighborhood of the point $x_1 = \ldots = x_k = y_1 = \ldots = y_n = 0$ breaks into a continuum sum of the nonoverlapping sets D having the properties:*

1) the integral curve of system (3.3.1) starting in one of the sets D stays therein until it leaves this neighborhood of the origin of coordinates,

2) between these sets and a sufficiently small neighborhood of the origin of coordinates in the k-dimensional space it is possible to establish a one-to-one correspondence in such a manner that if D corresponds to the set of numbers $(c_1,\ldots,c_k)$, then all the integral curves of system (3.3.1) starting in D satisfy the condition $x_s \to c_s, y_s \to u_s(c_1,\ldots,c_k)$ as $t \to +\infty$.

Proof: Take a sufficiently small neighborhood of the k-dimensional space. Choose some point in this neighborhood $(\bar{c}_1,\ldots,\bar{c}_k)$. Under the conditions of Theorem 44, system (3.3.1) has a family of integrals which may be represented as $c_s = x_s + f_s$ $(x_1,\ldots,x_k, y_1,\ldots,y_n)$. Each of these equalities defines a certain manifold in the $(n+k)$-dimensional space. Denote by $D(\bar{c}_1,\ldots,\bar{c}_k)$ the set of all common points of the manifolds $\bar{c}_s = x_s + f_s, s = 1,\ldots,k$. The collection of all sets $D(\bar{c}_1,\ldots,\bar{c}_k)$ covers a sufficiently small neighborhood of the origin of coordinates in the $(n+k)$-dimensional space. Indeed, let $(x_1^{(0)},\ldots,x_k^{(0)}, y_1^{(0)},\ldots,y_n^{(0)})$ be some point of this neighborhood. Set $c_s^{(0)} = x_s^{(0)} + f_s(x_1^{(0)},\ldots,x_k^{(0)}, y_1^{(0)},\ldots,y_n^{(0)})$. It follows from the properties of functions f_s that the point $c_1^{(0)},\ldots,c_k^{(0)}$ belongs to a sufficiently small neighborhood of the k-dimensional space. Therefore, it corresponds to the set $D(c_1^{(0)}, \ldots, c_k^{(0)})$; it is clear that the point $(x_1^{(0)}, \ldots, x_k^{(0)}, y_1^{(0)}, \ldots, y_n^{(0)})$ belongs to this set. Let $(X(t), Y(t))$ be the integral curve of system (3.3.1) starting at the point $(X^{(0)}, Y^{(0)})$. Then the following equalities hold along the integral curve involved

$$X_s(t) = f_s(x_1(t), \ldots, x_k(t), y_1(t), \ldots, y_n(t)) = c_s^{(0)}, \quad s = 1, \ldots, k.$$

Hence this integral curve is contained in the set $D(c_1^{(0)}, \ldots, c_k^{(0)})$. As is shown in Theorem 43, the zero solution of system (3.3.12) is stable. In this case, $\eta_j \to 0$ as $t \to +\infty$, $x_s \to c_s$ as $t \to +\infty$. Therefore, it follows from (3.3.3) that $y_j \to u_j(c_1, \ldots, c_k)$ as $t \to +\infty$ if $|c_s|, s = 1, \ldots, k$, are sufficiently small. From this and from the above reasoning it follows that any integral curve lying in the set $D(c_1, \ldots, c_k)$ is such that $x_s(t) \to c_s, y_j(t) \to u_j(c_1, \ldots, c_k)$ as $t \to +\infty$. ■

Note that the theorems formulated here apply to any system of $n + k$ differential equations with holomorphic right parts if the matrix of coefficients of the linear system constituting the first approximation has k zero eigenvalues which correspond to prime elementary divisors, and the remaining n eigenvalues of this matrix have negative real parts. This assertion follows from the fact that it is always possible to select the linear nonsingular transformation of the desired functions reducing such a system to the form (3.3.1).

Now we briefly outline the general case of the zero solution stability for a system as in (3.3.1). Let

$$X_s^{(0)} = \bar{X}_s(x_1, \ldots, x_k, 0, \ldots, 0) = X_s(x_1, \ldots, x_k, u_1, \ldots, u_n), \quad s = 1, \ldots, k,$$

$$Y_j^{(0)} = \bar{Y}_j(x_1, \ldots, x_k, 0, \ldots, 0) = -\sum_{s=1}^{k} \frac{\partial u_j}{\partial x_s} X_s^{(0)}, \quad j = 1, \ldots, n.$$

The functions $X_s^{(0)}$ and $Y_j^{(0)}$ are expanded as convergent series

$$X_s^{(0)} = \sum_{m=\mu}^{\infty} X_s^{(m)}, \quad Y_j^{(0)} = \sum_{m=\nu}^{\infty} Y_j^{(m)}, \quad s = 1, \ldots, k, \quad j = 1, \ldots, n, \quad \nu > \mu.$$

The zero solution of system (3.3.1) is stable (asymptotically stable) irrespective of the choice of the forms $X_s^{(m)}, m > \mu$, if the zero solution of the system

$$\frac{dx_s}{dt} = X_s^{(0)}, \quad s = 1, \ldots, k$$

is stable (asymptotically stable) [Malkin (1966)]. Thus, in the studies on the general case, it is possible to employ the results contained in section 1.5.

3. Suppose there is a linear system of differential equations

$$\frac{dx_s}{dt} = \sum_{i=1}^{n} a_{si} x_i, \quad s = 1, \ldots, n.$$

Natural Oscillations of Autonomous Systems

If the characteristic equation of this system has pure imaginary roots, then there are periodic oscillations in it. If there are several pure imaginary roots, where at least two $i\omega_1$ and $i\omega_2$ are incommensurable, then, apart from periodic oscillations the system also has almost periodic oscillations caused by a superposition of several periodic oscillations with incommensurable periods. The same phenomenon is observed in some nonlinear autonomous systems.

Consider the system of $n + 2k$ equations, whose linear approximation has k pairs of pure imaginary roots, among which there are no multiples. Our reasoning will go along two directions: first, we will discuss the properties of stability of an equilibrium, and, second, we will study the analytic properties of periodic and almost periodic solutions depending on the choice of their defining initial conditions. Consider the system

$$\frac{d\bar{x}_s}{dt} = -\lambda_s \bar{y}_s + X_s(\bar{x}_1, \ldots, \bar{x}_k, \bar{y}_1, \ldots \bar{y}_k, \bar{z}_1, \ldots, \bar{z}_n),$$

$$\frac{d\bar{y}_s}{dt} = \lambda_s \bar{x}_s + Y_s(\bar{x}_1, \ldots, \bar{x}_k, \bar{y}_1, \ldots, \bar{y}_k, \bar{z}_1, \ldots, \bar{z}_n), \quad s = 1, \ldots, k, \quad (3.3.17)$$

$$\frac{d\bar{z}_j}{dt} = \sum_{i=1}^{n} r_{ji} \bar{z}_i + Z_j(\bar{x}_1, \ldots, \bar{x}_k, \bar{y}_1, \ldots, \bar{y}_k, \bar{z}_1, \ldots, \bar{z}_n), \quad j = 1, \ldots, n.$$

Assume that the functions X_s, Y_s, Z_j are expanded as convergent power series with respect to $\bar{x}_1, \ldots, \bar{x}_k, \bar{y}_1, \ldots, \bar{y}_k, \bar{z}_1, \ldots, \bar{z}_n$ containing no linear terms. The right-hand sides of system (3.3.17) are assumed to be real. The quantities $\lambda_1, \ldots, \lambda_k$ are positive and are such that

$$\sum_{i=1}^{k} \mu_i \lambda_i \neq 0 \quad (3.3.18)$$

for any integer $\mu_1, \ldots, \mu_k$, $\sum_{i=1}^{k} |\mu_i| \neq 0$. The system

$$\sum_{j=1}^{n} \frac{\partial \bar{x}_s}{\partial \bar{z}_j} \left(\sum_{i=1}^{n} r_{ji} \bar{z}_i + \bar{Z}_j \right) = -\lambda_s \bar{y}_s + \bar{X}_s, \quad (3.3.19)$$

$$\sum_{j=1}^{n} \frac{\partial \bar{y}_s}{\partial \bar{z}_j} \left(\sum_{i=1}^{n} r_{ji} \bar{z}_i + \bar{Z}_j \right) = \lambda_s \bar{x}_s + \bar{Y}_s, \quad s = 1, \ldots, k,$$

has the solution

$$x_s = u_s(\bar{z}_1, \ldots, \bar{z}_n), \quad y_s = v_s(\bar{z}_1, \ldots, \bar{z}_n),$$

which is holomorphic in a neighborhood of the point $\bar{z}_1 = \ldots = \bar{z}_n = 0$. We make a transformation

$$x_s = \bar{x}_s + u_s, \quad y_s = \bar{y}_s + v_s, \quad s = 1, \ldots, k, \quad (3.3.20)$$

in system (3.3.17):

$$\frac{d\bar{x}_s}{dt} = -\lambda_s \bar{y}_s + \bar{X}_s(\bar{x}_1, \ldots, \bar{x}_k, \bar{y}_1, \ldots, \bar{y}_k, \bar{z}_1, \ldots, \bar{z}_n),$$

$$\frac{d\bar{y}_s}{dt} = \lambda_s \bar{x}_s + \bar{Y}_s, \tag{3.3.21}$$

$$\frac{d\bar{z}_j}{dt} = \sum_{i=1}^{k}(p_{ji}\bar{x}_i + q_{ji}\bar{y}_i) + \sum_{i=1}^{n} r_{ji}\bar{z}_i + \bar{Z}_j.$$

Clearly, the functions $\bar{X}_s, \bar{Y}_s, \bar{Z}_j$ have the same properties as the original functions X_s, Y_s, Z_j and, additionally, $\bar{X}_s = \bar{Y}_s \equiv 0$ when $\bar{x}_s = \bar{y}_s = 0, s = 1, \ldots, k$. We make a replacement in system (3.3.21)

$$\bar{x}_s = r_s \cos \vartheta_s, \quad \bar{y}_s = r_s \sin \vartheta_s, \tag{3.3.22}$$

which results in the system

$$\frac{dr_s}{dt} = R_s, \quad s = 1, \ldots, k,$$

$$\frac{d\vartheta_s}{dt} = \lambda_s + \Theta_s, \tag{3.3.23}$$

$$\frac{d\bar{z}_j}{dt} = \sum_{i=1}^{n} r_{ji}\bar{z}_i + P_j(r_1, \ldots, r_k, \vartheta_1, \ldots, \vartheta_k, \bar{z}_1, \ldots, \bar{z}_n), \quad j = 1, \ldots, n,$$

where

$$R_s = \cos \vartheta_s \bar{X}_s(r_1 \cos \vartheta_1, \ldots, r_k \cos \vartheta_k, r_1 \sin \vartheta_1, \ldots, r_k \sin \vartheta_k, \bar{z}_1, \ldots, \bar{z}_n)$$
$$+ \sin \vartheta_s \bar{Y}_s(r_1 \cos \vartheta_1, \ldots, r_k \cos \vartheta_k, r_1 \sin \vartheta_1, \ldots, r_k \sin \vartheta_k, \bar{z}_1, \ldots, \bar{z}_n),$$

$$\Theta_s = \frac{\cos \vartheta_s \bar{Y}_s - \sin \vartheta_s \bar{X}_s}{r_s},$$

$$P_j = \bar{Z}_j(r_1 \cos \vartheta_1, \ldots, r_k \cos \vartheta_k, r_1 \sin \vartheta_1, \ldots, r_k \sin \vartheta_k, \bar{z}_1, \ldots, \bar{z}_n).$$

Evidently, the functions R_s and P_j are holomorphic in the neighborhood of $r_1 = \ldots = r_k = \bar{z}_1 = \ldots = \bar{z}_n = 0$. In this case $R_s \equiv 0$ for $r_1 = \ldots = r_k = 0$. The coefficients in the expansions of the functions R_s and P_j are periodic functions with respect to $\vartheta_i, i = 1, \ldots, k$. The functions Θ_s are assumed to have the same properties as R_s and P_j. Note that condition (3.3.18) allows system (3.3.21) to be transformed, whereupon the functions Θ_s are formally expanded as series in powers of $r_1, \ldots, r_k, \bar{z}_1, \ldots, \bar{z}_n$. We seek a solution to system (3.3.23) as the series

$$r_s = c_s + \sum_{m=2}^{\infty} r_s^{(m)}(\vartheta_1, \ldots, \vartheta_k, c_1, \ldots, c_k),$$

$$\bar{z}_j = \sum_{m=1}^{\infty} \bar{z}_j^{(m)}(\vartheta_1, \ldots, \vartheta_k, c_1, \ldots, c_k). \tag{3.3.24}$$

The functions $r_s^{(m)}$ and $\bar{z}_j^{(m)}$ are homogeneous forms for $c_1, \ldots, c_k$ to the power m with the periodic coefficients for $\vartheta_1, \ldots, \vartheta_k$ to be defined.

Natural Oscillations of Autonomous Systems

We set up equations to find solutions to a system (3.3.23) in the form of (3.3.24)

$$\sum_{i=1}^{k} \frac{\partial r_s}{\partial \vartheta_i}(\lambda_i - \Theta_i) = R_s, \quad s = 1, \ldots, k, \tag{3.3.25}$$

$$\sum_{i=1}^{k} \frac{\partial \bar{z}_j}{\partial \vartheta_i}(\lambda_i + \Theta_i) = \sum_{i=1}^{n} r_{ji}\bar{z}_i + P_i, \quad j = 1, \ldots, n.$$

Substituting the series (3.3.24) into the system (3.3.25) and equating the forms of the same power with respect to $c_1, \ldots, c_k$, we get the system of equations

$$\sum_{i=1}^{k} \frac{\partial r_s^{(m)}}{\partial \vartheta_i}\lambda_i = R_s^{(m)}, \quad s = 1, \ldots, k, \tag{3.3.26}$$

$$\sum_{i=1}^{k} \lambda_i \frac{\partial \bar{z}_j^{(m)}}{\partial \vartheta_i} = \sum_{i=1}^{n} r_{ji}\bar{z}_i^{(m)} + P_j^{(m)}, \quad j = 1, \ldots, n,$$

where the forms of $R_s^{(m)}$ and $P_j^{(m-1)}$ are determined if the functions $r_s^{(m_1)}$ and $\bar{z}_j^{(m_2)}$, $m_1 < m_2, m_2 < m - 1$, are found. Assume that functions (3.3.25) are defined from system (3.3.26) as periodic in $\vartheta_1, \ldots, \vartheta_k$ and constitute a solution to system (3.3.25) for sufficiently small $|c_s|$. Let the series (3.3.24) be constructed so that $r_s^{(m)} = 0$ for $\vartheta_1 = \ldots = \vartheta_k = 0, m \geq 2$.

In this special case we consider the stability problem for the zero solution of system (3.3.17).

Theorem 45 *If the system (3.3.23) has the set of restricted solutions (3.3.24), then the zero solution of system (3.3.17) is Lyapunov stable.*

Proof: In the system (3.3.23) we make a replacement

$$r_s = \rho_s + \sum_{m=2}^{\infty} r_s^{(m)}(\vartheta_1, \ldots, \vartheta_k, \rho_1, \ldots, \rho_k), \tag{3.3.27}$$

where $\rho_1, \ldots, \rho_k$ are the new desired functions:

$$\frac{d\rho_s}{dt} = \bar{R}_s,$$

$$\frac{d\vartheta_s}{dt} = \lambda_s + \bar{\Theta}_s, \tag{3.3.28}$$

$$\frac{dz_j}{dt} = \sum_{i=1}^{n} r_{ji}z_i + \bar{P}_j.$$

The system (3.3.28) has the solution set

$$\rho_s = c_s, \quad s = 1, \ldots, k, \quad z_j = 0, \quad j = 1, \ldots, n.$$

The system

$$\sum_{j=1}^{n} \frac{\partial \rho_s}{\partial \bar{z}_j}\left(\sum_{i=1}^{n} r_{ji}\bar{z}_i + \bar{P}_j\right) + \sum_{i=1}^{k} \frac{\partial \rho_s}{\partial \vartheta_i}(\bar{\lambda}_i + \bar{\Theta}_i) = R_s \tag{3.3.29}$$

has the solution set

$$\rho_s = c_s + F_s(\bar{z}_1, \ldots, \bar{z}_n, \vartheta_1, \ldots, \vartheta_k, c_1, \ldots, c_k), \quad s = 1, \ldots, k. \quad (3.3.30)$$

The functions F_s are expanded in convergent power series of $\bar{z}_1, \ldots, \bar{z}_n, c_1, \ldots, c_k$ for sufficiently small $|\bar{z}_j|$ and $|c_s|$, $F_s \equiv 0$ for $\bar{z}_1, \ldots, \bar{z}_n = 0$. Using equalities (3.3.30) we eliminate the quantities ρ_s from the third group of equations (3.3.28). Then we get the system to define the functions $\bar{z}_i$ from which it follows that for all sufficiently small $|c_s|$ and for any choice of continuous real functions $\vartheta_1(t), \ldots, \vartheta_k(t)$ we have that $\bar{z}_j(t) \to 0$ as $t \to +\infty$ uniformly in $c_1, \ldots, c_k$ if $|\bar{z}_j(0)|$ are sufficiently small. This and (3.3.30) suggest that $\rho_s \to c_s$ as $t \to +\infty$ and, therefore, the zero solution of system (3.3.17) is stable. ∎

Theorem 46 *For the system (3.3.23) to have the family of constrained solutions (3.3.24), it is necessary and sufficient that there exist k holomorphic integrals of system (3.3.21), as in (3.3.32), possessing the property (3.3.33).*

Proof: *Necessity.* Solve equalities (3.3.30) for $c_1, \ldots, c_k$:

$$c_s = \rho_s + \Phi_s(\vartheta_1, \ldots, \vartheta_k, \rho_1, \ldots, \rho_k, \bar{z}_1, \ldots, \bar{z}_n), \quad s = 1, \ldots, k. \quad (3.3.31)$$

In equalities (3.3.31) we pass to the quantities $r_1, \ldots, r_k$ and, using (3.3.22), we get

$$c_s^2 = \bar{x}_s^2 + \bar{y}_s^2 + \psi_s(\bar{x}_1, \ldots, \bar{x}_k, \bar{y}_1, \ldots, \bar{y}_k, \bar{z}_1, \ldots, \bar{z}_n), \quad s = 1, \ldots, k. \quad (3.3.32)$$

The functions ψ_s are expanded as convergent series in integral positive powers of quantities $\bar{x}_s, \bar{y}_s, \bar{z}_i$. So equalities (3.3.32) give k holomorphic integrals of system (3.3.21), here

$$\sqrt{\bar{x}_s^2 + \bar{y}_s^2 + \psi_s} = r_s + \bar{\psi}_s(\vartheta_1, \ldots, \vartheta_k, r_1, \ldots, r_k, \bar{z}_1, \ldots, \bar{z}_n), \quad (3.3.33)$$

where $\bar{\psi}_s$ are holomorphic in the neighborhood of the point $r_1 = \ldots = r_k = \bar{z}_1 = \ldots = \bar{z}_n = 0$.

Sufficiency is obtained by performing inverse transformations of the integrals in (3.3.33). ∎

Consider in some detail the case where all the functions $r_s^{(m)}$ and $\bar{z}_j^{(m)}$ are periodic. Using the functions (3.3.24), we eliminate variables $r_1, \ldots, r_k, \bar{z}_1, \ldots, \bar{z}_n$ from the second group of equations in (3.3.23). Then, to define the functions $\vartheta_1, \ldots, \vartheta_k$, we obtain the system

$$\frac{d\theta_s}{dt} = \lambda_s + \Phi_s(\vartheta_1, \ldots, \vartheta_k, c_1, \ldots, c_k), \quad s = 1, \ldots, k. \quad (3.3.34)$$

The functions Φ_s on the right-hand side of system (3.3.34) are expanded as series in integral positive powers of $c_1, \ldots, c_k$ containing no free terms in the latter. The coefficients in these series are finite trigonometric polynomials in $\vartheta_1, \ldots, \vartheta_k$. Introduce new independent variables by the formulas $\tau_s = th_s$, where h_s are series of the form

$$h_s = \lambda_s + \sum_{m=1}^{\infty} h_s^{(m)},$$

Natural Oscillations of Autonomous Systems

and $h_s^{(m)}$ are homogeneous forms of the quantities $c_1, \ldots, c_k$ of order m with the coefficients to be defined. The system (3.3.34) has a solution determined by the conditions $\vartheta_s = \vartheta_s^{(0)}, s = 1, \ldots, k$ with $t = 0$. This solution may be represented as series in powers of variables $t, c_1, \ldots, c_k$ and the coefficients of these series are determined as the known trigonometric polynomials in $\vartheta_1^{(0)}, \ldots, \vartheta_k^{(0)}$. Define the coefficients of the forms $h_s^{(m)}$ so that the quantities ϑ_s would be dependent on t only via the variables $\tau_1, \ldots, \tau_k$. We shall show that the functions $h_1, \ldots, h_k$ can always be chosen so that there be the above dependence. Indeed, consider the system of partial differential equations

$$\sum_{l=1}^{k} \frac{\partial \vartheta_s}{\partial \tau_i} h_i = \lambda_s + \Phi_s, \quad s = 1, \ldots, k. \tag{3.3.35}$$

We shall seek a solution to these equations as

$$\vartheta_s = \vartheta_s^{(0)} + \tau_s + T_s, \quad s = 1, \ldots, k. \tag{3.3.36}$$

Substituting functions (3.3.36) into system (3.3.35) we find

$$\sum_{i=1}^{k} \frac{\partial T_s}{\partial \tau_i} h_i = \lambda_s - h_s + \bar{\Phi}_s. \tag{3.3.37}$$

In equations (3.3.37) $\bar{\Phi}_s$ stand for the functions obtained from Φ_s by replacing the quantities with formulas (3.3.36). We shall show that the functions h_s can be chosen in such a manner that the functions T_s are uniquely defined from system (3.3.37) as series for $c_1, \ldots, c_k$ whose coefficients are trigonometric polynomials in $\tau_1, \ldots, \tau_k$. We shall seek a solution to system (3.3.37) as series $T_s = \sum_{m=1}^{\infty} T_s^{(m)}$, where $T_s^{(m)}$ is the form to the power m with respect to $c_1, \ldots, c_k$ with coefficients that are trigonometric polynomials in $\tau_1, \ldots, \tau_k$ and are to be defined. Substitute these series into equation (3.3.37) and, equating the forms on the left and right for $c_1, \ldots, c_k$ to the same power, obtain

$$\sum_{i=1}^{k} \frac{\partial T_s^{(m)}}{\partial \tau_i} \lambda_i = h_s^{(m)} + \bar{\Phi}_s^{(m)}, \quad s = 1, \ldots, k, \tag{3.3.38}$$

where $\bar{\Phi}_s^{(m)}$ is the known form to the power m for $c_1, \ldots, c_k$ whose coefficients are trigonometric polynomials in $\tau_1, \ldots, \tau_k$ if all the forms $T_s^{(m_1)}(m_1 < m)$ are found as mentioned above.

We shall consider system (3.3.38) in some detail. The functions $\bar{\Phi}_s^{(m)}$ are given as

$$\bar{\Phi}_s^{(m)} = \sum_{l_1+\ldots+l_k=m} c_1^{l_1} \ldots c_k^{l_k} R_s^{(m,l_1,\ldots,l_k)}. \tag{3.3.39}$$

The functions $R_s^{(m,l_1,\ldots,l_k)}$ are trigonometric polynomials of the form

$$R_s^{(m,l_1,\ldots,l_k)} = \sum R_s^{(m,l_1,\ldots,l_k,p_1,\ldots,p_k)} e^{i \sum_{j=1}^{k} p_j \tau_j} + R_{s0}^{(m,l_1,\ldots l_k)}. \tag{3.3.40}$$

Summation in the formula (3.3.40) extends to all integers $p_1, \ldots, p_k$ that are simultaneously nonzero and whose sum of absolute values does not exceed some finite limit $N(m)$. We shall seek a solution to the equations of (3.3.38) as

$$T_s^{(m)} = \sum_{l_1+\ldots+l_k=m} c_1^{l_1} \ldots c_k^{l_k}$$

$$\sum_{\sum_{j=1}^{k} |p_j| \leq N(m)} T_s^{(m,l_1,\ldots,l_k,p_1,\ldots,p_k)} e^{i \sum_{j=1}^{k} p_j \tau_j}. \tag{3.3.41}$$

Substituting series (3.3.41) into system (3.3.38), to define the coefficients, we obtain the system

$$i \sum_{j=1}^{k} p_j \lambda_j T_s^{(m,l_1,\ldots,l_k,p_1,\ldots,p_k)} = R_s^{(m,l_1,\ldots,l_k,p_1,\ldots,p_k)}.$$

In view of the rational independence of quantities $\lambda_1, \ldots, \lambda_k$ the coefficients of $T_s^{(m)}$ forms are determined uniquely. Hence, for the existence of a solution to system (3.3.38) as in (3.3.41) it is necessary and sufficient that the forms $h_s^{(m)}$ satisfy the relationship

$$h_s^{(m)} + \sum_{l_1+\ldots+l_k=m} c_1^{l_1} \ldots c_k^{l_k} R_{s0}^{(m,l_1,\ldots,l_k)} = 0.$$

The last relation for $m \geq 1$ allows the functions $h_1^{(m)}, \ldots, h_k^{(m)}$ to be uniquely defined, here $h_s^{(1)} \equiv 0$. So there is a unique system of functions $h_s = \lambda_s + \sum_{m=2}^{\infty} h_s^{(m)}$ such that ϑ_s may be represented as $\vartheta_s = \vartheta_s^{(0)} + \tau_s + T_s$. Using formulas (3.3.36), we eliminate the variables $\vartheta_1, \ldots, \vartheta_k$ from the functions of (3.3.22) and return to the original variables $\bar{x}_s, \bar{y}_s, \bar{z}_j$. Then we have that these functions may be represented as power series of $c_1, \ldots, c_k$ whose coefficients are trigonometric polynomials in $\tau_1, \ldots, \tau_k$. Introduce the notation $b_1 = \frac{h_2}{h_1}, \ldots, b_{k-1} = \frac{h_k}{h_1}$. If the quantities $c_1, \ldots, c_k, \vartheta_1^{(0)}, \ldots, \vartheta_k^{(0)}$ are selected so that $b_1, \ldots, b_{k-1}$ are rational, then the constructed functions $\bar{x}_s, \bar{y}_s, \bar{z}_j$ supply a periodic solution to the original system. If they include an irrational number, then the original system has the almost periodic solution determined by these quantities. Indeed, the functions $\bar{x}_s, \bar{y}_s, \bar{z}_j$ are expanded in power series of $c_1, \ldots, c_k$ whose coefficients are trigonometric polynomials in $\tau_1, \ldots, \tau_k$, where $\tau_s = th_s$. In the first case, the polynomials have one period in common, while in the second case there is no such period; hence the first case corresponds to periodic solutions, and the second to almost periodic solutions if the fall-out of terms with commensurable frequencies is not observed here. The same properties are exhibited by the functions $\bar{x}_s, \bar{y}_s, \bar{z}_j$, which constitute a solution to system (3.3.21), if they are defined by (3.3.22). In these functions we set $t = 0$ and denote by $x_j^{(0)}, y_j^{(0)}, z_j^{(0)}$ the series obtained. These quantities are series in $c_1, \ldots, c_k$ whose coefficients are trigonometric polynomials in $\vartheta_1^{(0)}, \ldots, \vartheta_k^{(0)}$. Using the first two series, we eliminate from the series h_s and $\bar{z}_j$ the quantities $c_1, \ldots, c_k, \vartheta_1^{(0)}, \ldots, \vartheta_k^{(0)}$ and denote by $\tilde{h}_s$ and $\tilde{z}_j$ the newly obtained functions. These functions are expanded in power series of $x_1^{(0)}, \ldots, x_k^{(0)}, y_1^{(0)}, \ldots, y_k^{(0)}$.

Natural Oscillations of Autonomous Systems

In this case, the functions $\tilde{z}_j$ constitute a solution to system of partial differential equations

$$\sum_{i=1}^{k} \frac{\partial \tilde{z}_j}{\partial \bar{x}_i}(-\lambda_i \bar{y}_i + \bar{X}_i) + \sum_{i=1}^{k} \frac{\partial \tilde{z}_j}{\partial \bar{y}_i}(\lambda_i \bar{x}_i + \bar{Y}_i) = \sum_{i=1}^{n} r_{ji} \bar{z}_i + \bar{Z}_j, \quad j = 1, \ldots, n. \quad (3.3.42)$$

Substitute the functions $\tilde{z}_j$ that are a solution to system (3.3.42) in the first two groups of equations (3.3.21) instead of $\bar{z}_1, \ldots, \bar{z}_n$. Then we get the system of equations

$$\frac{d\bar{x}_s}{dt} = -\lambda_s \bar{y}_s + \tilde{X}_s(\bar{x}_1, \ldots \bar{x}_k, \bar{y}_1, \ldots, \bar{y}_k), \quad (3.3.43)$$

$$\frac{d\bar{y}_s}{dt} = \lambda_s \bar{x}_s + \tilde{Y}_s(\bar{x}_1, \ldots \bar{x}_k, \bar{y}_1, \ldots \bar{y}_k), \quad s = 1, \ldots, k.$$

The functions $\tilde{h}_s, s = 1, \ldots, k$ are integrals of this system.

In summary, only those integral curves of system (3.3.21) enter into a set of periodic and almost periodic solutions for which the initial data of the functions $\tilde{z}_j$ are the functions $x_1^{(0)}, \ldots, x_k^{(0)}, y_1^{(0)}, \ldots, y_k^{(0)}$ defined by system (3.3.42).

Consider the $(k-1)$-dimensional space and isolate the following two sets therein: A is the set of points $(a_1, \ldots, a_{k-1})$ such that all coordinates are rational, and C is the set of points $(c_1, \ldots, c_{k-1})$ such that their coordinates include the irrational one. When the quantities $c_1, \ldots, c_k, \vartheta_1^{(0)}, \ldots, \vartheta_k^{(0)}$ are cancelled out, the functions $b_1, \ldots, b_{k-1}$ are represented as series for $x_1^{(0)}, \ldots, x_k^{(0)}, y_1^{(0)}, \ldots, y_k^{(0)}$. Denote by $F(a_1, \ldots, a_{k-1})$ the point set for the sphere of radius r

$$\sum_{i=1}^{k} x_i^2 + \sum_{i=1}^{k} y_i^2 \leq r^2 \quad (3.3.44)$$

such that $b_1 = a_1, \ldots, b_{k-1} = a_{k-1}$. The set $F(a_1, \ldots, a_{k-1})$ is closed. Denote by F the union of such closed sets for all points of the set A. Since the set A is countable, the set F is the set of class F_σ [Kantorovič and Akilov (1977)]. Denote by G the set which permits F to complete the closed sphere (3.3.44). We show that the set G is the set of class G_δ [Kantorovič and Akilov (1977)]. Indeed, consider the open sphere of radius $r_1 > r$ and the system of open sets which is an intersection of this sphere and the sets $F(a_1, \ldots, a_{k-1})$. The set G may then be obtained as an intersection of the countable number of these open sets and the closed sphere (3.3.44) and, therefore, is the type G_δ set [Kantorovič and Akilov (1977)].

Theorem 47 *If system (3.3.23) has the constrained solution set (3.3.24), then any integral curve of system (3.3.21) determined by the initial data $x_1^{(0)}, \ldots, x_k^{(0)}$, $y_1^{(0)}, \ldots, y_k^{(0)} \in F$ and $z_j^{(0)} = \tilde{z}_j(x_1^{(0)}, \ldots, x_k^{(0)}, y_1^{(0)}, \ldots, y_k^{(0)})$, $j = 1, \ldots, n$, is periodic, and any integral curve determined by the initial data $(x_1^{(0)}, \ldots, x_k^{(0)}, y_1^{(0)}, \ldots, y_k^{(0)}) \in G$, $z_j^{(0)} = \tilde{z}_j(x_1^{(0)}, \ldots, x_k^{(0)}, y_1^{(0)}, \ldots, y_k^{(0)})$, $j = 1, \ldots, n$, is almost periodic. The set F has no interior points, and the set G may also have no interior points. In this case, the set G a fortiori has no interior points for $k = 2$ and $b_1 \neq \lambda_2/\lambda_1$.*

Proof: As noted above, the periodicity or almost periodicity of integral curves of system (3.3.21) is dependent on the values of integrals $\tilde{h}_s, s = 1, \ldots, k$, of system (3.3.34). The integral curve is periodic if the quantities $b_1, \ldots, b_{k-1}$ are rational, or almost periodic if they include the irrational one. This condition, in turn, is related to the belonging of initial data to the set F or G, which proves the validity of the first part of Theorem 47.

We show that the set F has no interior points. Assume that there is the point a which is an interior point of the set F. Then there exists a sphere of a sufficiently small radius which is centered at this point and is lying entirely in the set F. At least one of the functions $b_1, \ldots, b_{k-1}$ must assume no less than two values (otherwise all of these functions will be constant), hence $b_i = \lambda_i/\lambda_1$, which is not possible, since these quantities include the irrational ones. Then, by the Weierstrass theorem, the function assuming two different rational values will assume irrational values, which contradicts the above assumption. Thus, F has no interior points. In the case of $k = 2$, it is proved in the same manner that the set G cannot have interior points except the case when $b_1 = \lambda_2/\lambda_1$. ∎

Example. Let the following system be given

$$\dot{x}_1 = -\lambda_1 y_1 - y_1(x_1^2 + y_1^2 + x_2^2 + y_2^2),$$

$$\dot{y}_1 = \lambda_1 x_1 + x_1(x_1^2 + y_1^2 + x_2^2 + y_2^2),$$

$$\dot{x}_2 = -\lambda_2 y_2 - y_2(x_1^2 + y_1^2),$$

$$\dot{y}_2 = \lambda_2 x_2 + x_2(x_1^2 + y_1^2),$$

$$\dot{z} = -z + x_1(x_2^2 + y_2^2).$$

This system has the constrained solution set

$$x_1 = x_1^0 \cos \tau_1 - y_1^0 \sin \tau_1, \quad y_1 = x_1^0 \sin \tau_1 + y_1^0 \cos \tau_1;$$

$$x_2 = x_2^0 \cos \tau_2 - y_2^0 \sin \tau_2, \quad y_2 = x_2^0 \sin \tau_2 + y_2^0 \cos \tau_2;$$

$$z = \frac{y_1^0(x_1^{0^2} + y_1^{0^2})(\lambda_1 + x_1^{0^2} + y_1^{0^2} + y_2^{0^2} + x_2^{0^2}) + x_1^0(x_1^{0^2} + y_1^{0^2})}{(1 + (\lambda_1 + x_1^{0^2} + y_1^{0^2} + x_2^{0^2} + y_2^{0^2})^2)} \cos \tau_1$$

$$+ \frac{x_1^0(x_1^{0^2} + y_1^{0^2})(\lambda_1 + x_1^{0^2} + y_1^{0^2} + x_2^{0^2} + y_2^{0^2}) + y_1^0(x_1^{0^2} + y_1^{0^2})}{(1 + (\lambda_1 + x_1^{0^2} + y_1^{0^2} + y_2^{0^2} + x_2^{0^2})^2)} \sin \tau_1,$$

where

$$\tau_1 = t(\lambda_1 + x_1^{0^2} + y_1^{0^2} + x_2^{0^2} + y_2^{0^2}), \quad \tau_2 = t(\lambda_2 + x_1^{0^2} + y_1^{0^2})^2.$$

Evidently, the quantities τ_1/t and τ_2/t are the integrals of the first four equations. The solution is periodic at the points where the function

$$b_1 = \frac{\lambda_2 + x_1^{0^2} + y_1^{0^2}}{\lambda_1 + x_1^{0^2} + y_1^{0^2} + x_2^{0^2} + y_2^{0^2}}$$

Forced Periodic and Almost Periodic Oscillations

assumes rational value; if, however, the function assumes the irrational value, then the solution is almost periodic. The four-dimensional space of variables (x_1, y_1, x_2, y_2) is partitioned into two sets F and G, each having no interior points. This testifies the presence of the almost periodic solution in arbitrarily small neighborhood of a point corresponding to the periodic solution and, conversely, periodic solutions start in an arbitrarily small neighborhood of the point corresponding to the almost periodic solution.

Remark. Examples may be provided for the systems where the functions $h_1, \ldots, h_k$ are dependent, and even where only one integral is involved in formation of the series describing the constrained family of solutions.

3.4 Forced Periodic and Almost Periodic Oscillations

1. As shown in section 1.2, in the stationary linear system, whose proper oscillations are damped, the external perturbing periodic or almost periodic force gives rise to a unique periodic or, respectively, almost periodic forced oscillation which is global-asymptotically stable. In other words, if there is given the nonhomogeneous linear system

$$\dot{X} = \mathbf{A}X + Q(t), \qquad (3.4.1)$$

where $\mathbf{A}$ is the real constant matrix whose eigenvalues have negative real parts, and $Q(t)$ the continuous periodic or almost periodic vector function, then system (3.4.1) has a unique periodic and, respectively, an almost periodic solution

$$X(t) = \int_{-\infty}^{t} e^{\mathbf{A}(t-\tau)} Q(\tau) d\tau. \qquad (3.4.2)$$

In this case, the solution (3.4.2) of system (3.4.1) is global-asymptotically stable.

The problem arises as to the conditions under which the nonlinear system has the same property as system (3.4.1). We first consider a system of n differential equations of the general form

$$\dot{X} = F(t, X), \qquad (3.4.3)$$

the right-hand sides of which are given for all $X \in E_n, t \in (-\infty, +\infty)$ and are real and continuous in a collection of all their variables. We shall assume that the vector-valued function $F(t, X)$ with any fixed X is an almost periodic function of t. The almost periodic vector function $F(t, X)$ is called almost periodic uniformly in any finite region $\|X\| \leq r$ if for any $\epsilon > 0$ and $r > 0$ it is possible to choose a number $L(r, \epsilon)$ such that in any interval $[\alpha, \alpha + L(r, \epsilon)]$ of the length $L(r, \epsilon)$ there is at least one number τ satisfying the inequality

$$\|F(t+\tau, X) - F(t, X)\| < \epsilon, \quad t \in (-\infty, +\infty), \quad \|X\| \leq r. \qquad (3.4.4)$$

Theorem 48 *For the vector function $F(t, X)$ to be the almost periodic function of t uniformly in X in any finite region $\|X\| \leq r$, it is necessary and sufficient* that $F(t, X)$ is the almost periodic function for any fixed $X \in E_n$ and the uniformly continuous function of X over any finite range of changes in X for $t \in (-\infty, +\infty)$.

Proof: We shall first show that the almost periodicity of the function $F(t, X)$ in t, which is uniform in any finite region $\|X\| \leq r$, implies its uniform continuity in X in any bounded region of changes in X for $t \in (-\infty + \infty)$. Indeed, let the bounded region G be given. Take r so large that $\|X\| \leq r$ for all $X \in G$. Choose $\epsilon > 0$ and determine $L(r, \epsilon)$ for it and for a number r. The vector-valued function $F(t, X)$ is uniformly continuous in $t \in [0, L], \|X\| \leq r$, hence for the given ϵ it is possible to select a number δ such that

$$\|F(t, X) - F(t, Y)\| < \epsilon \quad \text{for} \quad \|X - Y\| < \delta, \quad t \in [0, L], \quad \|X\| \leq r, \quad \|Y\| \leq r.$$

Now let t be any real number. Choose a number τ in the interval $[-t, -t + L]$, then $0 \leq t + \tau \leq L$. From this we get

$$\|F(t, X) - F(t, Y)\| \leq \|F(t, X) - F(t + \tau, X)\|$$
$$+ \|F(t, Y) - F(t + \tau, Y)\| + \|F(t + \tau, X) - F(t + \tau, Y)\| \leq 3\epsilon,$$
$$t \in (-\infty, +\infty), \quad \|X - Y\| \leq \delta, \quad \|X\| \leq r, \quad \|Y\| \leq r.$$

The necessity condition has thus been established.

We shall show that it is also sufficient. Let the region $\|X\| \leq r$ be given. Take $\epsilon > 0$ and choose $\delta > 0$ so that

$$\|F(t, X) - F(t, Y)\| \leq \epsilon \quad \text{for} \quad \|X - Y\| < \delta \ \ t \in (-\infty, +\infty), \quad \|X\| \leq r, \quad \|Y\| \leq r.$$

Cover the ball $\|X\| \leq r$ with the balls of radius 2δ taken in a finite number. Let $X_1, \ldots, X_k$ be the centers of these balls. The vector-valued function $F(t, X_j)$ is almost periodic in t for any $j = 1, \ldots, k$, hence for the given $\epsilon > 0$ it is possible to select $L(r, \epsilon)$ such that in every interval $[\alpha, \alpha + L]$ of length L it is possible to select a number τ which is the common ϵ-almost period of these functions for which the following inequalities hold

$$\|F(t + \tau, X_j) - F(t, X_j)\| \leq \epsilon, \quad t \in (-\infty, +\infty), \quad j = 1, \ldots, k.$$

Estimate the difference

$$\|F(t + \tau, X) - F(t, X)\| \leq \|F(t + \tau, X) - F(t + \tau, X_j)\|$$
$$+ \|F(t, X) - F(t, X_j)\| + \|F(t + \tau, X_j) - F(t, X_j)\| \leq 3\epsilon,$$

where X_j is the center of the radius 2δ ball, inside or on the boundary of which is located the point X. The last inequality holds for all $t \in (-\infty, +\infty), \|X\| \leq r$. Consequently, $F(t, X)$ is the almost periodic function of t uniformly in any finite region $\|X\| \leq r$. ∎

In what follows it is assumed that the right-hand side of system (3.4.3) is the uniformly continuous function of X in any finite region for $t \in (-\infty, +\infty)$. In

Forced Periodic and Almost Periodic Oscillations

other words, if $F(t, X)$ is regarded as a family of functions defined by parameter $t \in (-\infty, +\infty)$, then this family is equicontinuous in X over any bounded range. For short in formulation of theorems, we introduce the concept of convergence.

Definition 34. System (3.4.3) has the convergence property, if it has a unique almost periodic solution $X = \Phi(t)$ such that for every $\epsilon > 0$ it is possible to select $\delta > 0$ such that for $\|X^{(0)} - \Phi(t_0)\| < \delta$ we get $\|X(t, X^{(0)}, t_0) - \Phi(t)\| < \epsilon$ for all $t \geq t_0$ and, additionally $\|X(t, X^{(0)}, t_0) - \Phi(t)\| \to 0$ as $t - t_0 \to +\infty$ uniformly in $t_0 \in (-\infty, +\infty)$ in any finite region $\|X^{(0)}\| \leq r$. Here we denote by $X(t, X^{(0)}, t_0)$ the integral curve of system (3.4.3) passing through the point $(X^{(0)}, t_0)$. In what follows, we shall distinguish the periodic and almost periodic convergence. The former occurs where the right-hand side of (3.4.3) is the periodic function of period ω, with $\Phi(t)$ being also the periodic function of the same period. In other cases, convergence will be called almost periodic. Henceforth these concepts will be employed only when some particular case is to be identified. The system (3.4.1) has a unique almost periodic solution (3.4.2) which is global uniform-asymptotically stable. Therefore, system (3.4.1), as defined, has the convergence property. We shall find the conditions that are necessary and sufficient for system (3.4.3) to have the convergence property.

Theorem 49 *If the right-hand sides of system (3.4.3) satisfy the Lipschitz condition for X over any bounded range of X with the t-independent constant, then for system (3.4.3) to have the convergence property it is necessary and sufficient that the following conditions hold:*

1) any integral curve of system (3.4.3) $X(t, X^{(0)}, t_0)$ is bounded for all $t \geq t_0$;

2) for any $r > 0$ and every $\epsilon > 0$ it is possible to choose $\delta(r, \epsilon)$ such that for $\|X^{(0)} - Y^{(0)}\| < \delta(r, \epsilon)$ we get $\|X(t, X^{(0)}, t_0) - X(t, Y^{(0)}, t_0)\| < \epsilon$ for all $t \geq t_0$, and also $\|X(t, X^{(0)}, t_0) - X(t, Y^{(0)}, t_0)\| \to 0$ as $t - t_0 \to +\infty$ uniformly in $t_0 \in (-\infty, +\infty)$ and $\|X^{(0)}\| \leq r$, $\|Y^{(0)}\| \leq r$;

3) for each point $(X^{(0)}, t_0)$ and each $\epsilon > 0$ it is possible to choose two numbers L and T such that in each interval $[\alpha, \alpha, +L]$ there is a number τ satisfying the inequality $\|X(t+\tau, X^{(0)}, t_0) - X(t, X^{(0)}, t_0)\| < \epsilon'$ for all $t \geq t_0 + T$ and $t + \tau \geq t_0 + T$, where τ is an ϵ'-almost period of the function $F(t, X)$ in the region G containing considered integral curve $X(t, X^{(0)}, t_0)$, for all $t \geq t_0$. Note that $\epsilon' \to +0$ as $\epsilon \to +0$ and vice versa.

Proof: *Necessity.* If system (3.4.3) has the convergence property, then it has a unique almost periodic solution $X = \Phi(t)$. It is well known that any almost periodic function is bounded. Consequently, the vector-valued function $\Phi(t)$ satisfies the inequality

$$\|\Phi(t)\| \leq M < \infty, \quad t \in (-\infty, +\infty).$$

Then for any solution $X = X(t, X^{(0)}, t_0)$ with $t \geq t_0$ we have

$$\|X\| \leq \|\Phi\| + \|X - \Phi\|. \tag{3.4.5}$$

Since $\|X - \Phi\| \to 0$ as $t - t_0 \to +\infty$, there is a number $T > 0$ such that $\|\Phi - X\| \leq 1$ for all $t \geq T + t_0$. The function $\|\Phi - X\|$ is continuous in $t \in [t_0, t_0 + T]$ and, therefore, is bounded by the number M_1. Let $\bar{M} = \max\{1, M_1\}$, then from inequality (3.4.5) we obtain $\|X(t, X^{(0)}, t_0)\| \leq M + \bar{M}$ for all $t \geq t_0$. Thus, condition 1) holds.

We shall show that condition 2) holds. Since $\|\Phi(t) - X(t, X^{(0)}, t_0)\| \to 0$ as $t - t_0 \to +\infty$ uniformly in $t_0 \in (-\infty, +\infty)$, $\|X^{(0)}\| \leq r$, for $\epsilon/2$ there are numbers T and δ_1 such that for $\|\bar{X}^{(0)} - \Phi(\bar{t}_0)\| < \delta_1$, there will be $\|\Phi(t) - X(t, \bar{X}^{(0)}, \bar{t}_0)\| < \epsilon/2$ for any $t \geq \bar{t}_0$, $\bar{t}_0 \in (-\infty, +\infty)$ and $\|X(t_0 + T, X^{(0)}, t_0) - \Phi(t_0 + T)\| < \delta_1$; therefore,

$$\|X(t, X^{(0)}, t_0) - \Phi(t)\| < \frac{1}{2}\epsilon \quad \text{for all} \quad t \geq T + t_0.$$

Consider the family of functions $X(t_0 + t, X^{(0)}, t_0)$, $t \in [0, T]$. This family is equicontinuous in $t_0 \in (-\infty, +\infty)$ with $\|X^{(0)}\| \leq r$ for $t \in [0, T]$. Indeed, set

$$Z = X(t + t_0, X^{(0)}, t_0) - X(t + t_0, Y^{(0)}, t_0).$$

The vector function Z then satisfies the equation

$$\dot{Z} = F(t + t_0, Z + X(t + t_0, Y^{(0)}, t_0)) - F(t + t_0, X(t + t_0, Y^{(0)}, t_0)) \quad (3.4.6)$$

and the initial condition $Z = Z^{(0)} = X^{(0)} - Y^{(0)}$ for $t = 0$. From system (3.4.6) we get

$$\|Z\| \leq \|Z^{(0)}\| e^{n\lambda T},$$

where λ is the Lipschitz constant for the components of vector F (see section 1.1). Choose a number δ so that $\delta e^{\lambda n T} < \epsilon$, then for $\|X^{(0)} - Y^{(0)}\| < \delta$, $\|X^{(0)}\| \leq r$, $\|Y^{(0)}\| \leq r$ we get

$$\|X(t, X^{(0)}, t_0) - X(t, Y^{(0)}, t_0)\| < \epsilon \quad (3.4.7)$$

for all $t \in [t_0, t_0 + T]$; with $t \geq t_0 + T$ we have

$$\|X(t, X^{(0)}, t_0) - X(t, Y^{(0)}, t_0)\|$$

$$\leq \|\Phi(t) - X(t, X^{(0)}, t_0)\| + \|\Phi(t) - X(t, Y^{(0)}, t_0)\| \leq \epsilon \quad (3.4.8)$$

for all $\|X^{(0)}\| \leq r$, $\|Y^{(0)}\| \leq r$, $t_0 \in (-\infty, +\infty)$. It follows from the inequalities (3.4.7) and (3.4.8) that

$$\|X(t, X^{(0)}, t_0) - X(t, Y^{(0)}, t_0)\| \leq \epsilon \quad \text{for} \quad t \geq t_0,$$

$$t_0 \in (-\infty, +\infty), \quad \|X^{(0)}\| \leq r, \quad \|Y^{(0)}\| \leq r, \quad \|X^{(0)} - Y^{(0)}\| < \delta.$$

Moreover, (3.4.8) implies that

$$\|X(t, X^{(0)}, t_0) - X(t, Y^{(0)}, t_0)\| \to 0 \quad \text{for} \quad t - t_0 \to +\infty, \quad \|X^{(0)}\| \leq r, \quad \|Y^{(0)}\| \leq r.$$

Now we shall show that condition 3) is also satisfied. Take any point $(X^{(0)}, t_0)$ and construct the integral curve $X(t, X^{(0)}, t_0)$. Let $\epsilon > 0$, for $\epsilon/2$ there is a number T such that

$$\|X(t, X^{(0)}, t_0) - \Phi(t)\| < \epsilon/2 \quad \text{for all} \quad t \geq T + t_0.$$

From the almost periodicity condition for the vector-valued function $\Phi(t)$ we choose for $\epsilon/2$ a number L such that in each interval $[\alpha, \alpha + L]$ there is τ such that $\|\Phi(t+\tau) - \Phi(t)\| < \epsilon/2$. For such numbers τ we get

$$\|X(t+\tau, X^{(0)}, t_0) - X(t, X^{(0)}, t_0)\| \leq \|X(t+\tau, X^{(0)}, t_0) - \Phi(t+\tau)\|$$

Forced Periodic and Almost Periodic Oscillations

$+\|X(t, X^{(0)}, t_0) - \Phi(t)\| + \|\Phi(t+\tau) - \Phi(t)\| < \epsilon$ for all $t \geq t_0 + T$, $t + \tau \geq t_0 + T$.

Sufficiency. We first show that under the conditions of Theorem 49 system (3.4.3) has an almost periodic solution $X = \Phi(t)$. Take the sequence of positive values $\epsilon_k \to +0$ as $k \to \infty$ and the point $X^{(0)}$. Construct the integral curve passing through it for $t = t_0 = 0$. By condition 3), it is then possible to select sequences T_k and L_k such that in each interval $[\alpha, \alpha + L_k]$ there is a number τ_k satisfying the inequality

$$\|X(t + \tau_k, X^{(0)}, 0) - X(t, X^{(0)}, 0)\| \leq \epsilon_k \quad \text{for all} \quad t \geq T_k, \quad t + \tau_k \geq T_k.$$

The sequence τ_k may be chosen so that $\tau_k \geq T_k + \Theta_k$, where $\Theta_k \to +\infty$ and $X_k^{(0)} = X(\tau_k, X^{(0)}, 0) \to \Phi_0$ as $k \to \infty$. The latter is possible because the integral curve $X(t, X^{(0)}, 0)$ is bounded for $t \geq 0$ and, therefore, all points $X_k^{(0)}$ are in the bounded region. Thus, from the sequence $X_k^{(0)}$ it is possible to choose a convergent sequence which is also denoted by $X_k^{(0)}$. We shall show that the solution of system (3.4.3), $X = \Phi(t)$, passing through the point Φ_0 for $t = 0$, is almost periodic. To do this, consider the functions $X_k = X(t + \tau_k, X^{(0)}, 0)$. Clearly, they satisfy the system of equations

$$\dot{X} = F_k(t, X) = F(t + \tau_k, X)$$

and the initial condition $X = X_k^{(0)}$ for $t = 0$. Since τ_k is an ϵ'_k-almost period of the function $F(t, X)$ in some finite region G, the relationship $F_k(t, X) \to F(t, X)$ as $k \to \infty$ holds in this region uniformly in $t \in (-\infty, +\infty)$. From the choice of the sequence $X_k^{(0)}$ we have $X_k^{(0)} \xrightarrow[k \to \infty]{} \Phi_0$. We show then that for any choice of the quantities $T > 0$ and $\epsilon > 0$ it is possible to select a number $K(T, \epsilon)$ such that the relationship $\|X_k - \Phi\| \leq \epsilon$ for $t \in [-T, T]$ holds for $k > K$. Indeed, the function $Z = X_k - \Phi$ satisfies the equation

$$\dot{Z} = F_k(t, Z + \Phi) - F(t, \Phi) \tag{3.4.9}$$

and the initial condition $Z = X_k^{(0)} - \Phi_0$ for $t = 0$. We rewrite equation (3.4.9) as

$$\dot{Z} = F(t, Z + \Phi) - F(t, \Phi) + R,$$

where $R = F_k(t, Z + \Phi) - F(t, Z + \Phi)$. Hence we have that

$$\|Z\| \leq \delta_1 e^{\lambda n T} + \frac{\delta_2}{\lambda n}, \tag{3.4.10}$$

where, as above, λ is the Lipschitz constant for the components of vector F in the region G,

$$\delta_1 = \|X_k^{(0)} - \Phi_0\|, \quad \delta_2 = \sup_{\substack{x \in G \\ t \in (-\infty, +\infty) \\ X = Z + \Phi}} \|F_k(t, Z + \Phi) - F(t, Z + \Phi)\|,$$

G is a fixed bounded region where the initially chosen integral curve $X = X(t, X^{(0)}, 0)$ for all $t \geq 0$ is located. It follows from inequality (3.4.10) that the number $K(T, \epsilon)$

exists, since the numbers δ_1 and δ_2 are arbitrarily small for a sufficiently large k. Choose $\bar{\epsilon} > 0$. Then, by condition 3), its corresponding numbers $\bar{L}$ and $\bar{T}$ are found to be such that in any interval $[\alpha, \alpha + \bar{L}]$ it is possible to select τ such that

$$\|X(t+\tau, X^{(0)}, 0) - X(t, X^{(0)}, 0)\| < \bar{\epsilon} \quad \text{with all} \quad t + \tau \geq \bar{T}, \quad t \geq \bar{T}.$$

Estimate the magnitude

$$\|\Phi(t+\tau) - \Phi(t)\| \leq \|X_k(t+\tau) - \Phi(t+\tau)\|$$

$$+ \|\Phi(t) - X_k(t)\| + \|X_k(t+\tau) - X_k(t)\|. \tag{3.4.11}$$

The third summand in inequality (3.4.11) may be rewritten as

$$\|X_k(t+\tau) - X_k(t)\| = \|X(t+\tau+\tau_k, X^{(0)}, 0) - X(t+\tau_k, X^{(0)}, 0)\|.$$

Hence it is possible to choose a number $\bar{k}$, so that

$$\|X_k(t+\tau) - X_k(t)\| < \bar{\epsilon} \quad \text{for} \quad t \in [-T, T], \quad |\tau| \leq L,$$

where L is any sufficiently large number. Indeed, the last inequality holds for $t + \tau + \tau_k \geq \bar{T}$ and $t + \tau_k \geq \bar{T}$. Therefore, it suffices to choose $\bar{k}$ such that $\Theta_k - T - L > \bar{T}, k \geq \bar{k}$. If $t + \tau$ and $t \in [-T, T]$, then it is possible to select $K(\bar{\epsilon}, T)$ such that each of the two first summands on the right-hand side of (3.4.11) does not exceed $\bar{\epsilon}$ for $k \geq K(\bar{\epsilon}, T)$.

Let $K(\bar{\epsilon}, T) = \max\{K(\bar{\epsilon}, T), \bar{k}\}$. Then for $k \geq k_0$ from (3.4.11) we obtain the inequality $\|\Phi(t+\tau) - \Phi(t)\| \leq 3\bar{\epsilon}$ which holds for $t \in [-T, T], t + \tau \in [-T, T]$. Since both sides of the last inequality are k-independent, we have that it holds for all $t \in (-\infty, +\infty)$ and for any τ chosen above. Thus, the solution $X = \Phi(t)$ of system (3.4.3) is almost periodic. If in condition 2) we now take $X = \Phi(t)$ as one of the solutions, then this solution has the stability property, viz., for every $\epsilon > 0$ there is $\delta > 0$ such that for $\|X^{(0)} - \Phi(t_0)\| < \delta$ we get $\|X(t, X^{(0)}, t_0) - \Phi(t)\| < \epsilon$ for all $t \geq t_0$. Moreover, $\|X - \Phi\| \to 0$ as $t - t_0 \to +\infty$ uniformly in $t_0 \in (-\infty, +\infty), \|X^{(0)}\| \leq r$, where $r > 0$ is an arbitrary but fixed number. Hence it follows that $\Phi(t)$ is the unique almost periodic solution of system (3.4.3) and, therefore, this system has the convergence property. ∎

Remark. We show that in the periodic convergence case, condition 3) of Theorem 49 is unnecessary. Indeed, suppose that the right-hand side of system (3.4.3) is the t-periodic vector function with the period 2π, and the system involved has the periodic convergence property. Then there exists a 2π-periodic solution of this system $\Phi(t)$. The first two conditions of Theorem 49 are satisfied since the necessity of these conditions is proved in the manner described above. We shall show that these conditions are also sufficient. Indeed, take the solution $X = X(t, X^{(0)}, 0)$. Then any vector-valued function $X_k = X(t + 2k\pi, X^{(0)}, 0)$ is also a solution to system (3.4.3) with initial condition $X_k^{(0)} = X(2k\pi, X^{(0)}, 0)$ at $t = 0$. Choose a subsequence n_k such that $X_{n_k}^{(0)} \to \Phi_0$, and show that the integral curve $X = \Phi(t)$ of system

Forced Periodic and Almost Periodic Oscillations 181

(3.4.3) passing through the point Φ_0 at $t = 0$ is a 2π-periodic solution of this system. Since $X(t + 2\pi, X^{(0)}, 0)$ is a solution to system (3.4.3), the condition 2) implies the relationship

$$\|X(t + 2\pi, X^{(0)}, 0) - X(t, X^{(0)}, 0)\| \to 0 \quad \text{as} \quad t \to +\infty. \tag{3.4.12}$$

Consider the integral curves $X(t, X_{n_k}^{(0)}, 0)$ having the property $X \to \Phi_0$ as $n_k \to \infty, t = 0$. Also, we shall show that $X \to \Phi_0$ as $n_k \to \infty, t = 2\pi$. If we take the relationship (3.4.2) for $t = 2\pi n_k$, then

$$\|X(2\pi, X_{n_k}^{(0)}, 0) - X_{n_k}^{(0)}\| \to 0 \quad \text{as} \quad n_k \to \infty;$$

hence

$$\lim_{n_k \to \infty} X(2\pi, X_{n_k}^{(0)}, 0) = \Phi_0.$$

Next, from the theorem of continuity in initial data it follows that $\Phi(0) = \Phi(2\pi) = \Phi_0$. Hence, $\Phi(t)$ is a periodic solution of system (3.4.3). ∎

2. Theorem 49 is qualitative in its nature. Below are given sufficient conditions for system (3.4.3) to possess the convergence property. These conditions are sufficient for the conditions of Theorem 49 to be satisfied, and the entire investigation is carried out on this basis.

Definition 35. The function $V(X, Y, Z, t)$ given for all $X, Y, Z \in E_n, t \in (-\infty, +\infty)$, is real and continuous, and is called positive-definite (negative-definite) in the region $\|Z\| \leq \rho$ if for every $r > 0$ there is a function $V_r(Z)$ real continuous and such that $V_r(Z) > 0$ for $Z \neq 0, V_r(0) = 0, V(X, Y, Z, t) \geq V_r(Z)$ ($V(X, Y, Z, t) \leq -V_r(Z)$) for $\|Z\| \leq \rho, \|X\| \leq r, \|Y\| \leq r, t \in (-\infty, +\infty)$ and $V(X, Y, 0, t) \equiv 0$.

Definition 36. The continuous real function $V(X, Y, Z, t)$ given for all $X, Y, Z \in E_n, t \in (-\infty, +\infty)$, admits the infinitesimal upper limit in the region $\|Z\| \leq \rho$ if for $r > 0$ there is a function $U_r(Z)$ given for $\|Z\| \leq \rho$, which is real, continuous, and is such that $U_r(Z) > 0$ for $Z \neq 0, U_r(0) = 0$,

$$|V(X, Y, Z, t)| \leq U_r(Z) \quad \text{for} \quad \|X\| \leq r, \quad \|Y\| \leq r, \quad \|Z\| \leq \rho, \quad t \in (-\infty, +\infty).$$

Theorem 50 *Suppose the following assumptions hold:*

1. There exists the function $V_1(X, t)$ satisfying the conditions: a) V_1 is given for all $t \in (-\infty, +\infty), X \in E_n$, and is real and continuous; b) $V_1(X, t) \to +\infty$ as $\|X\| \to +\infty$ uniformly in $t \in (-\infty, +\infty)$; c) there exists the continuous total derivative of V_1 by system (3.4.3) which is nonpositive for $\|X\| \geq r_1, t \in (-\infty, +\infty)$; d) the function V_1 is uniformly bounded from above for all $t \in (-\infty, +\infty)$ in any finite region $r_1 \leq \|X\| \geq r$.

2. There exists the function $V(X, Y, Z, t)$ satisfying the conditions: a) the function $V(X, Y, Z, t)$ is positive-definite and admits the infinitesimal upper limit in the region $\|Z\| \leq \rho$; b) the function $V(X, Y, Z, t)$ continuously differentiable with respect to the vector components X, Y, Z, t and its partial derivatives are uniformly bounded in any finite region of variation in the variables X, Y, Z for all $t \in (-\infty, +\infty)$; c) the function

$$W = \frac{\partial V(X, Y, Z, t)}{\partial t} + \sum_{j=1}^{n} \frac{\partial V}{\partial x_j} f_j(t, X) + \sum_{j=1}^{n} \frac{\partial V}{\partial y_j} f_j(t, Y)$$

$$+\sum_{j=1}^{n}\frac{\partial V}{\partial z_j}(f_j(t,Z+Y)-f_j(t,Y))$$

is negative-definite in any finite region $\|Z\| \leq \rho$. Then the system (3.4.3) has the convergence property.

Proof: The proof involves successive verifications of the conditions to Theorem 49. We shall first show that condition 1) of Theorem 49 is met. Take the number $r > r_1 > 0$ and denote

$$m = \sup_{\substack{r_1 \leq \|X\| \leq r \\ t \in (-\infty, +\infty)}} V_1(X,t).$$

Let $(X^{(0)}, t_0)$ be a point such that $\|X^{(0)}\| \in [r_1, r]$. Then along the integral curve $X = X(t, X^{(0)}, t_0)$ we get $\frac{dV_1}{dt} \leq 0$, whence

$$V_1(X(t, X^{(0)}, t_0), t) \leq V_1(X^{(0)}, t_0) \leq m. \qquad (3.4.13)$$

The inequality (3.4.13) holds during the entire time of motion for $t \geq t_0$ as long as the integral curve $X(t, X^{(0)}, t_0)$ stays outside the region $\|X\| \leq r_1$. It follows from property b) of the function V_1 that there exists a number m_2 such that $\|X\| \leq m_2$. The latter inequality holds for any integral curves of system (3.4.3) starting in the region $r_1 \leq \|X^{(0)}\| \leq r$. Indeed, otherwise there exist sequences of points $X_k^{(0)}$ and numbers t_{0k}, t_k, such that

$$\|X(t_k, X_k^{(0)}, t_{0k})\| \to +\infty \quad \text{as} \quad k \to \infty, \quad r_1 \leq \|X_k^{(0)}\| \leq r.$$

Then we have that $V_1(X(t_k, X_k^{(0)}, t_{0k}), t_k) \to +\infty$ as $k \to \infty$, and this is not possible, because the inequality $V_1(X(t_k, X_k^{(0)}, t_{0k}), t_k) \leq m$ holds. Hence, all the solutions of system (3.4.3) are uniformly bounded for $t \geq t_0$ if $\|X^{(0)}\| \leq r$.

Now we shall show that condition 2) of Theorem 49 is satisfied. Let $X^{(0)}$ and $Y^{(0)}$ be two points of the region $\|X\| \leq r$. Construct two integral curves $X(t, X^{(0)}, t_0)$ and $X(t, Y^{(0)}, t_0)$ of system (3.4.3) and the vector function $Z = Z(t, Z^{(0)}, t_0) = X(t, X^{(0)}, t_0) - X(t, Y^{(0)}, t_0)$, $Z^{(0)} = X^{(0)} - Y^{(0)}$. Clearly, the vector-valued function Z is the integral curve for the system

$$\dot{Z} = F(t, Z + X(t, Y^{(0)}, t_0)) - F(t, X(t, Y^{(0)}, t_0)) \qquad (3.4.14)$$

and satisfies the condition $Z = Z^{(0)}$ for $t = t_0$. Choose $\epsilon > 0$ and introduce the notation

$$\Lambda = \inf_{\substack{\|X\| \leq m_r, \|Y\| \leq m_r \\ \|Z\| = \epsilon, t \in (-\infty, +\infty)}} V(X, Y, Z, t).$$

For the number Λ, determine $\delta > 0$ so that $V < \Lambda$ for $\|Z\| < \delta, t \in (-\infty, +\infty), \|X\| \leq m_r, \|Y\| \leq m_r$. We shall show that $Z(t, Z^{(0)}, t_0)$ satisfies the condition $\|Z\| < \epsilon$ for $t \geq t_0$ and $\|Z^{(0)}\| < \delta, t_0 \in (-\infty, +\infty)$. Indeed, $\frac{dV}{dt} < 0$. Hence

$$V(X, Y, Z, t) < V(X^{(0)}, Y^{(0)}, Z^{(0)}, t_0) < \Lambda,$$

Forced Periodic and Almost Periodic Oscillations

where $V(X, Y, Z, t)$ is the value of the function V computed for $X = X(t, X^{(0)}, t_0)$, $Y = X(t, Y^{(0)}, t_0)$, $Z = Z(t, Z^{(0)}, t_0)$. Assume that $\|Z\| = \epsilon$ for some $t \geq t_0$, then we get $V \geq \Lambda$, but this is not possible since for all $t \geq t_0$ we get $V < \Lambda$. Consequently, $\|Z\| < \epsilon$ for all $t \geq t_0$, if $\|Z^{(0)}\| < \delta < \epsilon$.

We shall now show that $\|Z\| \to 0$ as $t - t_0 \to +\infty$ uniformly in $t_0 \in (-\infty, +\infty)$, $\|X^{(0)}\| \leq r, \|Y^{(0)}\| \leq r$. Suppose this is not the case, then it is possible to indicate sequences $X_k^{(0)}, Y_k^{(0)}, t_{0k}, t_k$, and a number $\bar{\epsilon} > 0$ such that $\|Z(t, Z^{(0)}, t_{0k})\| \geq \bar{\epsilon}$ for $t \in [t_{0k}, t_k]$ and $t_k - t_{0k} \to +\infty$ as $k \to \infty$. Set

$$-\mu = \sup_{\substack{\bar{\epsilon} \leq \|Z\| \leq 2m_r \\ \|X\| \leq m_r, \|Y\| \leq m_r \\ t \in (-\infty, +\infty)}} W,$$

then we get the inequality $\frac{dV}{dt} < -\mu$, whence

$$V \leq V(X_k^{(0)}, Y_k^{(0)}, Z_k^{(0)}, t_{0k}) - \mu(t - t_{0k}). \tag{3.4.15}$$

This inequality also holds for $t = t_k$. So the right-hand side of inequality (3.4.15) assumes negative values that are arbitrarily large in modulus, while the left-hand side is positive. The obtained contradiction shows that condition 2) of Theorem 49 is satisfied.

It remains to show that condition 3) is also satisfied. Take the point $X^{(0)}$ and construct the integral curve of system (3.4.3), $X(t, X^{(0)}, t_0)$. Denote the vector-valued function $X(t, X^{(0)}, t_0)$ by Y, and the vector-valued function $X(t + \tau, X^{(0)}, t_0)$ by X. Clearly, the vector function X satisfies the system of equations

$$\dot{X} = F(t + \tau, X)$$

and the initial conditions $X = X(\tau + t_0, X^{(0)}, t_0)$ with $t = t_0$. Let $Z(t) = X - Y$, then the vector function Z satisfies the equation

$$\dot{Z} = F(t + \tau, X) - F(t, Y) = F(t + \tau, Z + Y) - F(t, Y)$$

$$= F(t, Z + Y) - F(t, Y) + R, \tag{3.4.16}$$

where $R = F(t + \tau, Z + Y) - F(t, Z + Y)$. Calculate the vector function $V(X, Y, Z, t)$ on the vector functions $X(t), Y(t), Z(t)$ and find its total derivative with respect to t:

$$\frac{dV}{dt} = \frac{\partial V}{\partial t} + \sum_{i=1}^{n} \left(\frac{\partial V}{\partial x_i} F_i(t + \tau, X) + \frac{\partial V}{\partial y_i} F_i(t, Y) \right.$$

$$\left. + \frac{\partial V}{\partial z_i} (F_i(t, Z + Y) - F_i(t, Y) + R_i) \right)$$

$$= W(X, Y, Z, t) + + \sum_{i=1}^{n} \left(\frac{\partial V}{\partial x_i} (F_i(t + \tau, X) - F_i(t, X)) + \frac{\partial V}{\partial z_i} R_i \right), \tag{3.4.17}$$

where $R_1, \ldots, R_n$ are the components of R from system (3.4.16). Choose $\epsilon > 0$ and denote

$$\lambda = \inf_{\substack{t \in (-\infty, +\infty) \\ \|Z\| = \epsilon, X \in G, Y \in G}} V,$$

where G is the region in which the integral curve $X = X(t, X^{(0)}, t_0)$ for $t \geq t_0$ is located. For a number λ there is $\delta > 0$ such that

$$V < \lambda \quad \text{for} \quad \|Z\| < \delta, \quad t \in (-\infty, +\infty), \quad X \in G, \quad Y \in G, \quad \delta < \epsilon.$$

Set

$$-\beta = \sup_{\substack{X \in G, Y \in G \\ t \in (-\infty, +\infty) \\ \delta \leq \|Z\| \leq d_G}} W,$$

where d_G is the diameter of the region G, $d_G = \sup_{X \in G, Y \in G} \|X - Y\|$. Let

$$\bar{M} = \sup_{\substack{X \in G, Y \in G \\ \|Z\| \leq d_G \, t \in (\infty, +\infty)}} \sqrt{\sum_{i=1}^{n} \left(\frac{\partial V}{\partial x_i}\right)^2 + \sum_{i=1}^{n} \left(\frac{\partial V}{\partial z_i}\right)^2}.$$

By the uniform almost periodicity of the function $F(t, X)$ for $X \in G$ and the function $F(t, Y + Z), Y + Z \in G$, using the number $\frac{\beta}{4\bar{M}}$, we find a number L such that in any interval $[\alpha, \alpha + L]$ there will be the number τ satisfying the inequality

$$\|F(t + \tau, X) - F(t, X)\| < \frac{\beta}{4\bar{M}} \quad \text{for} \quad t \in (-\infty, +\infty), \quad X \in G.$$

We shall assume that $\tau \geq 0$ (for $\tau < 0$ the reasoning is done analogously). With such a choice of the number τ, the expression under the summation sign in (3.4.17) may be estimated as follows:

$$\left|\sum_{i=1}^{n} \left(\frac{\partial V}{\partial x_i}(F_i(t + \tau, X) - F_i(t, X)) + \frac{\partial V}{\partial z_i} R_i\right)\right|$$

$$\leq \sqrt{\sum_{i=1}^{n} \left(\frac{\partial V}{\partial x_i}\right)^2} \|F(t + \tau, X) - F(t, X)\|$$

$$+ \sqrt{\sum_{i=1}^{n} \left(\frac{\partial V}{\partial z_i}\right)^2} \|F(t + \tau, Y + Z) - F(t, Y + Z)\| \leq \beta/2.$$

Consequently, from (3.4.17) we have that $dV/dt \leq -\beta/2$ for $\delta \leq \|Z\| \leq d_G$, $X \in G$, $Y \in G$, $t \in (-\infty, +\infty)$.

We show the existence of $T > 0$ such that for $t + \tau \geq T + t_0$ and $t \geq T + t_0$ we get $\|Z\| < \epsilon$. Let $\|Z\| > \delta$ throughout the motion, then

$$V \leq V(X^{(0)}, Y^{(0)}, Z^{(0)}, t_0) - \frac{\beta}{2}(t - t_0).$$

Forced Periodic and Almost Periodic Oscillations

The last inequality is not possible for all $t \geq t_0$, hence the number T is found to be such that
$$\|Z(t_0 + T)\| < \delta.$$
We shall show that $\|Z(t)\| < \epsilon$ for all $t \geq T + t_0$. Suppose there exists $\bar{t} > t_0 + T$ such that $\|Z(\bar{t})\| = \epsilon$. Then it is possible to select a number $\bar{\bar{t}}$ satisfying the conditions
$$\bar{\bar{t}} < \bar{t}, \quad \|Z\| = \delta \quad \text{and} \quad \|Z(t)\| > \delta \quad \text{for} \quad t \in (\bar{\bar{t}}, \bar{t}].$$
For $t \in [\bar{\bar{t}}, \bar{t}]$ the function V considered on $X(t), Y(t), Z(t)$ is the function of t. Denote it by $V(t)$ and then
$$V(\bar{\bar{t}}) < \lambda, \quad V(\bar{t}) \geq \lambda, \quad \frac{dV}{dt} < -\frac{1}{2}\beta \quad \text{for} \quad t \in (\bar{\bar{t}}, \bar{t}].$$
Thus, the function $V(t)$ is strictly decreasing in the interval $[\bar{\bar{t}}, \bar{t}]$, and hence $V(\bar{t}) < \lambda$. The obtained contradiction shows that the relationship $\|Z\| = \epsilon$ does not hold for any finite $\bar{t} \geq t_0 + T$. We have thus established that
$$\|X(t + \tau, X^{(0)}, t_0) - X(t, X^{(0)}, t_0)\| < \epsilon \quad \text{for} \quad t \geq T + t_0, \quad \tau + t \geq T + t_0.$$
Therefore, the condition 3) of Theorem 49 holds and the conditions of Theorem 50 are sufficient for system (3.4.3) to possess convergence property. ∎

Remark. Theorem 50 makes it possible to formulate the existence condition for the asymptotically stable almost periodic solution which is possibly not global-asymptotically stable. Denote
$$N_r = \sup_{\substack{t \in (-\infty, +\infty) \\ r_1 \leq \|X\| \leq r}} V_1$$
and find a number ρ_r having the property that
$$\inf_{\substack{t \in (-\infty, +\infty) \\ \|X\| \geq \rho_r}} V_1 > N_r,$$
where V_1 is the function which appears in the conditions of Theorem 50. If condition 2) of Theorem 50 holds for $\|Z\| \leq 2\rho_1$, $\rho_1 > \rho_r$ for some $r > r_1$, then system (3.4.3) has an almost periodic asymptotically stable solution. We show this. Take the point $X^{(0)}$ such that $\|X^{(0)}\| \leq r$. Then the integral curve $X(t, X^{(0)}, t_0)$ is bounded for all $t \geq t_0$, since throughout the time of motion the inequality $V_1 \leq N_r$ holds along the integral curve as long as this curve remains in the region $\|X\| \geq r_1$. When the integral curve leaves the region $\|X\| \leq \rho_r$, the above inequality is violated. If the region $\|X\| \leq \rho_r$ is taken as G, then it follows from the proof of Theorem 50 that the integral curve $X(t, X^{(0)}, t_0)$ satisfies property 3) of Theorem 49. So the proof of sufficiency for Theorem 49 suggests that system (3.4.3) has the almost periodic solution $X = \Phi(t)$ in the region $\|X\| \leq \rho_r$.

We shall now show that this solution is asymptotically stable uniformly in $t_0 \in (-\infty, +\infty)$. It follows from the proof of Theorem 50 that if we take $\epsilon > 0$ and select numbers λ and δ for it, using the relationships

$$\lambda = \inf_{\substack{\|X\|=\rho_1, \|Y\|\le\rho_1 \\ \|Z\|=\epsilon, t\in(-\infty,+\infty)}} V, \quad V < \lambda \text{ for } \|Z\| \le \delta,\ t \in (-\infty, +\infty),\ \|X\| \le \rho_1,\ \|Y\| \le \rho_1$$

then $\|X - Y\| \le \epsilon$ for $t \ge t_0$, $\|X^{(0)} - Y^{(0)}\| \le \delta$ and $\|X - Y\| \to 0$ as $t - t_0 \to +\infty$ uniformly in $t_0 \in (-\infty, +\infty)$, where $X = X(t, X^{(0)}, t_0), Y = X(t, Y^{(0)}, t_0)$ are two solutions to the system (3.4.3), $\|X^{(0)}\| < \rho_1$, $\|Y^{(0)}\| \le \rho_1$. ∎

Theorem 51 *Suppose the following assumptions are satisfied:*
1. There exists the function $Y_1(X,t)$ satisfying the conditions: a) V_1 is given for all $t \in (-\infty, +\infty), X \in E_n$, and is real and continuous, b) by system (3.4.3), there exists a continuous total t-derivative of V_1 satisfying the inequality $\frac{dV_1}{dt} < -\alpha$ for $\|X\| \ge r_1, \alpha > 0$, c) there exist two numbers r_2 and $r_3, r_3 > r_2 \ge r_1$ such that

$$\inf_{\substack{\|X\|=r_3 \\ t\in(-\infty,+\infty)}} V_1 > \sup_{\substack{\|X\|=r_2 \\ t\in(-\infty,+\infty)}} V_1,$$

d) is uniformly bounded below for $t \in (-\infty, +\infty), \|X\| \ge r_1$.
2. There exists the function $V(X, Y, Z, t)$ satisfying the conditions : a) the function V is positive-definite and admits the infinitesimal upper limit in the region $\|Z\| \le 2r_3$; b) the function V has continuous partial derivatives in all its variables, and these derivatives are the functions uniformly bounded in modulus for $t \in (-\infty, +\infty)$ in any finite region of variation in variables X, Y, Z; c) the function

$$W = \frac{\partial V}{\partial t} + \sum_{i=1}^{n} \left(\frac{\partial V}{\partial x_i} F_i(t, X) + \frac{\partial V}{\partial y_i} F_i(t, Y) + \frac{\partial V}{\partial z_i} (F_i(t, Y+Z) - F_i(t, Y)) \right)$$

is negative-definite in the region $\|Z\| \le 2r_3$.
The system (3.4.3) then has convergence property.

Proof: We shall first show that all the integral curves $X(t, X^{(0)}, t_0)$ starting in the region $r_1 \le \|X^{(0)}\| \le r$ fall within the region $\|X\| \le r_1$ during the time period not exceeding $T < +\infty$, where T is the constant independent of $X^{(0)}$ and t_0. Suppose the opposite is true, then there exist sequences $X_k^{(0)}, t_{0k}, t_k$ such that

$$\|X_k^{(0)}\| \in [r_1, r], \quad t_{0k} \in (-\infty, +\infty), \quad t_k \ge t_{0k}, \quad t_k - t_{0k} \to +\infty$$

as $k \to \infty$, and
$$\|X(t, X_k^{(0)}, t_{0k})\| > r_1 \quad \text{for} \quad t \in [t_{0k}, t_k].$$

Moreover, we have $V_1(X, t) \le -\alpha(t - t_{0k}) + V_1(X_k^{(0)}, t_{0k})$, where $V_1(X,t)$ is the function computed on the integral curve $X = X(t, X_k^{(0)}, t_{0k})$. From the last inequality we have that for $t = t_k$, $V_1 \to -\infty$ as $k \to \infty$, but this is not possible by item d) in condition 1 of Theorem 51. So the above assumption yields a contradiction, hence all the integral curves starting in the region $r_1 \leq \|X^{(0)}\| \leq r$ fall within the region $\|X\| \leq r_1$ for $t - t_0 \geq T$.

The integral curves starting on the surface $\|X\| = r_2$ cannot cut the surface $\|X\| = r_3$ with the increase in time. In fact, $V_1(X_1, t) < V_1(X^{(0)}, t_0)$ on any arc located in the region $r_2 \leq \|X\| \leq r_3$; therefore if the integral curve starting at $\|X^{(0)}\| = r_2$ falls on the surface $\|X\| = r_3$, we have a contradiction to item c) of condition 1 in Theorem 51. This means that all the integral curves starting in the region $r_1 < \|X^{(0)}\| \leq r$ stay in the region $\|X\| \leq r_3$ for all $t \geq T + t_0$.

It remains to show that the conditions 2) and 3) of Theorem 49 are satisfied. Take $\epsilon > 0$, denote

$$\lambda = \inf_{\substack{\|X\| \leq r_3, \|Y\| \leq r_3 \\ \|Z\| = \epsilon, t \in (-\infty, +\infty)}} V$$

and choose a number $\delta < \epsilon$ so that $V < \lambda$ for $\|Z\| \leq \delta, \|X\| \leq r_3, \|Y\| \leq r_3, t \in (-\infty, +\infty)$. Then all the integral curves starting in the region $\|Y^{(0)}\| \leq r_2, \|X^{(0)}\| \leq r_2$, satisfy the inequality $\|Z\| \leq \epsilon$ for $t \geq t_0$ if $\|X^{(0)} - Y^{(0)}\| < \delta$, and $\|Z\| \to 0$ as $t - t_0 \to +\infty$ uniformly in $t_0 \in (-\infty, +\infty), \|X^{(0)}\| \leq r_2; \|Y^{(0)}\| \leq r_2$. As before, X and Y are the integral curves of system (3.4.3): $X = X(t, X^{(0)}, t_0), Y = X(t, Y^{(0)}, t_0)$. The latter property is possessed by the integral curves X, Y starting in any region $\|X^{(0)}\| \leq r, \|Y^{(0)}\| \leq r$, because they, as shown before, inevitably fall within the region $\|X\| \leq r_1 \leq r_2$ at the expiration of time T. The set of functions

$$X(t_0 + t, X^{(0)}, t_0), \quad t \in [0, T], \quad \|X^{(0)}\| \leq r$$

equicontinuously in $X^{(0)}$ for $t_0 \in (-\infty, +\infty), t \in [0, T]$. Hence there is $\delta_1 > 0$ such that

$$\|X - Y\| < \epsilon \quad \text{for} \quad \|X^{(0)} - Y^{(0)}\| < \delta_1, \quad t_0 \in (-\infty, +\infty), \quad t \in [0, T].$$

If $\bar{\delta} = \min \delta, \delta_1$, then condition 2) of Theorem 49 is satisfied. Condition 3) is also met, since the region $\|X\| \leq r_3$ may be taken as the region G, and the integral curve starting on the surface $\|X\| = r_2$ as the original integral curve. ∎

3.5 Influence of External Perturbations on Stationary Modes

1. Consider the system of equations

$$\dot{X} = F(t, X), \tag{3.5.1}$$

where F is the vector-valued function given for all $t \in (-\infty, +\infty), X \in E_n$, being real and continuous in all its variables. We say that system (3.5.1) has the equilibrium $X = 0$ and its right-hand sides are 2π-periodic functions of t and continuously differentiable with respect to $x_1, \ldots, x_n$. Concurrent with system (3.5.1), the following system is described with due regard to external perturbation influences

$$\dot{X} = F(t, X) + R(t, X), \qquad (3.5.2)$$

where $R(t, X)$ is the vector-valued function given for all $t \in (-\infty, +\infty)$ in some neighborhood of $X = 0$, is real, continuous in X, satisfies the Lipschitz condition for $x_1, \ldots, x_n$ in this neighborhood and is such that $\|R(t, X)\| \leq am$, where a is a positive parameter, m is a constant. Also, we say that either $R(t, X)$ is 2π-periodic function in t or it has the period commensurable with 2π.

Theorem 52 *If the equilibrium $X = 0$ of system (3.5.1) is uniform-asymptotically stable, then there is a number $a_0 > 0$ such that system (3.5.2) for $0 \leq a \leq a_0$ has a periodic solution in the region $\|X\| \leq \delta(a)$, where $\delta(a)$ is a positive (with $a \neq 0$) continuous function, $\delta(0) = 0$. This periodic motion may degenerate into an equilibrium.*

Proof: The uniform asymptotic stability of the equilibrium $X = 0$ of system (3.5.1) implies the existence of two functions $V(X, t)$ and $W(X, t)$ having the following properties: $V(X, t)$ is positive-definite and admits the infinitesimal upper limit, $W(X, t)$ is negative-definite and is related to the function V by

$$W = \frac{\partial V}{\partial t} + \sum_{j=1}^{n} \frac{\partial V}{\partial x_j} f_j(t, X),$$

where $f_j(t, X)$ is the j-th component of the vector-valued function $F(t, X)$. The partial derivatives of the function V are continuous and uniformly bounded for $t \in (-\infty, +\infty), \|X\| \leq r, r = \text{const} > 0$. Let $\epsilon > 0$ be a sufficiently small number. Denote

$$\lambda = \inf_{\substack{\|x\| = \epsilon \\ t \in (-\infty, +\infty)}} V(X, t)$$

and choose a number $\delta < \epsilon$ such that $V < \lambda$ for $\|X\| < \delta, t \in (-\infty, +\infty)$. We construct the number δ_0 for the number δ as δ is constructed for ϵ. Let

$$-\beta = \sup_{\substack{\delta_0 \leq \|x\| \leq \epsilon \\ t \in (-\infty, +\infty)}} W(X, t).$$

Choose a number a_0 from the condition $-\beta + a_0 m m' < 0$, where

$$m' = \sup_{\substack{\|X\| \leq \epsilon \\ t \in (-\infty, +\infty)}} \sqrt{\sum_{j=1}^{n} \left(\frac{\partial V}{\partial x_j}\right)^2}.$$

We show that for any $a, 0 \leq a \leq a_0$ system (3.5.2) has a periodic solution in the region $\|X\| \leq \delta(a)$. In fact, any integral curve of system (3.5.2) starting in the region $\|X\| \leq \delta_0$ stays in the region $\|X\| \leq \delta$ for all $t \geq t_0$. And any integral curve starting in the region $\|X\| \leq \delta$ stays in the region $\|X\| < \epsilon$ for all $t \geq t_0$. We now show that

Influence of External Perturbations on Stationary Modes

there exists a number T such that all the integral curves $X = X(t, X_0, t_0)$ starting in the region $\delta_0 \leq \|X\| \leq \delta$ fall within the region $\|X\| \leq \delta_0$ for a particular $t \leq T + t_0$ and, therefore, stay for all $t \geq t_0 + T$ in the region $\|X\| < \delta$. Suppose the opposite is true, then there are sequences X_{0k}, t_{0k} and t_k such that $t_k - t_{0k} \to +\infty$ as $k \to \infty$, $\|X_{0k}\| < \delta$, and $\|X(t, X_{0k}, t_{0k})\| > \delta_0$ for $t \in [t_{0k}, t_k]$. On such arcs of the integral curves in system (3.5.2) we have

$$\frac{dV}{dt} = \frac{\partial V}{\partial t} + \sum_{j=1}^{n} \frac{\partial V}{\partial x_j}(f_j(t,X) + r_j(t,X)) = W + \sum_{j=1}^{n} \frac{\partial V}{\partial x_j} r_j(t,X) \leq -\beta + a_0 mm' < 0,$$

whence

$$V \leq V(X_{0k}, t_{0k}) - (\beta - a_0 mm') - (t - t_{0k}).$$

From the last inequality it follows that the function V must take negative values that are arbitrarily large in modulus. And this contradicts the positive definiteness of the function V, hence the number T with the above-mentioned properties exists. Choose a natural N such that $2\pi N > T$. Then the self-mapping $X(t_0 + 2\pi N, X_0, t_0)$ of the region $\|X_0\| \leq \delta$ satisfies all conditions of the Brauer theorem and hence there exists a fixed point $\bar{X}_0$ of this mapping such that $X(t_0 + 2\pi N, \bar{X}_0, t_0) = \bar{X}_0$. So the integral curve $X(t, \bar{X}_0, t_0)$ is a $2\pi N$-periodic motion of system (3.5.2). This motion is in the region $\|X\| \leq \epsilon$. The number a_0 is chosen to be dependent on ϵ, hence considering it as a monotone function of ϵ, we select ϵ as the function of a and denote it by $\delta(a)$. Thus, the obtained periodic solution is in the region $\|X\| \leq \delta(a)$. ∎

Remark. Investigations into behavior of the other integral curves of system (3.5.2) in the neighborhood of this periodic motion, and studies on the problem of its uniqueness without further assumptions for the functions F and R seem to be impossible. In other words, the result stated in Theorem 52 cannot be refined without further assumptions for the functions F and R.

We now assume that system (3.5.1) has a periodic mode

$$X = \Phi(t), \tag{3.5.3}$$

which, in particular, can be an equilibrium. In system (3.5.1) we replace the desired function by the formula

$$X = \Phi(t) + Y, \tag{3.5.4}$$

where Y is the new desired vector-valued function. To define it, we obtain the system

$$\dot{Y} = F(t, Y + \Phi(t)) - F(t, \Phi(t)) = \mathbf{P}(t)Y + H(t, Y), \tag{3.5.5}$$

where

$$\mathbf{P} = \{p_{jk}(t)\}, \quad p_{jk} = \frac{\partial f_j}{\partial x_k}\Big|_{X=\Phi(t)}.$$

The function $H(t, Y)$ satisfies a Lipschitz condition of the form

$$\|H(t, \bar{Y}) - H(t, \bar{\bar{Y}})\| \leq \lambda(h)\|\bar{Y} - \bar{\bar{Y}}\|, \quad t \in (-\infty, +\infty), \quad \|\bar{Y}\| \leq h, \quad \|\bar{\bar{Y}}\| \leq h, \tag{3.5.6}$$

where $\lambda(h)$ is a strictly monotonic continuous function which decreases to zero as $h \to +0$. Indeed,
$$H(t,Y) = F(t, Y + \Phi) - F(t, \Phi) - \mathbf{P}(t)Y,$$
therefore,
$$H(t,\bar{Y}) - H(t,\bar{\bar{Y}}) = F(t, \bar{Y} + \Phi) - F(t, \bar{\bar{Y}} + \Phi) - \mathbf{P}(t)(\bar{Y} - \bar{\bar{Y}})$$
$$= \left(\mathbf{Q}(t, \Phi + \bar{Y} + \vartheta(\bar{\bar{Y}} - \bar{Y})) - \mathbf{P}(t)\right)(\bar{Y} - \bar{\bar{Y}}),$$
where
$$\mathbf{Q} = \{q_{jk}\}, \quad q_{jk} = \frac{\partial f_j}{\partial x_k}\Big|_{X = \Phi + \bar{Y} + \vartheta(\bar{\bar{Y}} - \bar{Y})}.$$

Proceeding from the assumption that the partial derivatives of the function F with respect to $x_1, \ldots, x_n$ are continuous in the collection (t, X), we get $\|\mathbf{Q} - \mathbf{P}\| \to 0$ as $\|\bar{Y}\| \to 0$, $\|\bar{\bar{Y}}\| \to 0$, which implies the existence of the function $\lambda(h)$.

We say that the period of the function $\Phi(t)$ is commensurable with 2π. Then, by replacing the independent variable in system (3.5.5), we ensure that its right-hand sides are 2π-periodic functions of t. For the linear system
$$\dot{Y} = \mathbf{P}(t)Y \qquad (3.5.7)$$
we denote by $\lambda_1, \ldots, \lambda_n$ its characteristic indices.

Theorem 53 *Suppose the following conditions are satisfied:*

1) system (3.5.1) has the periodic solution (3.5.3) whose period is commensurable with 2π;

2) the characteristic indices of system (3.5.7) have nonzero real parts
$$Re\lambda_j < 0, \quad j = 1, \ldots, k, \quad Re\lambda_j > 0, \quad j = k+1, \ldots, n,$$

3) the vector-valued function $R(t, X)$ is given for $t \in (-\infty, +\infty)$ and is real and continuous in all its variables in some region $\bar{G} \subset E_n$ containing the periodic solution (3.5.3); in any bounded subregion $\bar{G} \subset G$ the function $R(t, X)$ is continuous in X uniformly for $t \in (-\infty, +\infty)$ and is an almost periodic function of t for any $X \in G$, satisfying also the Lipschitz condition
$$\|R(t, \bar{X}) - R(t, \bar{\bar{X}})\| \le aL\|\bar{X} - \bar{\bar{X}}\|, \quad \bar{X} \in G, \bar{\bar{X}} \in G \qquad (3.5.8)$$
and an inequality of the form
$$\|R(t, X)\| \le am \quad for \quad t \in (-\infty, +\infty), \quad X \in G, \qquad (3.5.9)$$
where a is a positive parameter. Then there exists the number a_0 and the function $\delta(a)$ given for $a \in [0, a_0]$ is continuous, strictly decreasing to zero as $a \to +0$ and such that system (3.5.2) has for any $a, 0 \le a \le a_0$, a unique almost periodic solution $X = \bar{X}(t)$ in the region $\|X - \Phi(t)\| \le \delta(a), t \in (-\infty, +\infty)$. In this case, a sufficiently small neighborhood of the solution $\bar{X}(t)$ for any $a \in [0, a_0]$ is filled with integral curves of two types, viz.: in this neighborhood there is a family of integral curves $X =$

Influence of External Perturbations on Stationary Modes

$X(t, c_1, \ldots, c_k)$ dependent on k arbitrary constants which stays in this neighborhood for all $t \geq t_0$ and satisfies the inequality

$$\|X(t, c_1, \ldots, c_k) - \bar{X}(t)\| \leq b\|C\|e^{-\bar{\lambda}(t-t_0)} \quad for \quad t \geq t_0, \quad \|C\| \leq c_0, \quad (3.5.10)$$

where $b, c_0, \bar{\lambda}$, are positive constants, and $\|C\| = \sqrt{\sum_{i=1}^{k} c_i^2}$. Any integral curve starting in the above-mentioned neighborhood at $t = t_0$ either enters into that family or leaves the neighborhood in a finite time. If $k = n$, then the solution $X = \bar{X}(t)$ is asymptotically stable in the sense of Lyapunov, for $k = 0$ any integral curve, as distinct from $X = \bar{X}(t)$ and starting at $t = t_0$ in its particular sufficiently small neighborhood, leaves this neighborhood in a finite time.

Proof: In (3.5.2) we replace the desired function X by (3.5.4). Then, to define the function Y, we obtain the system

$$\dot{Y} = \mathbf{P}(t)Y + H(t, Y) + R(t, Y + \Phi). \quad (3.5.11)$$

Concurrent with (3.5.11), we consider a system of integral equations

$$Y(t) = \int_{-\infty}^{+\infty} \mathbf{G}(t, \tau)\left(H(\tau, Y) + R(\tau, Y + \Phi(\tau))\right) d\tau = K(Y). \quad (3.5.12)$$

Denote by $C_{(-\infty,+\infty)}$ the set of continuous real functions given for $t \in (-\infty, +\infty)$ and satisfying the inequality $\|Y\|_c \leq h$, where $\|Y\|_c = \sup_{t \in (-\infty,+\infty)} \|Y(t)\|$. The operator $K(Y)$ converts the function $Z(t) = K(Y)$ which is real, continuous, given for $t \in (-\infty, +\infty)$ and satisfies the inequality

$$\|Z(t)\| \leq \gamma(h\lambda(h) + am), \quad (3.5.13)$$

where γ is a positive constant such that $\int_{-\infty}^{+\infty} \|\mathbf{G}(t,\tau)\| d\tau \leq \gamma$. The last integral is convergent, since the matrix $\mathbf{G}(t,\tau)$ satisfies the inequality $\|\mathbf{G}(t,\tau)\| \leq \gamma_1 e^{-\alpha|t-\tau|}$, where γ_1, α are the positive constants independent of t and τ.

Let $\bar{Y}, \bar{\bar{Y}}$, be two elements of the space $C_{(-\infty,+\infty)}$, and $\bar{Z}, \bar{\bar{Z}}$ their images with the map K. If $\|\bar{Y} - \bar{\bar{Y}}\|_c \leq l$, then

$$\|\bar{Z} - \bar{\bar{Z}}\| \leq \gamma(\lambda(h) + aL)l. \quad (3.5.14)$$

Choose numbers $a_0 > 0$ and h_0 so small that $\lambda(h_0) < \frac{1}{2}\beta, a_0 L < \frac{1}{2}\beta$, where $\beta \in (0,1)$. The operator K is then the contracting operator, viz. the inequality $\|K(\bar{Y}) - K(\bar{\bar{Y}})\|_c \leq \beta\|\bar{Y} - \bar{\bar{Y}}\|_c$ is satisfied. We now choose a as a function of h such that the operator $K(Y)$ converts $C_{(-\infty,+\infty)}$ into itself. To this end, as follows from inequality (3.5.13), it suffices to set $a = h(1 - \lambda(h))/m$. We say that h_0 is so small that $1 - \lambda(h_0) > 0$, and the number a_0 determined above is found from the formula $a_0 = (1 - \lambda(h_0))/m$. The earlier equality defines h as the function of a; denote it by $\delta(a)$. This function is continuous and strictly decreasing to zero as $a \to +0$. Thus, the operator $K(Y)$ converts the space $C_{(-\infty,+\infty)}$ into itself for any $a \in [0, a_0], h = \delta(a)$ and is the contracting operator. As $C_{(-\infty,+\infty)}$, we now take the set of all almost periodic vector-valued functions $Y(t)$ satisfying the condition $\|Y(t)\| \leq \delta(a)$ (a being fixed). Then it follows from the contracting map principle that there exists the unique almost periodic function $Y = \bar{Y}(t)$ satisfying the equation $\bar{Y} = K(\bar{Y})$ in view of the completeness of the space of almost periodic functions $C_{(-\infty,+\infty)}$. If, as before, we consider $C_{(-\infty,+\infty)}$ to be the space of all continuous real vector functions satisfying the condition $\|Y(t)\| \leq \delta(a)$, then in this space there is a unique fixed point of the operator K which has to be coincident with the above almost-periodic function $\bar{Y}(t)$ by uniqueness.

Any solution to system (3.5.11), which is given for $t \in (-\infty, +\infty)$, is real, continuous and satisfies the inequality $\|Y(t)\| \leq \delta(a)$, is a solution to the integral equation system (3.5.12) and, therefore, coincides with its unique solution $Y = \bar{Y}(t)$. Differentiating the identity $\bar{Y}(t) = K(\bar{Y})$, we have that $\bar{Y}(t)$ is also a solution to system (3.5.11). It follows from (3.5.4) that the function $X = \bar{X}(t) = \bar{Y}(t) + \Phi(t)$ is an almost periodic solution to system (3.5.2) which is located in the region $\|X - \Phi(t)\| \leq \delta(a)$ and is the unique solution of this kind for the system.

We clarify the behavior of integral curves of system (3.5.2) in the neighborhood of its solution $X = \bar{X}(t)$. To do this, in system (3.5.2) we make a replacement $Y = \bar{Y}(t) + U$, where U is a new desired vector-valued function. To define it, we obtain the system

$$\dot{U} = \mathbf{P}(t)U + Q(t, U), \qquad (3.5.15)$$

where $Q(t, U) = H(t, \bar{Y} + U) + R(t, \bar{X} + U) - H(t, \bar{Y}) - R(t, \bar{X})$. The vector-valued function $Q(t, U)$ vanishes at $U = 0$ and satisfies a Lipschitz condition of the form

$$\|Q(t, \bar{U}) - Q(t, \bar{\bar{U}})\| \leq \alpha \|\bar{U} - \bar{\bar{U}}\|, \qquad (3.5.16)$$

where α is some positive parameter. The inequality (3.5.16) holds for $\|\bar{U}\| \leq \sigma, \|\bar{\bar{U}}\| \leq \sigma$ and $t \in (-\infty, +\infty)$. The positive parameter α is a function of a and σ and assumes arbitrarily small values if a and σ are sufficiently small. Indeed,

$$Q(t, \bar{U}) - Q(t, \bar{\bar{U}}) = H(t, \bar{Y} + \bar{U}) - H(t, \bar{Y} + \bar{\bar{U}}) + R(t, \bar{X} + \bar{U}) - R(t, \bar{X} + \bar{\bar{U}});$$

hence

$$\|Q(t, \bar{U}) - Q(t, \bar{\bar{U}})\| \leq \|H(t, \bar{Y} + \bar{U}) - H(t, \bar{Y} + \bar{\bar{U}})\| + \|R(t, \bar{X} + \bar{U}) - R(t, \bar{X} + \bar{\bar{U}})\|$$

$$\leq \lambda(\sigma + \delta(a))\|\bar{U} - \bar{\bar{U}}\| + aL\|\bar{U} - \bar{\bar{U}}\| = \alpha\|\bar{U} - \bar{\bar{U}}\|,$$

where $\alpha = \lambda(\sigma + \delta(a)) + aL$. The numbers σ_0 and a_0 may be chosen in such a manner that in the region $\|U\| \leq \sigma_0$ there is a family of integral curves of system (3.5.15)

$$U = U(t, c_1, \ldots, c_k), \qquad (3.5.17)$$

which depends on k arbitrary constants, is determined for all $t \geq t_0$, is situated in that region and satisfies the inequality

$$\|U\| \leq b\|C\|e^{-\bar{\lambda}(t-t_0)} \quad \text{for} \quad \|C\| \leq c_0, \quad t \geq t_0,$$

where $b, \bar{\lambda}, c_0$ are positive constants. Every integral curve starting at $t = t_0$ in that region either belongs to the family (3.5.17) or leaves this neighborhood in a finite time. Availability of such a family follows from the results in section 1.3 that are applied to system (3.5.15). As to the statement about the behavior of the other integral curves of system (3.5.15) starting in the region $\|U\| \leq \sigma_0$ at $t = t_0$, its validity can be established as in section 2.5. Turning to the system (3.5.2), we have

$$X(t, c_1, \ldots, c_k) = \bar{X}(t) + U(t, c_1, \ldots, c_k). \quad (3.5.18)$$

Thus, system (3.5.2) has a unique almost periodic solution $X = \bar{X}(t)$ in the region $\|X\| \leq \delta(a)$. For any $a \in [0, a_0]$ it is possible to select a number σ_0 such that in the neighborhood $\|X - \bar{X}(t)\| \leq \sigma_0$ of this periodic solution there is the family of integral curves (3.5.18) satisfying the inequalities

$$\|X(t, c_1, \ldots, c_k) - \bar{X}(t)\| \leq b\|C\|e^{-\bar{\lambda}(t-t_0)}, \quad \|C\| \leq c_0, \quad t \geq t_0$$

and situated in that neighborhood for all $t \geq t_0$. The remaining integral curves starting in the above-mentioned neighborhood either enter into the family (3.5.18) or leave this neighborhood in a finite time. ∎

2. Consider the influences of external "finite amplitude" perturbations on the system possibly having an equilibrium which is global asymptotically stable.

Theorem 54 *Suppose the following conditions are satisfied:*
1) there exist two functions $V(t, X)$ and $W(t, X)$ given for $\|X\| \leq r, t \in (-\infty, +\infty)$, real and continuous, $V(t, X) \geq 0, V(t, X) \to +\infty$ as $\|X\| \to +\infty$ uniformly in $t \in (-\infty, +\infty)$; the function $V(t, X)$ is uniformly bounded for $t \in (-\infty, +\infty)$ in any region $0 < r \leq r' \leq \|X\| \leq r'' < +\infty$, is continuously differentiable in its variables and satisfies the relationship

$$\frac{\partial V}{\partial t} + \sum_{j=1}^{n} \frac{\partial V}{\partial x_j} f_j(t, X) = W(t, X);$$

2) the vector-valued function $R(t, X)$ in system (3.5.2) is given for $X \in E_n, t \in (-\infty, +\infty)$, is real and continuous, satisfies the Lipschitz condition in X and is bounded;
3) for each m it is possible to select a number $\bar{r} \geq r$ such that for any $r_1 \geq \bar{r}$ we get

$$W + m\sqrt{\sum_{j=1}^{n} \left(\frac{\partial V}{\partial x_j}\right)^2} < -\beta < 0 \quad for \quad t \in (-\infty, +\infty), \quad \bar{r} \leq \|X\| \leq r_1,$$

where β is a positive constant which may depend on r_1 for the given m. With any choice of the vector-valued function $R(t, X)$ satisfying condition 2) of this theorem, system (3.5.2) then has a periodic solution which possibly reduces to an equilibrium.

Proof: Suppose the vector-valued function $R(t, X)$ is given and there exists a positive constant m such that $\|R(t, X)\| \leq m, t \in (-\infty, +\infty), X \in E_n$. By condition 3), we choose a number $\bar{r}$ and take some number $r_1 \geq \bar{r}$. Denote $\lambda_1 = \sup\limits_{\substack{\|X\|=r_1 \\ t\in(-\infty,+\infty)}} V(t,X)$. Choose a number r_2 such that $V > \lambda_1$ for $t \in (-\infty,+\infty)$, $\|X\| \geq r_2 > r_1$. Set $\lambda_2 = \sup\limits_{\substack{\|X\|=r_2 \\ t\in(-\infty,+\infty)}} V(t,X)$. For λ_2 we select $r_3 > r_2$ such that $V(t,X) > \lambda_2$ for $\|X\| \geq r_3, t \in (-\infty,+\infty)$.

All integral curves of system (3.5.2) starting at $t = t_0$ in the region $\|X\| \leq r_1$ stay for $t \geq t_0$ in the region $\|X\| \leq r_2$. All integral curves of system (3.5.2) starting at $t = t_0$ in the region $\|X\| \leq r_2$ stay for $t \geq t_0$ in the region $\|X\| \leq r_3$. We establish the indicated property for the integral curves of system (3.5.2) starting, for example, in the region $\|X\| \leq r_1$. On each integral curve $X = X(t, X_0, t_0)$ starting in the region $\|X_0\| \leq r_1$ and falling within the region $r_1 \leq \|X\| \leq r_2$ it is possible to select a point corresponding to the time t_1 such that $\|X'\| = r_1$, where $X' = X(t_1, X_0, t_0)$. If some arc of this integral curve with $t_1 \leq t \leq t_2$ lies in the region $r_1 \leq \|X\| \leq r_2$, then along this region the inequality $V(t,X) < V(t_1, X')$ holds for $t \in (t_1, t_2)$, and then $V(t,X) \leq \lambda_1$. Assume that for some $t \in (t_1, t_2)$, $\|X(t, X_0, t_0)\| = r_2$, therefore for this point $V(t,X) > \lambda_1$. The obtained contradiction suggests that the above statement is valid.

We show that there exists a number T such that all the integral curves $X(t, X_0, t_0)$ starting at $t = t_0$ in the region $r_1 \leq \|X_0\| \leq r_2$ fall for some $t \in [t_0, t_0 + T]$ within the region $\|X\| \leq r_1$ and, therefore, for all $t \geq t_0 + T$ stay in the region $\|X\| \leq r_2$. In fact, if the integral curves of system (3.5.2) do not possess this property then there exist the sequences $X_0^{(k)}, t_0^{(k)}, t_k$ satisfying the conditions $t_k - t_0^{(k)} \to +\infty$ as $k \to \infty$, $r_1 \leq \|X_0^{(k)}\| \leq r_2$ and such that $\|X(t, X_0^{(k)}, t_0^{(k)})\| > r_1$ for $t \in [t_0^{(k)}, t_k]$. The following relationship holds along these arcs of integral curves

$$V(t,X) \leq V(t_0^{(k)}, X_0^{(k)}) - \beta(r_2)(t - t_0^{(k)}).$$

This suggests that the function V on the arcs of such integral curves must assume negative values that are arbitrarily large in modulus, which is not possible by the uniform boundedness of this function in the region $r_1 \leq \|X\| \leq r_2$. The obtained contradiction shows that there exists the number T possessing the above-indicated properties. Let ω be a natural number such that $2\pi\omega \geq T$. Then we have that the mapping $X(t_0 + 2\pi\omega, X_0, t_0)$ transforms the closed sphere $\|X\| \leq r_2$ into itself. Consequently, there exists a fixed point $\bar{X}_0$ of this mapping which satisfies the equation

$$\bar{X}_0 = X(t_0 + 2\pi\omega, \bar{X}_0, t_0).$$

Therefore $X(t, \bar{X}_0, t_0)$ is a periodic solution of system (3.5.2). ∎

Remark. Theorem 54 also remains valid for all vector-valued functions $R(t,X)$ for which condition 3) holds if m is replaced by $\|R(t,X)\|$.

Example 1. Consider a linear system $\dot{X} = \mathbf{P}(t)X$ with 2π-periodic continuous real coefficients and characteristic indices $\lambda_1, \ldots, \lambda_n$ having negative real parts. If $R(t,X)$ is an arbitrary real continuous function, which is 2π-periodic in t, satisfies the Lipschitz condition for X in any finite region and obeys the inequality

$$\|R(t, X)\| \leq \varphi(\|X\|) \|X\|,$$

where $\varphi(\|X\|) \to 0$ as $\|X\| \to +\infty$, then the system

$$\dot{X} = \mathbf{P}(t)X + R(t, X)$$

has a periodic solution possibly reducing to an equilibrium.

We define a positive-definite quadratic form of V as a solution to the equation

$$\frac{\partial V}{\partial t} + \sum_{j=1}^{n} \frac{\partial V}{\partial x_j} \sum_{i=1}^{n} p_{ji}(t)x_i = -\sum_{j=1}^{n} x_j^2 = W,$$

where $p_{ji}(t)$ are the elements of the matrix $\mathbf{P}(t)$. Then the functions V and W satisfy all conditions of Theorem 54.

Example 2. Consider the system

$$\frac{dx_s}{dt} = X_s^{(\mu)}, \quad \mu = 2k + 1, \quad k \geq 0, \quad s = 1, \ldots, n.$$

Assume that the zero solution of this system is asymptotically stable. Here $X_s^{(\mu)}$ are homogeneous forms of odd order μ with respect to $x_1, \ldots, x_n$ with constant real coefficients. If $R(t, X)$ is an arbitrary real continuous function, which is 2π-periodic in t, satisfies the Lipschitz condition for X in any finite region and obeys the inequality

$$\|R(t, X)\| \leq \varphi(\|X\|) \|X\|^{\mu},$$

where $\varphi(\|X\|) \to 0$ as $\|X\| \to +\infty$, then the system

$$\frac{dx_s}{dt} = X_s^{(\mu)} + r_s(t, x_1, \ldots, x_n), \quad s = 1, \ldots, n.$$

has a periodic solution possibly reducing to an equilibrium.

As shown in section 1.5, the asymptotic stability condition for the zero solution of the system having homogeneous right-hand sides implies the existence of two homogeneous functions V and W such that V is positive-definite and is of order $m - \mu + 1$, and W is negative-definite and is of order m, where m is an even number, $m > 2k$. The functions V and W satisfy all conditions of Theorem 54, since the function V can be regarded as continuously differentiable (see section 1.5), and its partial derivatives as satisfying the inequality $\left|\frac{\partial V}{\partial x_j}\right| \leq b\|X\|^{m-\mu}$. Then

$$\|R(t, X)\| \sqrt{\sum_{j=1}^{n} \left(\frac{\partial V}{\partial x_j}\right)^2} \leq b\varphi(\|X\|)\|X\|^m.$$

The function W satisfies the inequality $W < -a \|X\|^m$, where a and b are the positive constants independent of X.

Example 3. Consider an equation of the form $\dot{x} = f(x)$ which has a unique equilibrium that is global-asymptotically stable. Note that the equilibrium $x = 0$ in this equation is global-asymptotically stable if and only if $xf(x) < 0$ for $x \neq 0$. The question is: under what conditions the equation $\dot{x} = f(x) + p(t)$, where $p(t)$ is a 2π-periodic continuous real function of t, will have a periodic solution for any choice of the function $p(t)$? To this end, it is necessary and sufficient that $\sup_{x \leq 0} f(x) = +\infty$, $\inf_{x \geq 0} f(x) = -\infty$.

As the function V, we take the function x^2, then $W = 2xf(x)$. Given the function $p(t)$, there exist a finite number $m = \sup_{t \in [0, 2\pi]} |p(t)|$ and two numbers a and b such that $f(x) + m > 0$ for $x = a < 0$, $f(x) + m > 0$ for $x = b > 0$. All of the integral curves starting in the closed interval $[a, b]$ fall for $t \geq T + t_0$ within the closed interval $[a + \epsilon, b - \epsilon]$, where ϵ is sufficiently small, $\epsilon > 0$, and stay there for all $t \geq T + t_0$. Therefore, there exists a fixed point $\bar{x}_0$ of the mapping $x = x(t_0 + 2\pi N, x_0, t_0)$ of the closed interval $[a, b]$ into itself, $2\pi N \geq T$. Thus the solution $x = x(t, \bar{x}_0, t_0)$ is a periodic solution of the equation involved, where $\bar{x}_0 = x(t_0 + 2\pi N, \bar{x}_0, t_0)$. This solution does not reduce to the equilibrium if $p(t)$ is not a constant.

The necessity of the obtained conditions is established by the fact that, by definition, the equation $\dot{x} = f(x) + c$ must have at least one equilibrium for any $c \in (-\infty, +\infty)$. This means that the function $y = f(x)$ is to map a real axis $(-\infty, +\infty)$ onto itself, which in turn, is equivalent to the above-obtained conditions.

Remark. In our discussion of the last example, we used a somewhat different form of Theorem 54, viz., conditions 1) and 3) in Theorem 54 were replaced by the existence of functions V and W given for $\|X\| \geq r, t \in (-\infty, +\infty)$. In this case, the function V is bounded uniformly in t in every region $r \leq \|X\| \leq \bar{r} < +\infty$, is continuously differentiable with respect to $t, x_1, \ldots, x_n$, is related to W by

$$\frac{\partial V}{\partial t} + \sum_{s=1}^{n} \frac{\partial V}{\partial x_s} f_s(t, X) = W$$

and is such that for each $m > 0$ it is possible to select three numbers $r_1 < r_2 < r_3, r_1 \geq r$ such that the following inequalities hold

$$\sup_{\substack{\|X\|=r_1 \\ t \in (-\infty, +\infty)}} V(t, X) < \inf_{\substack{\|X\|=r_2 \\ t \in (-\infty, +\infty)}} V(t, X),$$

$$\sup_{\substack{\|X\|=r_2 \\ t \in (-\infty, +\infty)}} V(t, X) < \inf_{\substack{\|X\|=r_3 \\ t \in (-\infty, +\infty)}} V(t, X)$$

and

$$W + m\sqrt{\sum_{j=1}^{n} \left(\frac{\partial V}{\partial x_j}\right)^2} < -\beta < 0, \quad r_1 \leq \|X\| \leq r_3, \quad t \in (-\infty, +\infty),$$

where β is a constant which is possibly dependent on r_1, r_3. Under these assumptions the system (15.2) has a periodic solution which possibly reduces to the equilibrium under any choice of the vector-valued function $R(t, X)$ satisfying the condition 2) in Theorem 54.

Problem. Consider a system of n differential equations $\dot{X} = F(X)$ having a unique equilibrium $X = 0$ which is global-asymptotically stable. The question is: which of the necessary and sufficient conditions are required for the system $\dot{X} = F(X) + P(t)$ to have a periodic solution for any choice of a real continuous 2π-periodic vector-valued function $P(t)$. It is suggested to argue for or against the following points: the necessary and sufficient conditions are that the vector-valued function $Y = F(X)$ should map E_n onto itself. Of course, the Lipschitz condition is satisfied for both an equation and a system of equations in any finite region.

Chapter 4

Methods for Investigation and Construction of Stationary Modes

4.1 Elements of Recurrent Function Theory

1. As is shown in section 3.1, any compact minimal set of a dynamical system consists of recurrent motion trajectories. A.A. Andronov formulated the problem of finding mathematical techniques suitable for description of recurrent motions. This section deals with general properties of recurrent functions. Also, a class of recurrent functions having an ergodic property is constructed. This construction is based on utilization of one generalization of the Kronecker theorem with respect to simultaneous solutions for a system of inequalities. The generalization of the Kronecker theorem for a special case was made by H. Weyl, but this is new for the general case. Studies on recurrent motions of dynamical systems require mathematical techniques to describe these motions. The present section offers one possible method for construction of such mathematical techniques. In this case, we will discuss recurrent functions rather than recurrent motions of some dynamical system.

Definition 37. The function $f(t)$ given and continuous for $t \in (-\infty, +\infty)$ is called recurrent if for every $\epsilon > 0$ it is possible to select a number $L_\epsilon > 0$ such that in every interval $(\alpha, \alpha + L_\epsilon)$ of a real axis $\alpha \in (-\infty, +\infty)$ for any real number t there is a number τ_t satisfying the condition $\mid f(t+\tau_t) - f(t) \mid < \epsilon$. If the number τ for any $\epsilon > 0$ can be chosen as independent of t, then $f(t)$ is an almost periodic function in the sense of Bohr.

Denote by R_f the set of all recurrent functions.

Theorem 55 *If $f(t) \in R_f$, then the function $f(t)$ is bounded.*

Proof: Specify $\epsilon > 0$. Choose $L_\epsilon > 0$ according to Definition 37. Let $c = \sup\limits_{t \in [0, L_\epsilon]} f(t)$. Then $c < +\infty$ by the continuity of the function $f(t)$. Let t be any finite real number. According to Definition 37, we choose in the interval $(-t, -t + L_\epsilon)$ a number τ_t. Then we get $t + \tau_t \in [0, L_\epsilon]$, $\mid f(t + \tau_\epsilon) - f(t) \mid < \epsilon$. Hence we find that $\mid f(t) \mid < c + \epsilon$ for all $t \in (-\infty, +\infty)$. ∎

Theorem 56 *The set R_f is a complete space in terms of uniform convergence* *along real axis.*

Proof: Suppose a sequence of functions $f_n(t) \in R_f$ is given such that $f_n(t)$ uniformly in $t \in (-\infty, +\infty)$ converges to the function $f(t)$ as $n \to \infty$. We show that $f(t) \in R_f$. For a number $\frac{\epsilon}{3}$, by the uniform convergence it is possible to select n_0 such that

$$|f(t) - f_{n_0}(t)| < \epsilon/3 \text{ for } t \in (-\infty, +\infty).$$

By Definition 37, for the number $\frac{\epsilon}{3}$ it is possible to select a quantity $L_\epsilon > 0$ such that $|f_{n_0}(t+\tau_t) - f_{n_0}(t)| < \frac{\epsilon}{3}$, where τ_t is a certain quantity from the interval $(\alpha, \alpha + L_\epsilon)$ which corresponds to a given t. Estimate the quantity $|f(t+\tau_t) - f(t)|$:

$$|f(t+\tau_t) - f(t)| \leq |f(t+\tau_t) - f_{n_0}(t+\tau_t)| + |f(t) - f_{n_0}(t)| + |f_{n_0}(t+\tau_t) - f_{n_0}(t)| < \epsilon.$$

Consequently, the $t, \frac{\epsilon}{3}$-almost period τ_t of the function $f_{n_0}(t)$ is a t, ϵ-almost period of the function f; so $f \in R_f$. ∎

Theorem 57 *If the function $F(z)$ is uniformly continuous on the set Z of a complex plane and $f(t) \in R_f$ is a function such that the values of $f(t)$ are entirely contained in the set Z, then a superposition of $F(f(t))$ is a recurrent function, i.e. $F(f(t)) \in R_f$.*

Proof: By the uniform continuity of $F(z)$, for every $\epsilon > 0$ there is $\delta > 0$ such that for $|z_1 - z_2| < \delta$ there is $|F(z_1) - F(z_2)| < \epsilon$. For the number δ, according to Definition 37, for the function $f(t)$ we select a number $L_\delta > 0$, and in each interval $(\alpha, \alpha + L_\delta)$ we take the number τ_t corresponding to a fixed real number t and satisfying the inequality $|f(t+\tau_t) - f(t)| < \delta$. Then we get $|F(t+\tau_t) - F(f(t))| < \epsilon$, i.e. the t, δ-almost period τ_t of the function $f(t)$ is the t, ϵ-almost period of the function $F(f(t))$. Thus $F(f(t)) \in R_f$. ∎

Theorem 58 *Recurrent functions are not invariant under addition and the set R_f is not transitive.*

Proof: The function $\sin t$ is recurrent and the function $\varphi(t)$ defined by the equalities

$$\varphi(t) = \begin{cases} \sin t, & t \notin [0, \pi], \\ 0, & t \in [0, \pi] \end{cases}$$

is also recurrent. But the function $\varphi(t) = \sin t - \varphi(t)$ is not recurrent. Thus, recurrent functions are not invariant under addition.

Let $f_1 = \sin t$, $f_2 = \cos t$, $f_3 = \varphi(t)$. Then $(c_1 f_1 + c_2 f_2) \in R_f$ for any finite c_1, c_2, $(d_1 f_2 + d_2 f_3) \in R_f$ with any finite d_1, d_2, but $(l_1 f_1 + l_2 f_3) \in R_f$ not with all finite values l_1, l_2. For example, for $l_1 = 1, l_2 = -1$ we get a nonrecurrent function $\varphi(t)$. ∎

This theorem is very important, for it shows that the space R_f is nonlinear and, moreover, it cannot segregate into linear subspaces such that we might transfer into them all recurrent functions, whose linear combinations with a given function $f \in R_f$ would again be from R_f.

Definition 38. A set of real numbers E is called relatively dense if there is such number $L > 0$ such that in each interval $(\alpha, \alpha + L)$ of a real axis there exists at least one element of the set E.

Elements of Recurrent Function Theory

Take a collection of continuous real functions $p_i(t)$ that are given and continuous in $(+\infty, -\infty)$, and the system of inequalities

$$|p_j(t+\tau) - p_j(t)| < \delta (\mod 2\pi) \; j = 1, \ldots, n. \tag{4.1.1}$$

Theorem 59 *If in any sufficiently small $\delta > 0$ the set of simultaneous real solutions of inequalities (4.1.1) for any fixed t is relatively dense, then any function*

$$f(t) \sum_{k=1}^{n} c_k e^{ip_k(t)}$$

is recurrent.

Proof: Evaluate $|f(t+\tau) - f(t)|$:

$$|f(t+\tau) - f(t)| \leq \sum_{k=1}^{n} |c_k| \left| \left(e^{ip_k(t+\tau)} - e^{ip_k(t)}\right) \right|,$$

hence we obtain the inequality

$$|f(t+\tau) - f(t)| \leq \sum_{k=1}^{n} |c_k| \left| e^{i(p_k(t+\tau) - p_k(t))} - 1 \right| \leq 2 \sum_{k=1}^{n} |c_k| \sin \frac{\delta}{2}.$$

If δ in (4.1.1) is such that

$$\delta < \frac{\epsilon}{\sum_{i=1}^{n} |c_k|} \quad \text{and} \quad \delta < \pi,$$

then we get $|f(t+\tau) - f(t)| < \epsilon$, where the τ is a simultaneous real solution of system (4.1.1) corresponding to the fixed t and δ. ∎

2. Let us generalize Kronecker's theorem of simultaneous real solutions for linear inequalities. Consider N real functions $\gamma_j(t_1, \ldots, t_m), j = 1, \ldots, N$ given in the m-dimensional Euclidean space, and a collection of n linear forms of these functions with real coefficients

$$p_k = \sum_{j=1}^{n} \lambda_j^k \gamma_j, \; k = 1, \ldots, n.$$

Definition 39. The set E of points in E_m is called relatively dense if there is $T > 0$ such that in each closed cube

$$K(c_1, \ldots, c_m, T) = \{c_j \leq t_j \leq c_j + T, j = 1, \ldots, m\}$$

there is at least one point of the set E.

This brings up the question: under what conditions does the system of inequalities

$$|p_k(t_1, \ldots, t_m) - \Theta_k| < \delta(\mod 2\pi), k = 1, \ldots, n, \tag{4.1.2}$$

have a simultaneous real solution or a relatively dense set of solutions for any sufficiently small $\delta > 0$? The solution of this problem is based on the following important assertion: if the system of inequalities (4.1.2) is to have a simultaneous real solution for any sufficiently $\delta > 0$ it is necessary and sufficient that the function

$$F(t_1,\ldots,t_m) = \sum_{k=1}^n e^{i(p_k(t_1,\ldots,t_m)-\Theta_k)}$$

would possess the property

$$\sup_{E_m} |F| = n+1.$$

Consider the functional L which is specified on some class **K** of the fixed functions $f(t_1,\ldots,t_m)$ and has the property

$$|L(f)| \leq \sup_{E_m} |f(t_1,\ldots,t_m)|.$$

Denote by $h(c_1,\ldots,c_m,T)$ characteristic function of the set $K(c_1,\ldots,c_m,T)$. We say that the class **K** comprises various linear combinations of finite functions of the form

$$e^{i\sum_{k=1}^n m_k p_k} h(c_1,\ldots,c_m,T),$$

where $m_1,\ldots,m_n$ are arbitrary integers, and $c_1,\ldots,c_m,T$ are fixed. Also, we assume that the functional L is additive on these linear combinations.

Definition 40. We say that the functional L and the functions $\gamma_1,\ldots,\gamma_N$ satisfy:
1) the condition A if there are $c_1,\ldots,c_m$ such that

$$L\left(e^{i\sum_{k=1}^n m_k p_k} a h(c_1,\ldots,c_m,T)\right) \to 0 \text{ as } T \to +\infty \tag{4.1.3}$$

assuming

$$\sum_{k=1}^n m_k p_k \not\equiv 0$$

(a is any complex number);
2) condition B, if (4.1.3) holds for any real $c_1,\ldots,c_m$ uniformly with respect to $(c_1,\ldots,c_m) \in E_m$;
3) condition A_1 if

$$L(ah(\bar{c}_1,\ldots,\bar{c}_m,T)) \to k \text{ as } T \to +\infty; \tag{4.1.4}$$

4) condition B_1 if (4.1.4) holds for any real $c_1,\ldots,c_m$ uniformly in $(c_1,\ldots,c_m) \in E_m$.

Theorem 60 [Zubov (1970)] *The following assertions are valid:*
a) if system (4.1.2) for any sufficiently small $\delta > 0$ has a simultaneous real solution, then from the relationship

$$\sum_{k=1}^n m_k \Lambda_k = 0, \tag{4.1.5}$$

Elements of Recurrent Function Theory

where $m_1, \ldots, m_k$ are integers, and $\Lambda_1, \ldots, \Lambda_n$ are the N-dimensional vectors with coordinates $\lambda_1^k, \ldots, \lambda_N^k$, $k = 1, \ldots, n$, it follows that

$$\sum_{k=1}^{n} m_k \Theta_k = 0 (\bmod\ 2\pi); \qquad (4.1.6)$$

b) if conditions A and A_1 are satisfied and (4.1.5) implies (4.1.6), then the system of inequalities (4.1.2) has a simultaneous real solution for any sufficiently small $\delta > 0$;

c) if conditions B and B_1 are satisfied and (4.1.5) implies (4.1.6), then the set of simultaneous real solutions to system (4.1.2) with any sufficiently small $\delta > 0$ is relatively dense.

Proof: a) Suppose that for any sufficiently small $\delta > 0$ the system of inequalities (4.1.2) has at least one simultaneous real solution. Then we show that (4.1.5) implies (4.1.6). Fix $\delta > 0$. In this case, there are integers $m_1, \ldots, m_k$ such that inequalities of the following form hold simultaneously for some values of $t_1, \ldots, t_m, n$

$$-\delta < p_k(t_1, \ldots, t_m) - \Theta_k - 2\pi m_k < \delta.$$

Multiplying each of them by the integers $\mu_1, \ldots, \mu_k$, we obtain

$$-\mu_k \epsilon_k \delta < \mu_k p_k(t_1, \ldots, t_m) - \Theta_k \mu_k - \mu_k 2\pi m_k < \mu_k \delta \epsilon_k,$$

where $\epsilon_k = \text{sign}\ \mu_k$. Summing the inequalities we find that

$$-m\delta < \sum_{k=1}^{n}(-\mu_k \Theta_k - \mu_k 2\pi m_k) < m\delta.$$

The numbers $\mu_1, \ldots, \mu_k$ are assumed to be chosen in conformity with the inequality (4.1.5) employed here. With variations of δ in each inequality, only the integers m_k can vary. Therefore, for $\delta > 0$ we have that the fractional part of the number $\frac{1}{2\pi} \sum_{k=1}^{n} \mu_k \Theta_k$ is zero, which means that the equality (4.1.6) holds.

b) Suppose that conditions A and A_1 are satisfied and (4.1.5) implies (4.1.6). We show then that for any sufficiently small $\delta > 0$, the system (4.1.2) has a simultaneous real solution. The proof of this assertion is based on the following lemma.

Lemma 10 *System (4.1.2) has a simultaneous real solution for any sufficiently small $\delta > 0$ if and only if*

$$\sup_{E_m} |f(t_1, \ldots, t_m)| = 1 + n, \qquad (4.1.7)$$

where

$$f(t_1, \ldots, t_m) = 1 + \sum_{k=1}^{n} e^{i(p_k(t_1, \ldots, t_m) - \Theta_k)}.$$

Proof: If system (4.1.2) for any sufficiently small $\delta > 0$ has a simultaneous real solution $t_1, \ldots, t_m$, then for all of these values the following inequality holds

$$\text{Re}(e^{i(p_k(t_1, \ldots, t_m) - \Theta_k)}) > \cos \delta.$$

From this, and from the inequality $|f| \geq \operatorname{Re} f$ we derive $|f| \geq 1 + n\cos\delta$, whence it follows that
$$\sup_{E_m} |f(t_1, \ldots, t_m)| = 1 + n.$$

We now assume that the relationship (4.1.7) holds, and show that the inequality (4.1.2) has a simultaneous real solution for any sufficiently small $\delta > 0$. To prove this, we set
$$z_k = e^{i(p_k(t_1,\ldots,t_m) - \Theta_k)} = x_k + iy_k, \ x_k = \cos(p_k(t_1, \ldots, t_m) - \Theta_k),$$
$$y_k = \sin(p_k(t_1, \ldots, t_m) - \Theta_k), \ k = 1, \ldots, n.$$

Estimate $|f|^2$ in terms of x_k:
$$|f|^2 = \left(1 + \sum_{k=1}^n x_k\right)^2 + \left(\sum_{k=1}^n y^k\right)^2 \leq (n+1)\left(1 + \sum_{k=1}^n x_k^2\right) + n\left(\sum_{k=1}^n y_k^2\right)$$
$$\leq (n+1) + \sum_{k=1}^n x_k^2 + n\sum_{k=1}^n \left(x_k^2 + y_k^2\right) \leq n^2 + n\sum_{k=1}^n x_k^2 + 1.$$

Hence we get
$$|f|^2 \leq n^2 + n\sum_{k=1}^n x_k^2 + 1.$$

Follows from the last inequality, a difference of $|f|$ from $n+1$ is arbitrarily small if and only if a difference of the sum $\sum_{k=1}^n x_k^2$ from n is arbitrarily small. This is possible provided that a simultaneous difference of all $|x_k|$ from unity is arbitrarily small. From the equality
$$|f| = \sqrt{\left(1 + \sum_{k=1}^n x_k\right)^2 + \left(\sum_{k=1}^n y_k\right)^2}$$

it follows that $|f|$ is arbitrarily close to $n+1$ if and only if all x_k are simultaneously close to unity as desired. Consequently, $|f|$ is arbitrarily close to $n+1$ if and only if the cosine arguments satisfy the relation (4.1.2) for any sufficiently small $\delta > 0$. ∎

The proof of assertion b) in the theorem can be easily completed by using this lemma. Suppose that assertion b) is not true, then
$$\sup_{E_m} |f| = c < n+1.$$

Consider the function
$$z = 1 + \sum_{k=1}^n z^k$$

and its p-th degree
$$z^p = \sum A_{m_1,\ldots,m_n} z_1^{m_1} \ldots z_n^{m_n}.$$

Hence
$$f^p(t_1, \ldots, t_m) = \sum A_s e^{iq_s(t_1,\ldots,t_m)},$$

Elements of Recurrent Function Theory 205

where $A_s = \sum A_{m_1,\ldots,m_n} e^{i\varphi_s}$, and the summation is made over all those integers $m_1, \ldots, m_n$ which satisfy the relationship

$$q_s(t_1, \ldots, t_m) = \sum_{k=1}^{n} m_k p_k(t_1, \ldots, t_m).$$

The presence in A_s of a complex multiplier with unity in absolute value is interpreted as follows. If m_k' and m_k'' are two number systems satisfying (4.1.5), then the condition (4.1.5) with integers $m_k = m_k' - m_k''$ is satisfied and, by assumption, the relationship (4.1.6) with the same m_k is valid. Hence, the following equality holds

$$\sum_{k=1}^{n} m_k'' \Theta_k = \sum_{k=1}^{n} m_k' \Theta_k (\text{mod} 2\pi),$$

i.e. similar terms appearing in the function f^p have the same phases to the nearest integer multiple 2π. We multiply both sides of the equality $f^p = \sum A_s e^{iq_s}$ by e^{-iq_s}:

$$f^p e^{-iq_s} = A_s + \sum{}' A_\gamma e^{i(q_\gamma - q_s)}.$$

here the sum $\sum'$ is extended to cover all $\gamma \neq s$. Multiplying the last equality by the characteristic function $h(c_1, \ldots, c_m, T)$ and then applying the functional L, after passage to the limit we obtain

$$A_s = \lim_{T \to \infty} L(h(c_1, \ldots, c_m, T)) f^p e^{-iq_s}.$$

From the last inequality we find

$$|A_s| \leq c^p. \tag{4.1.8}$$

Hence we have

$$|f|^p \leq \sum |A_s| = (n+1)^p$$

and

$$|f|^p \leq \sum |A_s| \leq c^p f_p,$$

where f_p is the number of terms in the sum $\sum A_s e^{iq_s}$. Then we obtain

$$|f|^p \leq c^p z_p,$$

where z_p is the number of terms in the sum $\sum A_{m_1,\ldots,m_n} z_1^{m_1}, \ldots, z_n^{m_n}$. By induction, it is possible to establish that $z_p = (p+1)^n$. Thus, for all p we have

$$\sum |A_s| = (n+1)^p \leq (p+1)^n c^p,$$

whence

$$1 \leq (p+1)^n e^{-\alpha p},$$

where $\alpha = -\ln \frac{c}{n+1} > 0$, which is not possible. Therefore, our assumption is incorrect and assertion b) is valid.

c) We show that if (4.1.5) implies (4.1.6), then under conditions B and B_1 the system (4.1.2) has a relatively dense set of simultaneous real solutions for any sufficiently small $\delta > 0$. Availability of at least one real solution to system (4.1.2) for a sufficiently small $\delta > 0$ follows from b), hence it remains only to prove that the collection of real solutions constitutes for any sufficiently small $\delta > 0$ a relatively dense set in the space E_m. Take the opposite case. Then there exists a sufficiently small $\delta > 0$ for which it is possible to choose a sequence of points $(c_1^k, \ldots, c_m^k)$ and a sequence of numbers $T_k, k = 1, 2, \ldots$, such that in the closed cube $K(c_1^k, \ldots, c_m^k, T_k)$ there are no simultaneous solutions to system (4.1.2) and $T_k \to \infty$ as $k \to \infty$. Then it follows from system (4.1.2) that there exists a number $c < n + 1$ such that $\sup\limits_{\substack{K(c_1^k,\ldots,c_m^k,T_k)\\k=1,2,\ldots}} |f| = c$.

As in b), it follows from conditions B_1 and B that

$$A_s = \lim_{T_k \to +\infty} L(h(c_1^k, \ldots, c_m^k, T_k)) f^p e^{iq_s},$$

whence $|A_s| \leq c^p$. From this inequality we have that in each cube discussed above the following relationships are valid

$$|f|^p = \sum |A_s| = (n+1)^p,$$

$$|f|^p = \sum |A_s| \leq z_p c^p \leq (p+1)^n c^p.$$

Therefore, in each such cube the inequality $1 \leq (1+p)^n e^{-\alpha p}$ holds for all p, which is not possible. Thus, for any sufficiently small $\delta > 0$ the arrangement of solutions to system (4.1.2) in E_m is relatively dense. ∎

Corollary 1. If the relationship (4.1.5) holds only for $m_1 = \ldots = m_n = 0$, then under conditions A, A_1 and with an arbitrary choice of real quantities $\Theta_1, \Theta_2, \ldots, \Theta_n$, the system (4.1.2) has a simultaneous real solution for any sufficiently small $\delta > 0$. If conditions B, B_1 are satisfied, then the arrangement of these simultaneous real solutions in E_m is relatively dense.

The proof of this corollary is contained in the above reasonings. In this case, quantities A_s are of the form $A_{m_1,\ldots,m_n} e^{i\psi_s}$.

Definition 41. The vectors $\Lambda_1, \ldots, \Lambda_n$ in the Euclidean real space E_n are called rationally independent if $\sum\limits_{k=1}^{n} m_k \Lambda_k = 0$ implies $m_1 = \ldots = m_n = 0$, where $m_1, \ldots, m_n$ are integers.

Thus, Corollary 1 covers the case where the linear form coefficients p_k constitute a rationally independent system of vectors.

Corollary 2. If $\Theta_k = p_k(\tau_1, \ldots, \tau_m)$, where $\tau_1, \ldots, \tau_m$ are some fixed point in the space E_m, then the fulfillment of (4.1.4) entails the fulfillment of (4.1.6) and the inequality system (4.1.2) for any sufficiently small $\delta > 0$ and any finite real numbers $\tau_1, \ldots, \tau_m$ has a simultaneous real solution only if A and A_1 are satisfied. If, however, the conditions B and B_1 are satisfied, then the set of these simultaneous real solutions is relatively dense.

Remark 1. The continuity of functions $\gamma(t_1, \ldots, t_m)$ is not employed in Theorem 60. Therefore, they may be considered as arbitrary (real- and single-valued) and

Elements of Recurrent Function Theory

such that the functional L is determined on the functions of the above-mentioned class K.

Remark 2. In item b) of Theorem 60, we proved the existence of a simultaneous real solution in the part of space E_m determined by inequalities $t_k \geq \bar{c}_k, k = 1, \ldots, m$. Similarly, we may prove the existence of a simultaneous real solution in the part of space E_m determined by inequalities $t_k \leq \bar{c}_k, k = 1, \ldots, m$, if conditions A and A_1 are changed in a proper way. For example, if conditions B and B_1 are satisfied only in the part of space determined by inequalities $t_k \geq 0$, then in this part it is possible to guarantee a relatively dense distribution of simultaneous real solutions to inequalities (4.1.2).

Remark 3. In general, the functions $\gamma_j(t_1, \ldots, t_m)$ can be specified on some arbitrary set G in the space E_m, and the problem is to seek simultaneous real solutions of (4.1.2) that are the points of this set. Basically, Theorem 60 is extended to this case without any changes. The challenge is to determine the functional L in a proper way. For example, if the functions γ_j are dependent on one variable and G is a closed set along a real straight line, then redefining the functions γ_j on additional intervals by constant values, while preserving continuity on the right or on the left, we obtain the new functions defined over the entire E_m. If the conditions of Theorem 60 hold for them, then for the original functions there are simultaneous real solutions of inequalities (4.1.2) from G.

We now focus on a special choice of the functional L. Let $g(t_1, \ldots, t_m)$ be a finite function which is summable in the sense of Lebesgue. Set

$$L(g) = \int_{c_1}^{c_1+T_1} \int_{c_2}^{c_2+T_2} \cdots \int_{c_m}^{c_m+T_m} \frac{g(t_1, \ldots, t_m) dt_1 \ldots dt_m}{T_1 \ldots T_m},$$

where $T_k > 0$, c_k are chosen so that outside the parallelepiped $c_k \leq t_k \leq c_k + T_k, k = 1, \ldots, m$ we have $g = 0$, and on any $m-1$ face of this parallelepiped there is at least one point at which $g \neq 0$. When using this functional, the conditions A_1 and B_1 are satisfied automatically, the condition A reduces to fulfillment of the equality

$$\lim_{T \to +\infty} \frac{1}{T^m} \int_{\bar{c}_1}^{\bar{c}_1+T} \cdots \int_{\bar{c}_m}^{\bar{c}_m+T} e^{i \sum_{k=1}^{n} m_k p_k} dt_1 \ldots dt_m = 0 \quad (4.1.9)$$

for $\sum_{k=1}^{n} m_k p_k \not\equiv 0$ and condition B reduces to fulfillment of equality (4.1.9) for any real $c_1, \ldots, c_m$ uniformly in $(c_1, \ldots, c_m) \in E_m$. This condition will be denoted by C.

Henceforth we shall consider the case $m = 1$.

Example 1. Let $\gamma_j(t) = t^j$, then $p_k = \sum_{j=1}^{N} \lambda_j^k t^j$. We will show that the condition (4.1.9) and, moreover, the condition C hold for the above-introduced functional L.

Let q_s be a polynomial of degree not exceeding N:

$$q_s = \sum_{k=1}^{n} m_k p_k,$$

where $m_1, ..., m_k$ are integers. If the degree of q_s equals unity, then conditions (4.1.9) and C are satisfied explicitly. Let the q_s degree be greater than unity. Then it is possible to choose a segment $[-c, c]$ such that all real roots of the equation $q'_s = 0$ lie in this segment. The purpose of further discussion is to prove that the following quantity tends to zero uniformly in $D \in (-\infty, +\infty)$

$$\frac{1}{T} \int_D^{D+T} e^{iq_s} dt \text{ as } T \to +\infty.$$

For definiteness, we assume that $D < -c$. Then for a sufficiently large T, we can evaluate the quantity involved

$$\frac{1}{T} \int_D^{D+T} e^{iq_s} dt = \frac{1}{T} \left(\int_D^{-c} e^{iq_s} dt + \int_{-c}^{c} e^{iq_s} dt + \int_c^{D+T} e^{iq_s} dt \right),$$

viz.

$$\left| \frac{1}{T} \int_D^{D+T} e^{iq_s} dt \right| \leq \frac{2c + \left| \int_D^{-c} e^{iq_s} dt \right| + \left| \int_c^{D+T} e^{iq_s} dt \right|}{T}.$$

The last two integrals are taken from the formula of integration by parts:

$$\int_x^y e^{iq_s} dt = \left(\frac{e^{iq_s}}{iq'_s} \right) \Big|_x^y + \int_x^y e^{iq_s} \frac{q''_s}{i(q'_s)^2} dt.$$

Using this relationship, for $[x, y] \cap [-c_1 + c] = \Lambda$ (here Λ represents the empty set), we find that

$$\int_x^y e^{iq_s} dt \leq \frac{1}{|q'_s(x)|} + \frac{1}{|q'_s(y)|} + \gamma \left(|x|^{1-l} + |y|^{1-l} \right),$$

where l is the actual degree of the polynomial q_s, and γ is some constant independent of x and y. Thus,

$$\left| \frac{1}{T} \int_D^{D+T} e^{iq_s} dt \right| \leq \frac{1}{T} \left(2c + \frac{1}{|q'_s(D)|} + \frac{1}{|q'_s(-c)|} \right.$$

$$\left. + \gamma \left(|D|^{1-l} + |c|^{1-l} \right) + \frac{1}{|q'_s(c)|} + \frac{1}{|q'_s(D+T)|} + \gamma \left(|c|^{1-l} + |D+T|^{1-l} \right) \right).$$

It follows from this inequality that for $D < -c$ by $\epsilon > 0$ it is possible to choose a number $T_1 > 0$ such that for $T > T_1$ we obtain $\frac{1}{T} \left| \int_D^{D+T} e^{iq_s} dt \right| < \epsilon$. If $D \in [-c, +c]$, then, by applying similar reasoning and estimations, we find a number $T_2 > 0$ such that for $T > T_2$ we obtain $\frac{1}{T} \left| \int_D^{D+T} e^{iq_s} dt \right| < \epsilon$. If $D > c$, then we find similarly a number $T_3 > 0$ having the same property. Denoting $T_\epsilon = \max_{1 \leq j \leq 3} T_j$, we have that

$$\frac{1}{T} \left| \int_D^{D+T} e^{iq_s} dt \right| < \epsilon \text{ for } T > T_\epsilon, \ D \in (-\infty, +\infty).$$

Elements of Recurrent Function Theory 209

This implies condition C.

Thus, if in (4.1.2) $\gamma_j = t^j$, then system (4.1.2) has a relatively dense set of simultaneous real solutions for any sufficiently small $\delta > 0$ only if (4.1.5) implies (4.1.6). If, however, in (4.1.2), all other things being equal, $\Theta_k = p_k(\tau) = \sum_{j=1}^{N} \lambda_j^k \tau^j$, then for each $\tau \in (-\infty, +\infty)$ the system (4.1.2) with any sufficiently small $\delta > 0$ has a relatively dense set of simultaneous real solutions, since in this case the fulfillment of condition (4.1.5) is equivalent to that of condition (4.1.6) for any $\tau \in (-\infty, +\infty)$.

Example 2. Set $\gamma_j = t^{j/N}$, where $j = 1, ..., N$ and N is odd. Denote $p_k = \sum_{j=1}^{N} \lambda_j^k t^{j/N}$ and consider the system of inequalities (4.1.2) with the thus chosen functions p_k. We show that if the numbers λ_j^k are rationally independent, the system (4.1.2) then has a relatively dense set of simultaneous real solutions for any choice of real quantities Θ_k and for any sufficiently small $\delta > 0$.

In this case, relationship (4.1.5) holds for $m_1 = ... = m_n = 0$ only. Consequently, (4.1.5) implies (4.1.6) for any real numbers $\Theta_1, ..., \Theta_n$. It remains to show that the property B is satisfied, i.e., that $\lim_{T \to \infty} \frac{1}{T} \int_D^{D+T} e^{iq_s} dt = 0$ uniformly in $D \in (-\infty, +\infty)$, where q_s is some linear form of quantities γ_j containing $\gamma_N = t$ because of the rational independence of quantities λ_j^k. Replace $t = \tau^N$ in q_s. Then q_s becomes a polynomial of degree N in τ. Choose a segment $[-c, +c]$ such that all roots of the equation $q = 0$ lie in this segment. Assume that $D < -c$. Partition the integral $\int_D^{D+T} e^{iq_s} dt$ into three integrals by segments $[D, -c]$, $[-c, +c]$, $[c, D+T]$. To estimate the first and the third of these, we consider the integral $\int_x^y e^{iq_s} dt$. After the above replacement it transforms to the integral

$$\int_{x'}^{y'} e^{iq_s(\tau^N)} \tau^{N-1} N d\tau,$$

which is computed by parts:

$$\int_{x'}^{y'} e^{iq_s(\tau^N)} \tau^{N-1} N d\tau = N \left(\frac{e^{iq_s} \tau^{N-1}}{iq_s' \tau} \right) \Big|_{x'}^{y'} - \int_{x'}^{y'} e^{iq_s} \left(\frac{N \tau^{N-1}}{iq_s' \tau} \right)' d\tau.$$

The numerator of the fraction standing in the last integral has τ to the power not exceeding $2N - 4$, whereas the denominator has τ^{2N-2}. If $[x, y] \cap [-c, +c] = \Lambda$, then $\int_x^y e^{iq_s} dt$ can be estimated as follows:

$$\left| \int_x^y e^{iq_s} dt \right| \leq c_1 + c_2 \left(|x|^{-1/N} + |y|^{-1/N} \right),$$

where c_1 and c_2 are some constants that are independent of x and y. From this estimation we obtain

$$\left| \frac{1}{T} \int_D^{D+T} e^{iq_s} dt \right| \leq 2c + c_1 + c_2 \left(|D|^{-1/N} + |c|^{-1/N} \right) + c_1 + c_2 \left(|c|^{-1/N} + |D+T|^{-1/N} \right).$$

When $D \in [-c, +c]$ or $D > c$, similar estimations are valid, whence it follows that for $\epsilon > 0$ it is possible to choose a number $T_\epsilon > 0$ for which

$$\left| \frac{1}{T} \int_D^{D+T} e^{iq_a} dt \right| < \epsilon$$

when $T > T_\epsilon$. Hence condition C is satisfied. Thus, with this choice of functions p_k and an arbitrary choice of real constants $\Theta_1, \ldots, \Theta_n$, the system of inequalities (4.1.2) has a relatively dense set of simultaneous real solutions for any sufficiently small $\delta > 0$.

Example 3. Let $\gamma_j = t\omega_j(t)$, where ω_j are 2π-periodic continuous functions. Set

$$p_k = \sum_{j=1}^N \lambda_j^k \gamma_j = \varphi_k(t).$$

Denote by $\omega(t)$ a linear combination of functions φ_k with integer coefficients m_k. If for $t \in [0, 2\pi]$ the equation $\omega(t) = 0 \pmod{1}$ has a finite number of solutions that are possibly dependent on $m_1, \ldots, m_n$, and if (4.1.5) implies (4.1.6), then system (4.1.2) has a relatively dense set of simultaneous real solutions for $p_k = \varphi_k(t)$ and any sufficiently small $\delta > 0$.

The proof of this statement amounts to verifying condition C. We have

$$\frac{1}{T} \int_D^{D+T} e^{it\omega(t)} dt = \frac{1}{T} \int_0^T e^{i(t+D)\omega(t+D)} dt.$$

Setting $T = 2\pi(n+1)$, we consider the integral

$$\int_{2k\pi}^{2(k+1)\pi} e^{iD\omega(t+D)} e^{it\omega(t+D)} dt = \int_0^{2\pi} e^{iD\omega(t+D)} e^{it\omega(t+D)} e^{i2k\pi\omega(t+D)} dt,$$

whence

$$\frac{1}{T} \int_0^T e^{i(D+t)\omega(D+t)} dt = \frac{1}{T} \int_0^{2\pi} e^{iD\omega(t+D)} e^{it\omega(t+D)} \frac{e^{i2\pi(n+1)\omega(t+D)} - 1}{e^{i2\pi\omega(t+D)} - 1} dt.$$

Hence, we get

$$\left| \frac{1}{T} \int_D^{D+T} e^{it\omega(t)} dt \right| \le \frac{1}{T} \int_0^{2\pi} \left| \frac{e^{i2\pi(n+1)\omega(t+D)} - 1}{e^{i2\pi\omega(t+D)} - 1} \right| dt$$

$$= \frac{1}{T} \int_D^{D+2\pi} \left| \frac{e^{i2\pi(n+1)\omega(t)} - 1}{e^{i2\pi\omega(t)} - 1} \right| dt.$$

The denominator of the fraction under the integral has a finite number of roots on any 2π length interval. At these points the integrand fraction equals $n+1$.

Let $\epsilon > 0$. Choose $\delta < 0$ so small that $\sum \delta < \frac{\epsilon}{2}$, where the sum is extended to cover all zeros in the denominator of the integrand fraction. Let $\gamma(\delta)$ be the smallest value of $|e^{i2\pi\omega(t)} - 1|$ on S' (S' is the segment $[0, 2\pi]$ without intervals of length δ

Elements of Recurrent Function Theory

containing zeros of this function). Clearly, $\gamma(\delta) \to 0$ as $\delta \to 0$. Choose T so that $\frac{4\pi}{\gamma T} < \frac{\epsilon}{2}$. Then

$$\frac{1}{T}\int_D^{D+2\pi}\left|\frac{e^{i2\pi(n+1)\omega(t)}-1}{e^{i2\pi\omega(t)}-1}\right|dt \leq \int_S\left|\frac{e^{i2\pi(n+1)\omega(t)}-1}{e^{i2\pi\omega(t)}-1}\right|dt + \int_{S'}\left|\frac{e^{i2\pi(n+1)\omega(t)}-1}{e^{i2\pi\omega(t)}-1}\right|dt \leq \epsilon,$$

where S is the union of all intervals, each being of length δ and containing integrand denominator zeros. It is readily seen that

$$\lim_{T\to+\infty}\frac{1}{T}\int_D^{D+T}e^{it\omega(t)}dt = \lim_{n\to\infty}\frac{1}{2\pi(n+1)}\int_D^{D+2\pi(n+1)}e^{it\omega(t)}dt = 0$$

uniformly in $D \in (-\infty, +\infty)$, which completes the proof of the statement given above.

3. In what follows, we assume that the functions $\gamma_1(t), \ldots, \gamma_n(t)$ are real, linearly independent, continuous, given for $t \in (-\infty, +\infty)$, and satisfy condition B:

$$\lim_{T\to+\infty}\frac{1}{T}\int_D^{D+T}e^{iq(t)}dt = 0$$

uniformly in $D \in (-\infty, +\infty)$, where $q = \sum_{j=1}^{N}c_j\gamma_j \not\equiv 0$, $c_1, \ldots, c_N$ are any real constants. The set of all recurrent functions is a complete space in terms of uniform convergence along the entire real axis, but its study involves serious problems due to non-linearity and non-transitivity of R_f. This suggests that studies should be carried out on various classes of recurrent functions united by some common property. Such a property is represented by availability of common t, ϵ-almost periods τ_t with a fixed t.

Denote by $H^N(\gamma_1(t), \ldots, \gamma_N(t))$ the set of all recurrent functions $f(t)$ satisfying the following condition. For every $\epsilon > 0$ it is possible to select a collection of forms $P_{k\epsilon}(t)$ ($k \leq n_\epsilon$) with real coefficients that are linear to $\gamma_1, \ldots, \gamma_N$, and a number $\delta > 0$ such that all simultaneous real solutions of the system

$$|P_{k\epsilon}(t+\tau) - P_{k\epsilon}(t)| < \delta \pmod{2\pi}, \quad k \leq n_\epsilon,$$

with any fixed τ are the t, ϵ-almost periods of the function $f(t)$, i.e.

$$|f(t+\tau) - f(t)| < \epsilon.$$

Of course, the forms $P_{k\epsilon}$, their number and the quantity δ are, generally, dependent on the function f.

Theorem 61 [Zubov (1970)] *Suppose a recurrent function $f \in H^N(\gamma_1(t), \ldots, \gamma_N(t))$. Then there exists a countable sequence of real coefficient forms P_k, $k = 1, 2, \ldots$, linear in $\gamma_1(t), \ldots, \gamma_N(t)$, such that for every $\epsilon > 0$ it is possible to select numbers $N_\epsilon > 0$, $\delta > 0$, (N_ϵ is an integer) such that for every fixed t all simultaneous real solutions $\tau = \tau_t$ of the system*

$$|P_k(t+\tau) - P_k(t)| < \delta (mod\ 2\pi), \quad k \leq N_\epsilon,$$

are the t, ϵ-almost periods of the function $f(t)$.

Proof: If $f \in H^N(\gamma_1(t), \ldots, \gamma_N(t))$, then for $\epsilon_k > 0$, $\epsilon_k \underset{k \to \infty}{\longrightarrow} 0$ it is possible to select collection sequences of the forms $P_{l\epsilon_k}$ ($l \leq n_{\epsilon_k}$) and numbers δ_l such that every simultaneous real solution of the system

$$| P_{l\epsilon_k}(t+\tau) - P_{l\epsilon_k}(t) | < \delta_l (\mathrm{mod}\ 2\pi),\ l = 1, \ldots, n_\epsilon, \tag{4.1.10}$$

is a t, ϵ_k-almost period of the function $f(t)$. The functions $P_{l\epsilon_k}$ are labelled with natural numbers in increasing order of k, each being denoted by P_k. Choose $\epsilon > 0$. Find k such that $\epsilon_l < \epsilon$. Let $\delta = \delta_k$ and $n_\epsilon = \sum_{l \leq k} n_{\epsilon l}$. From Theorem 60 it then follows that with any fixed t the system of inequalities

$$| P_k(t+\tau) - P_k(t) | < \delta (\mathrm{mod}\ 2\pi)$$

has a relatively dense set of simultaneous real solutions. All these solutions satisfy the system (4.1.10). Therefore, for $l = k$ they are t, ϵ_k-almost periods, and hence t, ϵ-almost periods of $f(t)$. ∎

Corollary. The set $H^1(\gamma_1)$ with $\gamma_1 = t$ is the collection of all almost periodic Bohr functions.

In fact, as shown by N.N.Bogolyubov, if $f(t)$ is the almost periodic Bohr function, i.e. τ can be chosen as t-independent, then there exists only a countable sequence of real numbers λ_k such that for $\epsilon > 0$ it is possible to select $n_\epsilon > 0$, $\delta > 0$ (n_ϵ is an integer) for which all simultaneous real solutions of the system

$$| \lambda_k \tau | < \delta (\mathrm{mod}\ 2\pi),\ k \leq n_\epsilon,$$

are ϵ-almost periods of the functions $f(t)$, which by Theorem 61 means that $f(t) \in H^1(\gamma_1)$, where $\gamma_1 = t$.

Remark. Theorem 61 may be used as the basis for a new definition of the class $H^N(\gamma_1(t), \ldots, \gamma_N(t))$ which in some cases proves to be more convenient, for the ϵ determines here only δ and the number of the forms P_k, but not the forms themselves.

Theorem 62 [Zubov (1970)] *The space $H^N(\gamma_1(t), \ldots, \gamma_N(t))$ is linear and complete in terms of uniform convergence along the entire real axis.*

Proof: Suppose $f_n(t) \in H^N(\gamma_1(t), \ldots, \gamma_n(t))$ and f_n converge uniformly to f over $(-\infty, +\infty)$. Then $f(t)$ is a recurrent function and, moreover, $f \in H^N$. In fact, for $\frac{\epsilon}{3}$ it is possible to choose $n_{\epsilon/3}$ such that $| f(t) - f_n(t) | < \frac{\epsilon}{3}$ for $n \geq n_{\epsilon/3}$. Fix one of these n. By definition, for a given number $\frac{\epsilon}{3}$ we can choose $\delta > 0$ and $k_{\epsilon/3}$ such that simultaneous real solutions to the system of inequalities

$$| P_{l\frac{\epsilon}{3}}(t+\tau) - P_{l\frac{\epsilon}{3}}(t) | < \delta (\mathrm{mod}\ 2\pi),\ l \leq k_{\frac{\epsilon}{3}},$$

are $t, \frac{\epsilon}{3}$-almost periods of the function $f_n(t)$, i.e.

$$| f_n(t+\tau) - f_n(t) | < \epsilon/3.$$

Evaluate $|f(t+\tau) - f(t)|$:

$$|f(t+\tau) - f(t)| \leq |f(t+\tau) - f_n(t+\tau)| + |f(t) - f_n(t)| + |f_n(t+\tau) - f_n(t)| < \epsilon;$$

hence

$$f \in H^N(\gamma_1(t), \ldots, \gamma_N(t)).$$

We shall now show that the space $H^N(\gamma_1(t), \ldots, \gamma_N(t)) = H^N$ is linear. To do this, it suffices to show that if $f(t) \in H^N$, $g(t) \in H^N$, then $(f+g) \in H^N$, for it follows from the condition $f \in H^N$ that $cf \in H^N$, where c is an arbitrary complex number, and then any finite linear combination of H^N elements with complex coefficients belongs to the space H^N. We show that $(f+g) \in H^N$ if $f, g \in H^N$. By definition, for the number $\frac{\epsilon}{2}$ it is possible to select two collections of linear forms P_{fl} and P_{gl} and numbers δ_1 and δ_2 such that all simultaneous real solutions of the system

$$|P_{fl}(t+\tau) - P_{fl}(t)| < \delta_1 (\text{mod } 2\pi), \quad l \leq k_f, \qquad (4.1.11)$$

are $t, \frac{\epsilon}{2}$-almost periods of the function f, and all simultaneous real solutions of the system

$$|P_{gl}(\tau+t) - P_{gl}(t)| < \delta_2 (\text{mod} 2\pi), \quad l \leq k_g, \qquad (4.1.12)$$

are the $t, \frac{\epsilon}{2}$-almost periods of the function g. If in these two systems the numbers δ_1 and δ_2 are replaced by $\delta = \min(\delta_1, \delta_2)$, then we shall obtain a system of $(k_f + k_g)$ inequalities. Simultaneous real solutions of this system are, at the same time, the $t, \frac{\epsilon}{2}$-almost periods of the functions f and the $t, \frac{\epsilon}{2}$-almost periods of the function g. Hence, the function $h(t) = f(t) + g(t)$ satisfies the inequality

$$|h(t+\tau) - h(t)| < |f(t+\tau) - f(t)| + |g(t+\tau) - g(t)| < \epsilon,$$

i.e. $h(t) \in H^N(\gamma_1(t), \ldots, \gamma_N(t))$. ∎

Corollary. If $f(t) = e^{i\sum_{j=1}^{N} \lambda_j \gamma_j(t)}$, then $f \in H^N(\gamma_1, \ldots, \gamma_N)$.

In fact, we have the equalities

$$|f(t+\tau) - f(t)| = \left| e^{i\sum_{j=1}^{N} \lambda_j \gamma_j(t+\tau)} - e^{i\sum_{j=1}^{N} \lambda_j \gamma_j(t)} \right| = \left| e^{i\sum_{j=1}^{N} \lambda_j (\gamma_j(t+\tau) - \gamma_j(t))} - 1 \right|$$

$$= \sqrt{\left(\cos \sum_{j=1}^{N} \lambda_j (\gamma_j(t+\tau) - \gamma_j(t)) - 1\right)^2 + \sin^2 \sum_{j=1}^{N} \lambda_j (\gamma_j(t+\tau) - \gamma_j(t))}$$

$$= 2 \left| \sin \frac{1}{2} \sum_{j=1}^{N} \lambda_j (\gamma_j(t+\tau) - \gamma_j(t)) \right| < \epsilon,$$

if τ is chosen so that

$$\left| \sum_{j=1}^{N} \lambda_j (\gamma_j(t+\tau) - \gamma_j(t)) \right| < \delta (\text{mod } 2\pi),$$

where $\delta < \pi$ and $\delta < \epsilon$. Thus, the definition suggests that $f \in H^N(\gamma_1, \ldots, \gamma_N)$. By Theorem 62, we then have that any trigonometric polynomial of the form

$$f(t) = \sum_{k=1}^{N} c_k e^{i \sum_{j=1}^{N} \lambda_j^k \gamma_j(t)}$$

belongs to the space $H^N(\gamma_1, \ldots, \gamma_N)$. ∎

Denote by $G^N(\gamma_1, \ldots, \gamma_N)$ the collection of all such trigonometric polynomials. The above reasonings suggest that

$$G^N(\gamma_1, \ldots, \gamma_N) \subset H^N(\gamma_1, \ldots, \gamma_N).$$

Below we show that G^N is dense everywhere in H^N.

Theorem 63 *If the function $\varphi(z)$ is uniformly continuous on the set G of a complex plane and $f(t)$ is a recurrent function from $H^N(\gamma_1, \ldots, \gamma_N)$ with values in G, then a superposition of $\varphi(f(t))$ is a recurrent function from $H^N(\gamma_1, \ldots, \gamma_N)$.*

Proof: By uniform continuity of $\varphi(z)$, for $\epsilon > 0$ there is $\delta > 0$ such that $|\varphi(z_2) - \varphi(z_1)| < \epsilon$ for $|z_2 - z_1| < \delta$. For the number δ from the definition of space $H^N(\gamma_1, \ldots, \gamma_N)$ it is possible to choose numbers γ and a collection of functions $P_{l\delta}$ such that all simultaneous real solutions of the system $| P_{l\delta}(\tau + t) - P_{l\delta}(t) | < \gamma \pmod{2\pi}$, $l = 1, \ldots, k_\delta$, are the t, δ-almost periods of the function $f(t)$, i.e. $| f(t + \tau) - f(t) | < \delta$, and then

$$| \varphi(f(t)) - \varphi(f(t + \tau)) | < \epsilon.$$

Hence, all simultaneous real solutions of the system of inequalities are t, ϵ-almost periods of the function $\varphi(f(t))$, which implies that this function belongs to the space $H^N(\gamma_1, \ldots, \gamma_N)$. ∎

Corollary. *If f and $g \in H^N(\gamma_1, \ldots, \gamma_N)$, then $fg \in H^N(\gamma_1, \ldots, \gamma_N)$ as well.*

In fact, the function $\varphi(z) = z^2$ is uniformly continuous in any finite rectangle on the plane of a complex variable z. We show that if $f \in H^N(\gamma_1, \ldots, \gamma_N)$, then $f^2 \in H^2(\gamma_1, \ldots, \gamma_N)$. The function f is bounded, for $f \in R_f$. Hence, a rectangle G is found to contain all f values, and by Theorem 63, a superposition of the functions $\varphi = z^2$ and f is then contained in $H^N(\gamma_1, \ldots, \gamma_N)$. The product of two functions fg can be represented as $fg = \frac{(f+g)^2 - (f-g)^2}{4}$. Each term on the right-hand side of this equality is a member of $H^N(\gamma_1, \ldots, \gamma_N)$, and since $H^N(\gamma_1, \ldots, \gamma_N)$ is linear, $fg \in H^N(\gamma_1, \ldots, \gamma_N)$ as well. ∎

If it is not essential which of the functions $\gamma_j(t)$ form the basis for construction of the space $H^N(\gamma_1, \ldots, \gamma_N)$, then we denote this space by H^N bearing in mind that it is related to the fixed functions $\gamma_1, \ldots, \gamma_N$.

We will give a few examples of recurrent functions from different spaces H^N.

Example 4. Consider trigonometric polynomials in a real form:

$$a_0 + \sum_{k=1}^{n} \left(a_k \cos \left(\sum_{j=1}^{n} \lambda_j^k t^j \right) + b_k \sin \left(\sum_{j=1}^{n} \lambda_j^k t^j \right) \right).$$

Every possible polynomial of this form is in the space $H^N(t, t^2, \ldots, t^N)$. In particular, they can describe the increasing frequency oscillations. In this case, an increase in oscillation frequency may be caused by the fundamental tone itself with sufficiently large t's rather than by overtones. A complex analogue of such polynomials is provided by polynomials of the form

$$a_0 + \sum_{k=1}^{n} c_k e^{i \sum_{j=1}^{n} \lambda_j^k t^j}.$$

Example 5. Consider a trigonometric polynomial of the form

$$a_0 + \sum_{k=1}^{n} \left(a_k \cos \left(\sum_{j=1}^{n} \lambda_j^k t^{j/N} \right) + b_k \sin \left(\sum_{j=1}^{n} \lambda_j^k t^{j/N} \right) \right),$$

where N is odd, $\lambda_N^k \neq 0$ and $\sum_{k=1}^{n} m_k \lambda_N^k \neq 0$ for any set of integers $m_1, \ldots, m_N$ that are not equal to zero at a time. Every such trigonometric polynomial is in the space $H^N(t^{1/N}, t^{2/N}, \ldots, t)$.

Here an implication is that the functions $P_{k\epsilon}$ mentioned in the definition of the space H^N must be such that the inequalities appearing in the definition would have a relatively dense set of simultaneous real solutions for any sufficiently small $\delta > 0$, which is ensured by the presence of a term of the form $\lambda_N^k t$ in each of these functions. In this case the quantities λ_N^k, $k = 1, \ldots, n_\epsilon$ are rationally independent for each ϵ.

The trigonometric polynomials can describe, specifically, oscillations with frequencies varying with time, but approaching some constant values as $t \to +\infty$. A complex analogue of such polynomials is of the form

$$a_0 + \sum_{k=1}^{n} c_k e^{i \sum_{j=1}^{N} \lambda_j^k t^{j/N}}.$$

Example 6. Consider a trigonometric polynomial of the form

$$a_0 \sum_{k=1}^{n} \left(a_k \cos \left(t \sum_{j=1}^{N} \lambda_j^k \omega_j(t) \right) + b_k \sin \left(t \sum_{j=1}^{N} \lambda_j^k \omega_j(t) \right) \right),$$

where ω_j are 2π-periodic continuous real functions. The collection of all such polynomials is in the space $H^N(t\omega_1, \ldots, t\omega_N)$. Similar functions can describe oscillations with periodically varying (say, saw-tooth type) frequencies, i.e., the oscillations that are encountered in signal transmission problems.

4. In studies of some classes of functions or function spaces, strong emphasis is placed on the theory of approximation of these functions by the functions with a simpler structure. It is well known that the collection of all trigonometric polynomials G^N belongs to H^N and, moreover, $\bar{G}^N \subset H^N$, where $\bar{G}^N$ stands for the collection of all recurrent functions that are uniform limits for polynomials from G^N along the entire real axis. This suggests that $\bar{G}^N = H^N$. The following discussion shows that this equality is valid.

Theorem 61 relates each function $f(t) \in H^N$ to the sequence $P_1, P_2, \ldots$ of linear forms with real coefficients. We show that if these functions involve some of their finite linear combinations with integer coefficients, then a newly obtained system of functions will also correspond to the function $f(t)$ in terms of Theorem 61. Indeed, let $\epsilon > 0$. Then, by Theorem 61, it is possible to select a number $\delta > 0$ and an integer n_ϵ such that all simultaneous real solutions of the system

$$| P_k(t+\tau) - P_k(t) | < \delta (\text{mod } 2\pi), \ k \leq n_\epsilon, \qquad (4.1.13)$$

are t, ϵ-almost periods of the function $f(t)$. In parallel with (4.1.3), we consider the system of inequalities

$$| \bar{P}_k(t+\tau) - \bar{P}_k(t) | < (\text{mod } 2\pi), \ k \leq m, \qquad (4.1.14)$$

where $\bar{P}_k$ are linear combinations of the functions P_k with integer coefficients. It is readily seen that a simultaneous real solution of system (4.1.13) and (4.1.14) is the t, ϵ-almost period of $f(t)$. Thus, incorporation of functions $\bar{P}_k$ into functions P_k leads to another system of functions corresponding to the same function $f(t)$. Hence it follows that the trigonometric approximation polynomial has to be constructed from all kinds of linear combinations with integer coefficients of the functions P_k, $k = 1, 2, \ldots$, taken as a finite number and corresponding to the functions $f(t) \in H^N$.

Theorem 64 [Zubov (1970)] *If $f(t) \in H^N$, then for every $\epsilon > 0$ it is possible to select a trigonometric polynomial*

$$P_\epsilon = \sum_{k=1}^n c_k e^{i \sum_{j=1}^N m_j^k P_j(t)},$$

where $P_j(t)$ are the functions corresponding to $f(t)$, and m_j^k are some integers, so that $| f(t) - P_\epsilon(t) | < \epsilon$ for $t \in (-\infty, +\infty)$.

Proof: Consider the trigonometric polynomial

$$R_n = \gamma + \frac{\sum_{k=1}^n \left(e^{i(P_k(t) - P_k(x))} + e^{-i(P_k(t) - P_k(x))} \right)}{2n},$$

where $P_k(t)$ are linear forms of the functions $\gamma_1, \ldots, \gamma_N$ corresponding to a given function $f(t) \in H^N$, by Theorem 61, and γ is a sufficiently small positive variable. Denote by $H(x, n, \delta)$ the collection of all simultaneous real solutions to the system

$$| P_k(t) - P_k(x) | < \delta (mod 2\pi), \ k \leq n, \qquad (4.1.15)$$

and by $SH(x, n, \delta)$ the complement of the set $H(x, n, \delta)$ in $(-\infty, +\infty)$. Continuity of the functions $\gamma_1, \ldots, \gamma_n$ implies measurability of each of the sets $[-T, +T] \cap H(x, n, \delta)$ and $[-T, +T] \cap SH(x, n, \delta)$, where T is any finite positive number.

We obtain $| f(x) - f(t) | < \frac{\epsilon}{2}$ for $t \in H(x, n, \delta)$ provided that n and δ are chosen for the number $\frac{\epsilon}{2}$ according to Theorem 61. In what follows, the numbers n and δ

Elements of Recurrent Function Theory

are said to be chosen exactly in this manner. For $t \in SH(x,n,\delta)$, at least one of the inequalities of system (4.1.15) fails to hold, therefore

$$|R_n| \leq \gamma + \frac{\sum_{k=1}^{n} \cos(P_k(t) - P_k(x))}{n} \leq \gamma + \frac{n-1}{n} + \frac{\cos\delta}{n}.$$

Hence we have that

$$|R_n| \leq 1 - \beta \ (\beta \in (0,1)) \text{ if } t \in SH(x,n,\delta)$$

and a number γ is chosen such that $\gamma - \frac{1-\cos\delta}{n} < 0$. In what follows, we assume that γ is chosen exactly in this manner.

All possible functions of the form

$$\sum_{k=1}^{n} m_k P_k(t), \ n = 1, 2, \ldots,$$

where m_k are integers, constitute at most a countable set. Hence, they can be enumerated and denoted by $\pi_1, \pi_2, \ldots$. The function of the argument T

$$\frac{1}{2T} \int_{-T}^{+T} f(t) e^{i\pi_1(t)} dt$$

is bounded in the absolute value by the number $D = \sup_{t \in (-\infty, +\infty)} |f(t)|$.

Therefore, there exists a sequence $T_{k_1} \to +\infty \ (k_1 \to \infty)$, such that the sequence

$$\frac{1}{2T_{k_1}} \int_{-T_{k_1}}^{+T_{k_1}} f(t) e^{i\pi_1(t)} dt$$

has a limit as $k_1 \to \infty$. The sequence

$$\frac{1}{2T_{k_2}} \int_{-T_{k_2}}^{+T_{k_2}} f(t) e^{i\pi_2(t)} dt$$

is bounded. Hence there exists a subsequence T_{k_2} such that the sequence

$$\frac{1}{2T_{k_2}} \int_{-T_{k_2}}^{+T_{k_2}} f(t) e^{i\pi_2(t)} dt$$

has a limit as $k_2 \to \infty$. Proceeding in a similar manner, we obtain a countable set of sequences T_{k_j} such that for any fixed j there is

$$\lim_{k_j \to \infty} \frac{1}{2T_{k_j}} \int_{-T_{k_j}}^{+T_{k_j}} f(t) e^{i\pi_j(t)} dt.$$

Each of these sequences is a subsequence of the preceding sequence. Denoting by T_k a diagonal subsequence of the sequence T_{k_j}, we have that for any j there exists a limit

$$\lim_{k \to \infty} \frac{1}{2T_k} \int_{-T_k}^{+T_k} f(t) e^{i\pi_j(t)} dt.$$

We fix the sequence T_k, $k = 1, 2, \ldots$. Denote by R_n^l a free term of the trigonometric polynomial R_n^{2l}. It is stated that

$$R_n^l = \lim_{T \to +\infty} \frac{1}{2T} \int_{-T}^{+T} R_n^{2l} dt,$$

hence

$$R_n^l = \lim_{k \to \infty} \frac{1}{2T_k} \int_{-T_k}^{+T_k} R_n^{2l} dt. \qquad (4.1.16)$$

Indeed,

$$R_n^{2l} = \sum_m c_m e^{i(\pi_m(t) - \pi_m(x))},$$

where summation is taken over all m such that $\pi_m = \sum_{k=1}^{n} m_k P_k$, the m_k running through the values of term indices in the formula

$$\sum z_1^{m_1} \ldots z_n^{m_n} = 1 + \sum_{i=1}^{n} \left(z_i + \frac{1}{z_i}\right)^{2l};$$

furthermore,

$$\lim_{T \to +\infty} \frac{1}{2T} \int_{-T}^{+T} e^{i\pi_m(t)} dt = 0,$$

only if $\pi_m(t) \not\equiv 0$. This fact and the expansion of R_n^{2l} imply the formula (4.1.16) for R_n^l.

Set

$$P_n^l = \lim_{T_k \to +\infty} \frac{\frac{1}{2T_k} \int_{-T_k}^{+T_k} f(t) R_n^{2l} dt}{R_n^l} = \lim_{T_k \to +\infty} \frac{\frac{1}{2T_k} \int_{-T_k}^{T_k} f(t) \sum_m c_m e^{i(\pi_m(t) - \pi_m(x))} dt}{R_n^l},$$

whence

$$P_n^l = \frac{\sum_m c_m e^{-i\pi_m(x)}}{R_n^l} \lim_{k \to \infty} \frac{1}{2T_k} \int_{-T_k}^{+T_k} f(t) e^{i\pi_m(t)} dt. \qquad (4.1.17)$$

It follows from equality (4.1.17) that for any positive integer n, l the function P_n^l is a trigonometric polynomial from G^N. And (4.1.16) suggests that the following identity holds

$$f(x) = \lim_{k \to \infty} \frac{\frac{1}{2T_k} \int_{-T_k}^{+T_k} f(x) R_n^{2l} dt}{R_n^l}.$$

Evaluate the quantity $|f(x) - P_n^l(x)|$:

$$|f(x) - P_n^l(x)| = \left| \lim_{k \to \infty} \frac{\frac{1}{2T_k} \int_{-T_k}^{+T_k} (f(t) - f(x)) R_n^{2l} dt}{R_n^l} \right|$$

Elements of Recurrent Function Theory 219

$$\leq \overline{\lim}_{k\to\infty} \frac{\frac{1}{2T_k}\int_{-T_k}^{+T_k}|f(t)-f(x)|R_n^{2l}dt}{R_n^l}. \quad (4.1.18)$$

Partition $\frac{1}{2T_k R_n^l}\int_{-T_k}^{+T_k}|f(t)-f(x)|R_n^{2l}dt$ into two integrals with respect to $[-T_k, T_k] \cap H(x, n, \delta)$ and $[-T_k, T_k] \cap SH(x, n, \delta)$. Each of these integrals exists for any x because of the measurability of the set for which the integration is carried out. The first integral is evaluated as follows:

$$\frac{1}{2T_k R_n^l}\int_{[-T_k,T_k]\cap H(x,n,\delta)}|f(t)-f(x)|R_n^{2l}dt \leq \frac{\epsilon}{2}\frac{1}{2T_k R_n^l}\int_{[-T_k,T_k]\cap H(x,n,\delta)}R_n^{2l}dt$$

$$\leq \frac{\epsilon}{2}\frac{1}{2T_k R_n^l}\int_{-T_k}^{+T_k}R_n^{2l}dt \to \frac{\epsilon}{2} \text{ as } k\to\infty.$$

The second integral is evaluated as follows:

$$\frac{1}{2T_k R_n^l}\int_{[-T_k,T_k]\cap SH(x,n,\delta)}|f(t)-f(x)|R_n^{2l}dt$$

$$\leq 2D(1-\beta)^{2l}\frac{1}{R_n^l}\int_{[-T_k,T_k]\cap SH(x,n,\delta)}\frac{R_n^{2l}}{2T_k}dt \leq 2D(1-\beta)^{2l}\frac{1}{R_n^l},$$

where $D = \sup_{t\in(-\infty,+\infty)}|f(t)|$. In order to establish the behavior of the expression $\frac{1}{R_n^l}(1-\beta)^{2l}$ for a fixed n as $l\to\infty$, it is essential to evaluate R_n^l. We have that $R_n^l \to \infty$ as $l\to\infty$ for any fixed $n > 0$. Assuming that n has been chosen, we find the number l_0 that for $l > l_0$ we have $\frac{2D(1-\beta)^{2l}}{R_n^l} < \frac{\epsilon}{2}$. Then, from (4.1.18) it follows that

$$|f(x) - P_n^l(x)| < \epsilon.$$

Thus, we can take a trigonometric polynomial P_n^l as P_ϵ. ∎

Corollary. If $(t) \in H^N$, then there exists a sequence S_k of trigonometric polynomials

$$S_k = \sum_{j=1}^{k} c_{kj} e^{iP_{kj}},$$

such that $S_k \to f(t)$ as $k\to\infty$ uniformly in $t \in (-\infty, +\infty)$.

Let S_{k0} be a free term of these polynomials. Then S_{k0} has a limit as $k\to\infty$. Indeed, let $\epsilon_k \xrightarrow[k\to\infty]{} +0$ be a sequence of positive numbers. Then, by Theorem 64, it is possible to choose a sequence of trigonometric polynomials P_{nk} such that $|f(t) - P_{nk}| < \epsilon_k$. Set $S_k = P_{nk}$ and evaluate the integral:

$$\left|\frac{1}{2T_l}\int_{-T_l}^{T_l}(f(t)-S_k(t))dt\right| < \frac{1}{2T_l}\int_{-T_l}^{T_l}|f(t)-S_k(t)|dt < \epsilon_k.$$

The sequence $\frac{1}{2T_l} \int_{-T_l}^{T_l} f(t)dt$ has a limit. Hence, S_{k0} also has a limit as $k \to \infty$, since

$$\left| \frac{1}{2T_l} \int_{-T_l}^{T_l} f(t)dt - S_{k0} \right| < \epsilon_k \to +0 \text{ as } k \to \infty.$$

Theorem 65 [Zubov (1970)] *For any function $f(t) \in H^N$ there is a mean value $M_T(f) = \lim_{T \to +\infty} \frac{1}{2T} \int_{-T}^{T} f(t)dt$ and, moreover,*

$$M_T(f) = \lim_{T \to +\infty} \frac{1}{T} \int_a^{a+T} f(t)dt,$$

this limit being uniform in $a \in (-\infty, +\infty)$.

Proof: By Corollary of Theorem 64, for a fixed function $f(t) \in H^N$ there exists a sequence of trigonometric polynomials S_k such that $S_k \xrightarrow[k \to \infty]{} f$ and there exists $\lim_{k \to \infty} S_{k0}$, where S_{k0} are free terms of these trigonometric polynomials. Denote this limit by a_0. Then we have, uniformly in a,

$$\lim_{T \to +\infty} \frac{1}{T} \int_a^{a+T} S_k dt = S_{k0} \to a_0 \text{ as } k \to \infty.$$

Specify $\epsilon > 0$. For the number $\frac{\epsilon}{3}$ we choose k_0 such that

$$|S_{k0} - a_0| < \epsilon/3, \ |f(t) - S_k(t)| < \epsilon/3 \text{ for } k \geq k_0.$$

For the number $\frac{\epsilon}{3}$ we choose T_ϵ such that for $T > T_\epsilon$ the inequality

$$\left| \frac{1}{T} \int_a^{a+T} S_k dt - S_{k0} \right| < \epsilon/3$$

is satisfied uniformly in a. Evaluate $\frac{1}{T} \int_a^{a+T} f(t)dt - a_0$:

$$\left| \frac{1}{T} \int_a^{a+T} f(t)dt - a_0 \right| \leq \left| \frac{1}{T} \int_a^{a+T} (f(t) - S_k(t))\, dt \right|$$

$$+ \left| \frac{1}{T} \int_a^{a+T} S_k dt - S_{k0} \right| + |S_{k0} - a_0| < \epsilon,$$

only if $T > T_\epsilon$ and $k \geq k_0$. Hence we conclusively get

$$\left| \frac{1}{T} \int_a^{a+T} f(t)dt - a_0 \right| < \epsilon \text{ for } T \geq T_\epsilon.$$

This suggests that the mean value $M_T(f(t))$ exists and the equality

$$M_T(t) = \lim_{T \to +\infty} \frac{1}{T} \int_a^{a+T} f(t)dt$$

holds uniformly in $a \in (-\infty, +\infty)$.

Setting $T = 2T$ and $a = -T$, we obtain

$$M_T(f) = \lim_{T \to +\infty} \frac{1}{2T} \int_{-T}^{T} f(t)dt. \qquad \blacksquare$$

Corollary. The mean value M_T can be considered to be the functional given on the functions from H^N. This functional is homogeneous, additive, and continuous in terms of convergence in H^N.

In fact, the functional homogeneity follows from the expression

$$M_T(cf) = \lim_{T \to +\infty} \frac{1}{2T} \int_{-T}^{T} cf dt = c \lim_{T \to +\infty} \frac{1}{2T} \int_{T}^{T} f dt = cM_T(f).$$

Additivity follows from the fact that if f and g are two functions from H^N, then

$$M_T(f + g) = \lim_{T \to +\infty} \frac{1}{2T} \int_{-T}^{T} (f + g)dt = \lim_{T \to +\infty} \frac{1}{2T} \int_{-T}^{T} f dt$$

$$+ \lim_{T \to +\infty} \frac{1}{2T} \int_{-T}^{T} g dt = M_T(f) + M_T(g).$$

We shall prove continuity of M_T. Suppose $f_n \in H^N$ and f_n approaches f as $n \to \infty$ uniformly in $t \in (-\infty, +\infty)$. Then $f \in H^N$, and from additivity of the functional we have

$$\mid M_T(f_n) - M_T(f) \mid = \mid M_T(f_n - f) \mid = \left| \lim_{T \to +\infty} \frac{1}{2T} \int_{-T}^{T} (f_n - f) dt \right|$$

$$\leq \lim_{T \to +\infty} \frac{1}{2T} \int_{-T}^{T} \mid f_n - f \mid dt < \epsilon,$$

if it is chosen such that $\mid f - f_n \mid < \epsilon$. This exactly shows that the functional $f(t)$ is continuous.

Thus, M_T is a linear functional given on the space H^N. $\blacksquare$

4.2 Forced Multifrequency Oscillations

1. Consider a mechanical system which is subjected to several forces of some nature. Assume that these forces are periodic time functions, though having different frequencies, viz. $\omega_1, \ldots, \omega_k$. Suppose the motion of the mechanical system is described by the system of differential equations

$$\frac{dx_s}{dt} = f_s(x_1, \ldots, x_n, z_1, \ldots, z_k), \quad s = 1, \ldots, n, \quad z_j = \omega_j t, \quad j = 1, \ldots, k. \qquad (4.2.1)$$

We assume that the functions f_s are real, continuously differentiable with respect to all finite values of their arguments and, moreover, are 2π-periodic in each of the arguments $z_1, \ldots, z_k$. The question is: under what conditions the system (4.2.1) will have the solution

$$x_s = x_s(z_1, \ldots, z_k), \ s = 1, \ldots, n, \ z_j = \omega_j t, \ j = 1, \ldots, k \qquad (4.2.2)$$

with the property that the functions $x_s(z_1, \ldots, z_k)$ are 2π-periodic in each of the arguments $z_1, \ldots, z_k$. If such a solution (4.2.2) exists, then a question arises concerning its uniqueness, the method of its construction, and the behavior of other solutions to system (4.2.1) in a sufficiently small neighborhood of the solution (4.2.2).

Partial differential equations satisfied by the functions of (4.2.2) may serve as a starting point for solution of the problem involved. The differential equations can be derived as follows. If the solution (4.2.2) of system (4.2.1) exists and the functions of (4.2.2) are continuously differentiable with respect to variables $z_1, \ldots, z_k$, then taking the total derivative with respect t and considering the equalities $z_j = \omega_j t$, we obtain

$$\sum_{j=1}^{k} \omega_j \frac{\partial x_s}{\partial z_j} = f_s(x_1, \ldots, x_n, z_1, \ldots, z_k). \qquad (4.2.3)$$

From this it follows that the solution (4.2.2) of system (4.2.1) can be found by seeking solutions to system (4.2.3) that are 2π-periodic in each of the unknown variables $z_1, \ldots, z_k$. Introduce the vectors

$$X = (x_1, \ldots, x_n), \ Z = (z_1, \ldots, z_k), \ \omega = (\omega_1, \ldots, \omega_k), \ F = (f_1, \ldots, f_n).$$

Then system (4.2.1) can be written in vector form:

$$\frac{dX}{dt} = F(X, Z). \qquad (4.2.4)$$

Suppose system (4.2.4) is quasilinear. In other words, it is of the form

$$\frac{dX}{dt} = \mathbf{P}X + Q(Z) + \mu G(X, Z), \qquad (4.2.5)$$

where $\mathbf{P}$ is a real constant square matrix, and Q and G are the real n-dimensional vector-valued functions that are given for all finite values of their arguments, and are continuously differentiable. Furthermore, $Q(Z)$ and $G(X, Z)$ are assumed to be 2π-periodic functions with respect to each component of the vector Z. The number μ is a small parameter. The system (4.2.5) is considered for $Z = \omega t$.

Theorem 66 [Zubov (1970)] *Suppose that all eigenvalues of the matrix $\mathbf{P}$ have negative real parts. Then for any sufficiently large sphere $\|X\| \leq R$ it is possible to choose a number $\mu_0 > 0$ such that for any $|\mu| \leq \mu_0$ the system (4.2.5) has a unique solution in that sphere which is representable as*

$$X = X(Z), \ Z = \omega t. \qquad (4.2.6)$$

Here $X(Z)$ is a 2π-periodic vector-valued function with respect to each component of the vector Z. This solution is uniform-asymptotically stable in the sense of Lyapunov.

Proof: In this theorem
$$\|X\| = \sqrt{\sum_{s=1}^{n} x_s^2}.$$

1. We shall show that the linear system
$$\sum_{j=1}^{k} \omega_j \frac{\partial X}{\partial z_j} = \mathbf{P}X + Q(Z) \tag{4.2.7}$$

has a unique solution
$$X = X_0(Z), \tag{4.2.8}$$

which is 2π-periodic in each component of the vector Z. In this case
$$\max \|X_0(Z)\| \le c \max \|Q(Z)\|,$$

where c is a positive constant. For the system (4.2.7) we write the equations of characteristics
$$\frac{dz_j}{dt} = \omega_j, \ j = 1, \ldots, k \tag{4.2.9}$$
$$\frac{dX}{dt} = \mathbf{P}X + Q(Z).$$

The second group of equations, as in (4.2.9), is satisfied by the function (4.2.8) if the vector Z, as a function of the independent variable t, is determined from the first group of equations, as in (4.2.9), i.e. for $Z = \omega t + C$, where C is a vector with components $c_1, \ldots, c_k$ that are arbitrary constants. Thus, if the solution (4.2.8) exists, then the following identity holds
$$\frac{dX_0}{dt} = \mathbf{P}X_0 + Q(\omega t + C), \tag{4.2.10}$$

here X_0, as a function of the independent variable t, is bounded for $t \in (-\infty, +\infty)$. Multiplying both members of (4.2.10) by $e^{-\mathbf{P}\tau}$ and then integrating with respect to τ between the limits $-\infty$ and t, we obtain the equality
$$\int_{-\infty}^{t} e^{-\mathbf{P}\tau} \left(\frac{dX_0}{dt} - \mathbf{P}X_0 \right) d\tau = \int_{-\infty}^{t} e^{-\mathbf{P}\tau} Q(\omega \tau + C) d\tau.$$

Evaluating the integral on the left-hand side of the equality and considering that $e^{-\mathbf{P}\tau} X_0 \to 0$ as $\tau \to -\infty$, we obtain the relationship
$$e^{-\mathbf{P}t} X_0 = \int_{-\infty}^{t} e^{-\mathbf{P}\tau} Q(\omega \tau + C) d\tau.$$

Multiplying both sides of the last identity by $e^{\mathbf{P}t}$, subsequently replacing the vector C under the integral sign by the formula $C = Z - \omega t$ and substituting the integration variable τ by the formula $\tau - t = \xi$, we find
$$x_0 = \int_{-\infty}^{0} e^{-\xi \mathbf{P}} Q(\omega \xi + Z) d\xi. \tag{4.2.11}$$

The vector-valued function on the right-hand side of (4.2.11) is 2π-periodic in each component of the vector Z.

We shall show that the vector function in (4.2.11) satisfies (4.2.7) and, therefore, is the required solution (4.2.8). To do this, differentiate both sides of (4.2.11) with respect to z_j, with subsequent multiplication by ω_j; and summation over j;

$$\sum_{j=1}^{k}\omega_j\frac{\partial X_0}{\partial z_j} = \int_{-\infty}^{0} e^{-\xi\mathbf{P}}\sum_{j=1}^{k}\omega_j\frac{\partial Q(\omega\xi + Z)}{\partial z_j}d\xi = \int_{-\infty}^{0} e^{-\xi\mathbf{P}}\frac{d}{d\xi}Q(\omega\xi + Z)d\xi.$$

Computing the value of the last integral by parts, we obtain

$$\sum_{j=1}^{k}\omega_j\frac{\partial X_0}{\partial z_j} = Q(Z) + \mathbf{P}\int_{-\infty}^{0} e^{-\xi\mathbf{P}}Q(\omega\xi + Z)d\xi = Q(Z) + \mathbf{P}X_0.$$

Note that here the differentiation under the integral and application of the formula of integration by parts for the improper integral are legitimate by the absolute and uniform convergence of this integral and integrals obtained by differentiating a vector-valued function Q with respect to its variables. The vector-valued function in (4.2.11) is a unique solution to the system (4.2.7). Indeed, suppose there exist two solutions of the system (4.2.7): $X_1 = X_1(Z)$ and $X_2 = X_2(Z)$. Then the difference of these solutions $\bar{X} = X_1(Z) - X_2(Z)$ satisfies the equation

$$\sum_{j=1}^{k}\omega_j\frac{\partial X}{\partial z_j} = \mathbf{P}X.$$

Consequently, the function $\bar{X}(\omega t + C)$ satisfies the system $\frac{dX}{dt} = \mathbf{P}X$ and, additionally, is a restricted solution to this system for $t \in (-\infty, +\infty)$. Since the restricted solution for the last system can only be zero, then $\bar{X} = (\omega t + C) \equiv 0$, whence we have $X_1(C) = X_2(C)$. In view of the arbitrariness of the vector C, we get $X_1(Z) \equiv X_2(Z)$. The above reasonings also suggest that (4.2.11) is a restricted unique solution to (4.2.7). Estimate the solution (4.2.11). We have

$$\left\|e^{-\xi\mathbf{P}}Q(\omega\xi + Z)\right\| \leq \left\|e^{-\xi\mathbf{P}}\right\|\|Q(\omega\xi + Z)\| \leq a_1 e^{a_2\xi}\max\|Q(Z)\|,$$

where a_1 and a_2 are some positive constants, hence from (4.2.11) we get

$$\|X_0\| \leq \left\|\int_{-\infty}^{0} e^{-\xi\mathbf{P}}Q(\omega\xi + Z)d\xi\right\| \leq c\max\|Q(Z)\|$$

and, consequently,

$$\max\|X_0(Z)\| \leq c\max\|Q(Z)\|,$$

where $c = \frac{a_1}{a_2}$. The numbers a_1 and a_2 are determined by the eigenvalues of the matrix $\mathbf{P}$ and its elements.

2. In any sufficiently large sphere $\|X\| \leq R$ there is a unique solution to system (4.2.5) only if the parameter $|\mu| \leq \mu_0$, where μ_0 is a positive value (determined

Forced Multifrequency Oscillations

below). In order to prove this statement, we will construct a system of integral equations satisfied by the functions of (4.2.6). Suppose the functions in (4.2.6) satisfy the system (4.2.5) for $Z = \omega t$. Substitute (4.2.6) into (4.2.5):

$$\sum_{j=1}^{k}\omega_j \frac{\partial X}{\partial z_j} = \mathbf{P}X + Q(Z) + \mu G(X, Z). \qquad (4.2.12)$$

In system (4.2.12) G can be taken to be the known function of the vector Z. The earlier discussion then suggests that $X(Z)$ satisfies the identity

$$X(Z) = X_0(Z) + \mu \int_{-\infty}^{0} e^{-\xi \mathbf{P}} G(X(\omega \xi + Z), \omega \xi + Z) d\xi. \qquad (4.2.13)$$

We shall consider (4.2.13) to be a system of integral equations designed to define the functions of (4.2.6). This system can also be written as

$$X(Z) = X_0(Z) + \mu K(X), \qquad (4.2.14)$$

where K is the operator mentioned in (4.2.13). Take a sufficiently large number $R > 0$ such that $\max \|Q(Z)\| < \frac{R}{c}$, then $\max \|X_0(Z)\| < R$.

We shall construct a solution to the system (4.2.14) using the method of successive approximations. Take the vector function $X_0(Z)$ as a zero approximation,

$$X_1(Z) = X_0(Z) + \mu K(X_0)$$

as a first approximation, and

$$X_{m+1}(Z) = X_0(Z) + \mu K(X_m), \ m = 0, 1, \ldots$$

as an $(m+1)$ approximation. There exists a positive number μ_1 such that for any $|\mu| < \mu_1$ we get

$$\max \|X_m(Z)\| < R, \ m = 0, 1, \ldots.$$

Indeed, let $M = \max \|G(X, Z)\|$, where max is taken for $\|X\| \leq R$ and $z_j \in (0, 2\pi)$, $j = 1, \ldots, k$. Take μ_1 such that

$$\mu_1 > 0, \ \mu_1 < \frac{R - \max \|X_0(Z)\|}{cM}.$$

In the selection of a number μ_1 for the first approximation, we have

$$\|X_1\| \leq \|X_0\| + |\mu| \|K(X_0)\| \leq \max \|X_0\| + |\mu| cM < R,$$

if $|\mu| \leq \mu_1$. These estimates allow the proof to be continued by induction.

We show that there exists a positive number μ_2 such that the series

$$X_0 + (X_1 + X_0) + (X_2 - X_1) + \ldots + (X_{m+1} - X_m) + \ldots$$

converges uniformly and absolutely in any μ satisfying the inequality $|\mu| \le \mu_2$. Since all successive approximations are contained in the sphere $\|X\| < R$ and the function G is continuously differentiable in all finite values of its arguments, then there exists a positive constant L such that

$$\|G(X_m, Z) - G(X_{m-1}, Z)\| \le L\|X_m - X_{m-1}\| \le \max\|X_m - X_{m-1}\|.$$

Estimate the terms of the above-mentioned series:

$$\|X_{m+1} - X_m\| = |\mu|\,\|K(X_m) - K(X_{m-1})\| \le |\mu|\,cL\max\|X_m - X_{m-1}\|$$

whence

$$\max\|X_{m+1} - X_m\| \le |\mu|\,cL\max\|X_m - X_{m-1}\|.$$

From this inequality we obtain

$$\max\|X_{m+1} - X_m\| \le \gamma^m \max\|X_1 - X_0\|,$$

where $\gamma = |\mu|\,cL$.

Set $0 < \mu_2 < \min\{\mu_1, \frac{1}{cL}\}$. For all μ satisfying the condition $|\mu| \le \mu_2$, the successive approximations of X_m, $m = 0, 1, \ldots$ stay in the sphere $\|X\| < R$ and, as $m \to \infty$, converge uniformly and absolutely to the function $X(Z)$ which is 2π-periodic with respect to each component of the vector Z and satisfies the equation (4.2.14) since the equality $X_{m+1} = X_0 + \mu K(X_m)$ admits passage to the limit under the sign of operator K.

The solution constructed for equation (4.2.14) is a unique solution to this equation in the region $\|X\| < R$. Indeed, assume that in the region involved there are two solutions to the equation: $X_1 = X_1(Z)$ and $X_2 = X_2(Z)$. These functions then satisfy the identities

$$X_1(Z) = X_0(Z) + \mu K(X_1), \ \ X_2(Z) = X_0(Z) + \mu K(X_2).$$

By subtracting the equalities termwise and then making estimates, we obtain

$$\|X_1 - X_2\| \le |\mu|\,cL\max\|X_1 - X_2\|;$$

hence

$$\max\|X_1 - X_2\| \le |\mu|\,cL\max\|X_1 - X_2\|.$$

If $\max\|X_1 - X_2\| > 0$, then $|\mu|\,cL \ge 1$, which is not possible since $|\mu| \le \mu_2$, $\mu_2 cL < 1$. Thus, $\max\|X_1 - X_2\| = 0$, i.e. $X_1 \equiv X_2$.

We will show that the above vector function $X = X(Z)$ with $Z = \omega t$ satisfies the system (4.2.5). Substituting $X(Z)$ into equation (4.2.13), we obtain an identity. In this identity, we perform successive transformations by the formulas $Z = \omega t$, $\xi = \tau - t$. Then we have that the function $X(Z)$ with $Z = \omega t$ satisfies the identity

$$X = \int_{-\infty}^{t} e^{\mathbf{P}(t-\tau)}(Q(\omega\tau) + \mu G(X, \omega\tau))d\tau.$$

Forced Multifrequency Oscillations

Differentiating both sides with respect to t, we find that the function X satisfies the system (4.2.5).

3. There exists a positive number $\mu_3 > 0$ such that the solution (4.2.6) of equation (4.2.5) is uniform-asymptotically Lyapunov stable for all μ such that $|\mu| < \mu_3$. Indeed, let us construct a quadratic form V satisfying the partial differential equation

$$\sum_{s=1}^{n} \frac{dV}{dx_s} \sum_{j=1}^{n} p_{sj} x_j = -\sum_{s=1}^{n} x_s^2.$$

The quadratic form is then positive-definite.

Let X denote some solution to the system (4.2.5), and $X(Z)$ its solution (4.2.6). Denote by $\bar{X}$ the difference of these solutions and calculate the total derivative $V(\bar{X})$:

$$\frac{dV(\bar{X})}{dt} = \sum_{s=1}^{n} \frac{\partial v}{\partial \bar{x}_s} \cdot \frac{d\bar{x}_s}{dt}.$$

The vector function $\bar{X}$ satisfies the equation

$$\frac{d\bar{X}}{dt} = \mathbf{P}\bar{X} + \mu(G(X,Z) - G(X(Z), Z)) = \mathbf{P}\bar{X} + \mu \mathbf{B}\bar{X},$$

where $\mathbf{B}$ is the Jacobian matrix obtained by applying the finite increment formula to the difference of vector functions. Hence it follows that

$$\frac{dV}{dt} = -\sum_{s=1}^{n} \bar{x}_s^2 + \mu(\operatorname{grad} V \mathbf{B}\bar{X}) = W.$$

Choose $\mu_3 > 0$ such that

$$-b_1 \sum_{s=1}^{n} \bar{x}_s^2 \leq W \leq -b_2 \sum_{s=1}^{n} \bar{x}_s^2$$

for $\sum_{s=1}^{n} \bar{x}_s^2 < \epsilon^2$, where ϵ is a given value, and for all μ such that $|\mu| \leq \mu_3$. Here b_1 and b_2 are positive constants. Then, by the familiar theorems of motion stability theory, the uniform asymptotic stability is available for the solution (4.2.6) of the system (4.2.5). In this case, the following inequality holds for all $t \geq t_0$

$$\|\bar{X}\| \leq \alpha_1 \|\bar{X}(t_0)\| e^{-\alpha_2(t-t_0)},$$

where α_1 and α_2 are some positive constants. Set $\mu_0 = \min\{\mu_1, \mu_2, \mu_3\}$. This number satisfies all conditions in the theorem. ∎

2. We now consider the case where the matrix $\mathbf{P}$ has no eigenvalues with zero real parts.

Theorem 67 [Zubov (1970)] *Suppose among eigenvalues of the matrix* $\mathbf{P}$ *is σ with negative real parts and $n - \sigma$ with positive parts. Then, for a sufficiently large number $R > 0$, it is possible to indicate such number $\mu_0 > 0$ that the system (4.2.5) for $|\mu| \leq$*

μ_0 has a unique solution of the form (4.2.6) in the region $\|X\| < R$. There exists a family of integral curves that are dependent on σ arbitrary constants and approach indefinitely the solution (4.2.6) as $t \to +\infty$, and there exists a family of integral curves that are dependent on $n - \sigma$ arbitrary constants and approach indefinitely the solution (4.2.6) as $t \to -\infty$.

Proof: By applying a nonsingular linear transformation to the desired functions $x_1, \ldots, x_n$, the system (4.2.5) can be reduced to the form

$$\frac{dY_1}{dt} = \mathbf{P}_1 Y_1 + \mu G_1(Y_1, Y_2, Z) + Q_1(Z),$$

$$\frac{dY_2}{dt} = \mathbf{P}_2 Y_2 + \mu G_2(Y_1, Y_2, Z) + Q_2(Z), \qquad (4.2.15)$$

where Y_1 and Y_2 are respectively the vectors of dimensions σ and $n - \sigma$; $\mathbf{P}_1$ and $\mathbf{P}_2$ are respectively the constant real matrices of orders σ and $n - \sigma$. All eigenvalues of the matrix $\mathbf{P}_1$ have negative real parts and those of the matrix $\mathbf{P}_2$ have positive real parts. We seek a solution

$$Y_i = Y_i(Z), \ i = 1, 2, \ Z = \omega t \qquad (4.2.16)$$

to (4.2.15) such that the functions $Y_i(Z)$ are 2π-periodic in each component of the vector Z. As in the earlier proofs of the theorems, we set up the equations

$$Y_1(Z) = Y_{10} + \mu K_1(Y_1, Y_2), \qquad (4.2.17)$$

where

$$Y_{10} = \int_{-\infty}^{0} e^{-\xi \mathbf{P}_1} Q_1(\omega \xi + Z) d\xi,$$

$$K_1(Y_1, Y_2) = \int_{\infty}^{0} e^{-\xi \mathbf{P}_1} G_1(Y_1(\omega \xi + Z), Y_2(\omega \xi + Z), \omega \xi + Z) d\xi.$$

By a similar argument, we obtain the integral equation for the vector-valued function $Y_2(Z)$:

$$Y_2(Z) = Y_{20} + \mu K_2(Y_1, Y_2), \qquad (4.2.18)$$

where

$$Y_{20} = -\int_{0}^{+\infty} e^{-\xi \mathbf{P}_2} Q_2(\omega \xi + Z) d\xi,$$

$$K_2(Y_1, Y_2) = -\int_{0}^{+\infty} e^{-\xi \mathbf{P}_2} G_2(Y_1(\omega \xi + Z), Y_2(\omega \xi + Z), \omega \xi + Z) d\xi.$$

Using equations (4.2.17), (4.2.18), we construct a sequence of functions $Y_{im}(Z)$, $m = 0, 1, \ldots$. Determine the $(m+1)$ approximation in terms of the m approximation from the formula

$$Y_{im+1} = Y_{i0} + \mu K_i(Y_{1m}, Y_{2m}).$$

We can choose a number $\mu_1 > 0$ such that the vector-valued functions X_m corresponding to these approximations stay in the region $\|X\| < R$ for $|\mu| \leq \mu_1$. Next,

Forced Multifrequency Oscillations

it is possible to choose a number μ_2 ($\mu_1 \geq \mu_2 > 0$) such that the sequences of functions $Y_{im}(Z)$ uniformly converge to the functions $Y_i = Y_i(Z)$ as $m \to \infty$. These functions are 2π-periodic in each component of the vector Z, satisfy the system of integral equations (4.2.17), (4.2.18) and constitute a unique solution of this type for these equations in the region $\|X\| < R$. If the obtained functions are substituted into equations (4.2.17), (4.2.18), and $Z = \omega t$ and $\xi = \tau - t$ are set in the thus formed identities, then we get

$$Y_1 = \int_{-\infty}^{t} e^{\mathbf{P}_1(t-\tau)} Q_1(\omega\tau) d\tau + \mu \int_{-\infty}^{t} e^{\mathbf{P}_1(t-\tau)} G_1(Y_1, Y_2, \omega\tau) d\tau,$$

$$Y_2 = -\int_{t}^{+\infty} e^{\mathbf{P}_2(t-\tau)} Q_2(\omega\tau) d\tau - \mu \int_{t}^{+\infty} e^{\mathbf{P}_2(t-\tau)} G_2(Y_1, Y_2, \omega\tau) d\tau.$$

Differentiating these identities, we find that the functions $Y_i(Z)$ for $Z = \omega t$ satisfy system (4.2.15), hence the solution of integral equations gives the solution (4.2.16) of this system.

We show that in a sufficiently small neighborhood of the solution (4.2.16) there is a family of the integral curves of system (4.2.15) that are dependent on σ arbitrary constants and approach asymptotically the solution (4.2.16) as $t \to +\infty$. To this end, construct the system of integral equations satisfied by the integral curves of the above family:

$$Y_1 = e^{\mathbf{P}_1 t}(Y_1(0) + \alpha) + \mu \int_{0}^{t} e^{\mathbf{P}_1(t-\tau)} G_1(Y_1, Y_2, \omega\tau) d\tau + \int_{0}^{t} e^{\mathbf{P}_1(t-\tau)} Q_1(\omega\tau) d\tau,$$

$$Y_2 = -\mu \int_{t}^{+\infty} e^{\mathbf{P}_2(t-\tau)} G_2(Y_1, Y_2, \omega\tau) d\tau - \int_{t}^{+\infty} e^{\mathbf{P}_2(t-\tau)} Q_2(\omega\tau) d\tau, \qquad (4.2.19)$$

where $Y_1(0) = Y_1(t)$ at $t = 0$, and α is a vector with the components $\alpha_1, \ldots, \alpha_\sigma$ that are arbitrary constants and sufficiently small in the absolute value.

System (4.2.19) can also be written as

$$Y_1 = e^{\mathbf{P}_1 t}(Y_1(0) + \alpha) + L_1(Y_1, Y_2),$$

$$Y_2 = L_2(Y_1, Y_2), \qquad (4.2.20)$$

where L_1 and L_2 are the operators indicated in (4.2.19).

System (4.2.19) has the family of restricted solutions

$$Y_i = Y_i(t, \mu, \alpha), \ i = 1, 2,$$

for $t \geq 0$, $|\mu| \leq \mu_3$ and $\|\alpha\| < h_0$, where μ_3 and h_0 are sufficiently small positive constants. This statement can be established by applying successive approximation. Determine an $(m+1)$ approximation from the formulas

$$Y_{1m+1} = e^{\mathbf{P}_1 t}(Y_{1m+1}(0) + \alpha) + L_1(Y_{1m}, Y_{2m}),$$

$$Y_{2m+1} = L_2(Y_{1m}, Y_{2m}), \ m = 0, 1, \ldots, \qquad (4.2.21)$$

where $Y_{1m+1}(0) = Y_{1m+1}(t)$ at $t = 0$. The zero approximation is determined from the formulas

$$Y_{10} = e^{P_1 t}\alpha + Y_{10}(Z), \; Y_{20} = Y_{20}(Z) \text{ for } Z = \omega t.$$

The sequence of vector-valued functions determined from (4.2.21), with the conditions imposed on (4.2.19), converges as $m \to \infty$ for $t \in (0, +\infty)$. This is established by utilizing an estimation procedure as that developed in the proof of Theorem 66.

We show that any integral curve entering into a family, approaches asymptotically a stationary motion $Y_1 = Y_1(Z)$, $Y_2 = Y_2(Z)$, $Z = \omega t$, as time increases. Note that the stationary motion satisfies the system of equations (4.2.19) for $\alpha = 0$. Substituting the integral curves of the family into equation (4.2.19), we obtain identities. If we subtract termwise from these identities those derived from the same equations (4.2.19), when $\alpha = 0$ and stationary motions are substituted therein, and make estimates then we get

$$\|\eta_1(t)\| \le a_1 e^{-a_2 t}\|\alpha\| + |\mu| L \int_0^t a_1 e^{-a_2(t-\tau)}(\|\eta_1(\tau)\| + \|\eta_2(\tau)\|)d\tau,$$

$$\|\eta_2(t)\| \le |\mu| L \int_t^{+\infty} b_1 e^{b_2(t-\tau)}(\|\eta_1(\tau)\| + \|\eta_2(\tau)\|)d\tau, \qquad (4.2.22)$$

where $\eta_i = Y_i(t, \mu, \alpha) - Y_i(Z)$, $i = 1, 2$, $Z = \omega t$, L, a_i and b_i are some positive constants. Set $u_i(t) = \max_{\tau \ge t} \|\eta_i(t)\|$. The functions $u_i(t)$ will satisfy the inequalities

$$u_1(t) \le a_1 e^{a_2 t}\|\alpha\| + |\mu| L \int_0^t a_1 e^{-a_2(t-\tau)}(u_1(\tau) + u_2(\tau))d\tau,$$

$$u_2(t) \le |\mu| L \int_t^{+\infty} b_1 e^{b_2(t-\tau)}(u_1(\tau) + u_2(\tau))d\tau. \qquad (4.2.23)$$

From the last inequality we find $u_2(t) \le |\mu| b_3 u_1(t)$, where b_3 is some positive constant if $|\mu|$ is sufficiently small. Then the function $u_1(t)$ satisfies the inequality

$$u_1(t) \le a_1 e^{-a_2 t}\|\alpha\| + \lambda \int_0^t e^{-a_2(t-\tau)} u_1(\tau) d\tau, \qquad (4.2.24)$$

where $\lambda = |\mu|a_3$, the a_3 being some positive constant if $|\mu|$ is sufficiently small. Multiplying inequality (4.2.24) by $e^{(a_2-\lambda)t}$, we rewrite it as

$$\frac{d}{dt}\left(e^{-\lambda t} \int_0^t e^{a_2 \tau} u_1(\tau) d\tau\right) \le a_1 \|\alpha\| e^{-\lambda t}.$$

Integrating this inequality and then multiplying both sides by $e^{\lambda t}$, we find

$$\int_0^t e^{a_2 \tau} u_1(\tau) d\tau \le a_1 \|\alpha\| \frac{e^{\lambda t} - 1}{\lambda}.$$

By substituting the last expression into (4.2.24), we get

$$u_1(t) \le e^{a_2 t}\|\alpha\|(a_1 + e^{\lambda t} - 1),$$

whence it follows that $u_1 \to 0$, $u_2 \to 0$ as $t \to +\infty$ for all sufficiently small $|\mu|$, and these functions tend to zero exponentially. This exactly means that the integral curves of the family $Y_i(t, \mu, \alpha)$ asymptotically approach the stationary motion $Y_i(Z)$, $Z = \omega t$, $i = 1, 2$, as $t \to +\infty$. The existence of a family of integral curves that are dependent on $n - \sigma$ arbitrary constants and asymptotically approach the same stationary motion as $t \to -\infty$, is established analogously if t is replaced by $-t$ in the original system. ∎

4.3 Motions of Nonlinear Systems Determined by Boundary Conditions

1. Principles to many problems in modern science and engineering are studies of the properties of solutions to systems of differential equations satisfying some boundary conditions. The general form of such boundary conditions can be described as follows.

In the phase space of a system there are given two manifolds. We shall construct an integral curve for the system of differential equations which starts at a particular time t_1 on one manifold and reaches the other at a time t_2. It is easy to show that seeking such an integral curve can reduce to seeking a periodic motion of an auxiliary system of differential equations which is constructed from the original system of equations of motion. This enables utilization of the above-developed methods as applied to construction of periodic solutions in order to solve the problem involved. But it also seems worthwhile to expound a direct method of solving such problems which is called the sweep method. Below this method of solving boundary-value problems for ordinary differential equations is developed. The development is to extend the sweep method to the case of nonlinear problems.

We first turn to a boundary-value problem with one degree of freedom. Suppose that the following equation is given

$$\ddot{y} = f(t, y, \dot{y}) \tag{4.3.1}$$

and it is known that this equation has the solution

$$y = y(t), \; t \in [0, T] \tag{4.3.2}$$

satisfying the boundary conditions

$$\varphi(y(0), \dot{y}(0)) = 0, \tag{4.3.3}$$

$$\varphi(y(T), \dot{y}(T)) = 0. \tag{4.3.4}$$

We shall give the method for construction of the solution (4.3.2) to equation (4.3.1) satisfying the boundary conditions (4.3.3) and (4.3.4). In order to solve the problem involved, we introduce general solution of equation (4.3.1)

$$y = y(t, C_1, C_2). \tag{4.3.5}$$

Assume that the equations (4.3.5) and

$$\dot{y} = \frac{\partial}{\partial t} y(t, C_1, C_2) \tag{4.3.6}$$

are solvable for C_1 and C_2, so that

$$C_1 = g_1(t, y, \dot{y}), \quad C_2 = g_2(t, y, \dot{y}). \tag{4.3.7}$$

In this case, the functions $g_i(t, y, \dot{y})$, $i = 1, 2$, are known to be the first integrals of equation (4.3.1). We show that there exist two integrals of equation (4.3.1)

$$\xi = \xi(t, y, \dot{y}), \tag{4.3.8}$$

$$\eta = \eta(t, y, \dot{y}), \tag{4.3.9}$$

possessing the properties

$$\xi(0, y, \dot{y}) = \varphi(y, \dot{y}), \tag{4.3.10}$$

$$\eta(T, y, \dot{y}) = \psi(y, \dot{y}). \tag{4.3.11}$$

Indeed, using (4.3.5) and (4.3.6), we eliminate the quantities $y(0)$, $\dot{y}(0)$ from the expression for the function φ:

$$\varphi = \varphi\left(y(0, C_1, C_2), \frac{\partial y(0, C_1, C_2)}{\partial t}\right).$$

Now, eliminate the quantities C_1 and C_2 from the obtained expression using the formulas (4.3.7). Denote the result of this elimination by $\xi = \xi(t, y, \dot{y})$. This function is the integral of (4.3.1) since, by construction, it is the function of the first integrals (4.3.7). If we set $t = 0$, then we get $\xi(0, y, \dot{y}) = \varphi(y, \dot{y})$. Reasoning along similar lines, we substitute (4.3.5) and (4.3.6) into the function $\psi(y(T), \dot{y}(T))$ for $t = T$, eliminate C_1 and C_2 by the formula (4.3.7), denote the result of elimination by $\eta(t, y, \dot{y})$ and, finally, find that the function $\eta(t, y, \dot{y})$, is the integral equation (4.3.1) and satisfies the condition (4.3.11).

The above reasoning suggests that ξ and η are solutions of the equation

$$\frac{\partial \xi}{\partial t} + \frac{\partial \xi}{\partial y}\dot{y} + \frac{\partial \xi}{\partial \dot{y}} f(t, y, \dot{y}) = 0. \tag{4.3.12}$$

Note that any integral curve of equation (4.3.1) satisfying only the condition (4.3.3) the integral $\xi(t, y, \dot{y})$ retains constant value, here

$$\xi(t, y, \dot{y}) = 0. \tag{4.3.13}$$

Next, any integral curve which satisfies only the condition (4.3.4), the integral $\eta(t, y, \dot{y})$ also retains constant value, here

$$\eta(t, y, \dot{y}) = 0. \tag{4.3.14}$$

This enables us to propose several methods for solution of the above-stated problems of seeking the solution (4.3.2) to equation (4.3.1) satisfying the boundary conditions (4.3.3) and (4.3.4).

First method. The solution (4.3.2) with $t = T$ satisfies the system of equations

$$\xi(T, y, \dot{y}) = 0, \quad \psi(y, \dot{y}) = 0. \tag{4.3.15}$$

Suppose (4.3.15) has the solution $(\bar{y}, \dot{\bar{y}})$. Then, to obtain the solution (4.3.2), we integrate backwards equation (4.3.1) with respect to the time t with initial conditions

$$y(T) = \bar{y}, \quad \dot{y}(T) = \dot{\bar{y}}$$

or we integrate the equation $\xi(t, y, \dot{y}) = 0$ with the initial condition

$$y(T) = \bar{y}.$$

Second method. The solution (4.3.2) with $t = 0$ satisfies the system of equations

$$\varphi(y, \dot{y}) = 0, \quad \eta(0, y, \dot{y}) = 0. \tag{4.3.16}$$

Denote the solutions of system (4.3.16) by $\tilde{y}$ and $\dot{\tilde{y}}$. In order to construct the solution (4.3.2) of equation (4.3.1), we integrate equation (4.3.1) with initial conditions

$$y(0) = \tilde{y}, \quad \dot{y}(0) = \dot{\tilde{y}}$$

or the equation $\eta(t, y, \dot{y}) = 0$ with initial condition $y(0) = \tilde{y}$.

Third method. The solution (4.3.2) of equation (4.3.1) satisfies the conditions (4.3.3) and (4.3.4) simultaneously. Therefore, it also satisfies equations (4.3.13) and (4.3.14) simultaneously. Thus, the solution (4.3.2) can be found without integration from the system of equations

$$\xi(t, y, \dot{y}) = 0, \quad \eta(t, y, \dot{y}) = 0. \tag{4.3.17}$$

From the above methods of constructing the solution (4.3.2) it follows that the problem reduces to seeking a solution of equation (4.3.12) with initial conditions (4.3.10) and (4.3.11). Such solutions can be constructed by various computational and analytic methods.

We now give a computation algorithm for finding an approximate solution of boundary-value problems in quasilinear systems with one degree of freedom. Let the following equation be given

$$\ddot{y} + p(t)\dot{y} + g(t)y = r(t) + \mu f(t, y). \tag{4.3.18}$$

Here p, g, r are the continuous real functions given for $t \in [0, T]$, μ is a small parameter, and $f(t, y)$ is the continuous real function which satisfies the Lipschitz condition with respect to y:

$$\mid f(t, \bar{y}) - f(t, \bar{\bar{y}}) \mid \leq l \mid \bar{y} - \bar{\bar{y}} \mid,$$

where l is a positive constant. Assume that equation (4.3.18) has the solution

$$y = y(t), \tag{4.3.19}$$

satisfying the boundary conditions

$$y(0) = a_0 y(0) + b_0 \tag{4.3.20}$$

and

$$\dot{y}(T) = \alpha_0 y(T) + \beta_0, \tag{4.3.21}$$

where $a_0, b_0, \alpha_0, \beta_0$ are real constants. We shall describe a successive approximation method in order to seek the solution (4.3.19) of equation (4.3.18).

We suggest the plan for solution of the problem involved. Construct first the integral equation satisfied by the solution (4.3.19). Next, describe the successive approximation method designed to solve this integral equation. Finally, show that for sufficiently small μ's the approximations converge to the solution (4.3.19) of equation (4.3.18). So we first show that the function (4.3.19) satisfies an integral equation of the form

$$y = y_0(t) + \mu \int_0^t f(\tau, y, (\tau)) a_1(t, \tau) d\tau + \mu \int_T^t a_2(t, \tau) f(\tau, y(\tau)) d\tau. \tag{4.3.22}$$

To construct the function (4.3.22), assume that the function (4.3.19) has been found. Then the function $f(t, y(t))$ is the known function of time t. Denote it by $r_1(t)$. The function (4.3.19) satisfies the linear equation

$$\ddot{y} + p(t)\dot{y} + g(t)y = r(t) + \mu r_1(t) \tag{4.3.23}$$

and the boundary conditions (4.3.20) and (4.3.21). For the case involved the partial differential equation, (4.3.12) has the solutions

$$\xi = \dot{y} - a(t)y - b(t), \tag{4.3.24}$$

$$\eta = \dot{y} - \alpha(t)y - \beta(t), \tag{4.3.25}$$

and the sought solutions satisfy the equations

$$\xi = 0, \tag{4.3.26}$$

$$\eta = 0. \tag{4.3.27}$$

From (4.3.24) and (4.3.26) we have that the functions $a(t)$ and $b(t)$ satisfy the system of ordinary differential equations

$$\dot{a} + (a+p)a + g = 0, \quad \dot{b} + (a+p)b = r + \mu r_1 \tag{4.3.28}$$

and initial conditions

$$a(0) = a_0, \quad b(0) = b_0.$$

Similarly, the functions $\alpha(t)$ and $\beta(t)$ satisfy the system

$$\dot{\alpha} + (\alpha + p)\alpha + g = 0, \ \dot{\beta} + (\alpha + p)\beta = r + \mu r_1 \qquad (4.3.29)$$

and initial conditions

$$\alpha(T) = \alpha_0, \ \beta(T) = \beta_0.$$

From the second equation of system (4.3.28) we have

$$b(t) = b_0 e^{-\int_0^t (a(\tau)+p(\tau))d\tau} + \int_0^t e^{-\int_\tau^t (a(\Theta)+p(\Theta))d\Theta} (r(\tau) + \mu r_1(\tau))d\tau. \qquad (4.3.30)$$

Similarly, from the second equation of system (4.3.29) we have

$$\beta(t) = \beta_0 e^{-\int_T^t (\alpha(\tau)+p(\tau))d\tau} + \int_t^T e^{-\int_\tau^t (\alpha(\Theta)+p(\Theta))d\Theta} (r(\tau) + \mu r_1(\tau))d\tau. \qquad (4.3.31)$$

Subtracting (4.3.27) from (4.3.26) and dividing both sides of the obtained equality by $a(t) - \alpha(t)$, we find

$$y = -\frac{b(t) - \beta(t)}{a(t) - \alpha(t)}. \qquad (4.3.32)$$

Using (4.3.30) and (4.3.31), we have

$$y = y_0 + \mu \int_0^t a_1(t,\tau) r_1(\tau) d\tau + \mu \int_T^t a_2(t,\tau) r_1(\tau) d\tau, \qquad (4.3.33)$$

where

$$y_0 = -\frac{b_0 e^{-\int_0^t (a(\tau)+p(\tau))d\tau} + \int_0^t e^{-\int_\tau^t (a(\Theta)+p(\Theta))d\Theta} r(\tau)d\tau}{a(t) - \alpha(t)}$$

$$+ \frac{\beta_0 e^{-\int_T^t (\alpha(\tau)+p(\tau))d\tau} + \int_T^t e^{-\int_\tau^t (\alpha(\Theta)+p(\Theta))d\Theta} r(\tau)d\tau}{a(t) - \alpha(t)}, \qquad (4.3.34)$$

$$a_1(t,\tau) = -\frac{e^{-\int_\tau^t (a(\Theta)+p(\Theta))d\Theta}}{a(t) - \alpha(t)}, \qquad (4.3.35)$$

$$a_2(t,\tau) = \frac{e^{-\int_\tau^t (\alpha(\Theta)+p(\Theta))d\Theta}}{a(t) - \alpha(t)}. \qquad (4.3.36)$$

In (4.3.33) we replace the function $r_1(t)$ by an equality $r_1(\tau) = f(\tau, y(\tau))$. Then we have that the solution (4.3.19) of equation (4.3.18) actually satisfies the integral equation (4.3.22), where the functions $y_0(t)$, $a_1(t,\tau)$ and $a_2(t,\tau)$ are defined by the formulas (4.3.34), (4.3.35) and (4.3.36).

We write equation (4.3.22) in operator form:

$$y = y_0 + \mu K(y),$$

where $K(y)$ is an integral operator on the right-hand side of (4.3.22). Let us construct a sequence of functions $y_0(t)$, $y_1(t)$, ... by

$$y_n(t) = y_0(t) + \mu K(y_{n-1}), \quad n = 1, 2, \ldots. \tag{4.3.37}$$

There exists $\mu_0 > 0$ such that for $|\mu| < \mu_0$ the series $y_0 + (y_1 - y_0) + (y_2 - y_1) + \ldots$ converges absolutely and uniformly to the solution (4.3.19) of equation (4.3.18) only if $y_0(t)$, $a_1(t, \tau)$ and $a_2(t, \tau)$ are continuous functions of their arguments. Below we give a more exact formulation of this statement.

2. We now consider the case of several degrees of freedom. Suppose that a system of n ordinary differential equations is given as

$$\frac{dy_s}{dt} = f_s(t, y_1, \ldots, y_n), \quad s = 1, \ldots, n. \tag{4.3.38}$$

Assume that there exists a solution to this system

$$y_s = y_s(t), \quad t \in [0, T], \quad s = 1, \ldots, n, \tag{4.3.39}$$

which satisfies the boundary conditions

$$\varphi_i(y_1(0), \ldots, y_n(0)) = 0, \quad i = 1, \ldots, k; \tag{4.3.40}$$

$$\psi_j(y_1(T), \ldots, y_n(T)) = 0, \quad j = 1, \ldots, n - k. \tag{4.3.41}$$

The method of constructing such a solution is in order.
Consider the partial differential equation

$$\frac{\partial V}{\partial t} + \sum_{s=1}^{n} \frac{\partial V}{\partial y_s} f_s = 0. \tag{4.3.42}$$

Construct a solution to equation (4.3.42)

$$\xi_i = \xi_i(t, y_1, \ldots, y_n), \quad i = 1, \ldots, k, \tag{4.3.43}$$

$$\eta_j = \eta_j(t, y_1, \ldots, y_n), \quad j = 1, \ldots, n - k, \tag{4.3.44}$$

satisfying the initial conditions

$$\xi_i(0, y_1, \ldots, y_n) = \varphi_i(y_1, \ldots, y_n), \quad i = 1, \ldots, k, \tag{4.3.45}$$

$$\eta_j(T, y_1, \ldots, y_n) = \psi_j(y_1, \ldots, y_n), \quad j = 1, \ldots, n - k. \tag{4.3.46}$$

Utilizing (4.3.43) and (4.3.44), the solution (4.3.39) can be sought by various methods.
First method. Find the solution $\bar{y}_1, \ldots, \bar{y}_n$ to the system of equations

$$\xi_i(T, y_1, \ldots, y_n) = 0, \quad i = 1, \ldots, k,$$

$$\psi_j(y_1, \ldots, y_n) = 0, \quad j = 1, \ldots, n - k. \tag{4.3.47}$$

Motions of Nonlinear Systems

In order to find the solution (4.3.39), the system (4.3.38) is integrated backwards in time subject to
$$y_s(T) = \tilde{y}_s, \quad s = 1, \ldots, n.$$

Second method. Find the solution $\tilde{y}_1, \ldots, \tilde{y}_n$ to the system of equations
$$\varphi_i(y_1, \ldots, y_n) = 0, \quad i = 1, \ldots, k,$$
$$\eta_j(0, y_1, \ldots, y_n) = 0, \quad j = 1, \ldots, n-k. \tag{4.3.48}$$

In order to construct the solution (4.3.39), the system (4.3.38) is integrated with initial conditions
$$y_s(0) = \tilde{y}_s, \quad s = 1, \ldots, n.$$

Third method. Since the solution (4.3.39) satisfies the conditions (4.3.40) and (4.3.41), the following equalities hold
$$\xi_i(t, y_1, \ldots, y_n) = 0, \quad i = 1, \ldots, k,$$
$$\eta_j(t, y_1, \ldots, y_n) = 0, \quad j = 1, \ldots, n-k. \tag{4.3.49}$$

The equalities (4.3.49) hold along any integral curve satisfying (4.3.40) and (4.3.41). Consequently, the required solution (4.3.39) may be sought among them without integrating system (4.3.38). We employ the equalities (4.3.49) to construct a computational algorithm, which enables us to determine the solution (4.3.39) of system (4.3.38) to a previously given precision when the boundary conditions are linear and system (4.3.38) is quasilinear. Let

$$f_s(t, y_1, \ldots, y_n) = \sum_{i=1}^{n} p_{si}(t) y_i + q_s(t) + \mu g_s(t, y_1, \ldots, y_n),$$

$$\varphi_i = \sum_{s=1}^{n} a_{is} y_s + b_i; \quad \psi_j = \sum_{s=1}^{n} \alpha_{js} y_s + \beta_j.$$

Then equation (4.3.38) and boundary conditions (4.3.40) and (4.3.41) can be written in vector-matrix form:

$$\frac{dY}{dt} = \mathbf{P}(t)Y + Q(t) + \mu G(t, Y), \quad \mathbf{A}_0 Y(0) + B_0 = 0, \quad \mathbf{\Gamma}_0 Y(T) + \Delta_0 = 0, \tag{4.3.50}$$

where $Y = (y_1, \ldots, y_n)$, $Q(t)$, $G(t, Y)$ are vector-valued functions of their independent variables, B_0, Δ_0 the constant vectors, $\mathbf{P}(t)$, $\mathbf{A}_0$ and $\mathbf{\Gamma}_0$ the matrices of suitable dimensions. In order to derive the integral equations satisfied by the solution (4.3.39), the vector-valued function $G(t, Y)$ will be treated only as the time function considering that it has been computed on the solution (4.3.39) of interest to us. In the case under study, the solutions (4.3.43) and (4.3.44) to equation (4.3.42) are linear functions, viz.

$$\Xi = \mathbf{A}(t)Y + B(t), \quad H = \mathbf{\Gamma}(t)Y + \Delta(t),$$

where
$$\Xi = (\xi_1, \ldots, \xi_k), \quad H = (\eta_1, \ldots, \eta_{n-k}),$$

the matrix $\mathbf{A}(t)$ and the vector $B(t)$ satisfying the system of equations

$$\frac{d\mathbf{A}}{dt} + \mathbf{A}P = \mathbf{0}, \qquad (4.3.51)$$

$$\frac{dB}{dt} + \mathbf{A}Q' = 0, \qquad (4.3.52)$$

$$\mathbf{A}(0) = \mathbf{A}_0, \qquad (4.3.53)$$

$$B(0) = B_0. \qquad (4.3.54)$$

Similarly, the matrix $\mathbf{\Gamma}(t)$ and the vector $\Delta(t)$ satisfy the system

$$\frac{d\mathbf{\Gamma}}{dt} + \mathbf{\Gamma}P = \mathbf{0}, \qquad (4.3.55)$$

$$\frac{d\Delta}{dt} + \mathbf{\Gamma}Q' = 0, \qquad (4.3.56)$$

$$\mathbf{\Gamma}(T) = \mathbf{\Gamma}_0, \qquad (4.3.57)$$

$$\Delta(T) = \Delta_0; \qquad (4.3.58)$$

here

$$Q' = Q(t) + \mu G(t, Y(t)).$$

Integrating equation (4.3.52) with initial condition (4.3.54), we find

$$B(t) = B_1(t) - \mu \int_0^t \mathbf{A}(\tau)G(\tau, Y(\tau))d\tau,$$

$$B_1(\tau) = B_0 - \int_0^t \mathbf{A}(\tau)Q(\tau)d\tau;$$

and, similarly,

$$\Delta(t) = \Delta_1(t) - \mu \int_T^t \mathbf{\Gamma}(\tau)G(\tau, Y(\tau))d\tau,$$

$$\Delta_1(t) = \Delta_0 - \int_T^t \mathbf{\Gamma}(\tau)Q(\tau)d\tau.$$

From (4.3.49) we have

$$\mathbf{A}(t)Y + B(t) = 0, \qquad (4.3.59)$$

$$\mathbf{\Gamma}(t)Y + \Delta(t) = 0. \qquad (4.3.60)$$

Multiplying from the left (4.3.59) by $\mathbf{A}^*(t)$, and (4.3.60) correspondingly by $\mathbf{\Gamma}^*(t)$, and then adding term by term and multiplying from the left by the matrix $\mathbf{C}(t) = (\mathbf{A}^*\mathbf{A} + \mathbf{\Gamma}^*\mathbf{\Gamma})^{-1}$, we obtain

$$Y(t) = Y_0(t) + \mu \int_0^t \mathbf{A}_1(t,\tau)G(\tau, Y(\tau))d\tau + \mu \int_T^t \mathbf{A}_2(t,\tau)G(\tau, Y(\tau))d\tau, \qquad (4.3.61)$$

where $Y_0(t) = -\mathbf{C}(t)(\mathbf{A}^*B_1(t) + \mathbf{\Gamma}^*\Delta_1(t))$, $\mathbf{A}_1(t,\tau) = \mathbf{C}(t)\mathbf{A}^*(t)\mathbf{A}(\tau)$, $\mathbf{A}_2(t,\tau) = \mathbf{C}(t)\mathbf{\Gamma}^*(t)\mathbf{\Gamma}(\tau)$. Denote by $Z(t)$ the fundamental system of solutions for linear equations

$$\frac{dZ}{dt} + Z\mathbf{P} = 0, \tag{4.3.62}$$

where $\mathbf{Z}(0) = \mathbf{E}$, $\mathbf{E}$ is a unit matrix, then

$$\mathbf{A}(t) = \mathbf{A}_0\mathbf{Z}(t), \mathbf{\Gamma}(t) = \mathbf{\Gamma}_0\mathbf{Z}^{-1}(T)\mathbf{Z}(t).$$

Hence we have

$$\mathbf{A}^*(t)\mathbf{A}(t) + \mathbf{\Gamma}^*(t)\mathbf{\Gamma}(t) = \mathbf{Z}^*(t)\left(\mathbf{A}_0^*\mathbf{A}_0 + (\mathbf{Z}^{-1}(T))^*\mathbf{\Gamma}_0^*\mathbf{\Gamma}_0\mathbf{Z}^{-1}(T)\right)\mathbf{Z}(t),$$

which implies that for the matrix $\mathbf{C}(t)$ to exist it is necessary and sufficient that the constant matrix enclosed in parentheses be nonsingular.

Theorem 68 [Zubov (1970)] *Suppose the following conditions are satisfied: 1) the right-hand sides of the system of differential equations (4.3.50) given for $t \in [0, T]$, $\|Y - Y_0\| \leq r$, are real and continuous, and the functions $g_s(t, Y)$ satisfy the Lipschitz condition with respect to $y_1, \ldots, y_n$;*
2) the matrix $\mathbf{A}_0^*\mathbf{A}_0 + (\mathbf{Z}^{-1}(T))^*\mathbf{\Gamma}_0^*\mathbf{\Gamma}_0\mathbf{Z}^{-1}(T)$ is nonsingular. Then there exists $\mu_0 > 0$ such that the integral equation (4.3.61) has a unique solution which satisfies the condition $\|Y(t) - Y_0(t)\| \leq m|\mu|, |\mu| < \mu_0$. Here r and m are positive constants and $\|Y(t) - Y(t)_0(t)\| = \sqrt{\sum_{s=1}^{n}(y^s - y_{s0})^2}$.

Proof: Denote by K the integral operator appearing in equation (4.3.61). Then this equation can be written as

$$Y = Y_0 + \mu K(Y). \tag{4.3.63}$$

Construct a system of vector-valued functions

$$Y_n = Y_0 + \mu K(Y_{n-1}), \ n = 1, 2, \ldots.$$

We determine routinely that there exists a number $\mu_0 > 0$ such that for $|\mu| < \mu_0$ we get $\|Y_n - Y_0\| \leq |\mu|m < r$, and hence construction of all approximations proves to be possible. Next, as the convergence proof for the Picard theorem, we find with the help of the Lipschitz condition that the series

$$Y_0 + (Y_1 - Y_0) + (Y_2 - Y_1) + \ldots$$

converges absolutely and uniformly to the solution $Y(t)$ of the integral equation (4.3.61), which satisfies the original differential equations and boundary conditions. ∎

Academician V.V. Novozhilov gave the author a few hints about the possibility of constructing successive approximations for quasilinear systems with linear boundary conditions.

3. We now consider the general case of unseparated boundary conditions. Suppose that the following system of differential equations is given

$$\frac{dy_s}{dt} = f_s(t, y_1, \ldots, y_n) \tag{4.3.64}$$

and in the interval $t \in [0, T]$ the system (4.3.64) has the solution

$$y_s = y_s(t), \quad s = 1, \ldots, n, \tag{4.3.65}$$

satisfying the boundary conditions

$$\varphi_i(Y(t_0), Y(t_1), \ldots, Y(t_{k+1})) = 0, \quad i = 1, \ldots, n, \tag{4.3.66}$$

where $t_0 = 0$, $t_{k+1} = T$, $t_0 < t_1 < \ldots < t_{k+1}$, $Y(t) = (y_1, \ldots, y_n)$. A method is required for seeking the solution (4.3.65) of system (4.3.64) which satisfies the condition (4.3.66). Suppose the solution (4.3.65) has been found:

$$Y = Y(t);$$

set

$$Y_j(t) = Y(t_j + \alpha_j t), \quad j = 1, \ldots, k+1. \tag{4.3.67}$$

The vector-valued functions (4.3.67) satisfy the equations

$$\frac{dY_j}{dt} = F_j(t, Y_j), \tag{4.3.68}$$

where $F_j(t, Y_j) = \alpha_j F(t_j + \alpha_j t, Y_j)$. The vector-valued function F has the components $f_s(t, Y)$ that are the right-hand sides of (4.3.64). Moreover, the vector-valued functions (4.3.67) satisfy the initial conditions

$$Y_j(0) = Y(t_j), \quad j = 1, \ldots, k+1. \tag{4.3.69}$$

Now, if the system (4.3.64) has the solution (4.3.65) satisfying the condition (4.3.66), then the system of $n(k+2)$ equations (4.3.64), (4.3.68) has the solution (4.3.65) which satisfies the boundary conditions

$$\varphi_i(Y(0), Y_1(0), \ldots, Y_{k+1}(0)) = 0, \quad i = 1, \ldots, n. \tag{4.3.70}$$

We now turn to constant quantities $\alpha_1, \ldots, \alpha_{k+1}$. Choose these quantities so that at a given time $\bar{t} \in [0, T]$ the following relationships hold

$$Y(\bar{t}) = Y_j(\bar{t}), \quad j = 1, \ldots, k+1. \tag{4.3.71}$$

For (4.3.71) to hold, we have to set $\alpha_j = (\bar{t} - t_j)/\bar{t}$. With such a choice of α_j quantities for system (4.3.64), (4.3.68) on the interval $[0, \bar{t}]$ there arises a boundary value problem with divided boundary conditions (4.3.70), (4.3.71). Consequently, on this interval it is possible to apply the above-developed theory. However, we can also proceed in a straight-forward manner, i.e. consider the partial differential equation

$$\frac{\partial V}{\partial t} + \sum_{s=1}^{n} \frac{\partial V}{\partial y_s} f_s + \sum_{s=1}^{n} \sum_{j=1}^{k+1} \frac{\partial V}{\partial y_{sj}} f_{sj} = 0, \qquad (4.3.72)$$

where f_{sj} are components of the vectors F_j, and y_{sj} are components of the vectors Y_j. Let us construct n solutions for equation (4.3.72)

$$\xi_i = \xi_i(t, Y, Y_1, \ldots, Y_{k+1}), \quad i = 1, \ldots, n, \qquad (4.3.73)$$

with initial conditions

$$\xi_i(0, Y, Y_1, \ldots, Y_{k+1}) = \varphi_i(Y, Y_1, \ldots, Y_{k+1}). \qquad (4.3.74)$$

On the sought solution, the functions (4.3.73) satisfy the equations

$$\xi_i(t, Y, Y_1, \ldots, Y_{k+1}) = 0. \qquad (4.3.75)$$

In order to find the sought solution at the point $\bar{t}$, it suffices to set $t = \bar{t}$ in equalities (4.3.75) and conditions (4.3.71). Then equations (4.3.75) enable definition of the vector-valued function $\bar{Y}(t)$. If the quantity $\bar{t} \in [0, T]$ is now assumed to be arbitrary, then the above procedure makes it possible to find the solution (4.3.65) of system (4.3.64) without further integration.

Consider in some detail the case of quasilinear system with undivided linear boundary conditions. Suppose the boundary conditions (4.3.66) are of the form

$$\sum_{j=0}^{k+1} \mathbf{A}_{j0} Y(t_j) + B_0 = 0, \qquad (4.3.76)$$

where $\mathbf{A}_{j0}$ are real square matrices with constant elements, and B_0 is the vector with constant real components. Let the original system (4.3.64) be written in vector form

$$\frac{dY}{dt} = \mathbf{P}(t) Y + Q(t) + \mu F(t, Y). \qquad (4.3.77)$$

Then system (4.3.68) becomes

$$\frac{dY_j}{dt} = \mathbf{P}_j(t) Y_j + Q_j(t) + \mu F_j(t, Y_j), \quad j = 1, \ldots, k+1, \qquad (4.3.78)$$

where $\mathbf{P}_j(t) = \alpha_j \mathbf{P}(t_j + \alpha_j t)$, $Q_j(t) = \alpha_j Q(t_j + \alpha_j t)$, $F_j(t) = \alpha_j F(t_j + \alpha_j t, Y_j)$. Assume that the solution (4.3.65) has been found. Then, using this solution, it is possible to compute the vector-valued functions F_j and F:

$$R(t) = F(t, Y(t)), \quad R_j(t) = F_j(t, Y_j(t)), \quad j = 1, \ldots, k+1, \qquad (4.3.79)$$

where $Y_j(t)$ are the vector-valued functions defined by the formula (4.3.67). If in the systems (4.3.77) and (4.3.78) the nonlinear terms are replaced by the formula

(4.3.79), then we get the linear equations satisfied by the functions (4.3.65), (4.3.67). Equation (4.3.72), made up for the linear systems obtained, has the solution (4.3.73) that is linear in components of the vectors $Y, Y_1, \ldots, Y_{k+1}$. Denote by Ξ the vector whose components are the above functions ξ_i. Then

$$\Xi = \sum_{j=0}^{k+1} \mathbf{A}_j(t) Y_j + B(t). \tag{4.3.80}$$

The matrix $\mathbf{A}_j(t)$ and the vector $B(t)$ satisfy the linear system of differential equations

$$\frac{d\mathbf{A}_j}{dt} + \mathbf{A}_j \mathbf{P}_j = 0, \tag{4.3.81}$$

$$\frac{dB}{dt} + \sum_{j=0}^{k+1} \mathbf{A}_j(Q_j + \mu R_j) = 0. \tag{4.3.82}$$

In formulas (4.3.80) - (4.3.82)

$$Y_0(t) = Y(t), \ \mathbf{P}_0(t) = \mathbf{P}(t), \ Q_0(t) = Q(t), \ R_0(t) = R(t).$$

The matrices $\mathbf{A}_j(t)$ and the vector $B(t)$ satisfy the initial conditions

$$\mathbf{A}_j(0) = \mathbf{A}_{j0}, \ B(0) = B_0.$$

Denote by $\mathbf{Z}(t)$ the fundamental system of solutions for

$$\frac{d\mathbf{Z}}{dt} + \mathbf{Z}\mathbf{P} = 0, \tag{4.3.83}$$

where $\mathbf{Z}(0) = \mathbf{E}$, $\mathbf{E}$ being a unit matrix. The matrices $\mathbf{A}_j(t)$ can be represented then as

$$\mathbf{A}_j(t) = \mathbf{A}_{j0} \mathbf{Z}^{-1}(t_j) \mathbf{Z}(t_j + \alpha_j t), \ j = 0, \ldots, k+1. \tag{4.3.84}$$

Integrating equation (4.3.82), we find

$$B(t) = B_0 - \int_0^t \sum_{j=0}^{k+1} \mathbf{A}_j(\tau) Q_j(\tau) d\tau - \mu \int_0^t \sum_{j=0}^{k+1} \mathbf{A}_j(\tau) R_j(\tau) d\tau. \tag{4.3.85}$$

From (4.3.75) we have that the sought solution satisfies the equation system

$$\sum_{j=0}^{k+1} \mathbf{A}_j(t) Y_j + B(t) = 0. \tag{4.3.86}$$

Set $t = \bar{t}$ in (4.3.86). Using further the relationship (4.3.71), from (4.3.86) we obtain

$$\left(\sum_{j=0}^{k+1} \mathbf{A}_j(\bar{t}) \right) Y(\bar{t}) = -B(\bar{t}). \tag{4.3.87}$$

Assume that the matrix $\sum_{j=0}^{k+1} \mathbf{A}_j(\bar{t})$ is nonsingular. Denote its inverse by $\mathbf{C}(\bar{t})$, then from (4.3.87) we have

$$Y(\bar{t}) = \mathbf{C}(\bar{t})\left(\sum_{j=0}^{k+1}\int_0^{\bar{t}}\mathbf{A}_j(\tau)Q_j(\tau)d\tau - B_0\right) + \mu\mathbf{C}(\bar{t})\sum_{j=0}^{k+1}\int_0^{\bar{t}}\mathbf{A}_j(\tau)R_j(\tau)d\tau. \quad (4.3.88)$$

Replace in (4.3.88) the functions $R_j(\tau)$ by their expressions in terms of the functions F_j and introduce in the j-th summand the integration variable $\Theta = t_j + \alpha_j\tau$. Then formula (4.3.88) becomes

$$Y(\bar{t}) = Y^0(\bar{t}) + \mu\sum_{j=0}^{k+1}\int_{t_j}^{\bar{t}}\mathbf{A}_j(\bar{t},\Theta)F(\Theta,Y(\Theta))d\Theta, \quad (4.3.89)$$

where

$$\mathbf{A}_j(\bar{t},\Theta) = \mathbf{C}(\bar{t})\mathbf{A}_{j0}\mathbf{Z}^{-1}(t_j)\mathbf{Z}(\Theta),$$

$$Y^0(\bar{t}) = \sum_{j=0}^{k+1}\int_{t_j}^{\bar{t}}\mathbf{A}_j(\bar{t},\Theta)Q(\Theta)d\Theta - \mathbf{C}(\bar{t})B_0. \quad (4.3.90)$$

Equation (4.3.89) can be treated as an integral equation which serves to define the vector-valued function (4.3.65) since the quantity $\bar{t}$ in (4.3.89) can now be viewed as the independent variable $\bar{t} \in [0,T]$. Note that the vector-valued function $Y^0(t)$ defined by equation (4.3.90) yields a solution to the problem for $\mu = 0$. Denote by $K(Y)$ an integral operator appearing on the right-hand side of (4.3.89). Then equation (4.3.89) can be represented as

$$Y(t) = Y^0(t) + \mu K(Y). \quad (4.3.91)$$

Set

$$Y^N(t) = Y^0(t) + \mu K(Y^{N-1}), \quad N = 1, 2, \ldots.$$

It turns out that under broad assumptions about the right-hand sides of the original system of differential equations, the successive approximations of $Y^0, Y^1, \ldots$ uniformly converge to the solution (4.3.65) of system (4.3.77) satisfying the boundary conditions (4.3.76). This means that the following theorem is valid.

Theorem 69 [Zubov (1970)] *Suppose that the following conditions hold: 1) the right-hand sides of equation system (4.3.77) are given for $t \in [0,T]$, $\|Y^0(t) - Y\| \leq r$, are real and continuous, and satisfy Lipschitz conditions in Y,*

2) the matrix $\sum_{j=0}^{k+1}\mathbf{A}_{j0}\mathbf{Z}^{-1}(t_j)$ is nonsingular. Then there exists a positive μ_0 that any μ, $|\mu| < \mu_0$ system (4.3.77) has the continuous solution (4.3.65) which satisfies the boundary conditions (4.3.76), the solution being unique.

Proof: If μ_0 is selected so that $\mu_0 m_1 < r$, where m_1 and r are positive constants, then by the continuity of operator K

$$\|Y^N - Y^0\| \leq |\mu|m_1, \quad N = 1, 2, \ldots.$$

Now, from Lipschitz conditions we have

$$\|Y^N - Y^{N-1}\| \leq |\mu| m_2 \|Y^{N-1} - Y^{N-2}\|,$$

where $N = 2, 3, \ldots$. If the quantity μ_0 is additionally selected so that $\mu_0 m_2 < 1$, then the series

$$Y^0 + (Y^1 - Y^0) + (Y^2 - Y^1) + \cdots$$

converges uniformly and absolutely in $t \in [0, T]$ to the continuous solution of (4.3.77). This solution satisfies the boundary conditions (4.3.76). The uniqueness of this solution is established routinely. ∎

We shall present solvability conditions for the boundary value problem. In the above case the matrix

$$\mathbf{A}(t) = \sum_{j=0}^{k+1} \mathbf{A}_j(t) \tag{4.3.92}$$

is nonsingular. Equation (4.3.87) can be rewritten as

$$\mathbf{A}(t)Y(t) = B_1 + \mu B_2, \tag{4.3.93}$$

where

$$B_1 = \int_0^t \sum_{j=0}^{k+1} \mathbf{A}_j(\tau) Q_j(\tau) d\tau - B_0 = \sum_{j=0}^{k+1} \int_{t_j}^t \mathbf{A}_{j0} \mathbf{Z}^{-1}(t_j) \mathbf{Z}(\tau) Q(\tau) d\tau - B_0, \tag{4.3.94}$$

$$B_2 = \int_0^t \sum_{j=0}^{k+1} \mathbf{A}_j(\tau) R_j(\tau) d\tau = \sum_{j=0}^{k+1} \int_{t_j}^t \mathbf{A}_{j0} \mathbf{Z}^{-1}(t_j) \mathbf{Z}(\tau) F(\tau, Y(\tau)) d\tau. \tag{4.3.95}$$

We now suppose that the matrix (4.3.92) is singular. Then, as shown below, the rank of this matrix is t-independent. Denote it by k.

Theorem 70 [Zubov (1970)] *The system (4.3.77) with $\mu = 0$ has the continuous solution satisfying condition (4.3.76) if and only if the constant vector Λ satisfying the relationship $\Lambda^* \tilde{\mathbf{A}} = 0$ also satisfies the condition $\Lambda^* \tilde{B}_1 = 0$, where*

$$\tilde{\mathbf{A}} = \sum_{j=0}^{k+1} \mathbf{A}_{j0} \mathbf{Z}^{-1}(t_j), \quad \tilde{B}_1 = B_0 + \sum_{j=0}^{k+1} \int_0^{t_j} \mathbf{A}_{j0} \mathbf{Z}^{-1}(t_j) \mathbf{Z}(\tau) Q(\tau) d\tau.$$

Proof: If the conditions of the theorem are satisfied, the rank of matrix $\mathbf{A}(t)$ coincides with that of the augmented matrix obtained from $\mathbf{A}(t)$ by adding the $(n+1)$ column equal to $B_1(t)$. Then the linear system of equations obtained from (4.3.93) for $\mu = 0$, has a solution that is dependent on $n - k$ arbitrary constants. We show that the ranks of the above matrices coincide when the conditions of the theorem hold. The matrix $\mathbf{A}(t)$ defined by the expression (4.3.92) can be represented as

$$\mathbf{A}(t) = \sum_{j=0}^{k+1} \mathbf{A}_{j0} \mathbf{Z}^{-1}(t_j) \mathbf{Z}(t) = \tilde{\mathbf{A}} \mathbf{Z}(t).$$

Motions of Nonlinear Systems

Since the matrix $\mathbf{Z}(t)$ is nonsingular, then the rank of the matrix $\mathbf{A}(t)$ coincides with that of the constant matrix $\tilde{\mathbf{A}}$. Let Λ be some constant vector such that $\Lambda^*\tilde{\mathbf{A}} = 0$, then $\Lambda^*\mathbf{A}(t) = 0$. Calculate the scalar product of the vector Λ on $B_1(t)$:

$$\Lambda^* B_1(t) = \Lambda^* \left(\sum_{j=0}^{k+1} \int_{t_j}^{t} \mathbf{A}_{j0}\mathbf{Z}^{-1}(t_j)\mathbf{Z}(\tau)Q(\tau)d\tau \right)$$

$$= \Lambda \left(\sum_{j=0}^{k+1} \int_{0}^{t} \mathbf{A}_{j0}\mathbf{Z}^{-1}(t_j)\mathbf{Z}(\tau)Q(\tau)d\tau \right) - \Lambda^* \tilde{B}_1.$$

We show that the following equality holds

$$\Lambda^* B_1(t) = -\Lambda^* \tilde{B}_1.$$

Indeed, we get

$$\Lambda^* \sum_{j=0}^{k+1} \int_{0}^{t} \mathbf{A}_{j0}\mathbf{Z}^{-1}(t_j)\mathbf{Z}(\tau)Q(\tau)d\tau = \Lambda^* \tilde{\mathbf{A}} \int_{0}^{t} \mathbf{Z}(\tau)Q(\tau)d\tau = 0.$$

Thus, under the condition of the theorem, from $\Lambda^*\mathbf{A}(t) = 0$ it follows that $\Lambda^*B(t) = 0$. This exactly means that the ranks of the matrix $\mathbf{A}(t)$ and the augmented matrix coincide. ∎

If the matrix (4.3.92) is of rank k, then there necessarily exist $n - k$ linear independent vectors Λ_i satisfying the condition $\Lambda_i^*\mathbf{A}(t) = 0$. Assume that the conditions of Theorem 70 have been satisfied and, therefore, there exists the solution of system (4.3.77) satisfying the boundary conditions (4.3.76). Substituting this solution into equation (4.3.93), we obtain a system of identities. Multiplying from the left the obtained identities by Λ_i^*, we find

$$\Lambda_i^* B_2(Y) = 0, \quad i = 1, \ldots, n - k. \tag{4.3.96}$$

Thus, if system (4.3.77) has a solution satisfying conditions (4.3.76), then this solution necessarily satisfies equations (4.3.93) and (4.3.96). Transform equations (4.3.96):

$$\Lambda_i^* \sum_{j=0}^{k+1} \int_{t_i}^{t} \mathbf{A}_{j0}\mathbf{Z}^{-1}(t_j)\mathbf{Z}(\tau)F(\tau, Y(\tau))d\tau$$

$$= \Lambda_i^* \tilde{\mathbf{A}} \int_{0}^{t} \mathbf{Z}(\tau)F(\tau, Y(\tau))d\tau - \Lambda_i^* \sum_{j=0}^{k+1} \mathbf{A}_{j0}\mathbf{Z}^{-1}(t_j)\mathbf{Z}(\tau)F(\tau, Y(\tau))d\tau$$

$$= -\Lambda_i^* \sum_{j=0}^{k+1} \int_{0}^{t_j} \mathbf{A}_{j0}\mathbf{Z}^{-1}(t_j)\mathbf{Z}(\tau)F(\tau, Y(\tau))d\tau.$$

From this it follows that equations (4.3.96) can be represented as

$$\Lambda_i^* \sum_{j=0}^{k+1} \int_{0}^{t_j} \mathbf{A}_{j0}\mathbf{Z}^{-1}(t_j)\mathbf{Z}(\tau)F(\tau, Y(\tau))d\tau = 0. \tag{4.3.97}$$

Denote by $Y^0 = Y^0(t, C_1, \ldots, C_{n-k})$ a solution to equation (4.3.93) when $\mu = 0$. Substitute this solution into the integrand appearing in (4.3.97). The resulting functions of arbitrary constants $C_1, \ldots, C_{n-k}$ will be denoted by g_i:

$$g_i(C_1, \ldots, C_{n-k}) = \Lambda_i^* \sum_{j=0}^{k+1} \int_0^{t_j} \mathbf{A}_{j0} \mathbf{Z}^{-1}(t_j) \mathbf{Z}(\tau) F(\tau, Y^0) d\tau. \qquad (4.3.98)$$

Theorem 71 [Zubov (1970)] *Suppose that 1) the conditions of Theorem 70 are satisfied;*

2) there exist real constants $C_{10}, \ldots, C_{n-k0}$ which satisfy the equations

$$g_i = 0, \ i = 1, \ldots, n-k, \qquad (4.3.99)$$

and are such that the functional determinant of functions $g_1, \ldots, g_{n-k}$ for variables $C_1, \ldots, C_{n-k}$ is nonzero at the point $C_{10}, \ldots, C_{n-k0}$. Then there exists a positive number μ_0 such that for any $\mu, |\mu| \leq \mu_0$ there is a continuous solution to the system (4.3.77) satisfying the boundary conditions (4.3.76). This continuous solution can be obtained as a limit of successive approximations.

Proof: Denote by $Y^0(t)$ the solution of system (4.3.93) with $\mu = 0$ which corresponds to the quantities $C_{10}, \ldots, C_{n-k0}$. Define the solution of the system

$$\mathbf{A}(t)Y^1 = B_1(t) + \mu B_2(Y^0). \qquad (4.3.100)$$

This solution depends on $n - k$ arbitrary constants

$$Y^1 = Y^1(t, C_1, \ldots, C_{n-k}). \qquad (4.3.101)$$

Select arbitrary constants $C_1, \ldots, C_{n-k}$ so that the functions (4.3.101) would satisfy equations (4.3.97). Denote by $Y^1(t)$ the solution of system (4.3.100) corresponding to these values of arbitrary constants. Proceeding in this manner, we can define the N-th approximation as a solution to the equation

$$\mathbf{A}(t)Y^N = B_1(t) + \mu B_2(Y^{N-1}), \qquad (4.3.102)$$

which satisfies the relationships derived from (4.3.97) by substituting thereto the functions Y^N instead of Y.

This method of constructing successive approximations enables us to find uniquely the vector-valued functions Y^N, $N = 1, 2, \ldots$, considering that the arbitrary constants $C_1, \ldots, C_{n-k}$ are determined at each step as solutions to an equation of the form (4.3.99) situated in a small neighborhood of the point $C_{10}, \ldots, C_{n-k0}$. Convergence proofs for these successive approximations utilize differentiability of the right-hand sides of system (4.3.77). We assume that the right-hand sides of system (4.3.77) given for $t \in [0, T]$, $\|Y^0(t) - Y\| \leq r$, are real, continuous and continuously differentiable. Here, as before, r is some positive constant, and $Y^0(t)$ is the solution of equations (4.3.93) for $\mu = 0$ corresponding to the selected values of arbitrary constants $C_{10}, \ldots, C_{n-k0}$. Denote by C the $n - k$-vector whose components are arbitrary constants $C_1, \ldots, C_{n-k}$. Then from equations (4.3.102) we can find

$$Y^N(t) = \mathbf{H}_1(t)B_1(t) + \mu \mathbf{H}_2(t)B_2(Y^{N-1}) + \mathbf{H}_3(t)C^N, \qquad (4.3.103)$$

where $\mathbf{H}_1(t)$, $\mathbf{H}_2(t)$, $\mathbf{H}_3(t)$ are the matrices determined in the solution of equations (4.3.102). Substituting the functions $Y^N(t)$ into equality (4.3.97), we have that the constant vector C^N satisfies the equations

$$g_i(\mathbf{H}_1(t)B_1(t) + \mu\mathbf{H}_2(t)B_2(Y^{N-1}) + \mathbf{H}_3(t)C^N) = 0, \ i = 1,\ldots,n-k. \quad (4.3.104)$$

System (4.3.104) has the solution C^0 for $\mu = 0$. The functional determinant of these functions is nonzero at the point C^0. This implies a continuously differentiable function $C^N(\mu)$ which satisfies (4.3.104) and is such that $C^N(\mu) = C^0$ for $\mu = 0$.

Evaluate $\|Y^N - Y^{N-1}\|$. From (4.3.103) we have

$$\|Y^N - Y^{N-1}\| \leq m_1\|C^N - C^{N-1}\| + m_2|\mu|\,\|Y^{N-1} - Y^{N-2}\|.$$

Writing equalities (4.3.103) for the number $N-1$, subtracting termwise from (4.3.103) and, finally, applying the finite increment formula, we find

$$\|C^N - C^{N-1}\| \leq m_3|\mu|\,\|Y^{N-1} - Y^{N-2}\|,$$

whence

$$\|Y^N - Y^{N-1}\| \leq m_4|\mu|\,\|Y^{N-1} - Y^{N-2}\|.$$

It can be easily shown that a positive number μ_0 can be chosen so that for $|\mu| \leq \mu_0$ the quantities m_i, $i = 1,2,3,4$ are the positive constants independent of μ, and construction of all successive approximations is also possible. If μ_0 is additionally chosen so that $m_4\mu_0 < 1$, then the successive approximations of Y^0, $Y^1,\ldots$ converge uniformly for $t \in [0,T]$ to a continuous solution of system (4.3.77) which satisfies conditions (4.3.76). ∎

If in (4.3.76) we set $k = 0$, $\mathbf{A}_{00} = \mathbf{E}, \mathbf{A}_{10} = -\mathbf{E}$ and $B_0 = 0$, where $\mathbf{E}$ is a unit matrix, then we get the so-called periodic boundary conditions $Y = (0) = Y(T)$. For such type conditions the method of constructing integral equations in one particular case was indicated by A.M. Lyapunov.

4.4 Application of Method of Successive Approximations for Finding Stationary Modes

1. Consider the system

$$\dot{X} = \mathbf{P}(t)X + \Phi(t) + \mu F(t,X,\mu) \quad (4.4.1)$$

and

$$\dot{X} = \mathbf{P}(t)X + \Phi(t). \quad (4.4.2)$$

We assume that $\mathbf{P}(t)$ is the matrix function which is real, continuous, and 2π-periodic in t. Denote by $\lambda_1,\ldots,\lambda_n$ characteristic indices of the system

$$\dot{X} = \mathbf{P}(t)X. \quad (4.4.3)$$

Consider the case where $\Phi(t)$ is a real continuous 2π-periodic vector. Suppose we need to find a $2\pi N$-periodic solution to (4.4.2). In section 3.2 we have established that in the resonance-type case such a solution necessarily reduces to a 2π-periodic solution. Hence we will seek a 2π-periodic solution.

Take the non-resonance-type case. Suppose that among $\lambda_1, \ldots, \lambda_n$ there are no pure imaginary quantities of the form $\pm ki$, $k = 0, 1, \ldots$. Then, as follows from section 1.2, the system (4.4.2) has a unique 2π-periodic solution $X = X^{(0)}(t)$. Denote by $\mathbf{Y}(t)$ the matrix of a fundamental system of solutions for (4.4.3), $\mathbf{Y}(0) = \mathbf{E}$. Here the j-th column of the matrix $\mathbf{Y}(t)$ is assumed to be the solution of system (4.4.3) corresponding to a characteristic index λ_j. Denote by $\mathbf{H}(t, \tau)$ the matrix $\mathbf{Y}(t)\mathbf{Y}^{-1}(\tau)$. Then the above periodic solution of system (4.4.2) is contained in its general solution

$$X = \mathbf{Y}(t)X_0 + \int_0^t \mathbf{H}(t, \tau)\Phi(\tau)d\tau. \quad (4.4.4)$$

We find a constant vector X_0 corresponding to a periodic solution. The periodic solution of system (4.4.2) satisfies the relationship $X(0) = X(2\pi)$, hence we find

$$X_0 = \mathbf{Y}(2\pi)X_0 + \int_0^{2\pi} \mathbf{H}(2\pi, \tau)\Phi(\tau)d\tau$$

and then

$$X_0 = (\mathbf{E} - \mathbf{Y}(2\pi))^{-1} \int_0^{2\pi} \mathbf{H}(2\pi, \tau)\Phi(\tau)d\tau. \quad (4.4.5)$$

From (4.4.4) and (4.4.5) follows the expression for the periodic solution of system (4.4.2)

$$X^{(0)}(t) = \mathbf{Y}(t)(\mathbf{E} - \mathbf{Y}(2\pi))^{-1} \int_0^{2\pi} \mathbf{H}(2\pi, \tau)\Phi(\tau)d\tau + \int_0^t \mathbf{H}(t, \tau)\Phi(\tau)d\tau. \quad (4.4.6)$$

We now assume that the vector-valued function $F(t, X, \mu)$ is given for $t \in (-\infty, +\infty)$, $X \in G$, $\mu \in [0, \mu_1]$, and is real and continuous in all of its arguments. Also, it is 2π-periodic with respect to t and satisfies in X the Lipschitz condition

$$\|F(t, \bar{X}, \mu) - F(t, \bar{\bar{X}}, \mu)\| \leq L\|\bar{X} - \bar{\bar{X}}\|,$$

$$\bar{X}, \bar{\bar{X}} \in G, \ t \in (-\infty + \infty), \ \mu \in [0, \mu_1], \quad (4.4.7)$$

where L is a positive constant, and G some region of the E_n which contains a trajectory of the periodic solution (4.4.6) to system (4.4.2). Then it is possible to select a number $\mu_0 > 0$ such that for any μ, $0 \leq \mu \leq \mu_0$ the system (4.4.1) has a periodic solution in the region

$$\|X - X^{(0)}(t)\| \leq \delta(\mu),$$

where $\delta(\mu)$ is a continuous function which has positive values at $\mu \neq 0$ and is strictly decreasing from $\delta_0 = \delta(\mu_0)$ to zero as $\mu \to +0$. Here the solution tends to $X^{(0)}(t)$ as $\mu \to +0$. The system (4.4.1) has no other periodic solutions in the region $\|X - X^{(0)}(t)\| \leq \delta_0$ for any μ, $0 \leq \mu \leq \mu_0$. Indeed, suppose that the system (4.4.1) has

Application of Method of Successive Approximations

a periodic solution $X = X(t)$ whose trajectory is in G. Then, by substituting this solution into the system (4.4.1), we obtain an identity. Consider this identity for $t = \tau$, multiply on the left by $\mathbf{H}(t,\tau)$ and integrate with respect to τ from zero to t so that

$$\int_0^t \mathbf{H}(t,\tau)\dot{X}(\tau)d\tau = \int_0^t \mathbf{H}(t,\tau)(\mathbf{P}(\tau)X(\tau) + \Phi(\tau) + \mu F(\tau, X, \mu))d\tau.$$

Integrating it by parts and utilizing the relationship

$$\frac{d}{d\tau}\mathbf{H}(t,\tau) = -\mathbf{H}(t,\tau)\mathbf{P}(\tau),$$

we get

$$X(t) - \mathbf{H}(t,0)X(0) = \int_0^t \mathbf{H}(t,\tau)(\Phi(\tau) + \mu F(\tau, X, \mu))d\tau,$$

and hence

$$X(t) = \mathbf{Y}(t)X_0 + \int_0^t \mathbf{H}(t,\tau)(\Phi(\tau) + \mu F(\tau, X, \mu))d\tau. \quad (4.4.8)$$

Determine the value of a constant vector X_0 using the periodicity conditions of function $X(t)$, $X(0) = X(2\pi)$ and identity (4.4.8):

$$X_0 = \mathbf{Y}(2\pi)X_0 + \int_0^{2\pi} \mathbf{H}(2\pi,\tau)(\Phi(\tau) + \mu F(\tau, X, \mu))d\tau,$$

whence

$$X_0 = (\mathbf{E} - \mathbf{Y}(2\pi))^{-1}\int_0^{2\pi} \mathbf{H}(2\pi,\tau)(\Phi(\tau) + \mu F(\tau, X, \mu))d\tau. \quad (4.4.9)$$

From (4.4.8) and (4.4.9) we find

$$X(t) = \mathbf{Y}(t)(\mathbf{E} - \mathbf{Y}(2\pi))^{-1}\int_0^{2\pi} \mathbf{H}(2\pi,\tau)(\Phi(\tau) + \mu F(\tau, X, \mu))d\tau$$

$$+ \int_0^t \mathbf{H}(t,\tau)(\Phi(\tau) + \mu F(\tau, X, \mu))d\tau.$$

Considering (4.4.6), reduce the latter identity to a final form

$$X(t) = X^{(0)}(t) + \mu\left(\mathbf{Y}(t)(\mathbf{E} - \mathbf{Y}(2\pi))^{-1}\int_0^{2\pi} \mathbf{H}(2\pi,\tau)F(\tau, X, \mu)d\tau\right.$$

$$\left.+ \int_0^t \mathbf{H}(t,\tau)F(\tau, X, \mu)d\tau\right). \quad (4.4.10)$$

Thus, each 2π-periodic solution of system (4.4.1) satisfies identically the relationship (4.4.10). Now, we consider (4.4.10) to be a system of integral equations designed to define the unknown function $X(t)$. Each continuous solution of (4.4.10), whose trajectory lies in G, is a 2π-periodic solution of system (4.4.1). Indeed, let the system (4.4.10) have a continuous solution $X = X(t)$ with the trajectory lying in G. By substituting it in (4.4.10), we get the identity. Let us differentiate this identity with respect to t:

$$\frac{d}{dt}X = \frac{d}{dt}X^{(0)}(t) + \mu \left(\frac{d}{dt}\mathbf{Y}(t)(\mathbf{E} - \mathbf{Y}(2\pi))^{-1}\int_0^{2\pi} \mathbf{H}(2\pi,\tau)F(\tau,X,\mu)d\tau \right.$$

$$\left. + F(t,X,\mu) + \int_0^t \frac{d}{dt}\mathbf{H}(t,\tau)F(\tau,X,\mu)d\tau\right) = \mathbf{P}(t)X^{(0)}(t) + \Phi(t)$$

$$+\mu\left(\mathbf{P}(t)\mathbf{Y}(t)(\mathbf{E}-\mathbf{Y}(2\pi))^{-1}\int_0^{2\pi}\mathbf{H}(2\pi,\tau)F(\tau,X,\mu)d\tau\right.$$

$$\left. + F(t,X,\mu) + \int_0^t \mathbf{P}(t)\mathbf{H}(t,\tau)F(\tau,X,\mu)d\tau\right) = \mathbf{P}(t)X(t) + \Phi(t) + \mu F(t,X,\mu).$$

Hence it follows that $X(t)$ satisfies the system (4.4.1). We now show that the function $X(t)$ has a 2π-period, i.e. the relationship $X(t) = X(t+2\pi)$ is satisfied. Since $X(t)$ is a solution to (4.4.1), then it suffices to verify the equality $X(0) = X(2\pi)$. From (4.4.10) we have

$$X(0) = X^{(0)}(0) + \mu(\mathbf{E}-\mathbf{Y}(2\pi))^{-1}\int_0^{2\pi}\mathbf{H}(2\pi,\tau)F(\tau,X,\mu)d\tau,$$

$$X(2\pi) = X^{(0)}(2\pi) + \mu\left(\mathbf{Y}(2\pi)(\mathbf{E}-\mathbf{Y}(2\pi))^{-1}\int_0^{2\pi}\mathbf{H}(2\pi,\tau)F(\tau,X,\mu)d\tau\right.$$

$$\left. + \int_0^{2\pi}\mathbf{H}(2\pi,\tau)F(\tau,X,\mu)d\tau\right).$$

Subtracting these equalities termwise, we find

$$X(0) - X(2\pi) = \mu\left((\mathbf{E}-\mathbf{Y}(2\pi))^{-1}(\mathbf{E}-\mathbf{Y}(2\pi))\int_0^{2\pi}\mathbf{H}(2\pi,\tau)Fd\tau\right.$$

$$\left. - \int_0^{2\pi}\mathbf{H}(2\pi,\tau)Fd\tau\right) = 0.$$

Consequently, seeking 2π-periodic solution for the system (4.4.1) reduces to seeking continuous solutions for the system of integral equations (4.4.10).

Denote the right-hand side of system (4.4.10) by $X^{(0)}(t) + \mu R(X)$. As a zero approximation, we take the function $X^{(0)}(t)$, and as a $(p+1)$ approximation take the function

$$X^{(p+1)}(t) = X^{(0)}(t) + \mu R(X^{(p)}). \qquad (4.4.11)$$

Denote by γ a positive constant satisfying the inequality

$$\|\mathbf{Y}(t)\|\,\|(\mathbf{E}-\mathbf{Y}(2\pi))^{-1}\|\,\|\mathbf{H}(2\pi,\tau)\| + \|\mathbf{H}(t,\tau_1)\| \leq \gamma \text{ for } \tau_1, t, \tau \in [0, 2\pi],$$

and by $m(\delta)$ a positive constant satisfying the relationship

$$\|F(t,X,\mu)\| \leq m(\delta) \text{ for } t \in (-\infty,+\infty),\ \mu \in [0,\mu_1],\ \|X - X^{(0)}(t)\| \leq \delta.$$

If the parameter μ is selected in such a manner that $0 \leq \mu \leq \mu_2$, where $\mu_2 = \frac{\delta}{2\pi\gamma m(\delta)}$, then all approximations of $X^{(p+1)}(t)$ are in the region $\|X - X^{(0)}(t)\| \leq \delta$. Indeed, from (4.4.11) we have the inequality

$$\|X^{(p+1)} - X^{(0)}(t)\| \leq \mu \left(\|\mathbf{Y}(t)\| \, \|(\mathbf{E} - \mathbf{Y}(2\pi))^{-1}\| \int_0^2 \|\mathbf{H}(2\pi,\tau)\| \, \|F\| d\tau \right.$$
$$\left. + \int_0^t \|\mathbf{H}\| \, \|F\| d\tau \right) \leq \mu\gamma \int_0^{2\pi} \|F(\tau, X^{(p)}(\tau), \mu)\| d\tau \leq 2\pi\mu\gamma m(\delta) \leq \delta,$$

which is obtained on the assumption that $\|X^{(p)} - X^{(0)}(t)\| \leq \delta$. Consequently, the same inequality is satisfied by $X^{(p+1)}(t)$, etc. If the number δ is taken to be so small that the δ-neighborhood of the trajectory $X^{(0)}(t)$ belongs to G, then for all μ, $0 \leq \mu \leq \mu_2$ it is possible to construct successive approximations by the formula (4.4.11).

We now focus on the convergence problem for the series

$$X^{(0)}(t) + \sum_{p=0}^{\infty} \left(X^{(p+1)} - X^{(p)} \right). \tag{4.4.12}$$

For this purpose, evaluate $\|X^{(p+1)} - X^{(p)}\|$. From (4.4.11) we derive the inequality

$$\|X^{(p+1)} - X^{(p)}\| \leq \gamma L \mu \int_0^{2\pi} \|X^{(p)}(\tau) - X^{(p-1)}(\tau)\| d\tau,$$

whence

$$\sup_{t\in[0,2\pi]} \|X^{(p+1)}(t) - X^{(p)}(t)\| \leq 2\pi\gamma\mu L \sup_{t\in[0,2\pi]} \|X^{(p)}(t) - X^{(p-1)}(t)\|$$
$$\leq (2\pi\gamma\mu L)^p \sup_{t\in 0,2\pi} \|X^{(1)}(t) - X^{(0)}(t)\|. \tag{4.4.13}$$

If we set

$$m = \sup_{\substack{t\in(-\infty,+\infty) \\ X=X^{(0)}, \mu\in[0,\mu_1]}} \|F(t,X,\mu)\|, \quad \text{then} \quad \sup_{t\in[0,2\pi]} \|X^{(1)}(t) - X^{(0)}(t)\| \leq 2\pi m\gamma\mu.$$

Let $\mu_0 < \min\left(\mu_2, \frac{1}{2\pi\gamma L}\right)$. Then for any $\mu \in [0,\mu_0]$ the series (4.4.12) converges for $t \in [0,2\pi]$, since the series composed of its term norms $\sum_{p=0}^{\infty} \|X^{(p+1)} - X^{(p)}\|$, is majorized by the convergent series

$$\sum_{p=0}^{\infty} (2\pi\gamma\mu L)^p 2\pi m\gamma\mu. \tag{4.4.14}$$

Denote by $X(t)$ the sum of series (4.4.12). Then from (4.4.14) and (4.4.12) follows the inequality

$$\|X(t) - X^{(0)}(t)\| \leq \frac{2\pi\gamma\mu m}{1 - 2\pi\gamma\mu L} = \delta(\mu).$$

We show that the obtained periodic solution is unique. Assume that in the region $\|X - X^{(0)}(t)\| \leq \delta$, there is at least for one $\mu \in [0, \mu_0]$, a periodic solution of (4.4.1) other than the obtained $X = \bar{X}(t)$. Then, substituting both solutions into (4.4.10), we get two identities:

$$X(t) = X^{(0)}(t) + \mu R(X) \text{ and } \bar{X}(t) = X^{(0)}(t) + \mu R(\bar{X}).$$

Subtracting these identities termwise and estimating the norm on the left-hand side, we find the relationship

$$\sup_{t \in [0, 2\pi]} \|X(t) - \bar{X}(t)\| \leq 2\pi \gamma \mu L \sup_{t \in [0, 2\pi]} \|X(t) - X(\bar{t})\|.$$

Then $2\pi\gamma\mu L \geq 1$, and hence $\mu > \mu_0$. The obtained contradiction shows that

$$\sup_{t \in [0, 2\pi]} \|X(t) - \bar{X}(t)\| = 0 \text{ and } X(t) = \bar{X}(t).$$

We now observe how the successive approximations (4.4.11) shall be constructed without passing from (4.4.1) to (4.4.10). To this end, the equality (4.4.11) is differentiated termwise:

$$\frac{dX^{(p+1)}}{dt} = \mathbf{P}(t)X^{(p+1)} + \Phi(t) + \mu F(t, X^{(p)}, \mu). \tag{4.4.15}$$

Thus, seeking successive approximations actually amounts to seeking a 2π-periodic solution to system (4.4.15), since (4.4.11) can be viewed as the formula intended for notation of the sought periodic solution of (4.4.15). It now remains to find an estimate of the $(p+1)$ approximation accuracy by approximating the sought solution $X(t)$:

$$\|X(t) - X^{(p+1)}(t)\| \leq \|X^{(p+2)} - X^{(p+1)}\| + \|X - X^{(p+2)}\| \leq \sum_{k=p+1}^{\infty} \|X^{(k+1)} - X^{(k)}\|.$$

Considering (4.4.13), we obtain the required estimate

$$\|X(t) - X^{(p+1)}(t)\| \leq \sum_{k=p+1}^{\infty} \|X^{(k+1)} - X^{(k)}\|$$

$$\leq \sum_{k=p+1}^{\infty} (2\pi\gamma\mu L)^k 2\pi\gamma m = \frac{(2\pi\gamma\mu L)^{p+1} 2\pi\gamma\mu m}{1 - 2\pi\gamma\mu L}. \tag{4.4.16}$$

2. We now describe the successive approximation method for seeking the almost periodic solutions of system (4.4.1). Assume that the vector-valued function $\Phi(t)$ is almost periodic and that the real parts of characteristic indices of the system (4.4.3) are nonzero, i.e.

$$\text{Re}(\lambda_j) < 0, \ j = 1, \ldots, k, \ \text{Re}(\lambda_j) > 0, \ j = k+1, \ldots, n. \tag{4.4.17}$$

Introduce the matrix

$$\mathbf{G}(t,\tau) = \mathbf{Y}(t)\mathbf{G}_1\mathbf{Y}^{-1}(\tau) \text{ for } t \geq \tau,$$

$$\mathbf{G}(t,\tau) = \mathbf{Y}(t)\mathbf{G}_2\mathbf{Y}^{-1}(\tau) \text{ for } t < \tau,$$

$$\mathbf{G}_1 = \begin{pmatrix} \mathbf{I}_k & 0 \\ 0 & 0 \end{pmatrix}, \quad \mathbf{G}_2 = \begin{pmatrix} 0 & 0 \\ 0 & -\mathbf{I}_{n-k} \end{pmatrix},$$

where $\mathbf{I}_k$ and $\mathbf{I}_{n-k}$ are the unit matrices of orders k and $n-k$, respectively. Under the condition (4.4.17), the system (4.4.2) has the unique almost periodic solution

$$X^{(0)}(t) = \int_{-\infty}^{+\infty} \mathbf{G}(t,\tau)\Phi(\tau)d\tau. \tag{4.4.18}$$

The fact that the vector-valued function (4.4.18) is the almost periodic solution of system (4.4.3) is implied by what follows. The uniqueness of such a solution is established in section 1.2. We assume that the right-hand sides of system (4.4.1) are given for $t \in (-\infty, +\infty)$, $\mu \in [0, \mu_1]$, $X \in G$, where G is some region of the space E_n containing the closure of the trajectory described by the vector-valued function (4.4.18), are real, continuous in all of their arguments, almost periodic in t for any fixed $X \in G$ and satisfy the Lipschitz condition

$$\|F(t,\bar{X},\mu) - F(t,\bar{\bar{X}},\mu)\| \leq L\|\bar{X} - \bar{\bar{X}}\|. \tag{4.4.19}$$

Then it is possible to indicate some small $\mu_0 > 0$ that for all $\mu \in [0, \mu_0]$ the system (4.4.1) has a periodic solution in the region $\|X - X^{(0)}(t)\| \leq \delta(\mu)$, where $\delta(\mu)$ is some continuous real function given for $\mu \in [0, \mu_0]$ and strictly decreasing from δ_0 to zero for $\mu > 0$, $\delta_0 = \delta(\mu_0)$. In this case, for any $\mu \in [0, \mu_0]$ the system (4.4.1) has no other almost periodic solutions satisfying the inequality $\|X - X^{(0)}(t)\| \leq \delta_0$.

We shall demonstrate this. Suppose the system (4.4.1) has the almost periodic solution $X = X(t)$; then, substituting it in the system (4.4.1), we obtain an identity. Considering this identity for $t = \tau$, multiply both sides from the left by the matrix $\mathbf{G}(t,\tau)$ and integrate the obtained identity with respect to τ first between the limits $-\infty$ and t and then between the limits t and $+\infty$:

$$\int_{-\infty}^{t} \mathbf{G}(t,\tau)\dot{X}(\tau)d\tau = \int_{-\infty}^{t} \mathbf{G}(t,\tau)(\mathbf{P}(\tau)X(\tau) + \Phi(\tau) + \mu F(\tau, X, \mu))d\tau;$$

integrating it by parts, we find

$$\mathbf{G}(t, t-0)X(t) + \int_{-\infty}^{t} \mathbf{G}(t,\tau)\mathbf{P}(\tau)X(\tau)d\tau$$

$$= \int_{-\infty}^{t} \mathbf{G}(t,\tau)(\mathbf{P}(\tau)X(\tau) + \Phi(\tau) + \mu F(\tau, X, \mu))d\tau.$$

Similarly, we have

$$-\mathbf{G}(t, t+0)X(t) + \int_{t}^{+\infty} \mathbf{G}(t,\tau)\mathbf{P}(\tau)X(\tau)d\tau$$

$$= \int_t^{+\infty} \mathbf{G}(t,\tau)(\mathbf{P}(\tau)X(\tau) + \Phi(\tau) + \mu F(\tau, X, \mu))d\tau.$$

Adding termwise these identities and considering (4.4.18), we finally obtain

$$X(t) = X^{(0)}(t) + \mu \int_{-\infty}^{+\infty} \mathbf{G}(t,\tau)F(\tau, X, \mu)d\tau. \tag{4.4.20}$$

Thus, every almost periodic solution $X = X(t)$ of system (4.4.1), whose trajectory lies in the region G, satisfies the identity (4.4.20). We regard (4.4.20) as a system of integral equations. Then every restricted continuous real solution $X = X(t)$ of (4.4.20), whose trajectory is in G, is an almost periodic solution of system (4.4.1). Indeed, let $X = X(t)$ be such a solution to system (4.4.20). Substituting it in (4.4.20) and differentiating the obtained identity with respect to t, we have the equality

$$\frac{dX}{dt} = \frac{d}{dt}\int_{-\infty}^{t} + \frac{dX^{(0)}}{dt} + \frac{d}{dt}\int_{t}^{+\infty} = \mathbf{G}(t, t-0)\mu F(t, X, \mu)$$

$$+ \int_{-\infty}^{t} \mathbf{P}(t)\mathbf{G}(t,\tau)\mu F(\tau, X, \mu)d\tau - \mathbf{G}(t, t+0)\mu F(t, X, \mu)$$

$$+ \int_{t}^{+\infty} \mathbf{P}(t)\mathbf{G}(t,\tau)\mu F(\tau, X, \mu)d\tau + \mathbf{P}(t)X^{(0)}(t) + \Phi(t) = \mathbf{P}(t)X(t) + \Phi(t) + \mu F(t, X, \mu).$$

It remains to show that the vector-valued function $X(t)$ is almost periodic. For this purpose, we calculate its value at the point $t + \vartheta$:

$$X(t + \vartheta) = X^{(0)}(t + \vartheta) + \mu \int_{-\infty}^{+\infty} \mathbf{G}(t + \vartheta, \tau)F(\tau, X(\tau), \mu)d\tau$$

$$= X^{(0)}(t + \vartheta) + \mu \int_{-\infty}^{+\infty} \mathbf{G}(t + \vartheta, \tau + \vartheta)F(\tau + \vartheta, X(\tau + \vartheta), \mu)d\tau.$$

Evaluate $\|X(t + \vartheta) - X(t)\|$. From (4.4.20) follows the inequality

$$\|X(t + \vartheta) - X(t)\| \leq \|X^{(0)}(t + \vartheta) - X^{(0)}(t)\|$$

$$+ \mu \left\| \int_{-\infty}^{+\infty} \mathbf{G}(t + \vartheta, \tau + \vartheta)F(\tau + \vartheta, X(\tau + \vartheta), \mu)d\tau - \int_{-\infty}^{+\infty} \mathbf{G}(t, \tau)F(\tau, X(\tau), \mu)d\tau \right\|$$

$$\leq \|X^{(0)}(t+\vartheta) - X^{(0)}(t)\| + \mu \int_{-\infty}^{+\infty} \|\mathbf{G}(t+\vartheta, \tau+\vartheta) - \mathbf{G}(t,\tau)\| \, \|F(\tau+\vartheta, X(\tau+\vartheta), \mu)\| d\tau$$

$$+ \mu \int_{-\infty}^{+\infty} \|\mathbf{G}(t,\tau)\| \, \|F(\tau+\vartheta, X(\tau+\vartheta), \mu) - F(\tau, X(\tau+\vartheta), \mu)\| d\tau$$

$$+ \mu \int_{-\infty}^{+\infty} \|\mathbf{G}(t,\tau)\| \, \|F(\tau, X(\tau+\vartheta), \mu) - F(\tau, X(\tau), \mu)\| d\tau. \tag{4.4.21}$$

Recall that the fundamental system of solutions $\mathbf{Y}(t)$ for the system (4.4.3) is of the form $\mathbf{Y}(t) = \mathbf{Y}^{(0)}(t)e^{\mathbf{A}t}$, where $\mathbf{Y}^{(0)}(t)$ is a 2π-periodic matrix. Without loss of generality it may be assumed that the matrix $\mathbf{A} = \begin{pmatrix} \mathbf{A}_1 & 0 \\ 0 & \mathbf{A}_2 \end{pmatrix}$, where $\mathbf{A}_1$ and $\mathbf{A}_2$

Application of Method of Successive Approximations

are respectively the matrices of orders k and $n-k$ having eigenvalues $\lambda_1, \ldots, \lambda_k$ and $\lambda_{k+1}, \ldots, \lambda_n$. From this representation of matrix $\mathbf{Y}(t)$ we find that

$$\mathbf{G}(t+\vartheta, \tau+\vartheta) = \mathbf{Y}^{(0)}(t+\vartheta)e^{\mathbf{A}(t+\vartheta)}\mathbf{G}_1 e^{-\mathbf{A}(\tau+\vartheta)}\left(\mathbf{Y}^{(0)}(\tau+\vartheta)\right)^{-1}$$

$$= \mathbf{Y}^{(0)}(t+\vartheta)\begin{pmatrix} e^{\mathbf{A}_1(t-\tau)} & 0 \\ 0 & 0 \end{pmatrix}\left(\mathbf{Y}^{(0)}(\tau+\vartheta)\right)^{-1} \quad \text{for } t \geq \tau,$$

and if $t < \tau$, then

$$\mathbf{G}(t+\vartheta, \tau+\vartheta) = \mathbf{Y}^{(0)}(t+\vartheta)\begin{pmatrix} 0 & 0 \\ 0 & e^{-\mathbf{A}_2(t-\tau)} \end{pmatrix}\left(\mathbf{Y}^{(0)}(\tau+\vartheta)\right)^{-1}.$$

Considering these relationships, we get the inequality

$$\|\mathbf{G}(t+\vartheta, \tau+\vartheta) - \mathbf{G}(t,\tau)\| \leq \|\mathbf{Y}^{(0)}(t+\vartheta) - \mathbf{Y}^{(0)}(t)\|\left\|\begin{pmatrix} e^{\mathbf{A}_1(t-\tau)} & 0 \\ 0 & 0 \end{pmatrix}\right\|$$

$$\times \|(\mathbf{Y}^{(0)}(t+\vartheta))^{-1}\| + \|\mathbf{Y}^{(0)}(t)\|\left\|\begin{pmatrix} e^{\mathbf{A}_1(t-\tau)} & 0 \\ 0 & 0 \end{pmatrix}\right\|$$

$$\times \|(\mathbf{Y}^{(0)}(\tau+\vartheta))^{-1} - (\mathbf{Y}^{(0)}(\tau))^{-1}\| \quad \text{for } t \geq \tau. \qquad (4.4.22)$$

We have a similar inequality for $t > \tau$. It follows from these inequalities that

$$\|\mathbf{G}(t+\vartheta, \tau+\vartheta) - \mathbf{G}(t,\tau)\| \leq C_1\gamma_1 e^{-\alpha(t-\tau)},$$

where γ_1, α are some positive constants, and

$$C_1 = \sup_{t\in(-\infty,+\infty)} \|\mathbf{Y}^{(0)}(t+\vartheta) - \mathbf{Y}^{(0)}(t)\|.$$

Set

$$C_2 = \sup_{\substack{t\in(-\infty,+\infty) \\ X\in\bar{G}}} \|F(t+\vartheta, X, \mu) - F(t, X, \mu)\|,$$

where $\bar{G} \subset G$ is the bounded region containing the trajectory of the solution $X = X(t)$ under consideration;

$$C_3 = \sup_{t\in(-\infty,+\infty)} \|X^{(0)}(t+\vartheta) - X^{(0)}(t)\|,$$

$$C = \sup_{t\in(-\infty,+\infty)} \|X(t+\vartheta) - X(t)\|,$$

$$m = \sup_{\substack{t\in(-\infty,+\infty) \\ X\in\bar{G}}} \|F(t, X, \mu)\|.$$

From (4.4.21) and (4.4.22) we find

$$\|X(t+\vartheta) - X(t)\| \leq C_3 + \mu C_1 m\gamma + \mu L\gamma C + \mu C_2\gamma, \qquad (4.4.23)$$

where γ is a positive quantity such that

$$\int_{-\infty}^{+\infty} \|\mathbf{G}(t,\tau)\| d\tau \leq \gamma \text{ and } \int_{-\infty}^{+\infty} \gamma_1 e^{-\alpha(t-\tau)} d\tau \leq \gamma.$$

From the almost periodicity condition of functions F, $X^{(0)}$ and from the periodicity of matrix $\mathbf{P}(t)$ for every $\epsilon > 0$ it is possible to select a number $L(\epsilon)$ such that in any interval of the length $L(\epsilon)$ there is a number for which $C_1 \leq \epsilon$, $C_2 \leq \epsilon$, $C_3 \leq \epsilon$. From (4.4.23) we then derive the inequality

$$C(1 - \mu L\gamma) \leq \epsilon(1 + \mu\gamma m + \mu\gamma).$$

We assume that the number μ_1 is such that $\mu_1 \leq \frac{1}{\gamma L}$. Then for all $\mu \in [0, \mu_1]$ we have $C \leq \epsilon' = \frac{\epsilon(1 + \mu\gamma m + \mu\gamma)}{1 - \mu L\gamma}$. Thus, the required almost periodicity is established.

To seek a restricted continuous solution for system (4.4.20), we will apply the successive approximation method. For this purpose, we take as a zero approximation the function (4.4.18) and set

$$X^{(p+1)}(t) = X^{(0)}(t) + \mu \int_{-\infty}^{+\infty} \mathbf{G}(t,\tau) F(\tau, X^{(p)}(t), \mu) d\tau, \ p \geq 0, \quad (4.4.24)$$

and

$$m(\delta) = \sup_{\substack{t \in (-\infty, +\infty) \\ \|X - X^{(0)}\| \leq \delta \\ \scriptstyle \mu \in [0, \mu_1]}} \|F(t, X, \mu)\|, \quad \mu_2 = \frac{1}{\gamma m(\delta)}.$$

For all $\mu \in [0, \mu_2]$ any one of the approximations then satisfies the inequality $\|X^{(p+1)}(t) - X^{(0)}(t)\| \leq \delta$, $p = 0, 1, \ldots$. Since δ is selected so that the condition $\|X - X^{(0)}(t)\| \leq \delta$ implies $X \in G$, we have that the formula (4.4.24) determines successive approximations.

We show that the series

$$X^{(0)}(t) + \sum_{p=0}^{\infty} \left(X^{(p+1)}(t) - X^{(p)}(t) \right) \quad (4.4.25)$$

converges to the almost periodic function $X(t)$:

$$\|X^{(p+1)}(t) - X^{(p)}(t)\| \leq \mu \int_{-\infty}^{+\infty} \|\mathbf{G}(t,\tau)\| \|F(\tau, X^{(p)}, \mu) - F(\tau, X^{(p-1)}, \mu)\| d\tau$$

$$\leq \mu L \int_{-\infty}^{+\infty} \|\mathbf{G}(t,\tau)\| \|X^{(p)} - X^{(p-1)}\| d\tau \leq (\mu L\gamma)^p m,$$

where $m = \sup_{t \in (-\infty, +\infty)} \|X^{(1)}(t) - X^{(0)}(t)\| \leq \gamma\mu m_0$, $m_0 = m(0)$. From this it follows that the series (4.4.25) converges for $\mu < \frac{1}{\gamma L}$. Choose the number μ_0 so that $\mu_0 <$

$\min\{\mu_1, \mu_2, \frac{1}{\gamma L}\}$. Then we have that for any $\mu \in [0, \mu_0]$ the system (4.4.1) has the almost periodic solution $X = X(t)$ satisfying the inequality

$$\|X(t) - X^{(p)}(t)\| \leq \frac{\mu \gamma m_0}{1 - L\gamma \mu}.$$

In this case, the p approximation determined by the formula (4.4.24) approximates $X(t)$ to an accuracy of

$$\|X(t) - X^{(p)}(t)\| \leq \frac{\mu \gamma m_0}{1 - \mu \gamma L}(\mu \gamma L)^p.$$

The uniqueness of the obtained almost periodic solution in the region $\|X - X^{(0)}(t)\| \leq \delta$ is established just as in the case of periodic solutions. Now we may ask: how are the successive approximations (4.4.24) related to the original system (4.4.1)? On the one hand, the earlier considerations suggest that $X^{(p)}$ is an almost periodic vector-valued function and, on the other, differentiating (4.4.24) with respect to t, we have the relationship

$$\frac{dX^{(p+1)}}{dt} = \mathbf{P}(t)X^{(p+1)} + \Phi(t) + \mu F(t, X^{(p)}, \mu). \tag{4.4.26}$$

Thus, the construction of successive approximations reduces to that of almost periodic solutions for the system (4.4.26), where we should set $X^{(p)} = X^{(0)}$ for $p = 0$.

4.5 Quantitative Stability Criteria for Stationary Modes in Automatic Control Systems

1. The stability problem most commonly arises in studies on the behavior of integral curves in the neighborhoods of the equilibrium position and periodic modes. This section offers one of the possible numerical solutions of the stability problem for the equilibrium position and periodic orbits when the solution of the stability problem reduces to investigation in the behavior of solutions of homogeneous differential equations.

Suppose that an automatic control system is described by the differential equation system

$$\dot{X} = F(X, t, \mu), \tag{4.5.1}$$

where $\mu = (\mu_1, \ldots, \mu_k)$ is a collection of some real parameters varying within a region Ω. The parameters must be chosen so that the equilibrium position $X = \bar{X}$ of the system (4.5.1) would be asymptotically stable. In (4.5.1), we replace the required function X by the formula $X = \bar{X} + Y$. Then, isolating in the following system the terms that are linear in Y

$$\dot{Y} = F(Y + \bar{X}, t, \mu),$$

we obtain the system of linear differential equations

$$\dot{Y} = \mathbf{P}Y. \tag{4.5.2}$$

Assume that the elements of matrix **P** are t-independent and are merely the functions of μ, and that the nonlinear terms in the function $F(\bar{X} + Y, t, \mu)$ are of higher order than the first one. Then the equilibrium position $X = \bar{X}$ in (4.5.1) is asymptotically stable if and only if the zero solution of system (4.5.2) is asymptotically stable. As shown in section 1.2, the zero solution of (4.5.2) is asymptotically stable if and only if all the eigenvalues of matrix **P** have negative real parts. In order to find quantitative description of this necessary and sufficient condition for asymptotic stability, it is common practice to employ the Hurwitz condition [Faddeev and Faddeeva (1963)], or an equivalent, which has extensive applications in the automatic control theory. To realize each of such criteria, it is essential to find a characteristic equation. In other words, we have to evaluate the determinant

$$|\mathbf{P} - \lambda \mathbf{E}| = 0. \qquad (4.5.3)$$

After evaluating the determinant, equation (4.5.3) may be written as

$$\lambda^n + P_1 \lambda^{n-1} + \ldots + P_n = 0,$$

where P_j are the known functions of μ. To obtain such parameter combinations, for which the zero solution of system (4.5.2) is asymptotically stable, the quantities $P_1, \ldots, P_n$ should obey Routh-Hurwitz inequalities [Faddeev and Faddeeva (1963)]

$$g_i(P_1, \ldots, P_n) > 0. \qquad (4.5.4)$$

Note that the determinant order (4.5.3) in real systems can be rather high. This shows that utilization of the criterion (4.5.4) entails derivation of a characteristic equation involving a large amount of work for evaluation of the determinant (4.5.3). Application of computers for evaluation of such determinants involves considerable machine time. Similar remarks also apply to the determinants appearing in Hurwitz conditions. In this paragraph, attempts are made to depart from this traditional method of solving the stability problem and the method is given for numerical solution of this problem with the help of digital computers. The latter method partly eliminates the above-mentioned drawbacks of the traditional method and allows the computational process to be automated to a greater degree.

The central idea of the proposed method is as follows. Let a region G be given on the plane of a complex variable λ, and a circle of unit radius with the center at $\rho = 0$ be given on the plane of a complex variable ρ. Assume that necessary and sufficient conditions are to be found for all eigenvalues of matrix **P** to be in the region G. Denote by P_G the collection of all matrices of order n whose eigenvalues are in G, and by B_ρ the set of all matrices **B** of order n whose eigenvalues are inside a unit circle with the center at the point $\rho = 0$. Also, assume that there exists a one-to-one correspondence of two sets P_G and B_ρ. In other words, if $\mathbf{P} \in P_G$, then $\mathbf{H}(\mathbf{P})$ is a matrix of order n such that $\mathbf{H}(\mathbf{P}) \in B_\rho$. Conversely, if **B** is a matrix of family B_ρ, then $\mathbf{H}^{-1}(\mathbf{B})$ is a matrix of order n such that $\mathbf{H}^{-1}(\mathbf{B}) \in P_G$. Here $\mathbf{H}^{-1}$ is the inverse mapping of **H**. For $\mathbf{B} \in B_\rho$, it is necessary and sufficient that $\mathbf{B}^k \to \mathbf{0}$ as $k \to \infty$. Indeed, if **B** takes a diagonal canonical form, then there is such nonsingular matrix

Quantitative Stability Criteria for Stationary Modes

C that $\mathbf{C}^{-1}\mathbf{R}\mathbf{C} = \mathbf{B}$, where $\mathbf{R}$ is the diagonal matrix along whose principal diagonal are arranged eigenvalues $\rho_1, \ldots, \rho_n$ of the matrix $\mathbf{B}$. From this relationship we find $\mathbf{B}^k = \mathbf{C}^{-1}\mathbf{R}^k\mathbf{C}$, whence it follows that $\mathbf{B}^k \xrightarrow[k \to \infty]{} \mathbf{0}$ if and only if $\mathbf{R}^k \xrightarrow[k \to \infty]{} \mathbf{0}$. The latter is possible if and only if $\rho_j^k \to 0$, $k \to \infty$, $j = 1, \ldots, n$, because $\mathbf{R}^k$ is the diagonal matrix along whose principal diagonal are the arranged numbers $\rho_1^k, \ldots, \rho_n^k$. In the general case, this statement is established in the same manner. This property of matrices of the family B_ρ implies the following statement: for $\mathbf{P} \in P_G$ it is necessary and sufficient that $(\mathbf{H}(\mathbf{P}))^k \to \mathbf{0}$ as $k \to \infty$.

We now discuss this idea in some detail from the point of view of automatic control theory. It is well known that the most popular requirements placed upon the linear system (4.5.2) when automatic control systems are designed, are the requirements for asymptotic stability with a given margin as well as the requirements for restriction of oscillation ability in the control system loop. In other words, the problem of seeking parameters from the condition of asymptotic stability with a given margin can be formulated as follows: to indicate all such combinations H_0 of parameters μ for which all eigenvalues of matrix $\mathbf{P}$ are in the half-plane $\mathrm{Re}\,\lambda < \alpha$. Denote by $g(\alpha)$ the region determined by this inequality. Then

$$\rho = 1 + 2(\lambda - 1 - \alpha)^{-1} \tag{4.5.5}$$

maps $g(\alpha)$ onto unit radius circle and, therefore, H_0 involve only those parameters for which $\mathbf{B}^k \to \mathbf{0}$ as $k \to \infty$, where

$$\mathbf{B} = \mathbf{E} + 2(\mathbf{P} - (\alpha + 1)\mathbf{E})^{-1}. \tag{4.5.6}$$

In control system designs, constraints are also imposed on oscillation ability in control loops. Such constraints require all eigenvalues of matrix $\mathbf{P}$ to be within the half-strip $g(\alpha, \beta)$ determined by the inequalities

$$\mathrm{Re}\,\lambda < \alpha, \ |\,\mathrm{Im}\,\lambda\,| < \beta. \tag{4.5.7}$$

For all eigenvalues of matrix $\mathbf{P}$ to be within $g(\alpha, \beta)$, it is necessary and sufficient that they lie simultaneously in three half-planes:

$$\mathrm{Re}\,\lambda < \alpha, \ |\,\mathrm{Im}\,\lambda\,| < \beta, \ |\,\mathrm{Im}\,\lambda\,| > -\beta.$$

Each of these half-planes will be mapped onto a unit circle by the formulas

$$\rho = 1 + 2(\lambda - \alpha - 1)^{-1}, \tag{4.5.8}$$

$$\rho = 1 + 2(-i\lambda - \beta - 1)^{-1}, \ \rho = 1 + 2(i\lambda - 1 - \beta)^{-1}. \tag{4.5.9}$$

For the eigenvalues of matrix $\mathbf{P}$ to be within $g(\alpha, \beta)$ it is necessary and sufficient that the following relationships hold

$$\mathbf{B}^k \to \mathbf{0}, \ \mathbf{B}_1^k \to \mathbf{0}, \ \mathbf{B}_2^k \to \mathbf{0} \ \text{as} \ k \to \infty,$$

where
$$\mathbf{B}_1 = \mathbf{E} + 2(-i\mathbf{P} - (\beta+1)\mathbf{E})^{-1}, \tag{4.5.10}$$
$$\mathbf{B}_2 = \mathbf{E} + 2(i\mathbf{P} - (\beta+1)\mathbf{E})^{-1}. \tag{4.5.11}$$

Note that if the matrix $\mathbf{P}$ has real elements, then the matrices $\mathbf{B}_1$ and $\mathbf{B}_2$ are adjoint. Consequently, in the above-mentioned criterion we are to consider only two matrices: either $\mathbf{B}$ and $\mathbf{B}_1$ or $\mathbf{B}$ and $\mathbf{B}_2$.

The oscillation ability constraints are often relaxed in the form, i.e. it is assumed that the matrix $\mathbf{P}$ may have eigenvalues with pure imaginary parts that are large in the modulus, if the real parts of these eigenvalues are also negative quantities large enough in the modulus. This requirement is stated analytically as follows. The parameter set H_0 is to be chosen from Ω in such a manner that all eigenvalues of their corresponding matrices $\mathbf{P}$ would be in the region $g(\alpha, \beta, \varphi)$ formed by the intersection of the half-plane Re$\lambda < \alpha$ and the 2φ angle with the vertex at the point β on the real axis, the angle being symmetric with respect to the real axis and containing the infinite interval of negative real semi-axis. For all eigenvalues of the matrix $\mathbf{P}$ to be in the region $g(\alpha, \beta, \varphi)$, it is necessary and sufficient that at the same time they would be within three half-planes, at the intersection of which the region $g(\alpha, \beta, \varphi)$ is produced. This implies the analytic criterion which is as follows. We find the functions mapping the above-mentioned half-planes onto the unit circle:

$$\rho = 1 + 2(\lambda - \alpha - 1)^{-1}, \tag{4.5.12}$$
$$\rho = 1 + 2(ie^{-i\varphi}\lambda - i\beta e^{-i\varphi} - 1)^{-1}, \tag{4.5.13}$$
$$\rho = 1 + 2(-ie^{i\varphi}\lambda + i\beta e^{i\varphi} - 1)^{-1}. \tag{4.5.14}$$

Then all eigenvalues of the matrix $\mathbf{P}$ lie in $g(\alpha, \beta, \varphi)$ if and only if the following relations hold
$$\mathbf{B}^k \to 0, \ \mathbf{B}_3^k \to 0, \ \mathbf{B}_4^k \to 0 \text{ as } k \to \infty,$$
where
$$\mathbf{B}_3 = \mathbf{E} + 2(ie^{-i\varphi}\mathbf{P} - (i\beta e^{-i\varphi} + 1)\mathbf{E})^{-1},$$
$$\mathbf{B}_4 = \mathbf{E} + 2(-ie^{i\varphi}\mathbf{P} + (i\beta e^{i\varphi} - 1)\mathbf{E})^{-1}.$$

If the matrix $\mathbf{P}$ is real, the matrix $\mathbf{B}_4$ is adjoint with respect to $\mathbf{B}_3$. Therefore, in this case the matrices $\mathbf{B}$ and $\mathbf{B}_3$, or $\mathbf{B}$ and $\mathbf{B}_4$, are to be considered. Note that the operation involving some matrix inversion is encountered in all the above-mentioned criteria. This is known to result in machine time losses and, therefore, affect adversely the proposed method in economical respects.

2. We discuss another approach avoiding, in the general case, the need for matrix inversion in handling the problem of stability and oscillation ability in the control loop. On the plane of a complex variable λ, consider the circle of radius r with center at the point $a = \alpha - r$. For all eigenvalues of the matrix $\mathbf{P}$ to lie in the region $g(\alpha)$, it is necessary and sufficient that there be number $r > 0$ such that all eigenvalues of the matrix $\mathbf{P}$ lie inside the circle of radius r with center at the point a. The function mapping this circle into the radius circle $\rho = \frac{\lambda - a}{r}$ has a simple form. This reasoning suggests that all eigenvalues of the matrix $\mathbf{P}$ lie in the region $g(\alpha)$ if and only if $\mathbf{D}^k \to 0$ as $k \to \infty$, where

$$\mathbf{D} = \frac{\mathbf{P} - a\mathbf{E}}{r} = \mathbf{E} + \frac{\mathbf{P} - \alpha\mathbf{E}}{r}.$$

On the plane of a complex variable λ we construct three circles with radius r and centers at the points $a = \alpha - r$, $a_1 = \alpha + i(r - \beta)$, $a_2 = \alpha + i(\beta - r)$. The eigenvalues of matrix $\mathbf{P}$ are all in the region $g(\alpha, \beta)$ if and only if there is a number $r > 0$ such that all the eigenvalues of matrix $\mathbf{P}$ simultaneously lie inside these three circles, which in turn holds if and only if the matrices $\mathbf{D}^k, \mathbf{D}_1^k, \mathbf{D}_2^k$ simultaneously tend to a zero matrix as $k \to \infty$, where

$$\mathbf{D}_1 = \frac{\mathbf{P} - a_1\mathbf{E}}{r}, \quad \mathbf{D}_2 = \frac{\mathbf{P} - a_2\mathbf{E}}{r}.$$

On the plane of λ, we construct a new system of radius r circles with centers at the points

$$a = \alpha - r, \ a_3 = \alpha - r\sin\varphi + i((\beta - \alpha)\operatorname{tg}\varphi - r\cos\varphi),$$

$$a_4 = \alpha - r\sin\varphi - i((\beta - \alpha)\operatorname{tg}\varphi - r\cos\varphi).$$

All eigenvalues of the matrix $\mathbf{P}$ lie in the region $g(\alpha, \beta, \varphi)$ if and only if there is $r > 0$ such that all eigenvalues of the matrix $\mathbf{P}$ simultaneously lie inside the radius r circles with centers at the points a, a_3, a_4. The latter is possible if and only if the martices $\mathbf{D}^k$, $\mathbf{D}_3^k$ and $\mathbf{D}_4^k$ simultaneously approach the zero matrix as $k \to \infty$.

The above approach was tested by digital computers and showed its usefulness. In this approach, the principal process affords matrix involution. The point is whether the exponent of matrix $\mathbf{B}^k$ tends to the zero matrix as $k \to \infty$. In the general case, this problem is handled when the traces of n successive exponents of this matrix are considered. Indeed, evaluate the determinant $|\mathbf{B} - \rho\mathbf{E}|$ and derive the characteristic equation

$$\rho^n + a_1\rho^{n-1} + \ldots + a_n = 0. \tag{4.5.15}$$

The coefficients of this equation are expressed in terms of the traces to successive exponents of matrix $\mathbf{B}$ by the familiar formulas

$$a_j = \varphi_j(S_1, \ldots, S_n), \ j = 1, \ldots, n,$$

where S_k is the trace of matrix $\mathbf{B}^k$; thus, e.g.,

$$a_1 = -S_1, \ a_2 = \frac{S_1^2 - S_2}{2}, \ldots.$$

In equation (4.5.15), replace the sought variable ρ by the formula $\rho = \frac{\lambda+1}{\lambda-1}$. Then it passes to an equation of the form

$$\lambda^n + p_1\lambda^{n-1} + \ldots + p_n = 0,$$

where p_j are the known linear-fractional functions of $a_1, \ldots, a_n$, since, after the above-mentioned substitution, (4.5.15) may be written as

$$\sum_{j=0}^{n} a_j(\lambda+1)^{n-j}(\lambda-1)^j = 0, \quad a_0 = 1.$$

By collecting the terms with equal exponents (powers) λ and then reducing the equation to a normal form, we obtain the required expression for the quantities p_j. Utilizing further the expressions of $a_1, \ldots, a_n$ through the traces of successive exponents of matrix $\mathbf{B}$ and the Hurwitz condition composed for quantities $p_1, \ldots p_n$, we derive the inequalities

$$\Psi_j(S_1, \ldots, S_n) > 0, \qquad (4.5.16)$$

that are the necessary and sufficient conditions for all eigenvalues of the matrix $\mathbf{B}$ to lie inside a unit circle. So the conditions (4.5.16) are necessary and sufficient for $\mathbf{B}^k \to \mathbf{0}$ as $k \to \infty$. They form the analytic foundation underlying the numerical solution of stability problem.

The finding of successive exponents of the matrix $\mathbf{B}$ calls for storage of the original matrix in computer memory. Sometimes this may cause a considerable inconvenience. Therefore it is proposed to find the successive exponents of matrices such as $\mathbf{B}, \mathbf{B}^2, \mathbf{B}^4, \mathbf{B}^8, \ldots$. This precludes the need for storage of the matrix $\mathbf{B}$ in the memory and makes it possible to use merely the exponents obtained therefrom, since every subsequent matrix is the square of the preceding one. Denoting by $\sigma_0, \sigma_1, \ldots, \sigma_{n-1}$, the traces of exponents of matrices $\mathbf{B}, \mathbf{B}^2, \ldots, \mathbf{B}^{2^{n-1}}$, we may find $a_1, \ldots, a_n$ in terms of $\sigma_0, \sigma_1, \ldots, \sigma_{n-1}$ and then $p_1, \ldots, p_n$ in terms of $a_1, \ldots, a_n$. As a result, applying the Hurwitz criterion, for the quantities $\sigma_0, \ldots, \sigma_{n-1}$ we find the conditions

$$\chi_j(\sigma_0, \ldots, \sigma_{n-1}) > 0,$$

that are necessary and sufficient for all eigenvalues of the matrix $\mathbf{B}$ to lie inside a unit circle. Note that such necessary and sufficient conditions are suitable at all times and, therefore, have no desired simplicity. In view of this, attempts were made to carry out, approximately, in some practical problems, numerical solutions of the stability problem, proceeding from the traces of several exponents of such matrices as $\mathbf{B}^{32}, \mathbf{B}^{64}, \mathbf{B}^{128}$, and considering that the trace of matrix $\mathbf{B}^{2^k}$ is the sum of all its eigenvalues raised to the 2^k powers:

$$\operatorname{Sp} \mathbf{B}^{2^k} = \sum_{i=1}^{k} \rho_i^{2^k}.$$

In this case, it turns out that the thus obtained results are very close to the true ones and require little machine time.

As another method, which may be proposed for numerical solution of the stability problem, we may refer to the possibility of using matrix norms. For $\mathbf{B}^k \to \mathbf{0}$ as $k \to \infty$, it is necessary and sufficient that there be such integer $k_0 > 0$ that the matrix $\mathbf{B}^{k_0}$ has the norm less than one (to be definite, the square root of the sum of modulus squares of matrix elements is regarded as the matrix norm). Based on this, computations can be carried out as follows. Specify a number k_0 and check the inequality

$$\left\|\mathbf{B}^{k_0}\right\| < 1. \qquad (4.5.17)$$

If the inequality (4.5.17) is satisfied, then $\mathbf{B}^k \to \mathbf{0}$ as $k \to \infty$. Otherwise, either a reverse conclusion is drawn approximately or no conclusion is drawn. The trace of matrix $\mathbf{B}^k$ or its norm is a function of the collection of parameters μ. Hence in the space Ω of parameter μ variations, a variety of directed search methods can be employed proceeding from any point μ_0 on the boundary of the asymptotic stability region with a given margin α or the region where constraints are imposed on oscillation ability. Gradient and Gauss methods refer to this category. Moreover, the known algorithms can also be utilized here to detour on the boundary of stability region. In the numerical solution of stability problem by the proposed method, exhaustive information can be obtained about the behavior of transients in automatic control systems. Also, thorough treatment of the obtained results shows that to reduce further the amount of computational work, the problem of stability and oscillation ability in control loops can be solved with some probability.

It is known that the general solution of linear system can be represented as $X = e^{\mathbf{P}t}X_0$, where X is the vector with coordinates $x_1, \ldots, x_n$, $X = X_0$ at $t = 0$. Assume that the number α is the larger real part of roots in equation (4.5.3). Then each coordinate of vector X can be written as

$$x_s = \sum_{i=1}^{n} y_{si}(t) x_i^0 e^{\alpha t}, \ s = 1, \ldots, n,$$

where $y_{si}(t)$ are the elements of matrix $e^{(\mathbf{P}-\alpha\mathbf{E})t}$. From these relationships we find that

$$\|x_s\| \leq \sqrt{\sum_{i=1}^{n} (x_i^0)^2} A_s e^{\alpha t}, \ s = 1, \ldots, n,$$

where $A_s = \sup\limits_{t \geq 0} \sqrt{\sum\limits_{i=1}^{n} y_{si}^2(t)}$, $s = 1, \ldots, n$. Note that these inequalities with such a choice of $A_1, \ldots, A_n$ are strict. The amplitudes of $A_1, \ldots, A_n$ characterize the quality of control system. Set $A = \max\{A_1, \ldots, A_n\}$ and $q = \ln \frac{A}{|\alpha|}$. The damping rate of transient processes in the automatic control system is described by q more adequately than by α. Therefore, q can be taken as the quality index of the system, here the smaller the q the higher is the quality.

Assume that the canonical structure of matrix $\bar{\mathbf{P}}$ is diagonal, then $\bar{\mathbf{P}} = \mathbf{S}^{-1}\mathbf{M}_1\mathbf{S}$, where $\mathbf{M}_1$ is a diagonal matrix. Consequently, the following equality holds

$$e^{\bar{\mathbf{P}}t} = \mathbf{S}^{-1}e^{\mathbf{M}_1 t}\mathbf{S}.$$

Denote by $\mathbf{M}_2$ the diagonal matrix produced from $\mathbf{M}_1$ when the left half-plane λ is transformed into a unit circle of the plane ρ. Then we get $\bar{\mathbf{B}} = \mathbf{S}^{-1}\mathbf{M}_2\mathbf{S}$, whence

$$\bar{\mathbf{B}}^N = \mathbf{S}^{-1}\mathbf{M}_2^N\mathbf{S} = \mathbf{S}^{-1}e^{N\mathbf{M}_3}\mathbf{S},$$

where $M_3 = \ln M_2$. These relationships imply that the matrices $\bar{\mathbf{B}}^N$ and $e^{\bar{\mathbf{P}}t}$ are closely related to each other. Denote

$$\bar{A} = \sup_{i,k=1,\ldots,n} \sqrt{\sum_{j=1}^{n} \left(b_{ij}^{(k)}\right)^2},$$

where $b_{ij}^{(k)}$ are the elements of matrix $\bar{\mathbf{B}}^{2^k}$. Set $\bar{q} = \frac{\bar{A}}{|\alpha|}$. The quantity $\bar{q}$ may be regarded as the total characteristic of system quality. After finding stability regions, it is good practice to isolate subregions, where the quantity $\bar{q}$ does not exceed a given limit.

To be noted is that the differential system

$$\frac{dx}{dt} = \mathbf{P}X$$

can be replaced by the finite difference system

$$X((k+1)h) = X(kh) + h\mathbf{P}X(kh) = (\mathbf{E} + h\mathbf{P})X(kh),$$

whence

$$X((k+1)h) = (\mathbf{E} + h\mathbf{P})^{k+1}X^0, \quad \text{where} \quad X^0 = X(0).$$

This solution of the finite difference system is as close to the solutions of the original system as desired. Attention is drawn to the fact that the matrix $(\mathbf{E} + \mathbf{P}h)^k$ coincides with the matrix $\mathbf{D}^k$ for zero stability margin $\alpha = 0$ and $h = \frac{1}{r}$. Thus, solving the stability problem by the proposed method, all transient processes can also be constructed over a wide range of variations in the independent variable.

Chapter 5

Oscillations in Nonlinear and Controlled Systems

5.1 Construction and Investigation of Stationary Self-oscillations in Automatic Control Systems

1. Consider some entity K whose state is described by a vector-valued function $X \in E_n$ determined from a system of equations

$$\dot{X} = \bar{F}(t, X). \tag{5.1.1}$$

Physically, the vector-valued function X is assumed to be a set of quantities $X = (x_1, \ldots, x_n)$ that are characteristics of departure of the entity K from a stationary mode. In order to eliminate or, at least, to restrict these departures, a control system is introduced. For this purpose, the entity K is connected to the measuring instruments required to control variations in the vector-valued function X. By the current values of the vector-valued function X, the measuring instruments determine a vector-valued function $Y \in E_m$ related to X and defined by differential equations

$$\dot{Y} = G(t, X, Y). \tag{5.1.2}$$

Some devices (controllers) are devised on the entity K, their characteristic being given by a vector-valued function $Z \in E_k$. The controllers Z influence variations in the vector-valued function X; so, in view of such influences, the differential equations describing the state of the entity K become

$$\dot{X} = F(t, X, Z). \tag{5.1.3}$$

In the general case, in equations (5.1.3) Z can be taken as an arbitrary function of time (the entity is controlled by an arbitrarily present program). To make the control of the entity K automatic, it is essential to match the value of Z to the current values of X. This is accomplished through the measuring instruments and the auxiliary device which determines Z in terms of Y. In the broad class of cases we have

$$\dot{Z} = H(t, X, Y, Z). \tag{5.1.4}$$

The equation systems (5.1.2)–(5.1.4), taken in their totality, are a system which serves to describe the process of automatic control of the entity K. The argument of X may appear in the function H in the pure form because among the measuring instruments are the ideal ones, i.e. those which give by the X values a completely defined function of these values. The function H itself is selected from the conditions imposed on the control properties. The system of equations (5.1.2)–(5.1.4) is commonly referred to as the indirect control system. By contrast, system (5.1.2)–(5.1.3) and

$$Z = H_1(t, X, Y) \tag{5.1.5}$$

may be called the direct control system.

Consider the problem of natural oscillations in the automatic control system provided with one indirect action control and described by the equations

$$\dot{\bar{X}} = \bar{\mathbf{A}}\bar{X} + \bar{B}Z,$$

$$\dot{Z} = f(\sigma), \qquad \sigma = \sum_{i=1}^{n-1} \bar{\gamma}_i \bar{x}_i + \bar{\gamma}_n Z, \tag{5.1.6}$$

where $\bar{X}$ is the $(n-1)$-vector, $\bar{\mathbf{A}}$ is a constant real square $(n-1)$-matrix, $\bar{B} \in E_{n-1}$ is a constant vector, $\bar{\gamma}_i$ are real constants, $i = 1, \ldots, n$, and f is some nonlinear function described below. Set $Z = x_n$, $X = (x_1, \ldots, x_n)$. Then system (5.1.6) can be rewritten in vector form:

$$\dot{X} = \mathbf{A}X + Nf(\bar{\Gamma}^* X), \tag{5.1.7}$$

where N is the n-dimensional column vector with zero components, except for the last one which is equal to unity, $\bar{\Gamma}^* = (\bar{\gamma}_1, \ldots, \bar{\gamma}_n)$. Apart from (5.1.7), we consider the system of equations

$$\dot{X} = \mathbf{A}X.$$

There exists a real nonsingular matrix $\mathbf{C}$ such that after transformation

$$X = \mathbf{C}Y \tag{5.1.8}$$

we have

$$\dot{Y} = \mathbf{C}^{-1}\mathbf{A}\mathbf{C}Y = \mathbf{A}_0 Y,$$

where $\mathbf{A}_0$ is the matrix which has a canonical real form. For real eigenvalues of the matrix $\mathbf{A}$ there are Jordan boxes in the matrix $\mathbf{A}_0$, and for complex eigenvalues of the matrix $\mathbf{A}$ of the form $a \pm ib$ there are slightly modified Jordan boxes which contain on the principal diagonal the second order matrices

$$\begin{pmatrix} a & -b \\ b & a \end{pmatrix}.$$

After transformation (5.1.8) the system (5.1.7) becomes

$$\dot{X} = \mathbf{A}_0 X + Bf(\Gamma^* X), \tag{5.1.9}$$

where $B = \mathbf{C}^{-1}N, \Gamma^* = \bar{\Gamma}^*\mathbf{C}$. Note that the desired vector-valued function in (5.1.9) is again denoted by X. If some component b_s of the vector B is zero and the s-th row of the matrix $\mathbf{A}$ contains only one element $\lambda_s \neq 0$, then any stationary natural oscillation of system (5.1.9) is in the hyperplane $x_s = 0$. Also, if $\lambda_s = 0$, then the above-mentioned stationary modes are on hyperplanes of the form $x_s = c_s$. Development of this argument, as applied to the rows of the matrix $\mathbf{A}_0$ corresponding to the adjoint eigenvalues, leads in seeking periodic solutions of system (5.1.9) to lowering its order if the vector B has zero components.

2. Assume that the function $f(\sigma)$ is of the form

$$\begin{aligned}
f(\sigma) &= 0 & \text{for} & \quad X \in S_0, \\
f(\sigma) &= \sigma - \sigma_1 & \text{for} & \quad X \in S_{-1}, \\
f(\sigma) &= \sigma - \sigma_2 & \text{for} & \quad X \in S_1, \\
f(\sigma) &= l_1 - \sigma_1 & \text{for} & \quad X \in S_{-2}, \\
f(\sigma) &= l_2 - \sigma_2 & \text{for} & \quad X \in S_2,
\end{aligned}$$

where $\sigma = \Gamma^* X$, and $S_0, S_1, S_{-1}, S_2, S_{-2}$ - are the sets in E_n determined respectively by $\sigma \in [\sigma_1, \sigma_2]$, $\sigma \in [\sigma_2, l_2]$, $\sigma \in [l_1, \sigma_1]$, $\sigma \in (l_2, +\infty)$, $\sigma \in (-\infty, l_1)$. The quantities $l_1, l_2, \sigma_1, \sigma_2$ are related by $l_1 < \sigma_1 \leq 0 \leq \sigma_2 < l_2$.

Consider the case where $\sigma_1 < 0, \sigma_2 > 0$, i.e. the effects of dead zone are allowed for. System (5.1.9) in the regions $S_0, S_1, S_{-1}, S_2, S_{-2}$ coincides respectively with the systems of linear equation

$$\dot{X} = \mathbf{A}_0 X, \tag{5.1.10}$$

$$\dot{X} = \mathbf{P}X - B\sigma_2, \tag{5.1.11}$$

$$\dot{X} = \mathbf{P}X - B\sigma_1, \tag{5.1.12}$$

$$\dot{X} = \mathbf{A}_0 X + B(l_2 - \sigma_2), \tag{5.1.13}$$

$$\dot{X} = \mathbf{A}_0 X + B(l_1 - \sigma_1), \tag{5.1.14}$$

where $\mathbf{P} = \mathbf{A}_0 + B\Gamma^*$. The basic idea underlying the investigation of natural oscillations in the system (5.1.9) is to study the behavior of integral curves of linear systems (5.1.10)–(5.1.14) in suitable regions and, subsequently, to unite the results of such studies by gluing together the integral curves in continuity when a passage is made from one region to another. All stationary modes of system (5.1.9) are divided into two groups (we are discussing the points of rest, the periodic and almost periodic solutions of system (5.1.9)). The "linear" stationary oscillation lying entirely in one of the regions $S_i, i = 0, \pm 1, \pm 2$ will be placed into one group. The "nonlinear" stationary oscillations, i.e. the stationary modes of system (5.1.9) lying in several regions S_i, or on the boundary of some two regions, will be placed in another group. Construction, detection and investigation of the neighborhoods of stationary modes of the first group are quite simple, while construction of stationary modes of the second group as well as detection and investigation of the neighborhoods of these modes are rather difficult to carry out.

Denote by $\lambda_1, \ldots, \lambda_n$ the eigenvalues of the matrix $\mathbf{A}$, and by $\mu_1, \ldots, \mu_n$ the eigenvalues of the matrix $\mathbf{P}$. If among $\lambda_1, \ldots, \lambda_n$ are the quantities with zero real parts, then the stationary modes that are other than $X = 0$ and are lying within the boundaries of dead zone appear in system (5.1.9). In this case, if among $\lambda_1, \ldots, \lambda_n$ are pure imaginary quantities, then a family of periodic solutions for (5.1.9) is within a dead zone. Also, if there are several such quantities, and among these are incommensurable quantities, then a family of almost periodic solutions is observed within a dead zone. For the zero roots there are the equilibrium positions of system (5.1.9) that are other than $X = 0$ and are lying in the dead zone. Each of the above-mentioned stationary modes is stable in the sense of Lyapunov if for the quantities $\lambda_1, \ldots, \lambda_k$ with zero real parts there are prime elementary divisors, and the remaining $\lambda_{k+1}, \ldots, \lambda_n$ have negative real parts. Otherwise, each stationary oscillation of system (5.1.9) lying in the dead zone is unstable in the sense of Lyapunov. If $\text{Re}\lambda_j \neq 0$, $j = 1, \ldots, n$, then system (5.1.9) has no stationary modes (or states) other than $X = 0$ located within a dead zone. Note that the matrix $\mathbf{A}_0$ always has at least one zero eigenvalue because the last row of the matrix $\mathbf{A}$ is zero. From this it follows that system (5.1.9) has a family of equilibrium positions in the region S_0. System (5.1.11) may not have equilibrium positions. Then it is the same with system (5.1.12) and, therefore, system (5.1.9) has no stationary modes lying within each of the regions S_1, S_{-1}. This case is possible only if among $\mu_1, \ldots, \mu_n$ there is at least one quantity $\mu_j = 0$. If all the eigenvalues $\mu_1, \ldots, \mu_n$ of the matrix $\mathbf{P}$ are nonzero, then each of the systems (5.1.11) and (5.1.12) has the rest point $X_1 = \mathbf{P}^{-1} B \sigma_2$ and $X_{-1} = \mathbf{P}^{-1} B \sigma_1$, respectively. If these rest points are in the regions S_1, S_{-1}, then system (5.1.9) has periodic solutions in these regions when among $\mu_1, \ldots, \mu_n$ are pure imaginary quantities. Also, if among eigenvalues are several pure imaginaries, then system (5.1.9) has almost periodic motions when at least two incommensurable numbers are present among the indicated eigenvalues. If the relationships

$$\Gamma^* X_1 \bar{\in} (\sigma_2, l_2), \quad \Gamma^* X_{-1} \bar{\in} (l_1, \sigma_1), \qquad (5.1.15)$$

are satisfied, then system (5.1.9) has no stationary modes lying entirely in the regions S_1, S_{-1}. System (5.1.13) may have no rest points. Then it is the same with system (5.1.14) and, therefore, system (5.1.9) has no stationary modes lying entirely in the regions S_2 and S_{-2}. If, however, the systems (5.1.13) and (5.1.14) have the rest points lying respectively in the regions S_2 and S_{-2}, then in the neighborhood of these points the system (5.1.9) possesses the same properties as within the boundaries of the dead zone (in the sense of the availability of stationary modes and the behavior of other motions in their neighborhoods). Thus, all stationary modes of (5.1.9) placed into the first group are computed and investigated without any additional problems involving nonlinearity of $f(\sigma)$.

We describe the method of finding periodic modes in (5.1.9) pertaining to the second group, viz. such periodic modes that are located simultaneously in several

Construction and Investigation of Stationary Self-oscillations

regions S_i. Assume that system (5.1.9) has a periodic solution whose trajectory has the points lying in each of the regions S_i and intersects each boundary of adjacent regions at two points only. Denote by $Z_1 - Z_8$ the successive intersection points for the trajectory of periodic solution

$$X = X(t) \tag{5.1.16}$$

with the planes $\Gamma^* X = l_1$, $\Gamma^* X = \sigma_1$, $\Gamma^* X = \sigma_2$, $\Gamma^* X = l_2$ corresponding to the instants $t = 0, t = t_1, \ldots, t = t_8$, so that $X(0) = Z_1 = X(t_8)$. Assume that the periodic solution (5.1.16) for $t \in (0, t_1)$ lies in the region S_{-1}, for $t \in (t_1, t_2)$ in the region S_0, for $t \in (t_2, t_3)$ in S_1, for $t \in (t_3, t_4)$ in S_2, ..., for $t \in (t_7, t_8)$ in S_{-2}, so that $X(0) = Z_1, X(t_1) = Z_2, \ldots, X(t_7) = Z_8, X(t_8) = Z_1$. From these relationships we have

$$Z_2 = e^{P t_1}(Z_1 - X_{-1}) + X_{-1}, \tag{5.1.17}$$

$$Z_3 = e^{A_0(t_2 - t_1)} Z_2, \tag{5.1.18}$$

$$Z_4 = e^{P(t_3 - t_2)}(Z_3 - X_1) + X_1, \tag{5.1.19}$$

$$Z_5 = e^{A_0(t_4 - t_3)}(Z_4 - X_2) + X_2, \tag{5.1.20}$$

$$Z_6 = e^{P(t_5 - t_4)}(Z_5 - X_1) + X_1, \tag{5.1.21}$$

$$Z_7 = e^{A_0(t_6 - t_5)} Z_6, \tag{5.1.22}$$

$$Z_8 = e^{P(t_7 - t_6)}(Z_7 - X_{-1}) + X_{-1}, \tag{5.1.23}$$

$$Z_1 = e^{A_0(t_8 - t_7)}(Z_8 - X_{-2}) + X_{-2}. \tag{5.1.24}$$

Each of the points $Z_1 - Z_8$ of equalities (5.1.17)–(5.1.24) can be expressed in terms of Z_1, and the equalities then become

$$Z_j = \mathbf{A}_j Z_1 + B_j, \quad j = 2, \ldots, 8, \tag{5.1.25}$$

$$Z_1 = \mathbf{D} Z_1 + F. \tag{5.1.26}$$

Assuming that the matrix $\mathbf{E} - \mathbf{D}$ is nonsingular, from (5.1.25) and (5.1.26) we find that

$$Z_j = C_j, \quad j = 1, \ldots, 8, \tag{5.1.27}$$

where C_j are the known functions of the quantities $t_1, \ldots, t_8$ to be defined. To appraise $t_1 - t_8$, from the condition that the points $Z_1 - Z_8$ lie on the planes delimiting the adjacent regions S_i we derive eight transcendental equations:

$$\Gamma^* C_1 = \Gamma^* C_8 = l_1, \quad \Gamma^* C_2 = \Gamma^* C_7 = \sigma_1,$$

$$\Gamma^* C_3 = \Gamma^* C_6 = \sigma_2, \quad \Gamma^* C_4 = \Gamma^* C_5 = l_2. \tag{5.1.28}$$

In the general case, these equations contain linear combinations of exponential functions whose coefficients are polynomials with trigonometric coefficients. After finding solutions to system (5.1.28), it is essential to construct the points $Z_j = C_j$ and establish that the arcs joining integral curves are actually lying in the regions prescribed to them. This being established, the system (5.1.9) has a periodic solution and, therefore, the system (5.1.28) together with the first equation of system (5.1.27) defines all possible periodic solutions of this type for (5.1.9). Similarly, equations are derived for definition of periodic solutions whose trajectories have no points lying in some of the regions S_i, but have the points lying in at least two different regions S_i.

Consider the case where the effects of a dead zone are absent or not allowed for, viz. $\sigma_1 = \sigma_2 = 0$. Here, in the regions

$$V_0 : l_1 \leq \sigma \leq l_2, \quad V_1 : \sigma \geq l_2, \quad V_{-1} : \sigma \leq l_1,$$

where $\sigma = \Gamma^* X$, $l_1 < 0 < l_2$, system (5.1.9) coincides respectively with the systems of equations

$$\dot{X} = \mathbf{P}X, \tag{5.1.29}$$

$$\dot{X} = \mathbf{A}_0 X + Bl_2, \tag{5.1.30}$$

$$\dot{X} = \mathbf{A}_0 X + Bl_1. \tag{5.1.31}$$

If the matrix $\mathbf{P}$ has no eigenvalues with zero real parts, then system (5.1.9) has no natural stationary oscillations lying wholly within V_0, except for $X = 0$. If, however, among $\mu_1, \ldots, \mu_n$ are the numbers with zero real parts, then system (5.1.9) has natural stationary oscillations in V_0, while for equilibria of (5.1.9) there are zero quantities μ_j, and for periodic motions there are pure imaginary μ_j. If among $\mu_1, \ldots, \mu_n$ are several pure imaginaries and, among them, at least two incommensurable quantities, then, aside from periodic motions, system (5.1.9) also has almost periodic motions. All stationary natural oscillations of (5.1.9) lying within V_0 are stable in the sense of Lyapunov if and only if among $\mu_1, \ldots, \mu_n$ there are no quantities with positive real parts, and for multiple eigenvalues with zero real parts there are prime elementary divisors. Assume that among the eigenvalues of the matrix $\mathbf{A}_0$ there are no zero numbers and, additionally, the following inequality is satisfied

$$\Gamma^* \mathbf{A}_0^{-1} B > -1. \tag{5.1.32}$$

System (5.1.9) then has no stationary natural oscillations whose trajectories would lie in one of the regions V_1 and V_{-1}. As an example, consider a linear system (5.1.30). It has a unique point of rest $X_1 = -\mathbf{A}_0^{-1} B l_2$ which, under condition (5.1.32), is in the half-space $\sigma < l_2$. Hence system (5.1.9) cannot have points of rest and periodic or almost periodic motions lying entirely in the region V_1; otherwise such motions would satisfy the relationship $\sigma > l_2$, which is not possible, since on any stationary motion there must be simultaneously

$$\lim_{t \to +\infty} \frac{1}{t} \int_0^t \sigma \, d\tau = 0, \quad \text{and} \quad \lim_{t \to +\infty} \frac{1}{t} \int_0^t \sigma \, d\tau \geq l_2 > 0.$$

Under condition (5.1.32) the rest point $X_{-1} = -\mathbf{A}_0^{-1} B l_1$ of system (5.1.31) lies entirely in the half-space $\sigma > l_1$. Hence system (5.1.9) also cannot have stationary modes whose trajectories are in the region V_{-1}. Thus, if (5.1.32) and the condition

$$\operatorname{Re}\mu_j < 0, \quad j = 1, \ldots, n, \qquad (5.1.33)$$

are satisfied, then system (5.1.9) has a unique equilibrium $X = 0$ that is asymptotically stable in the sense of Lyapunov, and has no stationary natural oscillations other than $X = 0$ whose trajectories are within one of the regions V_0, V_1, V_{-1}. If among $\lambda_1, \ldots, \lambda_n$ there are zero quantities, then systems (5.1.30) and (5.1.31) may generally have no rest points, which occurs if and only if for $\lambda_s = 0$ the component of vector B $b_s \neq 0$ for at least one s. In this case, system (5.1.9) also has no stationary natural oscillations whose trajectories lie wholly in the regions V_1 and V_{-1}. If $\lambda_s = 0$ implies $b_s = 0$, then systems (5.1.30) and (5.1.31) have the whole families of equilibrium positions that are dependent on as many arbitrary constants as the matrix $\mathbf{A}_0$ has zero eigenvalues (here the implication is that for zero eigenvalues there are prime elementary divisors; otherwise consideration merely becomes a little more complicated). Without loss of generality, it may be said that the first k of the eigenvalues of the matrix $\mathbf{A}_0$ are zero, $\lambda_j = 0$, $j = 1, \ldots, k$. System (5.1.9) then has stationary modes within the regions V_1 and V_{-1} if one of the numbers $\gamma_s \neq 0$, $s = 1, \ldots, k$, where γ_s are components of the vector Γ. If $\gamma_s = 0$ for all $s = 1, \ldots, k$, then system (5.1.9) has stationary motions within or on the boundary of V_1 and V_{-1} if and only if relationship (5.1.32) is disturbed, $\mathbf{A}_0$ and B are the quantities obtained from system (5.1.30) after deletion of the first k rows, and Γ is the $(n-k)$-vector with components $(\gamma_{k+1}, \ldots, \gamma_n)$.

We now turn to seeking the periodic solutions of system (5.1.9) whose trajectories have the points lying in various regions V_i. Assume that system (5.1.9) has the periodic solution $X = X(t)$ whose trajectory has the points lying in all the three regions V_i and is such that it cuts each of the planes $\sigma = l_1, \sigma = l_2$ at two points. Let Z_1, Z_4 be the points of intersection with the plane $\sigma = l_1$, Z_2, Z_3 be the points of intersection with the plane $\sigma = l_2$ and $t_1 - t_4$ are such instants of time that $X(0) = X(t_4) = Z_1$, $X(t_1) = Z_2$, $X(t_2) = Z_3$, $X(t_3) = Z_4$. Moreover, the arc of periodic solution for $t \in (0, t_1)$ lies in V_0, for $t \in (t_1, t_2)$ in V_1, etc. Then we get

$$Z_2 = e^{Pt_1} Z_1,$$

$$Z_3 = e^{A_0(t_2-t_1)}(Z_2 - X_1) + X_1, \quad Z_4 = e^{P(t_3-t_2)} Z_3, \qquad (5.1.34)$$

$$Z_1 = e^{A_0(t_4-t_3)}(Z_4 - X_{-1}) + X_{-1},$$

where $X_1 = -\mathbf{A}_0^{-1} B l_2$, and $X_{-1} = -\mathbf{A}_0^{-1} B l_1$ if $\mathbf{A}_0$ is a nonsingular matrix. If, however, $\mathbf{A}_0$ is a singular matrix and systems (5.1.30) and (5.1.31) have the points of rest, then X_1 and X_{-1} are some rest points of these systems. In the general case, the following formulas are to be applied

$$Z_3 = e^{A_0(t_2-t_1)}\left(Z_2 - \int_0^{t_1} \mathbf{A}_0(t_1-\tau)Bl_2 d\tau\right) + \int_0^{t_2} e^{A_0(t_1-\tau)}Bl_2 d\tau,$$

$$Z_1 = e^{A_0(t_1-t_3)}\left(Z_4 - \int_0^{t_3} \mathbf{A}_0(t_3-\tau)Bl_1 d\tau\right) + \int_0^{t_4} e^{A_0(t_4-\tau)}Bl_1 d\tau. \qquad (5.1.35)$$

Both in the particular (5.1.34) and in the general case (5.1.35), the vectors Z_j can be expressed in terms of Z_1. This will be done for the case of (5.1.34). We have

$$Z_2 = e^{Pt_1} Z_1, \quad Z_3 = e^{A_0(t_2-t_1)} e^{Pt_1} Z_1 - e^{A_0(t_2-t_1)} X_1 + X_1,$$

$$Z_4 = e^{P(t_3-t_2)} e^{A_0(t_2-t_1)} e^{Pt_1} Z_1 - e^{P(t_3-t_2)} e^{A_0(t_2-t_1)} X_1 + e^{P(t_3-t_2)} X_1, \quad (5.1.36)$$

$$Z_1 = e^{A_0(t_4-t_3)} e^{P(t_3-t_2)} e^{A_0(t_2-t_1)} e^{Pt_1} Z_1 - e^{A_0(t_4-t_3)} e^{P(t_3-t_2)} e^{A_0(t_2-t_1)} X_1$$
$$+ e^{A_0(t_4-t_3)} e^{P(t_3-t_2)} X_1 - e^{A_0(t_4-t_3)} X_{-1} + X_{-1}.$$

Considering that the points $Z_1 - Z_4$ are on the corresponding planes, from (5.1.36) we get the system of four equations

$$\Gamma^* Z_1 = \Gamma^* Z_4 = l_1, \quad \Gamma^* Z_2 = \Gamma^* Z_3 = l_2. \quad (5.1.37)$$

These equations together with the last equation in (5.1.36) serve to find all such periodic solutions of system (5.1.9). If the matrix

$$\mathbf{E} - e^{A_0(t_4-t_3)} e^{P(t_3-t_2)} e^{A_0(t_2-t_1)} e^{Pt_1}$$

is nonsingular, then we can eliminate Z_1 from system (5.1.37) and obtain the four-equation system to determine $t_1 - t_4$. After solving this system, from the last equation of system (5.1.36) we find the vector Z_1. Then t_4 is a period of the above periodic solution except for sign, and Z_1 is the point of its intersection with the plane $\sigma = l_1$. Of course, after solving the system we should see whether the points Z_1 and Z_2, Z_2 and Z_3, etc., can be joined by the arcs of integral curves lying wholly in their prescribed regions.

Conditions: a) the zero solution of system (5.1.29) is asymptotically stable, i.e. $\operatorname{Re}\mu_j < 0$, $j = 1, \ldots, n$;

b) the periodic solutions of system (5.1.9) are not available (which can be seen from investigation of transcendental algebraic equations of the above type);

c) $\gamma_s = 0$ for $\operatorname{Re}\lambda_s = 0$ are required for global asymptotic stability of the unique equilibrium of (5.1.9), $X = 0$.

3. We now turn to the case where the nonlinearity of $f(\sigma)$ in system (5.1.9) has the hysteresis-type loop, i.e. where the function $f(\sigma)$ is many-valued. This is possible in many cases, e.g. where the delay in relay operation is due to the release current being lower than the operating current. This phenomenon can be described by the two-valued characteristic $f(\sigma)$. Consider two single-valued piecewise-linear functions $f_1(\sigma)$ and $f_2(\sigma)$:

$$f_1(\sigma) = m_1 \text{ for } \sigma \leq l_2,$$
$$f_1(\sigma) = k_1(\sigma - l_2) + m_1 \text{ for } l_2 \leq \sigma \leq l'_2, \quad (5.1.38)$$
$$f_1(\sigma) = m_2 \text{ for } \sigma \geq l'_2,$$
$$f_2(\sigma) = m_2 \text{ for } \sigma \geq l_1,$$
$$f_2(\sigma) = k_2(\sigma - l_1) + m_1 \text{ for } l'_1 \leq \sigma \leq l_1, \quad (5.1.39)$$

$$f_2(\sigma) = m_1 \text{ for } \sigma \leq l'_1,$$

where $m_1 \leq 0 \leq m_2$, $m_1 < m_2$, $l_1 < 0 < l_2$. The quantities k_1 and k_2 can be arbitrary real numbers. However, to avoid complications in the analytic expressions required for investigation of the behavior of integral curves in system (5.1.9), these quantities are taken to be positive. Let

$$l'_2 = \frac{m_2 - m_1}{k_1} + l_2, \qquad l'_1 = \frac{m_1 - m_2}{k_2} + l_1.$$

Assume that for each value of σ the function $f(\sigma)$ in (5.1.9) takes two values: $f_1(\sigma)$ and $f_2(\sigma)$. Then $f(\sigma)$ is the single-valued function of the line segment $[l'_1, l'_2]$. Consider two systems:

$$\dot{X} = \mathbf{A}_0 X + B f_1(\sigma), \tag{5.1.40}$$

$$\dot{X} = \mathbf{A}_0 X + B f_2(\sigma). \tag{5.1.41}$$

We assume that any integral curve of system (5.1.9) starting at $t = 0$ in the region $\sigma > l'_2$ enters the region $\sigma \leq l'_2$ with time increase (decrease). Then in $\sigma \leq l'_2$ it should be extended as an integral curve of system (5.1.41) ((5.1.40), respectively). Next, if the integral curve of (5.1.9) starts at $t = 0$ in the region $\sigma < l'_1$ and enters the region $\sigma \geq l'_1$ as time increases (decreases), then in the region $\sigma \geq l'_1$ it must be extended as a solution of systems (5.1.40) ((5.1.41), respectively). This agreement is made with regard to a physical pattern of the phenomenon described in this case by system (5.1.9), and defines uniquely the behavior of the integral curves of system (5.1.9) appearing in the hysteresis loop zone. Note that the agreement may be opposite in nature if in the system we observe a lead rather than a delay, i.e. a forcing effect instead of a hysteresis effect. In the region $l'_1 \leq \sigma \leq l'_2$, two integral curves of system (5.1.9) pass through each point. One of the curves is a solution to system (5.1.40), and the other is a solution to system (5.1.41). Therefore, if the problem of behavior of system (5.1.9) is to be completely defined in this region, it is essential to indicate not only initial data of motion, but also which of the two values is assumed by $f(\sigma)$. The value of $f(\sigma)$ is determined by the actual position, say, of a relay lever. This condition, just as the above agreement, makes it possible to define uniquely all motions of system (5.1.9). To seek stationary modes of this system, consider the systems of linear equations

$$\dot{X} = \mathbf{A}_0 X + B m_1, \tag{5.1.42}$$

$$\dot{X} = \mathbf{A}_0 X + B m_2, \tag{5.1.43}$$

$$\dot{X} = \mathbf{P}_1 X + B(m_1 - k_1 l_2), \tag{5.1.44}$$

$$\dot{X} = \mathbf{P}_2 X + B(m_2 - k_2 l_1), \tag{5.1.45}$$

where $\mathbf{P}_1 = \mathbf{A}_0 + k_1 B \Gamma^*$, $\mathbf{P}_2 = \mathbf{A}_0 + k_2 B \Gamma^*$. From this it is apparent that both the motions in systems (5.1.40) and (5.1.41) and those in the original system (5.1.9) are adequately determined by integral curves of these linear systems.

Suppose the matrix $\mathbf{A}_0$ has no zero eigenvalues. Then each of the systems (5.1.42) and (5.1.43) has a unique equilibrium. If at least one of the following inequalities is satisfied

$$-\Gamma^*\mathbf{A}_0^{-1}Bm_1 \leq l_2, \tag{5.1.46}$$

$$-\Gamma^*\mathbf{A}_0^{-1}Bm_2 \geq l_1, \tag{5.1.47}$$

then system (5.1.9) has stationary modes. In this case, if one of the inequalities (5.1.46) and (5.1.47) is strictly satisfied and among $\lambda_1, \ldots, \lambda_n$ are pure imaginaries, then system (5.1.9) has a family of periodic solutions and even a family of almost periodic solutions when there are several imaginary eigenvalues of the matrix $\mathbf{A}_0$ and at least two of these are incommensurable. Each of such stationary motions of system (5.1.9) is stable in the sense of Lyapunov if and only if among $\lambda_1, \ldots, \lambda_n$ there are no eigenvalues with positive real parts, and for multiple imaginaries there are prime elementary divisors. If among $\lambda_1, \ldots, \lambda_n$ there are zero eigenvalues, then the systems (5.1.42), (5.1.43) have no rest points or have the whole family of such points. We assume that $\lambda_j = 0$, $j = 1, \ldots, k$ is the collection of all zero eigenvalues of the matrix $\mathbf{A}_0$ and to these correspond prime elementary divisors. Then system (5.1.9) has stationary modes in all cases where at least $\gamma_j \neq 0$, $j = 1, \ldots, k$. Otherwise system (5.1.9) continues to have stationary modes if one of the relationships (5.1.46), (5.1.47) is satisfied. Here it must be assumed that the matrix $\mathbf{A}_0$ is of order $n - k$, and the vectors B, Γ are of the form $(b_{k+1}, \ldots, b_n)$, $(\gamma_{k+1}, \ldots, \gamma_n)$.

We now turn to equations (5.1.44), (5.1.45). If each of the matrices $\mathbf{P}_1$ and $\mathbf{P}_2$ has no zero eigenvalues, then each of the systems (5.1.44), (5.1.45) has a unique equilibrium. If one of the following inequalities is satisfied

$$l_2 \leq -\Gamma^*\mathbf{P}_1^{-1}B(m_1 - k_1 l_2) \leq l_2', \tag{5.1.48}$$

$$l_1' \leq -\Gamma^*\mathbf{P}_2^{-1}B(m_2 - k_2 l_1) \leq l_1, \tag{5.1.49}$$

then system (5.1.9) has stationary modes. If at least one of these dual inequalities is strictly satisfied and among the eigenvalues corresponding to the matrix $\mathbf{P}_i$ are pure imaginaries, then system (5.1.9) has a family of periodic and possibly almost periodic stationary modes each of which is stable in the sense of Lyapunov if and only if the zero solution of a suitable linear system $\dot{X} = \mathbf{P}_i X$ is stable in the sense of Lyapunov. If among the eigenvalues of matrices $\mathbf{P}_i$ are zero quantities, then the above analysis of the matrix $\mathbf{A}_0$ can be applied to this case in full, thereby finding existence conditions for stationary modes in (5.1.9). It follows from the above discussion that when among the eigenvalues of the matrices $\mathbf{P}_1$, $\mathbf{P}_2$, $\mathbf{A}_0$ there are no zero quantities and none of the inequalities (5.1.46) – (5.1.49) is satisfied, the system (5.1.9) cannot have stationary modes lying in any of the regions $-\infty < \sigma \leq l_1'$, $l_1' \leq \sigma \leq l_1$, $l_1 \leq \sigma \leq l_2$, $l_2 \leq \sigma \leq l_2'$, $l_2' \leq \sigma < +\infty$, where $\sigma = \Gamma^*X$.

Denote these regions by $R_1 - R_5$. In the case where some of the above-mentioned matrices have zero eigenvalues it is possible to formulate the conditions under which there are no stationary modes of (5.1.9) in the regions $R_1 - R_5$. When such conditions are violated, the stationary modes lying within these regions appear in (5.1.9). As follows from the above, their construction and examination for stability amount to investigating the behavior of integral curves in one of the linear systems (5.1.42)–(5.1.45).

We derive equations for seeking the periodic solutions of (5.1.9) which have the points lying in each of the regions $R_1 - R_5$ and intersect only at two points each plane that is a common boundary of the adjacent regions R_i, R_{i+1}, $i = 1, \ldots, 4$. Suppose system (5.1.9) has a periodic solution $X = X(t)$ which, for definiteness, is assumed such that through the point Z_1 lying on the plane $\sigma = l'_1$ this solution enters into the region $\sigma > l'_1$ (as time increases) from the region $\sigma < l'_1$. Let t_1 be the instant of the first intersection of $X = X(t)$ with the plane $\sigma = l_2$ at the point Z_2 following intersection at Z_1, t_2 is the instant of intersection with the plane $\sigma = l'_2$ at the point Z_3, t_3 is the instant of intersecting $\sigma = l'_2$ at Z_4, t_4 is the instant of intersecting $\sigma = l_1$ at Z_5, t_5 is the instant of intersecting $\sigma = l'_1$ at Z_6, t_6 is the instant of intersecting $\sigma = l'_1$ at $Z_7 = Z_1$, so that $X(t_6) = X(0) = Z_1$. From this we find

$$Z_2 = e^{A_0 t_1} Z_1 + \int_0^{t_1} e^{A_0(t_1-\tau)} B m_1 d\tau, \tag{5.1.50}$$

$$Z_3 = e^{P_1(t_2-t_1)} Z_2 + \int_{t_1}^{t_2} e^{P_1(t_2-\tau)} B(m_1 - k_1 l_2) d\tau, \tag{5.1.51}$$

$$Z_4 = e^{A_0(t_3-t_2)} Z_3 + \int_{t_2}^{t_3} e^{A_0(t_3-\tau)} B m_2 d\tau, \tag{5.1.52}$$

$$Z_5 = e^{A_0(t_4-t_3)} Z_4 + \int_{t_3}^{t_4} e^{A_0(t_4-\tau)} B m_2 d\tau, \tag{5.1.53}$$

$$Z_6 = e^{P_2(t_5-t_4)} Z_5 + \int_{t_4}^{t_5} e^{P_2(t_5-\tau)} B(m_2 - k_2 l_1) d\tau, \tag{5.1.54}$$

$$Z_1 = Z_7 = e^{A_0(t_6-t_5)} Z_6 + \int_{t_5}^{t_6} e^{A_0(t_6-\tau)} B m_1 d\tau. \tag{5.1.55}$$

Formulas (5.1.50)–(5.1.55) enable all the vectors Z_j, $j = 2, \ldots, 7$, to be expressed in terms of the vector Z_1. Suppose that this has been realized, i.e.

$$Z_j = \mathbf{A}_j Z_1 + B_j, \quad j = 2, \ldots, 7. \tag{5.1.56}$$

The last equality, together with the equations

$$\Gamma^* Z_1 = \Gamma^* Z_6 = l'_1, \ \Gamma^* Z_2 = l_2, \Gamma^* Z_5 = l_1, \ \Gamma^* Z_3 = \Gamma^* Z_4 = l'_2 \tag{5.1.57}$$

determines all the required periodic solutions of (5.1.9). As follows from (5.1.50)–(5.1.55), the system obtained to determine periodic solutions can be slightly simplified by setting $Z_6 = Y_1$, $Z_2 = Y_2$, $Z_3 = Y_3$, $Z_5 = Y_4$. Suppose the time is reckoned along a periodic solution so that $t = 0$ corresponds to the point Y_1. Let τ_1 be the time length of the arc $Y_1 Y_2$, τ_2 the time length of the arc $Y_1 Y_3$, τ_3 the time length of the arc $Y_1 Y_4$, τ_4 the time length of the arc $Y_1 Y_5$; then

$$Y_2 = e^{A_0 \tau_1} Y_1 + \int_0^{\tau_1} e^{A_0(\tau_1-\tau)} B m_1 d\tau,$$

$$Y_3 = e^{P_1(\tau_2-\tau_1)}Y_2 + \int_{\tau_1}^{\tau_2} e^{P_1(\tau_2-\tau)}B(m_1 - k_1 l_2)d\tau, \tag{5.1.58}$$

$$Y_4 = e^{A_0(\tau_3-\tau_2)}Y_3 + \int_{\tau_2}^{\tau_3} e^{A_0(\tau_3-\tau)}Bm_2 d\tau,$$

$$Y_1 = e^{P_2(\tau_4-\tau_3)}Y_4 + \int_{\tau_3}^{\tau_4} e^{P_2(\tau_4-\tau)}B(m_2 - k_2 l_1)d\tau = Y_5.$$

It follows from the equalities (5.1.58) that all vectors Y_j can be expressed in terms of Y_1, $j = 2, \ldots, 5$. Now, we write four scalar equations from the condition for finding the points Y_j in suitable planes:

$$\Gamma^* Y_1 = l_1', \ \Gamma^* Y_3 = l_2', \ \Gamma^* Y_2 = l_2, \ \Gamma^* Y_4 = l_1. \tag{5.1.59}$$

Moreover, as shown above, from relationships (5.1.58) it is possible to derive the equation

$$Y_1 = \mathbf{A}_5 Y_1 + B_5. \tag{5.1.60}$$

Equations (5.1.59)–(5.1.60) represent a system of $(n + 4)$ equations which serves to seek all periodic solutions of (5.1.9), as required. In all cases, the quantities $\tau_1 - \tau_4$ can be found aside from the vector Y_1. This assertion is apparent, say, when the matrix $(\mathbf{E} - \mathbf{A}_5)$ is nonsingular. The vector Y_1 can then be eliminated from (5.1.59) using the equality $Y_1 = (\mathbf{E} - \mathbf{A}_5)^{-1} B_5$.

We shall now consider the case of (5.1.9), where $f(\sigma)$ also has a hysteresis loop, but of the other type. Set

$$f_1^0(\sigma) = m_1 \text{ for } \sigma < l_2, \ f_2^0(\sigma) = m_2 \text{ for } \sigma > l_1,$$

$$f_1^0(\sigma) = m_2 \text{ for } \sigma \geq l_2, \ f_2^0(\sigma) = m_1 \text{ for } \sigma \leq l_1.$$

Assume that in system (5.1.9), $f(\sigma)$ equals $f_1^0(\sigma)$ and $f_2^0(\sigma)$. This case of (5.1.9) will be regarded as limiting for the preceding case as $k_i \to \infty$, $i = 1, 2$. In this respect, the above agreement about the behavior of integral curves, on entering a hysteresis zone, remains valid. In each of the regions

$$-\infty < \sigma \leq l_1, \ l_1 \leq \sigma \leq l_2, \ l_2 \leq \sigma < +\infty$$

(they will be denoted by $M_1 - M_3$) there are no stationary modes if the following two conditions are satisfied:

$$-\Gamma^* \mathbf{A}_0^{-1} Bm_2 < l_1, \tag{5.1.61}$$

$$-\Gamma^* \mathbf{A}_0^{-1} Bm_1 > l_2. \tag{5.1.62}$$

If, however, one of these conditions is violated, then the system (5.1.9) has stationary modes within or on the boundary of the regions

$$\Gamma^* X \geq l_1, \tag{5.1.63}$$

$$\Gamma^* X \leq l_2, \tag{5.1.64}$$

Construction and Investigation of Stationary Self-oscillations

the stationary modes lying in the region (5.1.63) or (5.1.64) according to whether (5.1.61) or (5.1.62) is violated. If any of the inequalities (5.1.61)–(5.1.62) is violated in the sense of rigorous, and among eigenvalues of matrix $\mathbf{A}_0$ are pure imaginary quantities, then the system (5.1.9) has a family of periodic or even almost periodic solutions lying in a suitable region. Each of these stationary modes is stable in the sense of Lyapunov if and only if the zero solution of system $\dot{X} = \mathbf{A}_0 X$ is stable in the sense of Lyapunov. If the matrix $\mathbf{A}_0$ has zero eigenvalues, then on the strength of the earlier considerations it is possible to isolate the conditions under which in the regions $M_1 - M_3$ there are no stationary modes, and the stationary modes of (5.1.9) appear in the regions (5.1.63) and (5.1.64) when these conditions are violated.

We now construct equations for seeking periodic solutions that are nonlinear in nature. Assume that the system (5.1.9) has the periodic solution $X = X(t)$ whose orbit has the points lying in each of the regions $M_1 - M_3$, and this orbit intersects each of the planes $\sigma = l_1$, $\sigma = l_2$ at two different points only. Let the plane $\sigma = l_1$ intersect the above orbit at the points Z_1 and Z_4, while the plane $\sigma = l_2$ intersects it at Z_2, Z_3. For definiteness, suppose the periodic solution $X = X(t)$ enters the region $\sigma > l_1$ through the point Z_1 (as the time increases) from the region $\sigma < l_1$. If Z_2, Z_3 are the successive points of intersection of the periodic solution with the plane $\sigma = l_2$, and Z_4 is the point of intersection with $\sigma = l_1$, then setting

$$X(0) = Z_4,\ X(t_1) = Z_2,\ X(t_2) = Z_4,$$

we find

$$Z_2 = e^{A_0 t_1} Z_4 + \int_0^{t_1} e^{A_0(t_1 - \tau)} B m_1 d\tau, \qquad (5.1.65)$$

$$Z_4 = e^{A_0(t_2 - t_1)} Z_2 + \int_{t_1}^{t_2} e^{A_0(t_2 - \tau)} B m_2 d\tau.$$

Eliminate Z_2 from the last equation of (5.1.65) and denote Z_4 by Z, then

$$Z = e^{A_0 t_2} Z + \int_0^{t_1} e^{A_0(t_2 - \tau)} B m_1 d\tau + \int_{t_1}^{t_2} e^{A_0(t_2 - \tau)} B m_2 d\tau. \qquad (5.1.66)$$

Moreover, we have

$$\Gamma^* \left(e^{A_0 t_1} Z + \int_0^{t_1} e^{A_0(t_1 - \tau)} B m_1 d\tau \right) = l_2, \qquad (5.1.67)$$

$$\Gamma^* \left(e^{A_0 t_2} Z + \int_0^{t_1} e^{A_0(t_2 - \tau)} B m_1 d\tau + \int_{t_1}^{t_2} e^{A_0(t_2 - \tau)} B m_2 d\tau \right) = l_1. \qquad (5.1.68)$$

Equations (5.1.66)–(5.1.68) form the system of $(n+2)$ equations which serves to define all periodic solutions of system (5.1.9), as required. We will make a preliminary study of this system. If the matrix $\mathbf{A}_0$ has zero eigenvalues, then assuming that there are k pieces of $\lambda_j = 0$, $j = 1, \ldots, k$ and prime elementary divisors correspond to them, from (5.1.66) we find

$$t_1 b_j m_1 + (t_2 - t_1) b_j m_2 = 0,\ j = 1, \ldots, k.$$

If at least $b_j \neq 0$, then
$$t_2 = \alpha t_1,$$
where $\alpha = 1 - \frac{m_1}{m_2}$. We assume that $\mathbf{A}_0'$ is the matrix of order $(n-k)$ obtained from $\mathbf{A}_0$ by deleting the first k rows and the first k columns, and Z', Γ', B' are the vectors obtained from Z, Γ, B by deleting the first k components. The vector Z' may then be found from (5.1.66):

$$Z' = \left(\mathbf{E} - e^{A_0' t_2}\right)^{-1} \left(\int_0^{t_1} e^{A_0'(t_2-\tau)} B' m_1 d\tau + \int_{t_1}^{t_2} e^{A_0'(t_2-\tau)} B' m_2 d\tau\right). \quad (5.1.69)$$

Eliminating the vector Z' from equations (5.1.68) and (5.1.67) and then subtracting them termwise, we find

$$l_1 - l_2 = \Gamma'^* \left(\left(e^{A_0 t_2} - e^{A_0 t_1}\right)\left(\mathbf{E} - e^{A_0' t_2}\right)^{-1} \left(\int_0^{t_1} e^{A_0'(t_2-\tau)} B' m_1 d\tau\right.\right.$$
$$\left.+ \int_{t_1}^{t_2} e^{A_0'(t_2-\tau)} B' m_2 d\tau\right) + \int_0^{t_1} e^{A_0'(t_2-\tau)} B' m_1 d\tau + \int_{t_1}^{t_2} e^{A_0'(t_2-\tau)} B' m_2 d\tau$$
$$\left.- \int_0^{t_1} e^{A_0'(t_1-\tau)} B' m_1 d\tau\right). \quad (5.1.70)$$

In (5.1.70), we must set $t_2 = t_1 \alpha$. Then (5.1.70) is the transcendental equation intended to define the quantities t_1. After defining the quantity t_1, it is possible to find Z', Z from (5.1.69). In the case under study there may be the whole family of periodic solutions if at least one periodic solution of the desired type exists. If, however, all the quantities $b_1, \ldots, b_k$ are zero simultaneously, then relationships between t_1 and t_2 cannot be established. By eliminating the vector Z' from (5.1.67) and (5.1.68), we obtain two transcendental equations to determine the quantities t_1 and t_2. Each of these equations contains k arbitrary constants and the problem is not only to find the system parameters under which these equations have the required solutions, but also to find the values of arbitrary constants $c_1, \ldots, c_k$ which determine the manifold $Z_j = c_j$, $j = 1, \ldots, k$, containing the required periodic solution. When $\lambda_j \neq 0$, $j = 1, \ldots, n$, the formula (5.1.69) determines the vector Z. Eliminating it from (5.1.67) and (5.1.68), we get two transcendental equations for t_1, t_2 which contain only the parameters of (5.1.9).

5.2 Self-oscillations in Hysteresis Systems

1. Consideration of some control problems for various engineering installations often calls for investigation of the hysteresis systems of ordinary differential equations. In particular, central to this is the problem of stable or asymptotically stable periodic solutions of such systems, and the necessity to study conditions for their emergence. The relay hysteresis nonlinearity encountered in mathematical models for automatic control systems provides a physical description of the existing spatial lagging of elements or units in control systems [Lur'ye (1951)], [Aizerman (1966)], [Andronov, Witt, Khaikin (1959)], [Popov (1954)], [Popov and Pal'tov (1960)], [Tsypkin (1974)], [Zubov (1962a)], [Zubov (1966)], [Zubov (1975)]. Hysteresis is introduced into the controlled systems that are optimal in the sense of damping the functions or functionals, which ensures feasibility of program motions in these systems [Zubov (1966)]. In relay stabilization of program modes in controlled systems, hysteresis is used to perform small oscillations in some neighborhood of the modes [Zubov (1966)], [Zubov (1975)]. The presence of periodic oscillations in a hysteresis system makes it possible to discuss the quality of the relevant engineering installation and its behavior. Consideration is being given here to existence and stability of nonlinear periodic solutions in autonomous hysteresis systems of ordinary differential equations [Zubov (1975)].

Suppose there is a system

$$\dot{x} = \mathbf{A}x + c + (\mathbf{B}x + d)f(\sigma), \ \sigma = \Gamma^* x, \qquad (5.2.1)$$

where $x, c, d, \Gamma \in E_n, n \geq 2$, $\mathbf{A}, \mathbf{B}$ are $(n \times n)$-constant matrices, c, d, Γ are constant vectors, $*$ is the transposition sign, and $f(\sigma)$ is the hysteresis function,

$$f(\sigma) = \begin{cases} +1 & \sigma > -l, \\ -1 & \sigma < +l, \end{cases}$$

the constant $l > 0$. A solution to (5.2.1) is defined as follows.

Definition 42. For all given $t \geq 0$ and continuous in t, the function $x(t, x_0, \alpha)$ with its values in E_n ($x(0, x_0, \alpha) = x_0, \alpha = \pm 1$) is called the solution of system (5.2.1), if for any $T > 0$ there is a finite collection of numbers $t_0 = 0, t_1, \ldots, t_m = T$ ($t_k < t_{k+1}, k = 0, \ldots, m-1$) such that:
1) $\sigma(t_k, \sigma_0, \alpha) = (-1)^k \alpha l$, $k = 1, \ldots, m-1$,
2) $\sigma(t, \sigma_0, \alpha)(-1)^k \alpha > -l, t \in [t_k, t_{k+1}), k = 0, \ldots, m-1$,
3) $x(t, x_0, \alpha)$ is differentiable with respect to t for $t \in (t_k, t_{k+1})$,

$$\dot{x} = \dot{x}(t, x_0, \alpha) = \mathbf{A}x(t, x_0, \alpha) + c + (-1)^k \alpha (\mathbf{B}x(t, x_0, \alpha) + d), \ k = 0, \ldots, m-1,$$

where $\sigma(t, \sigma_0, \alpha) = \Gamma^* x(t, x_0, \alpha)$, $\sigma_0 = \Gamma^* x_0 = \sigma(0, \sigma_0, \alpha)$. The numbers $t_1, \ldots, t_{m-1}$ are called the switching moments of the solution $x(t, x_0, \alpha)$. Here we get $\alpha = 1$ for $\sigma_0 \geq l$, $\alpha = -1$ for $\sigma_0 \leq -l$, while for $|\sigma_0| < l$ we get $\alpha = 1$ or $\alpha = -1$.

Denote for $\epsilon > 0$, $x \in E_n$, $M \subset E_n$

$$\|x\| = \sqrt{x^* x}, \ \rho(x, M) = \inf_{y \in M} \|x - y\|, \ S(M, \epsilon) = \{y : y \in E_n, \rho(y, M) \leq \epsilon\}.$$

The matrix norm $\|\mathbf{A}\| = \sqrt{\sum\limits_{i,j=1}^{n} \{\mathbf{A}\}_{i,j}^2}$. Let $M(x_0, \alpha) = \{y : y \in E_n, y = x(t, x_0, \alpha), t \geq 0\} \subset E_n$ be an orbit of the solution $x(t, x_0, \alpha)$. We reformulate the stability definition for solutions of ordinary equation systems as applied to solutions of hysteresis systems.

Definition 43. The solution $x(t, x_0, \alpha)$ is called stable in Lyapunov's sense if for any $t_1 \geq 0, \epsilon > 0$ there is $\delta = \delta(t_1, \epsilon) > 0$ such that for any $x_1 \in S(x(t_1, x_0, \alpha), \delta)$ it is possible to choose $\beta = \beta(x_1)$, $\beta = \pm 1$ such that for all $t \geq 0$ we get $\|x(t, x_1, \beta) - x(t + t_1, x_0, \alpha)\| \leq \epsilon$. In addition, if $\|x(t, x_1, \beta) - x(t + t_1, x_0, \alpha)\| \to 0$ as $t \to +\infty$, then the solution $x(t, x_0, \alpha)$ is called asymptotically stable in Lyapunov's sense.

Definition 44. The solution $x(t, x_0, \alpha)$ is called orbitally stable if for any $\epsilon > 0$ there is $\delta = \delta(\epsilon) > 0$ such that for any $x_1 \in S(M(x_0, \alpha), \delta)$ it is possible to choose $\beta = \beta(x_1)$, $\beta = \pm 1$ such that for all $t \geq 0$ we get $x(t, x_1, \beta) \in S(M(x_0, \alpha), \epsilon)$. In addition, if $\rho(x(t, x_1, \beta), M(x_0, \alpha)) \to 0$ as $t \to +\infty$, then the solution $x(t, x_0, \alpha)$ is called orbit-asymptotically stable.

It can be easily shown that the stability of the solution $x(t, x_0, \alpha)$ in Lyapunov's sense implies its orbital stability, and the asymptotic stability of the solution $x(t, x_0, \alpha)$ in Lyapunov's sense implies its orbital asymptotic stability. Note that the orbit-asymptotically stable periodic solution is called a self-oscillation.

Definition 45. The orbit-asymptotically stable solution $x(t, x_0, \alpha)$ is said to have an asymptotic phase if there is $\delta > 0$ such that for any $x_1 \in S(M(x_0, \alpha), \delta)$ it is possible to choose numbers $\tau = \tau(x_1)$, $\beta = \beta(x_1)$, $\beta = \pm 1$ such that $\|x(t + \tau, x_1, \beta) - x(t, x_0, \alpha)\| \to 0$ as $t \to +\infty$.

We consider the existence problem for stable periodic solution of (5.2.1) and focus on periodic switching solutions only. Since system (5.2.1) is autonomous, periodicity of $x(t, x_0, \alpha)$ implies periodicity with the same period of the solution $x(t, x(\tau, x_0, \alpha), f(\sigma(\tau, \sigma_0, \alpha)))$ for any $\tau > 0$. When switching are employed, the time shifts normally allow one to ensure that for the periodic solution $x(t, x_0, \alpha)$ there would be $\sigma_0 = l, \alpha = 1$. Note that the periodic solution of system (5.2.1) has a constant number of switching in any span of time of the period length.

Definition 46. The periodic solution $x(t, x_0, 1)$ ($\sigma_0 = l$) is called periodic with two switching if there are $\tau_1 > 0$, $\tau_2 > 0$ such that the switching moments for the solution $x(t, x_0, 1)$ are the numbers $\tau_1 + pT$ and $(p+1)T$, where $T = \tau_1 + \tau_2, p = 0, 1, \ldots$, and $x(T, x_0, 1) = x_0$.

2. For $y \in E_n$, $t \in (-\infty, +\infty)$ we introduce the following notation:

$$X_1(y, t) = \exp(t(\mathbf{A} + \mathbf{B}))y + \int_0^t \exp((t-s)(\mathbf{A} + \mathbf{B}))(c + d)ds,$$

$$X_2(y, t) = \exp(t(\mathbf{A} - \mathbf{B}))y + \int_0^t \exp((t-s)(\mathbf{A} - \mathbf{B}))(c - d)ds.$$

From the above definitions it follows that the system (5.2.1) has a periodic solution with two switching if and only if the following system is solvable

$$\Gamma^* x_0 = l, \quad x_1 = X_1(x_0, t_1), \tag{5.2.2}$$

$$\Gamma^* x_1 = -l, \quad x_0 = X_2(x_1, t_2),$$

where $t_1 > 0$, $t_2 > 0$, $x_0, x_1 \in E_n$.

Theorem 72 *Let system (5.2.2) be solvable in the above sense and the following inequalities hold*

$$\Gamma^*((\mathbf{A} + \mathbf{B})x_1 + c + d) < 0, \quad \Gamma^*((\mathbf{A} - \mathbf{B})x_0 + c - d) > 0, \tag{5.2.3}$$

Self-oscillations in Hysteresis Systems 281

$$\|\mathbf{A}_2 \cdot \mathbf{A}_1\| < 1, \qquad (5.2.4)$$

where

$$\mathbf{A}_1 = \left(\mathbf{E} - \frac{((\mathbf{A}+\mathbf{B})x_1 + c + d)\Gamma^*}{\Gamma^*((\mathbf{A}+\mathbf{B})x_1 + c + d)}\right)\exp(t_1(\mathbf{A}+\mathbf{B})),$$

$$\mathbf{A}_2 = \left(\mathbf{E} - \frac{((\mathbf{A}-\mathbf{B})x_0 + c - d)\Gamma^*}{\Gamma^*((\mathbf{A}-\mathbf{B})x_0 + c - d)}\right)\exp(t_2(\mathbf{A}-\mathbf{B})).$$

Then the periodic solution $x(t, x_0, 1)$ of system (5.2.1) is orbit-asymptotically stable, has an asymptotic phase and is stable in Lyapunov's sense.

Proof: We introduce some designations. Let the symbol $C^0_{k,m}(Q)$ $\left(C^1_{k,m}(Q)\right)$ stand for the space of k-dimensional vector functions given on a dense in itself set $Q \subset E_m$ and continuous on Q in all their variables (having continuous partial derivatives of the first order on Q in all their variables). Introduce the following notation:

$$M_0 = M(x_0, 1), \; M_1 = \{y : y \in E_n, \; y = x(t, x_0, 1), \; t \in [0, t_1)\},$$

$$M_2 = \{y : y \in E_n, \; y = x(t, x_1, -1), \; t \in [0, t_2)\},$$

$$S_1(\epsilon) = S(\bar{M}_1, \epsilon) \cap \{y : y \in E_n, \; \Gamma^* y \geq -l\},$$

$$S_2(\epsilon) = S(\bar{M}_2, \epsilon) \cap \{y : y \in E_n, \; \Gamma^* y \leq l\},$$

$$R_1(\epsilon) = \{y : y \in E_n, \; \Gamma^* y = l, \; \|y - x_0\| \leq \epsilon\},$$

$$R_2(\epsilon) = \{y : y \in E_n, \; \Gamma^* y = -l, \; \|y - x_1\| \leq \epsilon\}, \; S'_1(\epsilon) = S_1(\epsilon) \backslash R_2(\epsilon),$$

$$S'_2(\epsilon) = S_2(\epsilon) \backslash R_1(\epsilon), \; L_1 = \|\mathbf{A}\| + \|\mathbf{B}\|, \; L_3 = \|c\| + \|d\|,$$

$$L_2 = \min\{\|\mathbf{A} - \mathbf{B}\|, \|\mathbf{A} + \mathbf{B}\|\}, \; L_4 = \max\{\|\mathbf{A} - \mathbf{B}\|, \|\mathbf{A} + \mathbf{B}\|\}.$$

Since $\left\|\mathbf{E} - \frac{(c+d)\Gamma^*}{\Gamma^*(c+d)}\right\|^2 \geq n - 1$, from condition (5.2.4) it follows that $L_4 > 0$. Next, for the relationship $\tilde{x} \in M_1$ ($\tilde{x} \in M_2$) it is everywhere implied that the relevant value $\alpha = 1$ ($\alpha = -1$). For any $\tilde{x} \in \bar{M}_1$, $\tilde{x} \in \bar{M}_2$ it is possible to choose $\tilde{t} = \tilde{t}(\tilde{x}) \in [0, t_1]$, $\tilde{t} = \tilde{t}(\tilde{x}) \in [0, t_2]$ such that $\tilde{x} = x(\tilde{t}, x_0, 1)$, $\tilde{x} = x(\tilde{t}, x_1, -1)$. Using periodicity of the solution $x(t, x_0, 1)$ given for $t \geq 0$, it is possible to determine $x(t, \tilde{x}, 1)$, $x(t, \tilde{x}, -1)$ for any $t < 0, \tilde{x} \in M_1$, $\tilde{x} \in M_2$.

Indeed, set

$$x(t, \tilde{x}, 1) = x(t + (t_1 + t_2)m, \tilde{x}, 1), \; x(t, \tilde{x}, -1) = x(t + (t_1 + t_2)m, \tilde{x}, -1),$$

where $m = E\left(-\frac{t}{t_1+t_2}\right) + 1$, $E(t)$ is an integral part of the real number t. Suppose there is the solution $x(t, x^*, \alpha)$, given for $t \geq 0$, and for a certain $t^* > 0$ we have

$$x_1^* = x(t^*, x^*, \alpha), \; \alpha^* = f(\sigma(t^*, \sigma^*, \alpha)),$$

where $\sigma^* = \Gamma^* x^*$. The solution $x(t, x_1^*, \alpha^*)$ can then be determined for any $t \geq -t^*$. That is, we set $x(t, x_1^*, \alpha^*) = x(t^* + t, x^*, \alpha)$.

We show that for a sufficiently small $\bar{\epsilon} > 0$ there are the functions

$$\tau_1(\tilde{y}) \in C^1_{1,n}(S_1(\bar{\epsilon})), \; \tau_2(\tilde{y}) \in C^1_{1,n}(S_2(\bar{\epsilon}))$$

implicitly given by the respective equalities

$$\Gamma^* X_1(\tilde{y}, \tau_1) + l = 0, \ \Gamma^* X_2(\bar{y}, \tau_2) - l = 0. \tag{5.2.5}$$

Indeed, let

$$\epsilon > 0, \ \tilde{y} \in S(\bar{M}_1, \epsilon), \ \tilde{x} \in \bar{M}_1, \|\tilde{y} - \tilde{x}\| \le \epsilon, \ \tilde{t} = \tilde{t}(\tilde{x}),$$

$$\bar{y} \in S(\bar{M}_2, \epsilon), \ \bar{x} \in \bar{M}_2, \ \|\bar{y} - \bar{x}\| \le \epsilon, \ \bar{t} = \bar{t}(\bar{x}).$$

For $\tilde{y} = \tilde{x}$, $\tau_1 = t_1 - \tilde{t}$, $\bar{y} = \bar{x}$, $\tau_2 = t_2 - \bar{t}$ the relationships (5.2.5) are then satisfied. It follows from conditions (5.2.3) that

$$\left.\frac{\partial}{\partial \tau_1}\Gamma^* X_1(\tilde{y}, \tau_1)\right|_{\substack{\tau_1 = t_1 - \tilde{t} \\ \tilde{y} = \tilde{x}}} = \Gamma^*((\mathbf{A} + \mathbf{B})x_1 + c + d) \ne 0,$$

$$\left.\frac{\partial}{\partial \tau_2}\Gamma^* X_2(\bar{y}, \tau_2)\right|_{\substack{\tau_2 = t_2 - \bar{t} \\ \bar{y} = \bar{x}}} = \Gamma^*((\mathbf{A} - \mathbf{B})x_0 + c - d) \ne 0,$$

and since

$$\Gamma^* X_1(\tilde{y}, \tau_1) \in C^1_{1,n+1}(E_{n+1}), \ \Gamma^* X_2(\bar{y}, \tau_2) \in C^1_{1,n+1}(E_{n+1})$$

and $\bar{M}_1$, $\bar{M}_2$ are closed, then the implicit function theorem [Fichtengol'c (1969)] implies the existence of a constant $\bar{\epsilon} > 0$ such that the functions $\tau_1(\tilde{y}) \in C^1_{1,n}(S_1, (\bar{\epsilon}))$, $\tau_2(\bar{y}) \in C^1_{1,n}(S_2(\bar{\epsilon}))$ are given so that

$$Y_1(\tilde{y}) = X_1(\tilde{y}, \tau_1(\tilde{y})) \in C^1_{n,n}(S_1(\bar{\epsilon})), \ Y_2(\bar{y}) = X_2(\bar{y}, \tau_2(\bar{y})) \in C^1_{n,n}(S_2(\bar{\epsilon}))$$

and

$$\Gamma^* Y_1(\tilde{y}) \equiv -l, \ \Gamma^* Y_2(\bar{y}) \equiv l$$

respectively for $\tilde{y} \in S_1(\bar{\epsilon})$, $\bar{y} \in S_2(\bar{\epsilon})$. The functions $Y_1(\tilde{y})$, $\tau_1(\tilde{y})$, $Y_2(\bar{y})$, $\tau_2(\bar{y})$ are defined and continuous on the closed sets $S_1(\bar{\epsilon}), S_2(\bar{\epsilon})$, respectively, and, hence are uniformly continuous there. Consequently, for any $\epsilon \in (0, \bar{\epsilon}]$ there are $\tilde{\delta} = \tilde{\delta}(\epsilon) \in (0, \epsilon]$, $\bar{\delta} = \bar{\delta}(\epsilon) \in (0, \epsilon]$ such that for all $\tilde{y}_1, \tilde{y}_2 \in S_1(\bar{\epsilon})$, $\bar{y}_1, \bar{y}_2 \in S_2(\bar{\epsilon})$, $\|\tilde{y}_1 - \tilde{y}_2\| \le \tilde{\delta}$, $\|\bar{y}_1 - \bar{y}_2\| \le \bar{\delta}$ there are

$$\|Y_1(\tilde{y}_1) - Y_1(\tilde{y}_2)\| \le \epsilon, \ |\tau_1(\tilde{y}_1) - \tau_1(\tilde{y}_2)| \le \epsilon,$$

$$\|Y_2(\bar{y}_1) - Y_2(\bar{y}_2)\| \le \epsilon, \ |\tau_2(\bar{y}_1) - \tau_2(\bar{y}_2)| \le \epsilon.$$

Since for all $\tilde{x} \in \bar{M}_1$, $\bar{x} \in \bar{M}_2$ we have $Y_1(\tilde{x}) = x_1$, $\tau_1(\tilde{x}) = t_1 - \tilde{t}(\tilde{x})$, $Y_2(\bar{x}) = x_0$, $\tau_2(\bar{x}) = t_2 - \bar{t}(\bar{x})$, then for any $\tilde{y} \in S_1(\tilde{\delta}(\epsilon))$, $\bar{y} \in S_2(\bar{\delta}(\epsilon))$, $\tilde{x} \in \bar{M}_1$, $\|\tilde{x} - \tilde{y}\| \le \tilde{\delta}(\epsilon)$, $\bar{x} \in \bar{M}_2$, $\|\bar{x} - \bar{y}\| \le \bar{\delta}(\epsilon)$ we get

$$\|Y_1(\tilde{y}) - x_1\| \le \epsilon, \ |\tau_1(\tilde{y}) + \tilde{t}(\tilde{x}) - t_1| \le \epsilon,$$

Self-oscillations in Hysteresis Systems

$$\|Y_2(\bar{y}) - x_0\| \leq \epsilon, \quad |\tau_2(\bar{y}) + \bar{t}(\bar{x}) - t_2| \leq \epsilon.$$

These relationships will be called the relationships (A) (or simply (A)). From (A) it follows that for any $y_0 \in R_1(\delta')$ we get

$$y_1 = Y_1(y_0) \in R_2(\bar{\delta}(\epsilon)), \quad y_2 = Y_2(y_1) \in R_1(\epsilon),$$

where $\delta' = \delta'(\epsilon) = \tilde{\delta}(\bar{\delta}(\epsilon)) \in (0, \epsilon]$. Since $Y(\bar{y}) = Y_2(Y_1(\bar{y})) \in C^1_{n,n}(S_1(\delta'))$, then $Y(y_0) \in C^1_{n,n}(R_1(\delta'))$, and hence the function $Y(y_0)$ is differentiable in $R_1(\delta')$. Consequently, for any $y_0 \in R_1(\delta')$ the following relationship is satisfied

$$y_2 - x_0 = \mathbf{D}(y_0 - x_0) + o(\|y_0 - x_0\|),$$

where $\frac{1}{\lambda}\|o(\lambda)\| \to 0$ as $\lambda \to +0$, and $\mathbf{D}$ is the functional matrix of the vector function $Y(y_0)$ calculated at the point $y_0 = x_0$ for the components of the vector y_0. Calculating this matrix, we get $\mathbf{D} = \mathbf{A}_2 \cdot \mathbf{A}_1$, and then it follows from (5.2.4) that $h = \|\mathbf{D}\| \in [0, 1)$. Because of this, there is a constant $\tilde{\epsilon} \in (0, \bar{\epsilon}]$, such that

$$\|o(\lambda)\| \leq \frac{\lambda}{2}(1 - h) \quad \text{for all} \quad \lambda \in [0, \tilde{\epsilon}].$$

Let $\tilde{\delta}' = \delta'(\tilde{\epsilon})$. Then for all $y_0 \in R_1(\tilde{\delta}')$ we get

$$\|y_2 - x_0\| \leq \frac{1}{2}(1 + h)\|y_0 - x_0\| < \|y_0 - x_0\|$$

and hence $y_2 = Y(y_0) \in R_1(\tilde{\delta}')$. From this it follows that there is a sequence

$$y_{2k} \in R_1(\tilde{\delta}'), \quad y_{2k+2} = Y(y_{2k}), \quad k = 0, 1, \ldots,$$

here $\|y_{2k} - x_0\| < h_1^k \|y_0 - x_0\|$, the constant $h_1 = (1 + h)/2 \in (0, 1)$, and hence $\|y_{2k} - x_0\| \to 0$ as $k \to \infty$.

Take arbitrary $\epsilon \in (0, \tilde{\delta}']$, $x' \in M_0$ and let $\delta \in (0, \epsilon]$, $\epsilon_1 \in (0, \tilde{\delta}']$. Then either $x' \in M_1$, or $x' \in M_2$. The values of δ, ϵ_1 will be improved in what follows. Let any $y' \in S(x', \delta)$. It follows from inequalities (5.2.3) that the hyperplanes $\Gamma^* x = l$, $\Gamma^* x = -l$ are transverse to the sets M_2, M_1 at the intersection points x_0, x_1. This implies the following statement. If $\Gamma^* y' \geq l$, then $y' \in S(x_0, \delta_1)$, and if $\Gamma^* y' \leq -l$, then $y' \in S(x_1, \delta_1)$, where $\delta_1 \geq \delta$, and by choosing a sufficiently small $\delta > 0$ a number δ_1 can be made as small as desired uniformly for all $x' \in M_0$ at once. Let

$$x'' = x' \text{ for } |\Gamma^* y'| < l, \quad x'' = x_0 \text{ for } \Gamma^* y' \geq l, \quad x'' = x_1 \text{ for } \Gamma^* y' \leq -l.$$

Then we have that

$$\|x'' - x'\| \leq 2\delta_1.$$

Since $\tilde{t}(\tilde{x}) \in C^0_{1,n}(\bar{M}_1)$, $\bar{t}(\bar{x}) \in C^0_{1,n}(\bar{M}_2)$ and $\bar{M}_1, \bar{M}_2$ are closed, $\tilde{t}(\tilde{x}), \bar{t}(\bar{x})$ are uniformly continuous on $\bar{M}_1, \bar{M}_2$ respectively. Let

$$t'' = \tilde{t}(x''), \; t' = \tilde{t}(x') \text{ for } x', x'' \in \bar{M}_1, \quad t'' = \bar{t}(x''), \; t' = \bar{t}(x') \text{ for } x', x'' \in \bar{M}_2.$$

Then the quantity $|t'' - t'|$ can be made arbitrarily small by choosing a sufficiently small δ_1, i.e. with the choice of a sufficiently small δ. Thus, $\|y' - x''\| \leq \delta_1$ and either $y' \in S_1'(\delta_1)$, $x'' \in M_1$ or $y' \in S_2'(\delta_1)$, $x'' \in M_2$. Assume that $\delta_1 \in (0, \delta'(\epsilon_1)]$. Then it follows from (A) that for $y' \in S_1'(\delta_1)$, $x'' \in M_1$ we get

$$y'' = x(\tau_1(y'), y', 1) = Y_1(y') \in R_2(\bar{\delta}(\epsilon_1)), \quad y_0 = x(\tau^*(y'), y', 1) = Y(y') \in R_1(\epsilon_1),$$

$$\tau^*(y') = \tau_1(y') + \tau_2(y''), \quad x^* = x(\tau^*(y'), x'', 1),$$

$$\tau_0(y') = \tau^*(y') + \tilde{t}(x'') - t_1 - t_2, \quad |\tau_0(y')| \leq \bar{\delta}(\epsilon_1) + \epsilon_1 \leq 2\epsilon_1,$$

and for $y' \in S_2'(\delta_1)$, $x'' \in M_2$ we get

$$y' \in S_2'(\bar{\delta}(\epsilon_1)), \quad y_0 = x(\tau^*(y'), y', -1) = Y_2(y') \in R_1(\epsilon_1), \quad \tau^*(y') = \tau_2(y'),$$

$$x^* = x(\tau^*(y'), x'', -1), \quad \tau_0(y') = \tau^*(y') + \tilde{t}(x'') - t_2, \quad |\tau_0(y')| \leq \epsilon_1.$$

In both cases $y_0 \in R_1(\epsilon_1)$, $|\tau_0(y')| \leq 2\epsilon_1$, $x^* = x(\tau_0(y'), x_0, 1)$. Since

$$\tau_1(\tilde{y}) \in C_{1,n}^1(S_1(\tilde{\delta}')), \quad \tau_2(\bar{y}) \in C_{1,n}^1(S_2(\bar{\delta}(\tilde{\epsilon}))), \quad Y_1(\tilde{y}) \in C_{n,n}^1(S_1(\tilde{\delta}')),$$

then

$$\tau_1(y_0) \in C_{1,n}^1(R_1(\tilde{\delta}')), \quad \tau_2(y_1) \in C_{1,n}^1(R_2(\bar{\delta}(\tilde{\epsilon}))), \quad Y_1(y_0) \in C_{n,n}^1(R_1(\tilde{\delta}'))$$

and hence the function $\tau^*(y_0) = \tau_1(y_0) + \tau_2(Y_1(y_0))$ is differentiable in $R_1(\tilde{\delta}')$. Consequently, for any $y_0 \in R_1(\epsilon_1)$ the following relationship holds

$$\tau_1(y_0) + \tau_2(Y_1(y_0)) - t_1 - t_2 = \Gamma_1^*(y_0 - x_0) + o(\|y_0 - x_0\|),$$

where $\frac{1}{\lambda} o(\lambda) \to 0$ as $\lambda \to +0$, and the vector $\Gamma_1 \in E_n$ is the gradient of the function $\tau^*(y_0)$ with respect to the components of the vector y_0 computed at the point $y_0 = x_0$. Computing this gradient, we get

$$\Gamma_1^* = -\left(\frac{\Gamma^* \exp(t_1(\mathbf{A} + \mathbf{B}))}{\Gamma^*((\mathbf{A} + \mathbf{B})x_1 + c + d)} + \frac{\Gamma^* \exp(t_2(\mathbf{A} - \mathbf{B}))}{\Gamma^*((\mathbf{A} - \mathbf{B})x_0 + c - d)} \mathbf{A}_1\right).$$

Let $L = 2\|\Gamma_1\|$ and suppose the constant $\epsilon_2 \in (0, \tilde{\delta}']$ is such that $|o(\lambda)| \leq \frac{1}{2}L$ for all $\lambda \in [0, \epsilon_2]$. Then for any $\epsilon_1 \in (0, \epsilon_2]$, $y_0 \in R_1(\epsilon_1)$ we get

$$|\tau_1(y_0) + \tau_2(Y_1(y_0)) - t_1 - t_2| \leq L \|y_0 - x_0\|.$$

Since $\epsilon_2 \in (0, \tilde{\delta}')$, then, as shown above, there is a sequence

$$y_{2k+1} = Y_1(y_{2k}) \in R_2(\bar{\delta}(\tilde{\epsilon})), \quad y_{2k+2} = Y(y_{2k}) \in R_1(\epsilon_1),$$

here

$$\|y_{2k} - x_0\| \leq h_1^k \|y_0 - x_0\| \leq h_1^k \epsilon_1,$$

Self-oscillations in Hysteresis Systems 285

$k = 0, 1, \ldots$, the constant $h_1 \in (0, 1)$. From this it follows that the series

$$T(y_0) = \sum_{k=0}^{\infty}(\tau_1(y_{2k}) + \tau_2(y_{2k+1}) - t_1 - t_2)$$

converges absolutely in $R_1(\epsilon_1)$, since

$$\sum_{k=0}^{\infty}|\tau_1(y_{2k}) + \tau_2(y_{2k+1}) - t_1 - t_2| \leq L \sum_{k=0}^{\infty} \|y_{2k} - x_0\|$$

$$\leq L \|y_0 - x_0\| \sum_{k=0}^{\infty} h_1^k = \frac{1}{1-h_1} L \|y_0 - x_0\|.$$

Again we have that

$$|T(y_0)| \leq \frac{1}{1-h_1} L \|y_0 - x_0\| \leq \frac{1}{1-h_1} L \epsilon_1,$$

$$T(y_{2m}) = \sum_{k=m}^{\infty}(\tau_1(y_{2k}) + \tau_2(y_{2k+1}) - t_1 - t_2),$$

$$|T(y_{2m})| \leq \frac{1}{1-h_1} L \|y_{2m} - x_0\| \leq \frac{1}{1-h_1} L h_1^m \|y_0 - x_0\| \leq \frac{1}{1-h_1} L h_1^m \epsilon_1,$$

$$m = 0, 1, \ldots.$$

Let $\alpha = 1$ for $x' \in M_1$, $\alpha = -1$ for $x' \in M_2$, $\beta = 1$ for $x'' \in M_1$, $\beta = -1$ for $x'' \in M_2$.

In order to prove the stability of the periodic solution $x(t, x_0, 1)$ in Lyapunov sense, it is necessary to evaluate $\|x(t, y', \beta) - x(t, x', \alpha)\|$ for all $t \geq 0$. We have that

$$\|x(t, y', \beta) - x(t, x', \alpha)\| \leq \|x(t, x', \alpha) - x(t, x'', \beta)\| + \|x(t, x'', \beta) - x(t, y', \beta)\|$$

for any $t \geq 0$,

$$\|x(t, x'', \beta) - x(t, y', \beta)\| \leq \|x(\tau, y_0, 1) - x(\tau - T(y_0), x_0, 1)\|$$

$$+\|x(\tau - T(y_0), x_0, 1) - x(\tau + \tau_0(y'), x_0, 1)\|$$

for all $t \geq \tau^*(y')$, $\tau \geq 0$, where $\tau = t - \tau^*(y')$. The solutions of system (5.2.1) on the finite interval $[0, \tau]$ with the constant value of f are continuously dependent on initial data. This follows from the inequality

$$\|x(t, x_0', \alpha') - x(t, x_0'', \alpha')\| \leq \exp(\tau L_1) \|x_0' - x_0''\|,$$

which holds for any $t \in [0, \tau]$ only if $f(\sigma(t, \sigma_0', \alpha')) = f(\sigma(t, \sigma_0'', \alpha')) = \alpha'$ for all $t \in [0, \tau]$, where $\sigma_0' = \Gamma^* x_0'$, $\sigma_0'' = \Gamma^* x_0''$. The above inequality will be called the relationship (B) (or simply (B)).

For all those real t from some neighborhood of zero, for which the solution $x(t, x_0', \alpha')$ has been determined and for which the value of $f(\sigma(t, \sigma_0', \alpha')) = \alpha'$ is constant, the following inequalities hold

$$\|x(t, x_0', \alpha') - x_0'\| \leq (\exp(|t| L_1) - 1) \left(\|x_0'\| + \frac{L_3}{L_2} \exp(|t| L_1)\right) \text{ for } L_2 > 0,$$

$$\|x(t,x_0',\alpha') - x_0'\| \leq (\exp(|t|L_1) - 1)\Big(\|x_0'\| + \frac{L_3}{L_4}\exp(|t|\,L_1)\Big) \text{ for } L_2 = 0, \|\mathbf{A} + \alpha'\mathbf{B}\| > 0,$$

$$\|x(t,x_0',\alpha') - x_0'\| \leq L_3|t| \quad \text{for} \quad \|\mathbf{A} + \alpha'\mathbf{B}\| = 0.$$

These relationships will be called the relationships (C) (or simply (C)). Since the set M_0 is bounded, then (C) implies the existence of $\tau' = \tau'(\epsilon) > 0$ such that for all $x_0' \in M_0$ there is $\|x(t,x_0',\alpha') - x_0'\| \leq \frac{1}{8}\epsilon$ for any real t such that $f(\sigma(t,\sigma_0',\alpha')) = \alpha'$, $|t| \leq \tau'$. It follows that

$$\|x(t+\tau,x_0',\alpha') - x(t,x_0',\alpha')\| \leq \frac{1}{4}\epsilon \quad \text{for all } t \geq 0, \ |\tau| \leq \tau'.$$

Since $x(t,x'',\beta) = x(t+t''-t',x',\alpha')$ for all $t \geq 0$, there is

$$\|x(t,x',\alpha) - x(t,x'',\beta)\| \leq \frac{1}{4}\epsilon \quad \text{for any } t \geq 0$$

only if $|t'-t''| \leq \tau'$. Since

$$|T(y_0) + \tau_0(y')| \leq \Big(\frac{1}{1-h_1}L + 2\Big)\epsilon_1,$$

for

$$0 < \epsilon_1 \leq \frac{(1-h_1)\tau'}{L + 2(1-h_1)},$$

there is $|T(y_0) + \tau_0(y')| \leq \tau'$ and hence

$$\|x(\tau - T(y_0), x_0, 1) - x(\tau + \tau_0(y'), x_0, 1)\| \leq \frac{\epsilon}{4} \quad \text{for all } \tau \geq 0.$$

Next, we assume that

$$0 < \epsilon_1 \leq \epsilon_3 = \min\{\epsilon_2, \frac{(1-h_1)\tau'}{L + 2(1-h_1)}\},$$

and $\delta > 0$ is so small that

$$|t'-t''| \leq \tau', \ \delta_1' \in (0, \delta'(\epsilon_1)\,].$$

We now show that for all sufficiently small δ we get

$$\|x(\tau,y_0,1) - x(\tau - T(y_0),x_0,1)\| \leq \frac{\epsilon}{2} \quad \text{for any } \tau \geq 0$$

and

$$\|x(t,x'',\beta) - x(t,y',\beta)\| \leq \frac{\epsilon}{2} \quad \text{for all } t \in [0,\tau^*(y')].$$

For arbitrary

$$\epsilon' \in [0,\bar{\epsilon}], \ \tilde{y} \in S_1'(\epsilon'), \ \tilde{x} \in M_1, \ \|\tilde{y} - \tilde{x}\| \leq \bar{\epsilon}', \ \bar{y} \in S_2'(\epsilon'), \ \bar{x} \in M_2, \ \|\bar{y} - \bar{x}\| \leq \epsilon'$$

we set
$$\tilde{\tau}_2 = \tilde{\tau}_2(\tilde{x}, \tilde{y}) = \max\{\tau_1(\tilde{y}), t_1 - \tilde{t}(\tilde{x})\},$$
$$\tilde{\tau}_1 = \tilde{\tau}_1(\tilde{x}, \tilde{y}) = \min\{\tau_1(\tilde{y}), t_1 - \tilde{t}(\tilde{x})\},$$
$$\tilde{\tau} = \tilde{\tau}(\tilde{x}, \tilde{y}) = \tilde{\tau}_2(\tilde{x}, \tilde{y}) - \tilde{\tau}_1(\tilde{x}, \tilde{y}),$$
$$\bar{\tau}_2 = \bar{\tau}_2(\bar{x}, \bar{y}) = \max\{\tau_2(\bar{y}), t_2 - \bar{t}(\bar{x})\},$$
$$\bar{\tau}_1 = \bar{\tau}_1(\bar{x}, \bar{y}) = \min\{\tau_2(\bar{y}), t_2 - \bar{t}(\bar{x})\},$$
$$\bar{\tau} = \bar{\tau}(\bar{x}, \bar{y}) = \bar{\tau}_2(\bar{x}, \bar{y}) - \bar{\tau}_1(\bar{x}, \bar{y}),$$
$$\tilde{d} = \tilde{d}(\tilde{y}) = \|Y_1(\tilde{y}) - x_1\|, \quad \bar{d} = \bar{d}(\bar{y}) = \|Y_2(\bar{y}) - x_0\|.$$

Then $\tilde{\tau}_1 \in [0, t_1]$, $\bar{\tau}_1 \in [0, t_2]$, $\tilde{\tau}(\tilde{x}, \tilde{y}) = |\tau_1(\tilde{y}) + \tilde{t}(\tilde{x}) - t_1|$, $\bar{\tau}(\bar{x}, \bar{y}) = |\tau_2(\bar{y}) + \bar{t}(\bar{x}) - t_2|$. For $t \geq 0$, we consider the solutions $x(t, \tilde{y}, 1), x(t, \tilde{x}, 1)$ ($x(t, \bar{y}, -1), x(t, \bar{x}, -1)$). By rotations, it means finite segments of the non-negative semi-axis $0 \leq t < +\infty$ such that

$$f(\sigma(t, \tilde{\sigma}, 1))f(\sigma(t, \tilde{\sigma}_0, 1)) = -1 \ (f(\sigma(t, \bar{\sigma}, -1))f(\sigma(t, \bar{\sigma}_0, -1)) = -1),$$

the ends of each of these segments corresponding to the intersection of the same hyperplane of the switching (or changeover) $\Gamma^* x = \pm l$ with the trajectories of relevant solutions. Here

$$\tilde{\sigma} = \Gamma^* \tilde{y}, \ \tilde{\sigma}_0 = \Gamma^* \tilde{x} \ (\bar{\sigma} = \Gamma^* \bar{y}, \ \bar{\sigma}_0 = \Gamma^* \bar{x}).$$

The length of the proper segment line will be referred to as the rotation time, and the distance between the points of trajectory of the solutions corresponding to the ends of that segment line will be called the rotation range. Since $\epsilon' \in (0, \bar{\epsilon}]$, for the solutions involved the first rotation exists, this being the segment line $[\tilde{\tau}_1, \tilde{\tau}_2]$ ($[\bar{\tau}_1, \bar{\tau}_2]$), and the numbers $\tilde{\tau}, \tilde{d}$ ($\bar{\tau}, \bar{d}$) are respectively the time and range of this rotation. To estimate the quantity

$$\tilde{r}(t) = \tilde{r}(t, \tilde{y}, \tilde{x}) = \|x(t, \tilde{y}, 1) - x(t, \tilde{x}, 1)\| \ (\bar{r}(t) = \bar{r}(t, \bar{y}, \bar{x}) = \|x(t, \bar{y}, -1) - x(t, \bar{x}, -1)\|)$$

outside rotations we will employ the relationship (B), and to estimate $\tilde{r}(t)$ ($\bar{r}(t)$) inside rotations we will use the relationships (A), (C). Since $\tilde{\tau}_1 \leq t_1$ ($\bar{\tau}_1 \leq t_2$), from (B) it follows that for all $t \in [0, \tilde{\tau}_1]$ ($t \in [0, \bar{\tau}_1]$) (before starting the first rotation) the quantity $\tilde{r}(t)$ ($\bar{r}(t)$) is arbitrarily small only if ϵ' is sufficiently small. From (A) we conclude that the time $\tilde{\tau}$ ($\bar{\tau}$) and the range $\tilde{d}$ ($\bar{d}$) of the first rotation are arbitrarily small if ϵ' is sufficiently small. The function $\tilde{d}(\tilde{y})$ ($\bar{d}(\bar{y})$) is defined and continuous in the closed set $S_1(\epsilon')$ ($S_2(\epsilon')$), and hence is bounded there. Because of this, for all $\tilde{y} \in S_1'(\epsilon')$ ($\bar{y} \in S_2'(\epsilon')$) the point $\tilde{y}' = Y_1(\tilde{y})$ ($\bar{y}' = Y_2(\bar{y})$) lies in the bounded set $\tilde{y}' \in R_2(\tilde{\epsilon}')$ ($\bar{y}' \in R_1(\bar{\epsilon}')$), where

$$\tilde{\epsilon}' = \max_{\tilde{y} \in S_1(\epsilon')} \tilde{d}(\tilde{y}) \quad \left(\bar{\epsilon}' = \max_{\bar{y} \in S_2(\epsilon')} \bar{d}(\bar{y})\right).$$

Then it follows from (C) that for all $t \in [\tilde{\tau}_1, \tilde{\tau}_2]$ ($t \in [\bar{\tau}_1, \bar{\tau}_2]$) (the first rotation) the quantity $\tilde{r}(t)$ ($\bar{r}(t)$) is arbitrarily small only if ϵ' is sufficiently small. Thus, for all

$t \in [0, \tilde{\tau}_2]$ $(t \in [0, \bar{\tau}_2])$ the quantity $\tilde{r}(t)$ $(\bar{r}(t))$ is arbitrarily small for a sufficiently small ϵ'. Note that the smallness for a small ϵ' of all the quantities involved is uniform in $\tilde{y}, \tilde{x}$ $(\bar{y}, \bar{x})$. We may also continue this discussion as applied to the time instants lying on the right of the first rotation end. To do this, we take

$$\tilde{y}_1 = x(\tilde{\tau}_2, \tilde{y}, 1), \quad \bar{x}_1 = x(\tilde{\tau}_2, \tilde{x}, 1), \quad \tilde{y}_1 = x(\bar{\tau}_2, \bar{y}, -1), \quad \bar{x}_1 = x(\bar{\tau}_2, \bar{x}, -1).$$

In this case, for any $\epsilon'' \in (0, \bar{\epsilon}]$ and for all sufficiently small $\epsilon' \in (0, \epsilon'']$ we get

$$\tilde{y}_1 \in S_2'(\epsilon''), \quad \bar{x}_1 \in M_2, \quad \|\tilde{y}_1 - \bar{x}_1\| \le \epsilon'' \quad (\tilde{y}_1 \in S_1'(\epsilon''), \quad \bar{x}_1 \in M_1, \quad \|\tilde{y}_1 - \bar{x}_1\| \le \epsilon'').$$

Consequently, the second rotation occurs, and over the entire span of time from the end of the first rotation to the end of the second rotation the quantity $\tilde{r}(t)$ $(\bar{r}(t))$ can be made as small as desired by choosing a sufficiently small ϵ''. Thus, over the entire span of time from the initial moment to the end of the second rotation the quantity $\tilde{r}(t)$ $(\bar{r}(t))$ can be made arbitrarily small for all

$$\tilde{y} \in S_1'(\epsilon'), \quad \tilde{x} \in M_1, \quad \|\tilde{y} - \tilde{x}\| \le \epsilon' \quad (\bar{y} \in S_2'(\epsilon'), \quad \bar{x} \in M_2, \quad \|\bar{y} - \bar{x}\| \le \epsilon')$$

only if ϵ' is chosen to be sufficiently small. Take $\epsilon' \in (0, \bar{\epsilon}]$ such that $\tilde{r}(t) \le \frac{\epsilon}{2}$ $(\bar{r}(t) \le \frac{\epsilon}{2})$ for all

$$\tilde{y} \in S_1'(\epsilon'), \quad \tilde{x} \in M_1, \quad \|\tilde{y} - \tilde{x}\| \le \epsilon' \quad (\bar{y} \in S_2'(\epsilon'), \quad \bar{x} \in M_2, \quad \|\bar{y} - \bar{x}\| \le \epsilon')$$

at any t in the time interval from the zero instant to the end of the second rotation. For all $\tau \ge 0$ we get

$$\|x(\tau, y_0, 1) - x(\tau - T(y_0), x_0, 1)\| \le \sup_{m \ge 0} N_m,$$

where

$$N_m = \max_{t \in [0, \tau^*(y_{2m})]} \|x(t, y_{2m}, 1) - x(t - T(y_{2m}), x_0, 1)\|, \quad m = 0, 1, \ldots.$$

Let m run through the values $0, 1, \ldots$. For the solutions $x(t, y_{2m}, 1)$, $x(t - T(y_{2m}), x_0, 1)$ the interval $[0, \tau^*(y_{2m})]$ then contains at least one rotation and two adjoint intervals between rotations. Since

$$\|y_{2m} - x_0\| \le \epsilon_1, \quad |T(y_{2m})| \le \frac{1}{1 - h_1} L\epsilon_1,$$

(C) implies the existence of $\epsilon_4 \in (0, \epsilon_3]$, such that for all $\epsilon_1 \in (0, \epsilon_4]$ we get

$$\tilde{y} \in S_1'(\epsilon'), \quad \tilde{x} \in M_1, \quad \|\tilde{y} - \tilde{x}\| \le \epsilon',$$

where

$$\tilde{y} = x(T(y_{2m}), y_{2m}, 1), \quad \tilde{x} = x_0 \quad \text{for } T(y_{2m}) > 0,$$
$$\tilde{y} = y_{2m}, \quad \tilde{x} = x(-T(y_{2m}), x_0, 1) \quad \text{for } T(y_{2m}) \le 0,$$

Self-oscillations in Hysteresis Systems

and for $T(y_{2m}) > 0$ we get

$$\|x(t, y_{2m}, 1) - x(t - T(y_{2m}), x_0, 1)\| \leq \frac{\epsilon}{2} \text{ for all } t \in [0, T(y_{2m})].$$

In this case, the segment line $[t^*(y_{2m}), \tau^*(y_{2m})]$ is contained in the time interval from the moment $t^*(y_{2m})$ to the end of the second rotation. Here

$$t^*(y_{2m}) = \max\{0, T(y_{2m})\}.$$

Then it follows from the choice of ϵ' that

$$\|x(t, y_{2m}, 1) - x(t - T(y_{2m}), x_0, 1)\| \leq \frac{\epsilon}{2} \text{ for all } t \in [0, \tau^*(y_{2m})],$$

and hence

$$N_m \leq \frac{\epsilon}{2}, \ m = 0, 1, \ldots.$$

Thus, for all $\tau \geq 0$ we get

$$\|x(\tau, y_0, 1) - x(\tau - T(y_0), x_0, 1)\| \leq \frac{\epsilon}{2}.$$

We now assume that $\epsilon_1 = \epsilon_4$. Since $\|y' - x''\| \leq \delta_1$ and either $y' \in S'_1(\delta_1)$, $x'' \in M_1$, or $y' \in S'_2(\delta_1)$, $x'' \in M_2$ and since the segment line $[0, \tau^*(y')]$ is contained in the time interval from the initial moment to the end of the second rotation, for $0 < \delta_1 \leq \epsilon'$ there is

$$\|x(t, x'', \beta) - x(t, y', \beta)\| \leq \frac{\epsilon}{2} \text{ for all } t \in [0, \tau^*(y')].$$

Because of this, for any $t \geq 0$ we get

$$\|x(t, y', \beta) - x(t, x', \alpha)\| \leq \epsilon$$

for all $\delta > 0$ such that

$$|t' - t''| \leq \tau', \ 0 < \delta_1 \leq \min\{\delta'(\epsilon_4), \epsilon'\}.$$

Thus, for arbitrary $\epsilon \in (0, \tilde{\delta}']$, $x' \in M_0$ there is $\delta \in (0, \epsilon]$ such that for any $y' \in S(x', \delta)$ it is possible to choose $\beta = \beta(y')$, $\beta = \pm 1$ such that for all $t \geq 0$ we get

$$\|x(t, y', \beta) - x(t, x', \alpha)\| \leq \epsilon.$$

Consequently, the periodic solution $x(t, x_0, 1)$ is stable in Lyapunov's sense, and hence is orbitally stable. Note that our proof has established independence of δ from x'. Therefore the Lyapunov stability of the periodic solution $x(t, x_0, 1)$ is uniform in the initially chosen time.

Proceeding from the same assumptions about the solutions $x(t, x', \alpha)$, $x(t, y', \beta)$, we shall prove the orbit-asymptotic stability of the periodic solution $x(t, x_0, 1)$ and the existence of its asymptotic phase. Let $t_0(y') = T(y_0) + \tau_0(y') + t' - t''$. Then we have that

$$0 \leq \rho(x(t + t_0(y'), y', \beta), M_0) \leq \|x(t, x', \alpha) - x(t + t_0(y'), y', \beta)\| \text{ for all } t \geq -t_0(y').$$

In this inequality we set $t = -t_0(y') + \tau^*(y') + \tau$, where $\tau \geq 0$. Since

$$x^* = x(\tau^*(y'), x'', \beta) = x(\tau_0(y'), x_0, 1), \quad x'' = x(t'' - t', x', \alpha),$$

we have

$$x(\tau^*(y') - \tau_0(y'), x'', \beta) = x_0,$$

and for all $\tau \geq 0$ we get

$$0 \leq \rho(x(\tau, y_0, 1), M_0) \leq \|x(\tau, y_0, 1) - x(\tau - T(y_0), x_0, 1)\|.$$

We show that

$$\|x(\tau, y_0, 1) - x(\tau - T(y_0), x_0, 1)\| \to 0 \text{ as } \tau \to +\infty.$$

As in the estimation of constants N_m, $m = 0, 1, \ldots$, we shall determine for a number $\epsilon^* \in (0, \tilde{\delta}']$ the corresponding quantities $\epsilon'^* \in (0, \bar{\epsilon}]$, $\epsilon_4^* \in (0, \epsilon_3]$. Since

$$\|y_{2m} - x_0\| \leq h_1^m \epsilon_4, \ |T(y_{2m})| \leq \frac{1}{1-h_1} L h_1^m \epsilon_4, \ m = 0, 1, \ldots,$$

the constant $h_1 \in (0, 1)$, for all $m \geq m(\epsilon^*) = E\left(\frac{\ln \epsilon_4^* - \ln \epsilon_4}{\ln h_1}\right) + 1$ we get

$$\|y_{2m} - x_0\| \leq \epsilon_4^*, \ |T(y_{2m})| \leq \frac{1}{1-h_1} L \epsilon_4^*.$$

As discussed above, we then conclude from the choice of ϵ'^*, ϵ_4^* that $N_m \leq \epsilon^*$ for all $m \geq m(\epsilon^*)$. Hence, for all

$$\tau \geq \tau(\epsilon^*) = T(y_0) - T\left(y_{2m(\epsilon^*)}\right) + (t_1 + t_2)m(\epsilon^*)$$

we get

$$\|x(\tau, y_0, 1) - x(\tau - T(y_0), x_0, 1)\| \leq \sup_{m \geq m(\epsilon^*)} N_m \leq \epsilon^*.$$

Thus

$$\|x(\tau, y_0, 1) - x(\tau - T(y_0), x_0, 1)\| \to 0 \text{ as } \tau \to +\infty$$

and hence

$$\rho(x(t, y', \beta), M_0) \to 0, \ \|x(t + t_0(y'), y', \beta) - x(t, x', \alpha)\| \to 0 \text{ as } t \to +\infty.$$

This exactly signifies the orbit-asymptotic stability of the periodic solution $x(t, x_0, 1)$ and the presence of its asymptotic phase. ∎

It should be noted that the research techniques discussed here can be applied to derivation of sufficient conditions for stability in Lyapunov's sense, orbit-asymptotic stability, and availability of asymptotic phase for the periodic switching solution of the system

$$\dot{x} = \mathbf{A}x + c + \mu g(x) + (\mathbf{B}x + d + \mu h(x))f(\sigma), \ \sigma = \Gamma^* x, \tag{5.2.6}$$

Self-oscillations in Hysteresis Systems

where $g(x)$, $h(x) \in C^1_{n,n}(E_n)$, the small parameter $\mu > 0$, and the remainder is as in system (5.2.1). When using the statements of Theorem 72, of special interest are the cases where the matrices $\mathbf{A}$, $\mathbf{B}$ are diagonal, which makes it possible to solve system (5.2.2) and write conditions (5.2.3), (5.2.4) explicitly.

3. Consider the existence of periodic solutions in hysteresis systems.

Theorem 73 *Suppose in system (5.2.1) $\mathbf{B} = \nu\mathbf{A}$, $|\nu| < 1$, all the eigenvalues of the matrix $\mathbf{A}$ have negative real parts $(\operatorname{Re}\lambda_i(\mathbf{A}) < 0, i = 1, \ldots, n)$ and the following inequalities hold*

$$\Gamma^* \mathbf{A}^{-1}(c+d) > (\nu+1)l, \quad \Gamma^* \mathbf{A}^{-1}(c-d) < (\nu-1)l. \tag{5.2.7}$$

Then in system (5.2.1) there is a periodic solution with two switchings lying in a bounded region of the phase space E_n.

Proof: Show that, under conditions imposed on system (5.2.1), any solution $x(t, x_0, \alpha)$ of this system for $t \geq 0$ has an integral sequence of switchings. Indeed, since

$$\operatorname{Re}\lambda_i((1+\nu)\mathbf{A}) < 0 \ (\operatorname{Re}\lambda_i((1-\nu)\mathbf{A}) < 0), \ i = 1, \ldots, n,$$

for the solution $x(t, x_0, 1)$ $(x(t, x_0, -1))$ of the system

$$\dot{x} = (1+\nu)\mathbf{A}x + c + d \ \ (\dot{x} = (1-\nu)\mathbf{A}x + c - d)$$

we get

$$\sigma(t, \sigma_0, 1) \to \frac{1}{1+\nu}\Gamma^*\mathbf{A}^{-1}(c+d) \ \left(\sigma(t, \sigma_0, -1) \to -\frac{1}{1-\nu}\Gamma^*\mathbf{A}^{-1}(c-d)\right) \text{ as } t \to +\infty,$$

where

$$\sigma(t, \sigma_0, \alpha) = \Gamma^* x(t, x_0, \alpha), \ \sigma_0 = \Gamma^* x_0, \ \alpha = \pm 1.$$

If follows from conditions (5.2.7) that

$$-\frac{1}{1+\nu}\Gamma^*\mathbf{A}^{-1}(c+d) < -l \ \left(-\frac{1}{1-\nu}\Gamma^*\mathbf{A}^{-1}(c-d) > l\right).$$

Since for the solution $x(t, x_0, 1)$ $(x(t, x_0, -1))$ of system (5.2.1) Definition 42 implies $\sigma_0 > -l$ $(\sigma_0 < l)$ and the function $\sigma(t, \sigma_0, \alpha)$ is continuous in t for $t \geq 0$, this suggests the availability of the first switching for the solution $x(t, \sigma_0, \alpha)$. This reasoning may also be continued for the time moments lying on the right of the first switching moment. To do this requires the first switching moment to be taken as initial. Thus the solution $x(t, x_0, \alpha)$ of system (5.2.1) for $t \geq 0$ has a sequence of switchings.

Let $\mathbf{V}$ be the $(n \times n)$-constant symmetric matrix that is a solution to Lyapunov's matrix equation $\mathbf{A}^*\mathbf{V} + \mathbf{V}\mathbf{A} = -\mathbf{E}$. The quadratic form $v = v(x) = x^*\mathbf{V}x$ is then positive-definite [Lyapunov (1950)]. We compute the derivative of v by system (5.2.1):

$$\frac{dv}{dt} = x^*\mathbf{V}((1+f\nu)\mathbf{A}x + c + fd) + (c^* + fd^* + (1+f\nu)x^*\mathbf{A}^*)\mathbf{V}x$$

$$= (1+f\nu)x^*(\mathbf{V}\mathbf{A} + \mathbf{A}^*\mathbf{V})x + x^*\mathbf{V}(c+fd) + (c+fd)^*\mathbf{V}x$$

$$= -(1 + f\nu) \|x\|^2 + 2x^* \mathbf{V}(c + fd),$$

where $f = f(\sigma(t, \sigma_0, \alpha))$. Then

$$\frac{dv}{dt} \leq a \|x\| - (1 - |\nu|) \|x\|^2,$$

where $a = 2 \|\mathbf{V}\| (\|c\| + \|d\|)$. Consequently, for all $\|x\| > b = \frac{a}{1-|\nu|}$ we get $\frac{dv}{dt} < 0$. Let

$$q = \max_{\|x\| \leq b} v(x), \quad Q = \{x : x \in E_n, v(x) \leq q\}.$$

We shall show that the set Q is invariant for the solutions of system (5.2.1), i.e. with all $x_0 \in Q$, $\alpha = \pm 1$ there is $x(t, x_0, \alpha) \in Q$ for any $t \geq 0$. Indeed, assuming the opposite we get $x' \in E_n$, $t' > 0$, $\alpha' = \pm 1$ such that $x' \in Q$, $x(t', x', \alpha') \bar{\in} Q$ and hence

$$u(0) = v(x') \leq q, \quad u(t') = v(x(t', x', \alpha')) > q,$$

where $u(t) = v(x(t, x', \alpha'))$. Since $u(t)$ is continuous in t for $t \geq 0$, there is $t_0 \in [0, t')$ such that $u(t_0) = q$, $u(t) > q$ for all $t \in (t_0, t']$. Let $t \in (t_0, t']$. The inequality $\|x(t, x', \alpha')\| > b$ holds, otherwise ($\|x(t, x', \alpha')\| \leq b$) and we get

$$u(t) \leq \max_{\|x\| \leq b} v(x) = q,$$

which contradicts the condition $u(t) > q$. From the fact that

$$\dot{u}(t) = \frac{dv}{dt}, \quad \|x(t, x', \alpha)\| > b,$$

follows the inequality $\dot{u}(t) < 0$ and hence

$$u(t) = u(t_0) + \int_{t_0}^{t} \dot{u}(\tau) d\tau \leq u(t_0) = q,$$

which contradicts the condition $u(t) > q$. The obtained contradiction proves the invariance of the set Q for the solutions of system (5.2.1).

Since any solution of system (5.2.1) for $t \geq 0$ has switchings, then the trajectory of any solution contained in Q for this system has common points with the hyperplanes $\Gamma^* x = \pm l$. Consequently, the following sets are nonempty

$$S_1 = Q \cap \{x : x \in E_n, \Gamma^* x = l\}, \quad S_2 = Q \cap \{x : x \in E_n, \Gamma^* x = -l\}.$$

Let $x_0 \in S_1$, and $t_1 = t_1(x_0)$, $t_2 = t_2(x_0)$ be the moments of the first and second switchings of the solution $x(t, x_0, 1)$, $0 < t_1 < t_2$. Since $x(t, x_0, 1) \in Q$ for all $t \geq 0$, we get

$$x(t_1, x_0, 1) \in S_2, \quad x(t_2, x_0, 1) \in S_1.$$

The operator $Z(x_0) = x(t_2(x_0), x_0, 1)$ then transfers the set S_1 into itself. Note that the set S_1 is holomorphic with respect to the $(n-1)$ sphere. Given in S_1, the operator $Z(x_0)$ is continuous. The validity of this statement follows from the proof of the preceding theorem. For the solutions $x(t, x_0, 1)$, $x(t, x_0', 1)$, where $x_0' \in S_1$, the quantity $\|Z(x_0) - Z(x_0')\|$ is the range of the second rotation and can, therefore, be made arbitrarily small by choosing a sufficiently small $\|x_0 - x_0'\|$. Thus, the continuous operator $Z(x_0)$ transfers into itself the set S_1 that is holomorphic with respect to the $(n-1)$ sphere. By the Ball-Brauer theorem [Kantorovič and Akilov (1977)], we then have that in S_1 there is a fixed point of the operator $Z(x_0)$, i.e. there is $x^* \in S_1$ such that $Z(x^*) = x^*$. Therefore,

$$x(t_1(x^*), x^*, 1) \in S_2, \quad x(t_2(x^*), x^*, 1) = x^*,$$

and hence the solution $x(t, x^*, 1)$ of system (5.2.1) lying in Q for all $t \geq 0$ is two-switching periodic. ∎

It should be noted that the techniques presented here make it possible to derive sufficient conditions for the existence of a two-switching periodic solution in system (5.2.6). Moreover, the proposed approach can be applied to hysteresis systems, where the quantity $l > 0$ is a small parameter. In this respect, we have the following result given here without any proof.

Theorem 74 *Let us consider the system*

$$\dot{x} = \mathbf{A}x + g(x) + (\nu \mathbf{A}x + h(x) + d)f(\sigma), \quad \sigma = \Gamma^* x, \tag{5.2.8}$$

where the functions $g(x)$, $h(x)$ satisfy the Lipschitz condition with respect to x in a neighborhood of the point $x = 0$, $\|g(x)\| = o(\|x\|)$, $\|h(x)\| = o(\|x\|)$ as $\|x\| \to 0$, and the remaining are as in system (5.2.1). Suppose that

$$\mathrm{Re}\lambda_i(\mathbf{A}) < 0, \; i = 1, \ldots, n, \; |\nu| < 1,$$

and the constant $(n \times n)$-symmetric matrix $\mathbf{V}$ is a solution to Lyapunov's matrix equation $\mathbf{A}^*\mathbf{V} + \mathbf{V}\mathbf{A} = -\mathbf{E}$. Then for $\Gamma = -\kappa \mathbf{V}d$ (the constant $\kappa > 0$) with all sufficiently small $l = 0$ in system (5.2.8) there is the two-switching periodic solution lying in a neighborhood of the point $x = 0$. This neighborhood can be made arbitrarily small by choosing a sufficiently small l. In this case, the period T of the above periodic solution is such that

$$\frac{1}{l}T \to \frac{4}{\kappa d^* \mathbf{V} d} \quad as \; l \to +0.$$

Remark. If the assumption $\mathrm{Re}\lambda_i(\mathbf{A}) < 0, i = 1, \ldots, n$ is abandoned and it is assumed that $\nu = 0$ and the vectors $d, \mathbf{A}d, \ldots, \mathbf{A}^{n-1}d$ are linearly independent, then the statements of Theorem 74 hold. In this case we should take the matrix $\mathbf{V}$ to be a solution to Lyapunov's matrix equation $(cd^* + \mathbf{A}^*)\mathbf{V} + \mathbf{V}(\mathbf{A} + dc^*) = -\mathbf{E}$, where the constant vector $c \in E_n$ is such that $\mathrm{Re}\lambda_i(\mathbf{A} + dc^*) < 0, i = 1, \ldots, n$.

5.3 Controlled Motion of Charged Particles in Magnetic Field

1. We set forth conditions for stability of integral manifolds (such conditions are given in section 2.2 for periodic orbits). These stability conditions are derived by reducing the stability problem of integral manifolds to the stability problem in coordinates of solutions to systems of differential equations.

Suppose the following system of differential equation is given

$$\dot{x} = f(x), \qquad (5.3.1)$$

where x is the vector of n dimensions, and $f(x)$ is the vector-valued function of the same dimension that is given for $x \in E_n$ and is real, continuous and continuously differentiable with respect to components of the vector x.

Definition 47. We say that

$$x = \varphi(\tau_1, ..., \tau_k) \qquad (5.3.2)$$

specifies the integral manifold M of system (5.3.1) if $x_0 \in M$ implies $x(t, x_0) \in M$ for $t \in (-\infty, +\infty)$, where $x(t, x_0)$ is the solution of system (5.3.1) satisfying the initial condition $x = x_0$ when $t = 0$.

The vector-valued function $\varphi(\tau_1, \ldots, \tau_k)$ is said to be given for $\tau \in E_k$, and is real, continuous and continuously twice differentiable with respect to components of the vector τ. Moreover, it is assumed that the vectors $\partial \varphi / \partial \tau_i$ are linearly independent with $j = 1, \ldots, k$ for all $\tau \in E_k$.

Definition 48. The integral manifold M of system (5.3.1) is called stable if for every $\epsilon > 0$ it is possible to choose $\delta(\epsilon) > 0$ such that for $\rho(x_0, M) < \delta$ we get $\rho(x(t, x_0)M) < \epsilon$ when $t \geq 0$. The stable manifold M is called asymptotically stable if $\delta(\epsilon) > 0$ can be chosen so that $\rho(x(t, x_0), M) \to 0$ as $t \to +\infty$.

Here the distance from the point x to the set M is denoted by $\rho(x, M)$:

$$\rho(x, M) = \inf_{y \in M} \|x - y\|, \ \|x - y\| = \sqrt{\sum_{s=1}^{n} (x_s - y_s)^2}.$$

Denote by $b_1(\tau), \ldots, b_r(\tau)$ the vector-valued functions which form the orthogonal basis for the orthogonal complement of the linear space stretched over the vectors $\partial \varphi / \partial \tau_i, j = 1, \ldots, k$. It is well known that the vectors $b_1, \ldots, b_r$ as given for $\tau \in E_k$, are real, continuous and continuously differentiable; here $k + r = n$.

Let

$$x = x(t, x_0) \qquad (5.3.3)$$

be a solution to system (5.3.1). Assume that there is a vector τ for which the following relationship holds

$$\rho(x(t, x_0), M) = \|x(t, x_0) - \varphi(\tau)\|. \qquad (5.3.4)$$

By the definition of distance, the components of vector τ then satisfy the equations

$$(x(t, x_0) - \varphi(\tau))^* \frac{\partial \varphi}{\partial \tau_j} = 0, j = 1, \ldots, k. \qquad (5.3.5)$$

We introduce the vector $\xi = (\xi_1, \ldots, \xi_r)$ and the matrix $\mathbf{B} = (b_1, \ldots, b_r)$. Set

$$x(t, x_0) - \varphi(\tau) = \mathbf{B}(\tau)\xi. \qquad (5.3.6)$$

Controlled Motion of Charged Particles

Formulas (5.3.5) and (5.3.6) are said to establish one-to-one correspondence between the components of vector x and the components of vectors τ and ξ. To define the vectors τ and ξ, we shall derive a system of differential equations. To this end, equalities (5.3.5) and (5.3.6) will be completely differentiated with respect to t, assuming that the vectors τ and ξ are time functions:

$$\left(f(\mathbf{B}\xi+\varphi) - \sum_{i=1}^{k}\frac{\partial\varphi}{\partial\tau_i}\frac{\partial\tau_i}{\partial t}\right)^{*}\frac{\partial\varphi}{\partial\tau_j} + \mathbf{B}\xi\sum_{i=1}^{k}\frac{\partial^2\varphi}{\partial\tau_i\partial\tau_j}\frac{\partial\tau_i}{\partial t} = 0, \; i = 1,\ldots,k, \qquad (5.3.7)$$

$$\mathbf{B}\frac{d\xi}{dt} = f(\mathbf{B}\xi+\varphi) - \sum_{i=1}^{k}\frac{\partial\varphi}{\partial\tau_i}\frac{\partial\tau_i}{\partial t} - \sum_{i=1}^{k}\frac{\partial\mathbf{B}}{\partial\tau_i}\xi\frac{\partial\tau_i}{\partial t}. \qquad (5.3.8)$$

Under the assumptions made for sufficiently small values of the components of vector ξ, equations (5.3.7) and (5.3.8) can be solved for $\dot{\tau}$ and $\dot{\xi}$:

$$\dot{\tau} = G(\tau,\xi), \; \dot{\xi} = H(\tau,\xi), \qquad (5.3.9)$$

where G and H are the known vector-valued functions of dimensions k and r, respectively. System (5.3.9) has a family of solutions

$$\tau = \tau(t,\tau_0), \; \xi = 0, \; \tau_0 \in E_k. \qquad (5.3.10)$$

Definition 49. The family (5.3.10) is assumed to be uniformly stable with respect to components of the vector ξ if for each $\epsilon > 0$ it is possible choose $\delta = \delta(\epsilon) > 0$ such that $\|\xi_0\| < \delta$ implies $\|\xi(t,\xi_0,\tau_0)\| < \epsilon$ for $t \geq 0$, $\tau_0 \in E_k$. In addition, if a number $\delta(\epsilon)$ can be chosen so that $\|\xi(t,\xi_0,\tau_0)\| \xrightarrow[\tau_0 \in E_k]{} 0$ as $t \to +\infty$, then the family (5.3.10) is called uniform-asymptotically stable with respect to components of the vector ξ.

Theorem 75 *For the integral manifold to be stable (asymptotically stable), it is necessary and sufficient that the family (5.3.10) be uniformly stable (uniform-asymptotically stable) with respect to components of the vector ξ.*

Definition 50. The function $V(\tau,\xi)$ is called positive-definite for components of the vector ξ uniformly in $\tau \in E_k$ if the following conditions are satisfied: 1) the function $V(\tau,\xi)$ given for $\tau \in E_k$, $\|\xi\| < \alpha$ is real and continuous, $V(\tau,0) = 0$, α is a positive constant, 2) for a sufficiently small $c_2 > 0$ it is possible to choose $c_1 > 0$ such that with $\|\xi\| > c_2$ we get $V(\tau,\xi) > c_1$ for $\tau \in E_k$.

Theorem 76 *Suppose there is a function $V(\tau,\xi)$ satisfying the conditions: 1) $V(\tau,\xi)$ is positive-definite for components of the vector ξ uniformly into $\tau \in E_k$, 2) the function $V(\tau,\xi) \to 0$ as $\xi \to 0$ uniformly into $\tau \in E_k$, 3) the total derivative of the function $V(\tau,\xi)$ computed by system (5.3.9)*

$$\frac{dV}{dt} = \frac{\partial V}{\partial \tau}G + \frac{\partial V}{\partial \xi}H = W(\tau,\xi),$$

is continuous and positive. Then the manifold M of system (5.3.1) is stable.

Theorem 77 *If there is the function $V(\tau,\xi)$ satisfying the first and second conditions of Theorem 75 and the function $W(\tau,\xi)$ is negative-definite for components* of the vector ξ uniformly into $\tau \in E_k$, then the integral manifold M is asymptotically stable.

Theorem 78 *If the vector-valued function φ is periodic with respect to each component of the vector τ, then for the manifold M to be asymptotically stable it is sufficient that there is the function $V(\tau, \xi)$ satisfying the conditions: 1)$V(\tau, \xi)$ is a periodic function with respect to components of the vector τ and is positive-definite with respect to components of the vector ξ for any fixed τ, 2) the function $W(\tau, \xi)$ is negative-definite with respect to components of the vector ξ for any fixed τ. Here it is assumed that the matrix $\mathbf{B}$ appearing in system (5.3.9) is periodic in components of the vector τ with the same period as functions φ and V.*

2. Consider the control problem for the motion of charged particles in a stationary magnetic field and indicate a possible structure of magnetic field which ensures the solution of the problem of focusing and acceleration of particle motion. We discuss the problem of charged particle motion in the magnetic field of dimensionless form:

$$\dot{x} = y, \tag{5.3.11}$$

$$\dot{y} = B \times y, \tag{5.3.12}$$

$$\mathrm{rot} B = I, \tag{5.3.13}$$

$$\mathrm{div} B = 0. \tag{5.3.14}$$

We shall find a dependence of the field $B = B(x)$ on the vector x components such that the motion of charged particles along the chosen axis, say the Ox_1 axis in Cartesian system $Ox_1x_2x_3$, is subjected to a focusing influence of the field B and is accelerated along this axis. In this formulation of the problem, the current I is a derivative and is determined in terms of B by equation (5.3.13). From equation (5.3.12) we have that the vector y keeps the square $y^2 = \mathrm{const}$ unaffected during the motion. For definiteness, we assume $y^2 = 1$. Introduce the vector-valued function $\eta = \eta(x)$ with components $\eta_1(x), \eta_2(x), \eta_3(x)$ as follows:

$$\eta_1(x) = (1 - \eta_2^2(x) - \eta_3^2(x))^{1/2},$$

where $\eta_2(x)$ and $\eta_3(x)$ are still arbitrary continuously differentiable functions satisfying the condition $\eta_2^2(x) + \eta_3^2(x) \leq 1$. Set

$$B = f(x)\eta(x) + \eta(x) \times \left(\frac{d\eta}{dx}\,\eta\right), \tag{5.3.15}$$

where $f(x)$ is an arbitrary function for which B satisfies condition (5.3.14). This condition leads to the equation

$$\frac{df}{dt} + f \mathrm{div}\eta + \mathrm{div} B' = 0, \tag{5.3.16}$$

where $B' = \eta \times \left(\frac{d\eta}{dx}\,\eta\right)$. The total derivative with respect to t in equation (5.3.16) is computed along integral curves of the ordinary system

$$\frac{dx}{dt} = \eta(x), \tag{5.3.17}$$

Controlled Motion of Charged Particles

so that (5.3.16) is a linear partial differential equation of the first order for variables x_1, x_2, x_3.

Theorem 79 *If the functions $\eta_2(x), \eta_3(x)$ are chosen so that the system (5.3.17) has a stable (or asymptotically stable) integral manifold $x_2 = x_3 = 0$, then the system (5.3.11) – (5.3.12) for the field chosen to be (5.3.15) has a conventionally stable (or asymptotically stable) equilibrium motion $x_2 = x_3 = 0$ with any choice of the function f satisfying condition (5.3.16).*

As an example, we set

$$\eta_2(x) = \lambda x_2 - \omega x_3, \quad \eta_3(x) = \omega x_2 + \lambda x_3,$$

where λ and ω are some functions of the vector x components.

Let $\lambda = \lambda(R)$, where $R = (x_2^2 + x_3^2)^{1/2}$ and ω is a nonvanishing scalar function. Then for every root of the equation $\lambda(R) = 0$ there is the integral manifold of system (5.3.17) which is asymptotically stable provided that $d\lambda/dR < 0$ or when an expansion of the function λ in the neighborhood of the root is of the form

$$\lambda(R) = g(R - R_1)^{2k+1}, \; g < 0, \; k \geq 0, \; \lambda(R_1) = 0.$$

System (5.3.11) – (5.3.12), under such a choice of functions η_2 and η_3 has condition-asymptotically stable integral manifolds $R = R_1$, so that particles are bunched about the above-mentioned manifolds.

Proof: The first step is to choose a vector field of velocities $\eta = \eta(x)$. For definiteness, we set $\eta^2(x) = 1$. If such a velocity field is produced by a magnetic stationary field $B(x)$, then the transverse component of this field B' is uniquely determined by

$$B' = \eta \times \left(\frac{d\eta}{dx}\,\eta\right).$$

The transverse component of the magnetic field is represented by

$$B - B' = f\eta.$$

In this case the transverse component does not affect the dynamics of particles arranged on the integral manifold $y = \eta(x)$. This, however, determines the flux along with the transverse component B'.

The second step in this proof is to establish that in the phase space of the system there is an integral manifold $\eta(x) - y = 0$. In other words, any particle having an initial position x_0 and an initial velocity $\dot{x}_0$ such that $\dot{x}_0 = \eta(x_0)$ for $t = 0$, remains on the manifold during the motion.

The third step is to consider the behavior of particles on the integral manifold $\eta(x) - y = 0$. This consideration completes the proof of Theorem 79. ∎

3. We now consider construction of an electromagnetic field which has a given equilibrium state and a focusing property in the neighborhood of this state. In the absence of an electric field component, Maxwell's equations apply only to components of the required magnetic field, and the current producing this field is a derivative. We find the structure of right-hand sides of the system of three differential equations

$$\dot{x} = \eta(x), \quad x = (x_1, x_2, x_3), \quad \eta = (\eta_1, \eta_2, \eta_3), \tag{5.3.18}$$

possessing the following properties: 1) system (5.3.18) has an integral manifold

$$F_1(x) = 0, \quad F_2(x) = 0, \tag{5.3.19}$$

2)
$$\eta^2 = \eta_1^2 + \eta_2^2 + \eta_3^2 = 1. \tag{5.3.20}$$

To solve this problem, introduce two orthogonal unit vectors

$$a_1 = \frac{\text{grad } F_1}{\sqrt{(\text{grad } F_1)^2}}, \quad a_2 = \frac{\text{grad } F_2 - \lambda \text{ grad } F_1}{\sqrt{(\text{grad } F_2 - \lambda \text{ grad } F_1)^2}},$$

where
$$\lambda = \frac{(\text{grad } F_1, \text{grad } F_2)}{(\text{grad } F_1)^2}.$$

Set $a_3 = a_1 \times a_2$. Then the right-hand side of (5.3.18) can be represented as

$$\eta = \sum_{i=1}^{3} a_i \varphi_i = \mathbf{A}\varphi, \tag{5.3.21}$$

where $\mathbf{A}$ is the matrix with columns a_1, a_2, a_3, and $\varphi = (\varphi_1, \varphi_2, \varphi_3)$. From condition (5.3.20) we have

$$\sum_{i=1}^{3} \varphi_i^2 = 1. \tag{5.3.22}$$

Since system (5.3.18) has the integral manifold (5.3.19), the following relationships hold

$$(\eta, \text{grad } F_1) = 0, \quad (\eta, \text{grad } F_2) = 0. \tag{5.3.23}$$

The relationships (5.3.23) hold subject to (5.3.20). From this it follows that

$$\varphi_1 = 0, \quad \varphi_2 = 0. \tag{5.3.24}$$

Let us assume that there are given two functions V and W: $V = V(x, F_1, F_2)$, $W = W(x, F_1, F_2)$ such that the relationship

$$\frac{dV}{dt} = W, \quad (\eta, \text{grad } V) = W \tag{5.3.25}$$

holds for system (5.3.18). Now introduce the vector

$$H = (H_1, H_2, H_3), \quad H_1 = \frac{(a_1, \text{grad } V)}{\mu},$$

Controlled Motion of Charged Particles

$$H_2 = \frac{(a_2, \operatorname{grad} V)}{\mu}, \quad H_3 = \frac{(a_3 \operatorname{grad} V)}{\mu},$$

where

$$\mu = \sqrt{(a_1, \operatorname{grad} V)^2 + (a_2, \operatorname{grad} V)^2 + (a_3, \operatorname{grad} V)^2}.$$

Then relationship (5.3.25) can be rewritten as

$$(H, \varphi) = \frac{W}{\mu}. \tag{5.3.26}$$

Introduce a unit vector G that is orthogonal to H. Put $Q = H \times G$, then $(G, \varphi) = \psi_1, (Q, \varphi) = \psi_2$. If $\xi = \left(\frac{W}{\mu}, \psi_1, \psi_2\right)$, then

$$\varphi = P^*\xi. \tag{5.3.27}$$

From condition (5.3.22) and equality (5.3.27) we get the relationship

$$\psi_1^2 + \psi_2^2 + \frac{W^2}{\mu^2} = 1. \tag{5.3.28}$$

This follows from the fact that the matrix $\mathbf{P} = \begin{pmatrix} H \\ G \\ Q \end{pmatrix}$ is orthogonal and $\mathbf{P}^{-1} = \mathbf{P}^*$, where $\mathbf{P}^*$ is the matrix transposed to $\mathbf{P}$. From condition (5.3.28) it follows that system (5.3.18) possessing the properties 1) and 2) satisfies condition (5.3.25) if and only if the following inequality holds

$$\frac{W^2}{\mu^2} \leq 1. \tag{5.3.29}$$

The functions F_1, F_2, W, V are said to satisfy condition (5.3.29). From condition (5.3.28) it also follows that

$$\psi_1 = \sqrt{1 - \frac{W^2}{\mu^2}} \sin \psi, \quad \psi_2 = \sqrt{1 - \frac{W^2}{\mu^2}} \cos \psi, \tag{5.3.30}$$

where $\psi = \psi(x, F_1, F_2)$ is an arbitrary function such that the equality $\varphi_1 = \varphi_2 = 0$ holds on the integral manifold. This means that the vector ξ is orthogonal to the first two rows of the matrix $\mathbf{P}$ and therefore this vector satisfies the relationship

$$\xi = \pm P_3, \tag{5.3.31}$$

where P_3 is the third column of the matrix $\mathbf{P}$, $P_3 = (H_3, G_3, Q_3)$. Finally we get

$$\eta = \mathbf{A}\mathbf{P}^*\xi, \tag{5.3.32}$$

where

$$\xi = \begin{pmatrix} \frac{W}{\mu} \\ \sqrt{1 - \frac{W^2}{\mu^2}} \sin \psi \\ \sqrt{1 - \frac{W^2}{\mu^2}} \cos \psi \end{pmatrix}.$$

Here the vector ξ satisfies condition (5.3.31) on the integral manifold (5.3.19).

Now consider stability of the integral manifold (5.3.19). Denote this manifold by M.

Theorem 80 [Zubov (1964)] *Suppose there exist two functions V and W satisfying the conditions:*

1) $V = 0$, $W = 0$ for $x \in M$, $V > \gamma > 0$, $W < -\gamma$ for $r(x, M) > \alpha$, where $\gamma = \gamma(\alpha) > 0$ for $\alpha > 0$,

2) V is continuously differentiable by (5.3.18) and the relationship $(\text{grad } V, \eta) = W$ holds.

Then the manifold (5.3.19) is asymptotically stable in Lyapunov's sense.

Remark. If the system (5.3.18) has the right-hand sides determined by the relationship (5.3.32), and the functions V, W satisfy conditions (5.3.29) and the conditions 1), 2) of Theorem 80, then the system (5.3.18) has the integral manifold (5.3.19), this being asymptotically stable in Lyapunov's sense. Here stability is assumed to be conventional.

We now turn to construction of the field $B(x)$ possessing the properties: a) the field $B(x)$ admits of equilibrium trajectories of charged particles, located on a manifold of the form (5.3.19); b) the field $B(x)$ is said to be focusing in the neighborhood of equilibrium motions. The equation of particle motion can be written as

$$\dot{x} = y, \ \dot{y} = B(x) \times y, \qquad (5.3.33)$$

$$\text{rot } B = I, \qquad (5.3.34)$$

$$\text{div } B = 0. \qquad (5.3.35)$$

Set

$$B = f\eta + \eta \times \left(\frac{d\eta}{dx} \eta\right), \qquad (5.3.36)$$

where f is an arbitrary function chosen so that the field (5.3.36) satisfies the condition (5.3.35). By substituting (5.3.36) in (5.3.35), we get

$$\dot{f} + f \text{ div } \eta + \text{div}\left(\eta \times \left(\frac{d\eta}{dx} \eta\right)\right) = 0. \qquad (5.3.37)$$

The total derivative in equation (5.3.37) of the function f is taken by the system of ordinary differential equations

$$\dot{x} = \eta(x). \qquad (5.3.38)$$

Theorem 81 *System (5.3.33) describing in dimensionless form the motion of particles in the field B has the equilibrium motion described by the equations*

$$F_1(x) = 0, \ F_2(x) = 0, \ y = \eta(x). \qquad (5.3.39)$$

This equilibrium motion is conditionally and asymptotically stable if the functions V and W appearing in $\eta(x)$ satisfy the conditions of Theorem 80.

4. Consider the problem of constructing a magnetic field, where a charged particle performs a given equilibrium motion. In the neighborhood of the given equilibrium

Controlled Motion of Charged Particles

motion, the required field has the focusing property or, to make it more precise, this equilibrium motion is orbit-asymptotically stable.

Let us assume that the equilibrium trajectory $\varphi = \varphi(\Theta)$ is given in some field $B = B(x)$. Determine the structure of this field, for which a given equilibrium motion is orbit-conditionally stable or conditionally and orbit-asymptotically stable. To this end, we introduce the system of equations

$$\dot{x} = \eta(x), \qquad (5.3.40)$$

possessing the property that the trajectory of equilibrium motion is that of system (5.3.40) and $\eta^2 = \eta_1^2 + \eta_2^2 + \eta_3^2 = 1$. Denote by $P(\Theta)$ a normal hyperplane of the curve $\varphi = \varphi(\Theta)$. If the trajectories of system (5.3.40) start on the hyperplane $P(0)$ at the time $t = 0$ in a sufficiently small neighborhood $\varphi(0)$, then under appropriate conditions imposed on the right-hand sides of system (5.3.40) they intersect the normal hyperplane $P(\Theta)$. In general, the time of intersection varies with each trajectory. Let $x = x(t, x_0)$ be one of such trajectories. Then the relationship

$$((x(t, x_0) - \varphi(\Theta)), \varphi'(\Theta)) = 0 \qquad (5.3.41)$$

holds for the time t of intersection of the hyperplane $P(\Theta)$. On the hyperplane $P(\Theta)$, we introduce the Cartesian system using the vectors $B_1(\Theta), B_2(\Theta)$. Here

$$(B_1(\Theta), B_2(\Theta)) = 0, \ B_1^2(\Theta) = B_2^2(\Theta) = 1;$$

then

$$x(t, x_0) = \varphi(\Theta) + \xi_1 B_1 + \xi_2 B_2. \qquad (5.3.42)$$

If the motion $x(t, x_0)$ is known, then formulas (5.3.41) and (5.3.42) determine uniquely the quantities Θ, ξ_2. The opposite statement is also true. Using system (5.3.40) and relationships (5.3.41) and (5.3.42), we find differential equations for the quantities Θ, ξ_1, ξ_2. By differentiating (5.3.41) with respect to t, we find

$$((x(t) - \varphi(\Theta)), \varphi''(\Theta))\dot{\Theta} + ((\eta(x) - \varphi'(\Theta)\dot{\Theta}), \varphi') = 0,$$

whence

$$\dot{\Theta} = \frac{(\eta(\varphi(\Theta) + \mathbf{B}\xi), \varphi')}{(\varphi')^2 - (\mathbf{B}\xi, \varphi'')}. \qquad (5.3.43)$$

Here the matrix with columns B_1, B_2 is denoted by $\mathbf{B}$, $\xi = (\xi_1, \xi_2)$. Similarly, by differentiating (5.3.42) we find $\eta(x) = \varphi'\dot{\Theta} + \mathbf{B}'\xi\dot{\Theta} + \mathbf{B}\dot{\xi}$. Multiply from the left the last equality by the matrix transposed to $\mathbf{B}$:

$$\dot{\xi} = \mathbf{B}^* \left(\eta(\varphi(\Theta) + \mathbf{B}\xi) - \frac{(\varphi' + \mathbf{B}'\xi)(\eta(\varphi(\Theta) + \mathbf{B}\xi), \varphi')}{(\varphi')^2 - (\mathbf{B}\xi, \varphi'')} \right). \qquad (5.3.44)$$

We write system (5.3.43) – (5.3.44) symbolically:

$$\dot{\Theta} = \zeta_1, \qquad (5.3.45)$$

$$\dot{\xi} = \zeta, \qquad (5.3.46)$$

where $\zeta = (\zeta_2, \zeta_3)$. For the equilibrium trajectory $\varphi = \varphi(\Theta)$ to be the trajectory of system (5.3.40) it is necessary and sufficient to satisfy the condition $\zeta = 0$ for $\xi = 0$. Assume that there are two functions $V(\Theta, \xi)$ and $W(\Theta, \xi)$, related by

$$\frac{dV}{dt} = W, \qquad (5.3.47)$$

where the total derivative is calculated by system (5.3.45), (5.3.46), then

$$(\text{grad } V, \zeta) = W - \frac{\partial W}{\partial \Theta} \zeta_1. \qquad (5.3.48)$$

From (5.3.48) we get

$$\zeta = \frac{\left(W - \frac{\partial V}{\partial \Theta}\zeta_1\right) \overset{*}{\text{grad}} V}{\|\text{grad } V\|^2} + \psi \bar{V}, \qquad (5.3.49)$$

where ψ is an arbitrary function of variables Θ and ξ, and $\bar{V}$ is the vector orthogonal to grad V. If the function ζ_1 is confined to positive constants at least in some neighborhood of the point $\xi = 0$, then the equilibrium $\xi = 0$ of system (5.3.46) is stable in Lyapunov's sense provided that the function V is positive-definite, and the function W is of constant sign. If, however, the function V is positive-definite and admits an infinitely small upper bound, and W is negative-definite, then the equilibrium $\xi = 0$ of system (5.3.46) is asymptotically stable. This statement becomes evident if time is eliminated from system (5.3.46) by equation (5.3.45). We now return to the original phase space x. Differentiating (5.3.42) with respect to t, we find

$$\dot{x} = \varphi' \zeta_1 + \mathbf{B}' \xi \zeta_1 + \mathbf{B} \zeta. \qquad (5.3.50)$$

Recall that in equality (5.3.50) the vector ξ and the quantity Θ are the known functions of the vector x, $\Theta = \Theta(x)$ being a solution to equation (5.3.41), and from (5.3.42) we have $\xi = \mathbf{B}^*(x - \varphi(\Theta))$. The functions ζ_1 and ψ are subject to the condition $\eta^2 = 1$, where

$$\eta = \zeta_1' \left(\varphi' + \mathbf{B}' \mathbf{B}^*(x - \varphi) - \frac{\frac{\partial V}{\partial \Theta} \text{grad}^* V}{\|\text{grad } V\|^2} \right)$$

$$+ \mathbf{B} \bar{V} \psi + \frac{\mathbf{B} W \text{grad}^* V}{\|\text{grad } V\|^2}. \qquad (5.3.51)$$

Denote by M an equilibrium trajectory $\varphi = \varphi(\Theta)$. To be noted is that M is an integral manifold. The equilibrium trajectory $\varphi(\Theta)$ will be called orbitally stable (orbit-asymptotically stable) if M is stable (asymptotically stable).

Theorem 82 *Suppose the following conditions are satisfied:*
1) the vector-valued function (5.3.49) has the property $\zeta = 0$ for $\xi = 0$,
2) the function V is positive-definite and admits of an infinitely small upper bound,

3) the function W is negative-definite and $\frac{dV}{dt} = W$. Then system (5.3.40) has the orbit-asymptotically stable trajectory $\varphi = \varphi(\Theta)$.

Remark. Here it is assumed that the trajectory $\varphi = \varphi(\Theta)$ is either a closed simple curve of homeomorphic circumference or a non-selfintersecting space curve given for all actual values of parameter Θ. The functions appearing on the right-hand side of (5.3.51) are required to satisfy the properties for which the right-hand sides of system (5.3.40) are continuous and satisfy the conditions of the uniqueness theorem.

Consider the motion of a charged particle in the phase space x under the action of the field $B = B(x)$:

$$\dot{x} = y, \qquad (5.3.52)$$

$$\dot{y} = B(x) \times y, \qquad (5.3.53)$$

$$\operatorname{rot} B = I, \qquad (5.3.54)$$

$$\operatorname{div} B = 0. \qquad (5.3.55)$$

Set

$$B = f\eta + \eta \times \left(\frac{d\eta}{dx}\,\eta\right), \qquad (5.3.56)$$

where $\eta = \eta(x)$ is determined by equality (5.3.51) and f is an arbitrary differentiable function which satisfies condition (5.3.55). This condition yields

$$(\operatorname{grad} f, \eta) + f \operatorname{div} \eta + \operatorname{div}\left(\eta \times \left(\frac{d\eta}{dx}\,\eta\right)\right) = 0. \qquad (5.3.57)$$

Equation (5.3.57) is a linear partial differential equation for the function f which can be integrated with the help of integrals of system (5.3.40). We assume that a solution to (5.3.57) has been found. Then the magnetic field B completely determines the current I and the motion of charged particles by equations (5.3.52), (5.3.53).

Theorem 83 *When the field is given by (5.3.56), the system (5.3.52) – (5.3.53) has an equilibrium motion $\varphi = \varphi(\Theta)$ which is conditionally and orbit asymptotically stable under conditions of Theorem 82.*

5. Consider the structure of a stationary magnetic field, where the surfaces are spanned by equilibrium motions of charged particles, and derive characteristics of the structure, which ensures conditional orbit-asymptotic stability of the equilibrium motions.

Consider the system of three equations

$$\dot{x} = \eta(x), \qquad (5.3.58)$$

assuming that it has an integral manifold $G(x) = 0$ and satisfies the condition $\eta^2 = 1$. We will clarify the properties of the vector-valued function $\eta(x)$ satisfying these two conditions. Denote by a_1, a_2, a_3 the Cartesian unit vectors chosen so that

$$a_1 = \frac{\operatorname{grad}^* G}{\|\operatorname{grad} G\|},$$

then
$$\eta(x) = a_1\varphi_1 + a_2\varphi_2 + a_3\varphi_3 = \mathbf{A}\varphi, \tag{5.3.59}$$
where $\mathbf{A}$ is the matrix with columns a_1, a_2, a_3, $\varphi = (\varphi_1, \varphi_2, \varphi_3)$. From the condition $\eta^2 = 1$ we get
$$\varphi^2 = 1. \tag{5.3.60}$$
Since system (5.3.58) has an integral manifold, the equality
$$(\operatorname{grad} G, \eta) = 0 \tag{5.3.61}$$
holds if $G = 0$; therefore we have $\varphi_1 = 0$ for $G = 0$. Denote by M the set of points of the phase space lying on the integral manifold $G = 0$. The integral manifold $G = 0$ will be called orbitally stable (orbit-asymptotically stable) if M is stable (asymptotically stable).

Assume that there are two functions $V(x)$ and $W(x)$ related by
$$(\operatorname{grad} V, \eta) = W. \tag{5.3.62}$$
We shall clarify further the properties of the vector-valued function $\eta(x)$ satisfying conditions (5.3.59) and (5.3.62). Set
$$H_1 = \frac{\mathbf{A}^* \operatorname{grad}^* V}{\|\mathbf{A}^* \operatorname{grad}^* V\|}$$
and construct two vectors H_2 and H_3 such that the vectors H_1, H_2, H_3 form the right-hand Cartesian system. Let $\mathbf{P}$ be the matrix whose columns are these vectors. Put
$$\mathbf{P}^*\varphi = \xi, \tag{5.3.63}$$
where $\xi = (\xi_1, \xi_2, \xi_3)$. Then it follows that
$$\xi_1 = \frac{W}{\|\operatorname{grad} V \mathbf{A}\|}. \tag{5.3.64}$$
From (5.3.63) we have $\varphi = \mathbf{P}\xi$ and therefore $\xi^2 = 1$. Hence we find $\xi_2 = \alpha \cos\psi$, $\xi_3 = \alpha \sin\psi$, where $\alpha = \sqrt{1 - \xi_1^2}$. The foregoing implies that the arbitrary function ψ and the functions V and W satisfy the relationships
$$(P_1, \xi) = 0 \text{ for } G = 0, \tag{5.3.65}$$
$$\xi_1^2 \leq 1, \tag{5.3.66}$$
where P_1 is the first row of the matrix $\mathbf{P}$. Under these conditions system (5.3.58) may be written as
$$\dot{x} = \eta(x) = \mathbf{AP}\xi. \tag{5.3.67}$$

Theorem 84 [Zubov (1964)] *Suppose the functions V and W possess the properties:*

1) $V = 0$ for $x \in M$, $V(x) > \gamma_1 > 0$ for $\rho(x, M) > \gamma_2 > 0$;

Controlled Motion of Charged Particles

2) $W = 0$ for $x \in M, W < -\gamma_1 < 0$ for $\rho(x, M) > \gamma_2$.

Then system (5.3.67) has an orbit-asymptotically stable integral manifold $G(x) = 0$.

Remark. Theorem 84 implies that the functions V and W and the functions ψ and G are such that the right-hand sides of system (5.3.67) satisfy conditions in the theorem of existence and uniqueness of solutions.

We now turn to the problem of motion of a charged particle in a stationary magnetic field. Define the magnetic field $B(x)$ by

$$B(x) = f\eta + \eta \times \left(\frac{d\eta}{dx}\eta\right). \qquad (5.3.68)$$

The function f, which is generally arbitrary, will be chosen so that Maxwell's equation be satisfied

$$\mathrm{div} B = 0. \qquad (5.3.69)$$

This equation leads to the relationship

$$(\mathrm{grad}\, f, \eta) + f \,\mathrm{div}\, \eta + \mathrm{div}\left(\eta \times \left(\frac{d\eta}{dx}\eta\right)\right) = 0. \qquad (5.3.70)$$

Equation (5.3.70) can be solved by using the integrals of system (5.3.67). In a dimensionless form, the equation of particle motion in this field can be represented as

$$\dot{x} = y, \quad \dot{y} = B \times y. \qquad (5.3.71)$$

Under conditions (5.3.69) the field (5.3.68) determines uniquely the spatial density of the current

$$\mathrm{rot} B = I. \qquad (5.3.72)$$

In this consideration, the density is an arbitrary quantity and is determined in terms of B as is shown.

Theorem 85 *If the conditions of Theorem 84 are satisfied and the field B is determined by the relationship (5.3.68), (5.3.69), then for any function f in system (5.3.71) there is an equilibrium manifold $G(x) = 0$, $y = \eta(x)$ which is conditionally and orbit-asymptotically stable.*

As an example, consider the system

$$\dot{x} = k\left(\frac{\lambda x}{R} - y\omega\right).$$

$$\dot{y} = k\left(\omega x + \frac{\lambda_1 y}{R}\right), \qquad (5.3.73)$$

$$\dot{z} = k((R - R_1)\mu + z\lambda),$$

where $\lambda_1 = \lambda(R - R_1 - \mu z)$. Here λ, ω, μ are some functions of the vector (x, y, z), $R = \sqrt{x^2 + y^2}$, R_1 is a positive constant, and k is the norming factor which is dependent on the vector (x, y, z) and satisfies the condition $\dot{x}^2 + \dot{y}^2 + \dot{z}^2 = 1$. Put $V = (R-R_1)^2 + z^2$, then

$$\frac{1}{2}\dot{V} = k\lambda V. \qquad (5.3.74)$$

Let us assume that λ is a function of the quantity V. If there is a positive number $\rho_1 < R_1$ such that $\lambda = -g_0(V_1 - \rho_1)^{2k+1} + \ldots$, where $V_1 = V^{1/2}$, then for $g_0 > 0$ and an integer $k \geq 0$ the system (5.3.73) has the set $V = \rho_1^2$ as an integral manifold. This integral manifold is orbit-asymptotically stable. Here ω and μ are said to be the completely defined functions assuming, say, positive values. If, however, $g_0 < 0$, the function λ for $V_1 < \rho_1$ remains negative and $\lambda = 0$ for $V = 0$, then system (5.3.73) has the orbit-asymptotically stable manifold $V = 0$ whose attraction region [Zubov (1962b)] is bounded by the surface $V = \rho_1$.

We take the right-hand sides of system (5.3.73) to be the function $\eta(x)$ appearing in (5.3.68). Then the motion of charged particle under the action of the field B is such that there is the equilibrium manifold $V = \rho_1^2$ which is conditionally and orbit-asymptotically stable. In the second case this manifold bounds the attraction region of the equilibrium orbit $V = 0$.

6. The method will be set forth for distribution of electric charges in the space which provides a solution to various problems arising from transport of beams. In particular, the method proposed for distribution of charges makes it possible to solve the problem of focusing and acceleration of electron beams.

Equations for motion of electrons in a stationary electric field can be written as

$$\frac{dX}{dt} = Y, \quad \frac{dY}{dt} = E(X), \qquad (5.3.75)$$

where $X = (x_1, x_2, x_3)$, $Y = (y_1, y_2, y_3)$ are the coordinates describing the phase state of a charged particle. The stationary electric field $E(X)$ satisfies the equations

$$\text{div} E = \rho, \quad \text{rot} E = 0, \qquad (5.3.76)$$

where $\rho = \rho(X)$ is the volume charge density. In forming electron beams, the problem is to find a distribution density ρ for which the beam has the required properties, in particular, it can be focused and accelerated.

In order to solve this problem, we specify a potential velocity field

$$\eta(X) = \nabla V(X). \qquad (5.3.77)$$

Set

$$E(X) = \left(\frac{d\eta(X)}{dX}\right)\eta, \qquad (5.3.78)$$

where $\left(\frac{d\eta(X)}{dX}\right) = \frac{D(\eta_1, \eta_2, \eta_3)}{D(x_1, x_2, x_3)}$. It follows from (5.3.77) that the vector-valued function $E(X)$ is potential and therefore rot $E(X) = 0$. From (5.3.78) we have that the volume charge density is defined by the formula

$$\rho(X) = \Delta \eta^2. \qquad (5.3.79)$$

Controlled Motion of Charged Particles

Theorem 86 *Suppose the following conditions are satisfied:*

1) the function $V(x_1, x_2, x_3)$ has a strict maximum with respect to the quantities x_2, x_3 at the point $x_2 = x_3 = 0$ for $x_1 \geq 0$;

2) the function $V(x_1, x_2, x_3) - V(x_1, 0, 0) \to 0$ as $x_2 \to 0, x_3 \to 0$ uniformly in $x_1 \geq 0$;

3) the function $(\nabla V)^2 \to +\infty$ as $x_1 \to +\infty, x_2^2 + x_3^2 \leq \epsilon$, where ϵ is a positive constant;

4) the following inequality holds

$$\frac{\partial}{\partial x_1}(V(x_1, x_2, x_3) - V(x_1, 0, 0)) + \frac{(\frac{\partial V}{\partial x_2})^2 + (\frac{\partial V}{\partial x_3})^2}{\frac{\partial V}{\partial x_1}} \geq \alpha(x_1) W(x_2, x_3),$$

where $W(x_2, x_3)$ is the function which is positive-definite with respect to x_2, x_3, and $\alpha(x_1)$ is a non-negative function such that

$$\int_0^{+\infty} \alpha(x_1) dx_1 = +\infty.$$

Then system (5.3.75) has an integral manifold $x_2 = 0$, $x_3 = 0$ which is conditional-asymptotically stable in Lyapunov's sense. In other words, if δ-beams of electrons start on the manifold

$$Y = \nabla V \tag{5.3.80}$$

at $t = 0$, then they remain on this manifold during the motion. In this case they are focused along the optic axis $x_2 = x_3 = 0$ and are accelerated at the same time.

Proof: We first ensure that if the volume charge density is chosen from formula (5.3.79), then system (5.3.75) has the integral manifold (5.3.80). We now eliminate the parameter t and consider the quantities x_2 and x_3 on the manifold (5.3.80) as the functions of x_1. As shown by studies, these functions describe the motions tending asymptotically to zero as $x_1 \to +\infty$. This implies conditional stability and focusing on δ-beams of electrons. It remains to be seen whether the focusing is accompanied by acceleration of charged particles, which follows immediately from the unbounded increase of the quantity $(\nabla V)^2$ along the optic axis. ∎

Example 1: Let $V = V_1(x_1) + V_2(x_2, x_3)$. All conditions of the theorem are satisfied if the function V_2 is negative-definite, and $V_1 = ax_1 + bx_1^2$, where a and b are positive constants.

Example 2: Put $V = V_1(x_1) + V_2(x_1, x_2, x_3)$, where the function V_2 is periodic in x_1, and negative-definite with respect to x_2 and x_3; $V_1 = ax_1 + p(x_1)$, where $p(x_1)$ is a periodic function of the same period as in V_2, and a is a positive constant such $a + \frac{dp}{dx_1} > 0$. If the function $(\frac{\partial V_2}{\partial x_2})^2 + (\frac{\partial V_2}{\partial x_3})^2$ is positive-definite, then all conditions of Theorem 86, except condition 2), are satisfied. Here the result is that the δ-beams of electrons on the integral manifold (5.3.80) are focused, but not accelerated unboundedly.

Remark 1. The relativistic approach does not introduce any additional problems into the finding of charge density. By employing the function V, we introduce into a charge density distribution some parameters which make it possible to alter various properties of beams, in particular, to reduce aberration.

Remark 2. Given the electric field E, the behavior of charged particles is analyzed by the well-known methods. It follows from the approach given here that these methods of analysis can be supplemented by constructing integral manifolds of the form (5.3.80), since the given electric field E and the partial differential equations of the first order make it possible to construct the function V related to the field E by (5.3.78).

7. We now turn to universality of electrodynamics equations and show that if the particle velocity field Maxwell's is given arbitrarily, then it is possible to construct the electromagnetic field producing the velocity field. This means that Maxwell's equation is universal.

Let us assume that a charged particle of mass m moves in an arbitrary field of forces. Then the equation of motion can be represented schematically as

$$\dot{X} = Y, \ (m\dot{Y}) = f(t, X, Y),$$

where $X = (x_1, x_2, x_3)$, $Y = (y_1, y_2, y_3)$ are coordinates and velocities of the particle.

Theorem 87 *There exist vectors $\hat{E}(t, X, Y)$ and $\hat{B}(t, X, Y)$ such that the field $f(t, X, Y)$ can be represented as*

$$f(t, X, Y) = \hat{E} + Y \times \hat{B}. \tag{5.3.81}$$

If the force field f is electromagnetic, then relationship (5.3.81) is assumed to hold for

$$\hat{E} = qE(t, X), \ \hat{B} = \frac{q}{c}B(t, X), \tag{5.3.82}$$

where q is a particle charge, c the velocity of light, $E(t, X)$ the electric field strength, $B(t, X)$ the magnetic field strength given by the relationships

$$\text{rot}E = -\frac{\partial B}{\partial t}, \ \text{rot}B = I + \frac{\partial E}{\partial t}, \ \text{div}B = 0, \ \text{div}E = \sigma. \tag{5.3.83}$$

Here I is the spatial current density, σ the charge density, the constants being omitted for simplicity. Given the density of current and that of charges, equations (5.3.83) determine the electromagnetic field, where the motion of a charged particle is described by equalities (5.3.82). Then the force field determined by (5.3.82) makes it possible to study the phase state of a charged particle. In particular, the particle velocity field is known.

The problem is whether there is another version of an electromagnetic field inducing a force field if the motion of a charged particle is given by an arbitrary velocity field

$$\dot{X} = \eta(X, t). \tag{5.3.84}$$

In other words, the question is whether it is possible to set up a current I and an electric charge density σ for which the charged particle travels in conformity with the given field of velocities.

Theorem 88 *For any field (5.3.84) there is a current density and a charges density determining the electromagnetic field under the action of which a charged particle has a given field of velocities.*

Proof: By a given velocity field (5.3.84), we derive equations for the magnetic field strength

$$\frac{dB}{dt} + \mathbf{P}B + \mathrm{rot}\left(\frac{\partial(m\eta)}{\partial t} + \left(\frac{d(m\eta)}{dX}\right)\eta\right) - \eta \mathrm{div} B = 0, \tag{5.3.85}$$

where the total derivative of vector B is taken by equations (5.3.84) and

$$\eta = (\eta_1, \eta_2, \eta_3), \quad \mathbf{P} = \begin{pmatrix} \mathrm{rot}\ i \times \eta \\ \mathrm{rot}\ j \times \eta \\ \mathrm{rot}\ k \times \eta \end{pmatrix}.$$

Here i, j, k are unit vectors of Cartesian stationary system.

Next, it is shown that system (5.3.85) has a unique solution $B = B(t, X)$ satisfying the condition $B(t, X) = \mathrm{rot} A(X)$ at $t = t_0$, where A is an arbitrary vector.

We first construct the electric field

$$E(t, X) = \frac{m}{q}\left(\frac{\partial(m\eta)}{\partial t} + \left(\frac{d(m\eta)}{dX}\right)\eta\right) + \frac{1}{c}B \times \eta,$$

and then the current density and the charge density.

It remains to establish that the obtained strength of magnetic and electric fields causes the particle to move in conformity with the velocity field (5.3.84). ∎

Theorem 89 *If the field (5.3.84) is linear and $m = $ const, then the field $E(t, X)$ can be constructed as linear and the field B as independent of spatial coordinates. Then the density of current I is a linear function of spatial coordinates when the charge density is independent of spatial coordinates.*

5.4 Phase Plane

1. If the autonomous system featuring one degree of freedom is given and the oscillations therein are described by the equations

$$\ddot{x} = f(x, \dot{x}), \tag{5.4.1}$$

then by introducing a new function $y = \dot{x}$ it is possible to pass from equation (5.4.1) to the first-order system of two differential equations

$$\dot{x} = y, \tag{5.4.2}$$

$$\dot{y} = f(x, y).$$

In order to represent geometrically the states of system (5.4.2), the $x0y$ plane is introduced, where a complete graphic pattern of oscillation propagation can be obtained. This plane is referred to as the phase plane for system (5.4.2) or (5.4.1).

We shall now discuss general concepts and results related to the behavior of integral curves of the system of two autonomous differential equations

$$\dot{x} = f_1(x, y), \qquad (5.4.3)$$

$$\dot{y} = f_2(x, y)$$

on a phase plane. Assume that the right-hand sides of system (5.4.3) given for $x \in (-\infty, +\infty)$, $y \in (-\infty, +\infty)$, are real, continuous, and satisfy Lipschitz conditions with respect to x, y in any bounded region G of the plane (x, y).

Denote by $f(p, t)$ the integral curve of system (5.4.3) passing at the time $t = 0$ through the point $p = (x_p, y_p)$. Recall that the point (a, b) is called a rest point or an equilibrium of system (5.4.3) when the equality $f_1(a, b) = f_2(a, b) = 0$ holds. Next, we assume that in the region G there is merely a finite number of equilibria. If the integral curve $f(p, t)$ has the property $f(p, t) = f(p, t + \omega)$, then system (5.4.3) is said to have a periodic solution whose graph is a closed curve on the plane. It will be referred to as an orbit or a closed trajectory. The point q is called $\omega(\alpha)$-limit for $f(p, t)$ if there is a sequence $t_n \to +\infty$ $(t_n \to -\infty)$ such that $f(p, t_n) \to q$ as $n \to \infty$. The set of points of the integral curve $f(p, t)$ for $t \geq 0$ will be called a positive semitrajectory and for $t \leq 0$ a negative semitrajectory and, finally, just a semitrajectory if the kind of semitrajectory is of no importance; the α and ω-limit points are called the limit points of the motion $f(p, t)$.

We shall note some basic facts relative to the limiting properties of integral curves in (5.4.3). If a positive semitrajectory has no equilibrium among its limit points, then either this semitrajectory is closed or it has a closed limit trajectory. If a positive semitrajectory has a closed limit trajectory, then the latter is its unique limit trajectory. If the limit points of a positive semitrajectory can include only a finite number of equilibria, then the entire set of limit points of this semitrajectory is made up of one equilibrium. Thus, under our restrictions, the positive semitrajectory can have a set of limit points of one of the following types: one equilibrium, one closed trajectory, and a set of integral trajectories some of which are equilibria while the other are trajectories which differ from equilibria and tend to equilibria as both $t \to +\infty$ and $t \to -\infty$. Note that in any closed trajectory there is at least one equilibrium. All the properties stated above (with explicit changes) also hold for negative semitrajectories.

We now turn to the behavior of integral curves of system (5.4.3) in the neighborhood of stationary modes of this system. The simplest stationary mode of system (5.4.3) is its rest point. All rest points are subdivided into two classes: the isolated and the nonisolated. The former are the rest points of system (5.4.3) for which it is possible to select their sufficiently small neighborhood that is free from other rest points of this system. We shall discuss the behavior of integral curves of system (5.4.3) in the neighborhood of the isolated rest point (a, b). Such points are first divided into two classes: Lyapunov stable and unstable. Recall that the point (a, b) is called Lyapunov stable if for every $\epsilon > 0$ it is possible to choose $\delta(\epsilon) > 0$ such that for $p_0 \in S(\delta)$ we get $f(p_0, t) \in S(\epsilon)$ for $t \geq 0$, whereby $S(\alpha)$ is denoted an

Phase Plane

α-neighborhood of the point (a,b), i.e. the collection of all points satisfying the inequality $(x-a)^2 + (y-b)^2 < \alpha^2$. The isolated stable points of rest (a,b) in turn are divided into two classes: asymptotic and nonasymptotic. The Lyapunov stable rest point (a,b) is called asymptotic or asymptotically stable in Lyapunov's sense if $\delta(\epsilon)$ can be chosen so that $f(p_0, t) \to (a,b)$ as $t \to +\infty$.

Theorem 90 *For the isolated rest point (a,b) to be stable in Lyapunov's sense, it is necessary and sufficient that there be no integral curve $f(p,t)$ of system (5.4.3) such that $p \neq (a,b)$ and $f(p,t) \to (a,b)$ as $t \to -\infty$. In addition, if it is to be asymptotically stable, it is necessary and sufficient that the additional condition is satisfied, i.e. existence of an integral curve $f(p,t)$, $p \neq (a,b)$ such that $f(p,t) \to (a,b)$ as $t \to +\infty$.*

Proof: *Necessity.* If the point (a,b) is Lyapunov stable, then for $\epsilon > 0$ we choose $\delta(\epsilon) > 0$ such that for $p_0 \in S(\delta)$ we get $f(p_0, t) \in S(\epsilon)$ for $t \geq 0$. Suppose there is such an integral curve $f(p,t)$ that $p \neq (a,b)$, $f(p,t) \to (a,b)$ as $t \to -\infty$. Then, setting $\epsilon = \sqrt{(x_p - a)^2 + (y_p - b)^2}$ and taking in conformance with it a number $\delta(\epsilon/2)$, we have that all integral curves starting in $S(\delta)$ stay in $S(\epsilon/2)$ for $t \geq 0$. By assumption, in $S(\delta)$ there is a point $q = f(p,T)$, $T < 0$ such that $f(q, -T) = p$ and therefore q is not in $S(\epsilon/2)$. The obtained contradiction shows that the necessity condition of the theorem has been established. The fact that for asymptotic stability there exists an integral curve which tends indefinitely to (a,b) as $t \to +\infty$, is evident.

Sufficiency. Lyapunov stability of the rest point (a,b) will be established if for every $\epsilon > 0$ it is possible to choose $\delta(\epsilon) > 0$ such that for $p_0 \in S(\delta)$ we get $f(p_0, t) \in S(\epsilon)$ with $t \geq 0$, or alternatively: all integral curves $f(p,t)$ starting on the circle ϵ with its center at (a,b) lie off $S(\delta)$ for $t \leq 0$. Assume that there is $\epsilon > 0$ for which there is no $\delta > 0$ with this property. Then it is possible to choose a sequence of numbers $\delta_k > 0$, $\delta_k \to +0$ as $k \to \infty$ and a sequence of points p_n lying on the circle of radius ϵ with its center at (a,b) for which there is a sequence of negative numbers t_n such that $f(p_n, t_n) \in S(\delta_n)$. Put $f(p_n, t_n) = p_n$. The sequence t_n is said to be chosen so that $f(p_n, t) \in S(\epsilon)$ for $t \in [t_n, 0)$. The integral curves $f(q_n, t)$ can either stay in $\overline{S(\epsilon)}$ when $t \leq 0$ or leave this neighborhood in the finite time. Without giving further consideration to the first opportunity, we assume that the integral curve $f(q_n, t)$ intersects the circle of radius ϵ with its center (a,b) at the point r_n. We will choose from the sequences p_n, r_n such subsequences that $p_n \to p$, $r_n \to r$ as $n \to \infty$ (to save notation the subsequences are designated by the same symbols). We show that the integral curve $f(p,t)$ for $t \leq 0$ remains in $\overline{S(\epsilon)}$. First establish that the sequence t_n has the property $t_n \to -\infty$ as $n \to \infty$. Otherwise there is a subsequence t_n such that $t_n > -T > -\infty$. For the numbers T and $\epsilon/2$ it is possible to choose such $\delta_1' > 0$ such that all the integral curves starting in $S(\delta_1')$ remain in $S(\epsilon/2)$ for $s \in [-T, T]$ (this follows from the continuity theorem for solutions in initial data since it is possible to take $x \equiv a$, $y \equiv b$ as one of the solutions). Then we may choose a number N such that $\delta_N < \delta_1'$. For such a choice of number N we have $f(q_n, -t_n) = p_n \notin S(\epsilon/2)$ and $f(q_n, -t_n) \in S(\epsilon/2)$ since $0 < -t_n \leq T$. This contradiction shows that the sequence t_n cannot have the subsequences bounded from below and hence $t_n \to -\infty$ as $n \to \infty$. Let us now assume that the integral curve $f(p,t)$ leaves $\overline{S(\epsilon)}$ at a certain $\bar{t} < 0$. Then, by the continuity theorem for solutions in initial data, all the integral curves starting in a sufficiency small neighborhood of the point p also leave $\overline{S(\epsilon)}$ at $t = \bar{t} < 0$. This contradicts the fact that in an arbitrarily small neighborhood of the point p there are points p_n such that the integral curves passing through them remain in $S(\epsilon)$ as long as desired for $t \in [t_n, 0)$. Thus, $f(p,t) \in \overline{S(\epsilon)}$ for $t \leq 0$. So the assumption that the point (a,b) is not stable in Lyapunov's sense leads to availability of the integral curve $f(p,t)$ which remains for $t \leq 0$ in $\overline{S(\epsilon)}$. Consequently, the α-limit set of this integral curve also lies in $\overline{S(\epsilon)}$ and may be composed either of a unique periodic solution $f(q,t)$ to system (5.4.3) or of an equilibrium point the integral curve with its two ends entering the equilibrium point. In this case, the second and third possibilities cannot be realized, since, by definition, there are no integral curves $f(p_0, t)$ such that $p_0 \neq (a,b)$, and $f(p_0, t) \to (a,b)$ as $t \to -\infty$. The first possibility is realized and hence the point (a, b) is lying within the periodic solution $f(q, t)$. Denote by δ the distance from this point to the orbit of the periodic solution. Then all the integral curves starting in $S(\delta)$ remain in $S(\epsilon)$ when $t \geq 0$. This contradicts the assumption that for this ϵ there is no $\delta > 0$ having such a property, and therefore the rest point (a, b) is stable in Lyapunov's sense.

It now remains to show that, under the additional condition stated by the theorem, the rest point is also asymptotically stable. Let us assume that $f(p_0, t) \to (a, b)$ as $t \to +\infty$, $p_0 \neq (a, b)$ and $\epsilon = \sqrt{(x_{p_0} - a)^2 + (y_{p_0} - b)^2}$. By the available stability, for such an ϵ, we select a number $\delta(\epsilon) > 0$ and show that all the integral curves $f(p, t)$ starting in $S(\delta)$ have the property $f(p,t) \to (a,b)$ as $t \to +\infty$. If this is not the case, there must exist an integral curve $f(p,t)$, $p \in S(\delta)$ such that $f(p,t) \not\to (a,b)$ as $t \to +\infty$. Here the ω-limit set of this integral curve is lying in $S(\epsilon)$ and is the orbit of the periodic solution $f(q,t)$, since the ω-limit set can neither be the rest point nor contain it (which follows from the above reasoning). The orbit of the periodic solution $f(q,t)$ breaks the plane into two regions: inner and outer. The inner region contains the point (a,b), the outer the point p_0 and therefore the integral curve $f(p_0, t)$ for $t > 0$ intersects the orbit of the periodic solution $f(q,t)$, which is not possible by the available uniqueness. The obtained contradiction shows that all the integral curves of system (5.4.3) starting in $S(\delta)$ not only stay in $S(\epsilon)$ when $t \geq 0$ but also tend indefinitely to (a,b) as $t \to +\infty$. ∎

The sufficiently small neighborhood of the asymptotically stable equilibrium (a, b) is as follows: all the integral curves $f(p,t)$ starting in it tend indefinitely to (a,b) as $t \to +\infty$, $f(p,t) \to (a,b)$ as $t \to +\infty$ and there is no integral curve $f(p,t)$, $p \neq (a,b)$, $f(p,t) \to (a,b)$ as $t \to -\infty$. In general, such a behavior of integral curves in the neighborhood of the asymptotically stable equilibrium is topologically equivalent to the behavior of integral curves of the system

$$\dot{x} = -x,$$

$$\dot{y} = -y.$$

The neighborhood of the stable equilibrium that is not asymptotically stable is more complicated in composition. If δ is chosen for a number ϵ such that for $p \in S(\delta)$

we get $f(p,t) \in S(\epsilon)$ with $t \geq 0$ and $\overline{S(\epsilon)}$ contains no other rest points, then any integral curve $f(p,t)$ is either a periodic solution to system (5.4.3), or an integral with the ends spiraling toward the periodic solution of system (5.4.3) (which is observed for a sufficiently small $\epsilon > 0$). It may turn out that all the integral curves starting in $S(\delta)$ are the periodic solutions of system (5.4.3). This behavior of integral curves in the neighborhood of the equilibrium is called the center. In general, such a behavior of the integral curves is topologically equivalent to the behavior of integral curves of the system of equations

$$\dot{x} = -y,$$
$$\dot{y} = x.$$

If, however, in an arbitrarily small neighborhood of the point (a,b) there are both periodic and spiral-type motions, this case is called the center-focus. Here the qualitative pattern is topologically equivalent to the behavior of integral curves of the system

$$\dot{x} = -xf(x^2 + y^2) - y,$$
$$\dot{y} = x - yf(x^2 + y^2)$$

in the neighborhood of the equilibrium $x = y = 0$, where $f(z)$ is the function given for $z \in [0,1]$, which is real, continuously differentiable, and is not identically zero in any arbitrarily small interval $[0, \delta]$, but vanishes at least twice in this interval for any $\delta > 0$.

2. The other stationary motions of system (5.4.3) are its periodic solutions. All the periodic solutions fall into two classes: isolated and non-isolated. The periodic motion $f(p,t)$ of system (5.4.3) is called isolated or a limit cycle if there is a sufficiently small neighborhood $S(M, \delta)$ of the orbit M of the periodic motion $f(p,t)$ containing no other periodic motions of this system. Here $S(M, \delta)$ is the set of points (x,y) satisfying the inequality $\rho((x,y), M) < \delta$, where

$$\rho((x,y), M) = \inf_{(x_1, y_1) \in M} \sqrt{(x - x_1)^2 + (y - y_1)^2}.$$

The behavior of integral curves of system (5.4.3) in the neighborhood of the orbit M of the isolated periodic solution may be classed with one of the three possible types.

The limit cycle is called stable if there is $\delta > 0$ such that for $p \in S(M, \delta)$ there will be $\rho(f(p,t), M) \to 0$ as $t \to +\infty$. The limit cycle is called unstable if there is $\delta > 0$ such that for $p \in S(M, \delta)$ there is $\rho(f(p,t), M) \to 0$ as $t \to -\infty$. The limit cycle is called semistable if there is $\delta > 0$ such that for $p \in S(M, \delta)$ there is $\rho(f(p,t), M) \to 0$ as $t \to +\infty$ or as $t \to -\infty$, and there are two points p_1, p_2 not lying on M such that $\rho(f(p_1, t)M) \to 0$ as $t \to +\infty$ and $\rho(f(p_2, t), M) \to 0$ as $t \to -\infty$.

In general, the behavior of integral curves in the neighborhood of the limit cycle is topologically equivalent to the behavior of integral curves of the system

$$\dot{x} = -xh(x^2 + y^2) - y, \tag{5.4.4}$$
$$\dot{y} = x - yh(x^2 + y^2)$$

in the neighborhood of its periodic solution $x = \cos t$, $y = \sin t$, where $h(z)$ is a continuously differentiable function that is given for $z \in [1-\delta, 1+\delta]$ and takes real values there. In this case, system (5.4.4) has a stable limit cycle $x^2 + y^2 = 1$ if $h(1) = 0$, $h(z) < 0$ for $z \in [1-\delta, 1)$ and $h(z) > 0$ for $z \in (1, 1+\delta]$. If the sign of the last inequalities is reversed, the limit cycle $x^2 + y^2 = 1$ will be unstable. If $h(z)$ retains the sign for $z \in [1-\delta, 1)$ and $z \in (1, 1+\delta]$, then $x^2 + y^2 = 1$ is a semistable limit cycle of system (5.4.4). The behavior of integral curves of (5.4.4) in the neighborhood of the non-isolated periodic solution is topologically equivalent to the behavior of integral curves of system (5.4.4) in the neighborhood of its periodic solution $x = \cos t$, $y = \sin t$ if $h(z)$ is taken to be an arbitrary continuously differentiable function such that in either of the intervals $[1-\delta_1, 1]$, $[1, 1+\delta_1]$ it vanishes at least twice for any sufficiently small $\delta_1 > 0$. Note that the stable limit cycle is also called self-oscillation.

The main problems confronting the theory in studies on systems of the form (5.4.3) are to establish the existence conditions for periodic solutions, specifically for limit cycles and self-oscillations, to construct these periodic solutions, and to investigate the behavior of the remaining integral curves in their neighborhood. All the results in § 2.2 relating to comparison of the notions of Lyapunov stability of periodic solutions and orbital stability apply to system (5.4.3) in full. Recall that the periodic solution $x = \varphi(t), y = \psi(t)$ of system (5.4.3) is referred to as Lyapunov stable if for every $\epsilon > 0$ it is possible to choose $\delta > 0$ such that for $\rho((x_0, y_0), (\varphi(0), \psi(0))) < \delta$ there is $\rho((x(t), y(t)), (\varphi(t), \psi(t))) < \epsilon$ for $t \geq 0$. If the periodic solution is Lyapunov stable and isolated, then it is a self-oscillation of system (5.4.3). If, however, the periodic solution of system (5.4.3) $x = \varphi(t), y = \psi(t)$ is a self-oscillation of this system, this does not mean that it is Lyapunov stable.

Example : Consider the system

$$\dot{x} = \frac{x(1-r^2)^3}{r^2} - y(1 + (1-r^2)^2), \qquad (5.4.5)$$

$$\dot{y} = x(1 + (1-r^2)^2) + \frac{y}{r^2}(1-r^2)^3, \quad r = \sqrt{x^2 + y^2}.$$

All calculations will be carried out in some detail, since they allow one to establish all the above statements.

Set $x = r\cos\varphi$, $y = r\sin\varphi$, where r and φ are the new required functions. Multiplying the first of the equations in (5.4.5) by x, the second by y, and adding together, we obtain

$$\frac{d}{dt}\left(r^2\right) = 2(1-r^2)^3.$$

Multiplying the second of the equations in (5.4.5) by x, the first by y, and subtracting termwise the second equation from the first, we obtain

$$x^2 \frac{d}{dt}\left(\frac{y}{x}\right) = (1 + (1-r^2)^2)r^2,$$

whence

$$r^2 \cos^2\varphi \frac{d}{dt}(\operatorname{tg}\varphi) = r^2 \frac{d\varphi}{dt} = r^2(1 + (1-r^2)^2).$$

Setting $1 - r^2 = z$, we find two equations

$$\frac{dz}{dt} = -2z^3, \quad \frac{d\varphi}{dt} = 1 + z^2.$$

By integrating these equations, we get

$$z = \frac{z_0}{\sqrt{1 + 4z_0^2 t}}, \quad \varphi = \varphi_0 + t + \frac{1}{4}\ln\left(1 + 4z_0^2 t\right);$$

hence

$$x = \sqrt{1 - \frac{z_0}{\sqrt{1 + 4z_0^2 t}}} \cos\left(\varphi_0 + t + \frac{1}{4}\ln\left(1 + 4z_0^2 t\right)\right);$$

$$y = \sqrt{1 - \frac{z_0}{\sqrt{1 + 4z_0^2 t}}} \sin\left(\varphi_0 + t + \frac{1}{4}\ln\left(1 + 4z_0^2 t\right)\right),$$

where $z_0 = 1 - x_0^2 - y_0^2$. When $z_0 = 0$, system (5.4.5) has the periodic solution $x = \cos(t+\varphi_0)$, $y = \sin(t+\varphi_0)$ which is a self-oscillation of (5.4.5). This periodic solution is unstable in Lyapunov's sense, however small z_0 might be, the phase difference of this periodic solution and any oscillation coming arbitrarily close to it increases beyond all bounds:

$$\varphi - \varphi_0 - t = \frac{1}{4}\ln\left(1 + 4z_0^2 t\right).$$

Thus, if the motion $x(t)$, $y(t)$ has a phase φ_0 at the initial instant $t = 0$ and starts in an arbitrarily small neighborhood of the point $(\cos\varphi_0, \sin\varphi_0)$, then in the course of time the distance $\rho((x(t), y(t)), (\cos(t + \varphi_0), \sin(t + \varphi_0)))$ exceeds the number 2 (diameter of the limit cycle) as many times as desired.

5.5 Oscillations in Autonomous System

1. Consider the system

$$\dot{x} = f(x, y), \qquad (5.5.1)$$

$$\dot{y} = q(x, y),$$

the right-hand sides of which satisfy the following conditions:

1. The functions $f(x,y)$ and $g(x,y)$ are given for $x \in (-\infty, +\infty)$, $y \in (-\infty, +\infty)$, are real and continuous, and satisfy the Lipschitz condition for variables (x, y) in any finite region of the plane (x, y).

2. The function $g(x, y)$ satisfies the inequality $xg(x, y) < 0, x \neq 0, g(0, y) \equiv 0$.

3. Equation $f(x, y) = 0$ defines on a plane the line breaking the plane (x, y) into two regions, one of which contains a positive semi-axis $Oy, y > 0$, and the other a negative semi-axis $Oy, y < 0$. The inequality $f > 0$ is satisfied in the first of the above regions, and $f < 0$ in the second. In other words, it is assumed that there are two functions $\varphi(s)$ and $\psi(s)$, given for $s \in (-\infty, +\infty)$, real, piecewise continuously

differentiable, and satisfying the condition $\dot{\varphi}^2(s) + \dot{\psi}^2(s) > 0$, where by $\dot{\varphi}$ and $\dot{\psi}$ are meant the left-hand and right-hand derivatives of these functions. The functions $\varphi(s)$ and $\psi(s)$ satisfy the equation $f(\varphi(s), \psi(s)) \equiv 0$. Moreover, each point (x_0, y_0) satisfying the equality $f(x_0, y_0) = 0$ can be associated with a number s_0 such that $x_0 = \varphi(s_0), y_0 = \psi(s_0)$.

4. The branch of the continuous line $f(x, y) = 0$ lying on the right half-plane $x > 0$ either has no points in the half-plane $x \geq c_1 > 0$ for a positive c_1 or the equation $f(x, y) = 0$ for all $x > c_2 > 0$ defines a unique continuous function $y = y_1(x)$, where c_2 is a sufficiently large positive number. Similarly, the branch of the continuous line $f = 0$ lying on the left half-plane $x < 0$ either has no points in the half-plane $x \leq d_1 < 0$ or there is a number $d_2 < 0$ sufficiently large in absolute value, the equation $f(x, y) = 0$ defining a unique continuous function $y = y_2(x)$ for $x \leq d_2$. The plots of the functions $y = y_1(x)$ for $x \geq c_2$ and $y = y_2(x)$ for $x \leq d_2$ coincide with the line defined by the equation $f(x, y) = 0$ for $x \geq c_2$ and $x \leq d_2$, respectively. From these conditions it follows that the equations $f(x, y) = 0$ and $g(x, y) = 0$ have a unique simultaneous real solution $x = y = 0$. Denote by G_1 the region confined between the positive semi-axis Oy and the right-hand branch of the line $f(x, y) = 0$, i.e. that portion of the line $f(x, y) = 0$ which lies on the right half-plane $x > 0$. Denote by G_2 the region confined between the right-hand branch of the line $f(x, y) = 0$ and the negative semi-axis Oy. Denote by G_3 the region lying between the negative semi-axis Oy and the left-hand branch of the line $f(x, y) = 0$. The remaining portion of the plane confined between the positive semi-axis Oy and the left-hand branch of the line $f(x, y) = 0$ will be denoted by G_4. If system (5.5.1) has a limit cycle, then it encompasses the origin of coordinates and hence there are the solutions of (5.5.1) intersecting in succession and without bound all the semi-axis and branches of the line $f(x, y) = 0$. If, however, the above limit cycle is unique, then any integral curve of system (5.5.1), except this limit cycle, intersects infinitely many times the coordinate axes at geometrically different points and comes arbitrarily close to the limit cycle as $t \to +\infty$ or $t \to -\infty$. This argument shows that it is essential to clarify the conditions under which the solutions of system (5.5.1) intersect all coordinate axes. To this end, we take an arbitrary point p in the region G_1 and construct the region G_p defined by the relationships $y < y_p, x > x_p, (x, y) \in G_1$, where x_p, y_p are the Cartesian coordinates of the point p. The integral curve $F(p, t)$ starting with $t = 0$ at the point p stays in the region G_p for $t \geq 0$ until it intersects the right-hand branch of the line $f(x, y) = 0$, for the inequalities $x > x_p, y < y_p$ hold along this integral curve for $t > 0$, since $f(x, y) > 0, g(x, y) < 0$ for $(x, y) \in G_1$. Two cases are possible here: the region G_p is either bounded or not. The first case is possible when the right-hand branch of the line $f(x, y) = 0$ does not enter the half-plane $x \geq c_1$, or when $y_p \leq y_1(x)$ for at least one $x > c_2$. In this case the integral curve $F(p, t)$ will necessarily abandon the region G_p. The second case is possible if the inequality $y_1(x) < y_p$ holds for all $x > c_2$. Then the integral curve $F(p, t)$ may either abandon the region G_p or remain therein. We are aiming at selecting the condition under which the integral curve $F(p, t)$ in the last case also abandons the region G_p. Along the integral curve of system (5.5.1) we

Oscillations in Autonomous System

have the equation
$$\frac{dy}{dx} = \frac{g(x,y)}{f(x,y)}. \tag{5.5.2}$$

Integrating this equation along the integral curve $F(p,t)$, we find
$$y(x) = y(c_2) + \int_{c_2}^{x} \frac{g(x,y)}{f(x,y)} dx. \tag{5.5.3}$$

If the integral curve $F(p,t)$ is to leave the region G_p, it is necessary and sufficient that for all $x \geq c_2$ the following relationship be satisfied
$$y(x) - y_1(x) > 0. \tag{5.5.4}$$

From (5.5.3) and (5.5.4) we have that, if the integral curve $F(p,t)$ is to leave the region G_p, it is necessary and sufficient that the x magnitude of $\int_{c_2}^{x} \frac{g(x,y)}{f(x,y)} dx - y_1(x)$ would not be strictly bounded below by the number $-y(c_2)$ for all $x \geq c_2$. In what follows the obtained condition will be denoted by A.

Take the point $q \in G_2$ and construct the region G_q determined by the relationship $y < y_q, x < x_q, (x,y) \in G_2$. By the additional assumption about $f(x,y)$, the point q may be chosen so that the line $y = y_q$ does not intersect the right-hand branch of the curve $f(x,y) = 0$ for all $x \in [0, x_q]$. Consider the integral curve $F(q,t)$. This integral curve stays in the region G_q until leaving it after intersection of the negative semi-axis Oy. In fact, along this integral curve we have $x < x_q, y < y_q$, since $f(x,y) < 0, q(x,y) < 0$ for $(x,y) \in G_2$. Along the integral curve $F(q,t)$ the following relationship is true
$$\frac{dx}{dy} = \frac{f(x,y)}{g(x,y)};$$

integrating it along $F(q,t)$, we get the equality
$$x(y) = x_q + \int_{y_q}^{y} \frac{f(x,y)}{q(x,y)} dy. \tag{5.5.5}$$

From (5.5.5) we find that for the integral curve $F(q,t)$ of system (5.5.1) to leave the region G_q it is necessary and sufficient that the magnitude of
$$\int_{y_q}^{y} \frac{f(x,y)}{q(x,y)} dy$$

would not be strictly bounded below by the number $-x_q$ for $y \leq y_q$. In what follows this condition will be denoted by B.

We now take the point $r \in G_3$ and construct the region G_r determined by the relationships $x < x_r, y > y_r, (x,y) \in G_3$. Construct the integral curve $F(r,t)$. This integral curve stays in the region G_r for $t > 0$ until it intersects the left-hand branch of the line $f(x,y) = 0$. Analysis of the behavior of integral curve $F(r,t)$ in the region G_r is similar to that of the behavior of $F(p,t)$ in the region G_p. Therefore, if in the case $y_r < y_2(x)$ for $x \geq d_2$ the integral curve $F(r,t)$ is to leave the region G_r for $t > 0$, it is necessary and sufficient that the magnitude of $\int_{d_2}^{x} \frac{f(x,y)}{g(x,y)} dx - y_2(x)$ would not be strictly bounded above for $x \leq d_2$ by the quantity $y(d_2)$. In what follows this condition will be denoted by C.

Finally, let us take a point s in the region G_4 and denote by G_s the region determined by the relationships $x > x_s, y > y_s, (x,y) \in G_4$. Under the additional assumption about $f(x,y)$, the point s is said to be chosen so that the straight line $y = y_s$ does not intersect the left-hand branch of the line $f(x,y) = 0$ for $x \in [x_s, 0]$. Construct the integral curve $F(s,t)$. If this integral curve for some $t > 0$ is to leave the region G_s, it is necessary and sufficient that for $y \geq y_s$ the integral $\int_{y_s}^{y} \frac{f(x,y)}{g(x,y)} dy$ would not be strictly bounded above by the number $-x_s$. In what follows this condition will be denoted by D. Note that all the integrals in conditions $A - D$ are evaluated along the integral curves of system (5.5.1) and, therefore, are to be regarded as curvilinear. Direct calculation of these integrals is possible only when the integral curves of system (5.5.1) are known. Therefore, in what follows these integrals will not be calculated directly. However, based on the functional properties of the right-hand sides of system (5.5.1), a number of sufficient and even necessary and sufficient conditions will be found to satisfy conditions $A - D$.

Theorem 91 *If a zero solution to the system (5.5.1) is stable in Lyapunov's sense after replacing the time reference direction t by $-t$ therein, and conditions $A - D$ are satisfied, then any solution to system (5.5.1) intersects indefinitely many times each coordinate semi-axis and each branch of the continuous line $f(x,y) = 0$ as time increases.*

Proof: Let us take the point p on a positive semi-axis Oy and release from it the integral curve $F(p,t), t \geq 0$. Then, by condition A, this integral curve will leave the region $G_p, x > 0, y < y_p, (x,y) \in G_1$. Because of Lyapunov stability of the zero solution of system (5.5.1) after replacing therein the independent variable t by $-t$, there is a number $\delta > 0$ (see section 2.1) such that the circle of radius δ with its center at the origin of coordinates does not contain any points of the integral curve $F(p,t), t \geq 0$. Therefore, this integral curve intersects the semi-axis $Ox, x > 0$ and falls within a region of the form G_q. Indeed, if the opposite is true, then the integral curve $F(p,t)$ stays in the region bounded by the straight lines $y = y_p, y = y_q$ and either $x = c_1$ or $x = \xi$, where ξ is the abscissa of the point of intersection of the integral curve $F(p,t)$ with the plot of the function $y = y_1(x)$. The region, where the integral curve $F(p,t)$ is lying, is bounded on the left by the portions of the Oy axis and the arc of the radius δ circle with its center at the origin of coordinates. In the region constructed there is then the rest point that is either ω-limit for the integral curve $F(p,t)$ or lies within the limit cycle which serves as an ω-limit trajectory for the integral curve $F(p,t)$. Since system (5.5.1) has a unique equilibrium $x = 0, y = 0$, the contradiction obtained shows that the integral curve $F(p,t)$ falls within the region G_q. Under condition B, it then intersects the negative semi-axis Oy, whence it falls within a region of the form G_r. From condition C and the above reasoning, we have by analogy that the integral curve $F(p,t)$ passes from the region G_r to a region of the form G_s, whereupon, based on condition D, it intersects the positive semi-axis Oy for the second time at the point $\bar{p}$. Applying the above reasoning to the point $\bar{p}$, we have that the integral curve $F(p,t)$ intersects each semi-axis of coordinates and each branch of the line $f(x,y) = 0$ infinitely many times as $t \to +\infty$. ∎

Corollary. If the point $\bar{p}$ of the second intersection of the integral curve $F(p,t)$ with the positive semi-axis Oy does not lie above the point p, i.e. if $y_{\bar{p}} \leq y_p$, then system (5.5.1) has a periodic solution provided that the conditions of the theorem are satisfied.

If $y_{\bar{p}} = y_p$, then the integral curve $F(p,t)$ is closed, and hence is a periodic solution to system (5.5.1). Therefore, it remains to discuss only the case where $y_{\bar{p}} < y_p$. Consider the region $\bar{G}$, whose boundary is represented by the arc of the integral curve $F(p,t)$ joining the points p and $\bar{p}$, the portion of the Oy axis joining the points p and $\bar{p}$, and the circle of radius δ with its center at the origin. The integral curve $F(\bar{p},t)$ for $t > 0$ is lying entirely in the region $\bar{G}$, since it cannot intersect the portion $p\bar{p}$, for it is intersected by the integral curves of system (5.5.1) as time increases only if they pass from the left to the right and cannot enter the circle of radius δ because of the assumed stability. The region $\bar{G}$ has no equilibrium, therefore the ω-limit set of the integral curve $F(\bar{p},t)$ is a periodic solution to (5.5.1) (see section 5.4). ∎

The above statements form the basis for finding specific conditions for the right-hand sides of (5.5.1), under which there exist the periodic solutions of system (5.5.1) or the limit cycles of this system. With this in mind, we discuss conditions $A - D$. If the point $p \in G_1$, then the integral curve $F(p,t)$ for $t > 0$ leaves the region G_1 if and only if condition A or the following condition is satisfied

$$\sup_{f(x,y)=0,\ x>0} y = +\infty. \tag{5.5.6}$$

We find some criteria for satisfying of condition A. To this end, consider the function

$$h_1(x,\delta) = \frac{g(x, y_1(x) + \delta)}{f(x, y_1(x) + \delta)}. \tag{5.5.7}$$

The function (5.5.7) given for $x > c_2$ and $\delta > 0$, is continuous and assumes negative values only. Suppose

$$\overline{\lim_{x \to +\infty}} y_1(x) = c > -\infty.$$

Denote by $h_{1\delta}$ the function

$$h_{1\delta} = \sup_{\alpha \in (0, \delta - y(x))} h_1(x, \alpha).$$

If for any $\delta > 0$

$$\int_{c_2}^{x} h_{1\delta}(x)dx \to -\infty \text{ as } x \to +\infty, \tag{5.5.8}$$

then condition A is satisfied. Indeed, if the integral curve $F(p,t)$ of system (5.5.1) satisfies the condition $F(p,0) = p, p \in G_1$, then the inequality $y < y_p$ holds for $t > 0$,

where $y = y(t)$ is the ordinate of the representative point $F(p,t)$ as long as this curve stays in G_1. Finding y as a function of x from system (5.5.1), we have $y(x) > y_p$. Hence, the inequality
$$\frac{g(x, y(x))}{f(x, y(x))} \le h_1 y_p(x)$$
holds. From this we find that the following inequality holds for any $\xi > c_2$ as long as the integral curve is determined and is lying in the region G_1
$$\int_\xi^x \frac{g}{f} dx - y_1(x) \le \int_\xi^x h_1 y_p dx - y_1(x). \tag{5.5.9}$$

Now assume that for some $p \in G_1$ the integral curve $F(p,t)$ stays in G_1 for $t > 0$. Then there is a sequence $x_k \xrightarrow[k \to \infty]{} +\infty$, such that $y_1(x_k) \xrightarrow[k \to \infty]{} c > -\infty$. On such a sequence we have
$$\int_\xi^{x_k} h_1 y_p dx - y_1(x_k) \to -\infty \text{ as } k \to \infty. \tag{5.5.10}$$

From (5.5.9) and (5.5.10) we have that the magnitude
$$\int_\xi^x \frac{g}{f} dx - y_1(x)$$
cannot be bounded below by any finite number. Consequently, condition A is satisfied and, contrary to the above assumption, the integral curve $F(p,t)$ cannot stay in the region G_1 for all $t > 0$. Thus, (5.5.8) entails satisfaction of condition A. If, however, it turns out that $y_1(x) \to -\infty$ as $x \to +\infty$, then condition A is satisfied when for any $\delta > 0$ the following relationship holds
$$\inf_{x > \xi} \int_\xi^x h_{1\delta}(x) dx - y_1(x) = -\infty. \tag{5.5.11}$$

Condition A follows directly from (5.5.11), (5.5.9), and from the above reasoning.

Since condition C is similar to condition A, then the criterion for its fulfillment will be given without proof. Assume that
$$h_3(x, \delta) = \frac{g(x, y_2(x) + \delta)}{f(x, y_2(x) + \delta)}.$$

This function is given for $x < d_2$ and $\delta < 0$ and takes negative values. If
$$\inf_{x < 0, \ f(x,y)=0} y = -\infty, \tag{5.5.12}$$
then all the integral curves starting in the region G_3 leave it as time increases. If condition (5.5.12) is disturbed, then for the integral curves starting in the region G_3 to leave it as time increases, it is necessary and sufficient that condition C is satisfied. Condition C is satisfied if
$$\sup_{x > \xi} \int_\xi^x (h_{3\delta}(x) dx - y_2(x)) = +\infty, \tag{5.5.13}$$

Oscillations in Autonomous System

where $h_{3\delta}(x) = \sup\limits_{\alpha \in (\sigma - y_2(x), 0)} h_3(x, \alpha)$. If $\overline{\lim\limits_{x \to -\infty}} y_2(x) = d < +\infty$, then condition (5.5.13) is satisfied if for any $\delta < 0$,

$$\int_\xi^x h_{3\delta}(x) dx \to +\infty \text{ as } x \to -\infty$$

holds. We now consider criteria for satisfying conditions B and D. Introduce the notation

$$h_{2x_q} = \inf\limits_{x \in (0, x_q)} \frac{f(x, y)}{g(x, y)}. \tag{5.5.14}$$

Function (5.5.14) is defined for all $y < y_q$, where y_q is sufficiently large in absolute value of the negative number. If

$$\int_{y_q}^y h_{2x_q}(y) dy \to -\infty \text{ as } y \to -\infty \text{ for any } x_q > 0,$$

then condition B is satisfied. Indeed, suppose condition B is not satisfied. Then an integral curve $F(q, t)$ for $t \geq 0$ stays in the region G_2. From system (5.5.1) we find that the following relationship holds along this integral curve

$$x(y) \leq x_q,$$

and hence

$$h_{2x_q} \leq \frac{f(x, y)}{g(x, y)}. \tag{5.5.15}$$

From (5.5.15) it follows that the integral $\int_{y_q}^y \frac{f}{g} dy$ is not bounded below by any finite number as $y \to -\infty$. Therefore, the assumption that condition B is not satisfied is not valid. Similarly, set

$$h_{4x_s} = \inf\limits_{x \in (x_s, 0)} \frac{f(x, y)}{g(x, y)}. \tag{5.5.16}$$

Function (5.5.16) is given for $y > y_s$, where y_s is a sufficiently large number, the function being continuous and positive since $s \in G_4$. If

$$\int_{y_s}^y h_{4x_q} dy \to +\infty \text{ as } y \to +\infty \text{ for any } x_q,$$

then condition D is satisfied.

We now derive conditions under which the Corollary and Theorem 91 are satisfied. These conditions will be obtained from estimation of the sign of the number $y_p - y_{\bar{p}}$ by the right-hand sides of system (5.5.1). Let us take arbitrary positive numbers k_1, k_2, l_1, l_2 and construct on the plane (x, y) the rectangle

$$-k_2 \leq x \leq k_1, \quad -l_2 \leq y \leq l_1. \tag{*}$$

Choose a point p_0 on the positive semi-axis Oy with the ordinate y_0, and construct the integral curve $F(p_0, t)$ of system (5.5.1) for $t \geq 0$. Our assumption about this integral curve is that with an increase in t it intersects the straight line $x = k_1$ first at the point q_0, then at the point p_0 and further intersects for the first time the negative semi-axis Oy at the point p_1 with the ordinate y_1. Then it intersects the straight line $x = -k_2$ at the points q_1, r_1 and the positive semi-axis Oy at the point p_2 with the ordinate y_2, then it again intersects the straight line $x = k_1$ at the points q_2, r_2 and, finally, the negative semi-axis Oy at the point p_3 with the ordinate y_3. Assume that the point p_2 is lying above the point p_0, i.e. $y_2 > y_0$ for all sufficiently large y_0. The point p_0 can then be chosen in such a manner that the arc of the integral curve $F(p_0, t)$ joining the points p_0 and p_3 has no points in common with the contour of the rectangle (*). Indeed, if this is not the case, then for each point p_0 with a sufficiently large y_0 it is possible to choose a point on the contour of the rectangle (*) lying simultaneously on the integral curve arc involved. By Bolzano-Cauchy principle, we then find a point s on the contour of the rectangle (*) that is limiting for similar points as $y_0 \to +\infty$. Under conditions of Theorem 91, the integral curve $F(s, t)$ will intersect for the first time the positive axis Oy at the point $\bar{p}$ when $t > 0$. By the integral continuity theorem, we have that the integral curves starting in a sufficiently small neighborhood of the point s also intersect the Oy axis in an arbitrarily small neighborhood of the point $\bar{p}$. This is inconsistent with the fact that in an arbitrarily small neighborhood of the point s there are points represented with the form p_2 with the ordinate $y_2 > y_0$ and are, therefore, lying as far from the point $\bar{p}$ as desired. The obtained contradiction shows that the above assertion is true.

The course of further reasoning will be described in brief. We will impose on the right-hand sides of system (5.5.1) additional condition that the relationship $y_0 < y_2$ or $y_3 < y_1$, what is the same, with all sufficiently large y_0 yields contradiction. In other words, the inverse inequality $y_0 \geq y_2$ holds for at least one y_0, which involves availability of a periodic solution in (5.5.1). Assume that the inequality $y_2 > y_0$ holds for any sufficiently large $y_0 > 0$. Choose the numbers k_1, k_2, l_1, l_2 large enough for the following relationships to be satisfied

$$k_1 > c_2, -k_2 < d_2, f(x, y) > 0 \text{ with } y > l_1,$$

$$f(x, y) < 0 \text{ with } y < -l_2 \text{ for } x \in [-k_2, k_1].$$

This is exactly what constitutes the additional assumption about the function f. We now choose y_0 so large that the arc of the integral curve $F(p_0, t)$ joining the points p_0 and p_3 would have no points common to the rectangle (*). Consider the function

$$V(x, y) = -\int_0^x g(x, 0) dx + \int_0^y f(0, y) dy.$$

Under the conditions imposed on the right-hand side of system (5.5.1), the function V will be position-definite. Introduce three functions

$$W = \frac{dV}{dt} = -f(x,y)g(x,0) + g(x,y)f(0,y), W_1 = \frac{W}{f}, W_2 = \frac{W}{g}.$$

Oscillations in Autonomous System

Calculate the increment of function $V(x,y)$ along the arc of integral curve $F(p_0, t)$ joining the points p_1 and p_3. This increment

$$V(p_3) - V(p_1) = \int_0^{y_3} f(0,y)dy - \int_0^{y_1} f(0,y)dy = \int_{y_1}^{y_3} f(0,y)dy > 0,$$

since the assumption has been made that $y_3 < y_1$ and $f(0,y) < 0$ for $y < 0$. On the other hand, the increment of the function V along the above-mentioned arc can be calculated with the help of contour integrals:

$$V(p_3) - V(p_1) = \oint_{p_1 p_3} \frac{dV}{dt} dt = \oint_{p_1 p_3} W dt$$

$$= \int_{q_1}^{r_1} W dt + \int_{q_2}^{r_2} + \int_{p_1}^{q_1} + \int_{r_1}^{p_2} + \int_{p_2}^{q_2} + \int_{r_2}^{p_3} W dt.$$

Turning to integrals of the other form, we perform substitution $dt = \frac{dy}{g}$ on the arcs of integral curve $F(p_0, t)$ joining the points q_1, r_1, q_2, r_2, and $dt = \frac{dx}{f}$ on the other arcs. To determine the increment of the function V, we obtain

$$V(p_3) - V(p_1) = \int_{q_1}^{r_1} W_2 dy + \int_{q_2}^{r_2} W_2 dy + \int_{p_1}^{q_1} W_1 dx$$

$$+ \int_{r_1}^{p_2} W_1 dx + \int_{p_2}^{q_2} W_1 dx + \int_{r_2}^{p_3} W_1 dx. \tag{5.5.17}$$

We make some assumptions about the right-hand sides of system (5.5.1) that facilitate evaluation of integrals standing on the right-hand side of (5.5.17). The function $W_1(x,y)$ is said to be bounded:

$$|W_1(x,y)| \leq m \text{ for } x \in [-k_2, k_1], \ y \in [-l_2, l_1]. \tag{5.5.18}$$

Assume that the function $W_2(x,y)$ satisfies the inequalities

$$W_2(x,y) \geq m_1(y) \text{ for } x \geq k_1, \tag{5.5.19}$$

$$W_2(x,y) \leq m_2(y) \text{ for } x \leq -k_2. \tag{5.5.20}$$

For the function $f(0,y)$, assume that

$$\int_{y_0}^{y_0+y} f(0,y)dy \to +\infty \text{ as } y \to +\infty \tag{5.5.21}$$

uniformly in $y_0 > 0$ and

$$\int_{y_0}^{y_0+y} f(0,y)dy \to +\infty \text{ as } y \to -\infty \tag{5.5.22}$$

uniformly in $y_0 < 0$. We say that the functions $m_1(y)$ and $m_2(y)$ given for $y \in (-\infty, +\infty)$, can be integrated in each finite integral (a, b) and have the properties

$$\inf_{\substack{b > l_1, \\ a < -l_2}} \int_a^b (m_2(y) - m_1(y))dy = -\infty, \tag{5.5.23}$$

$$\left|\int_a^b m_1(y)dy\right| \leq m(n) < +\infty \tag{5.5.24}$$

for any $a < b$ such that $b - a \leq \bar{n}$ and

$$\int_a^b (-m_2(y))dy < m_1 < +\infty \text{ for } b > a. \tag{5.5.25}$$

Conditions (5.5.18) — (5.5.25) impose constraints on the right-hand sides of system (5.5.1). In what follows it will be shown that in the general case these constraints are not contradictory.

We now establish that system (5.5.1) has a periodic solution when conditions of Theorem 91 and relationships (5.5.18) — (5.5.25) are satisfied. It will be shown that the last four integrals on the right-hand side of (5.5.17) are bounded in absolute value by a constant L, whatever the choice of a point p_0 might be. This property of integrals will be established by reference to one of them. Evaluate the integral in the arc p_2q_2:

$$\left|\int_{p_2q_2} W_1(x,y)dx\right| \leq \int_0^{k_1} |W_1(x,y)|dx \leq mk_1.$$

Set $L = 2m(k_1 + k_2)$, then

$$\left|\int_{p_1}^{q_1} W_1 dx + \int_{r_1}^{p_2} W_1 dx + \int_{p_2}^{q_2} W_1 dx + \int_{r_2}^{p_3} W_1 dx\right| \leq L. \tag{5.5.26}$$

We shall show that variations in the ordinate along any arc of the integral curve $F(p_0, t)$ lying in any strip $x \in [-k_2, 0]$ or $x \in [0, k_1]$, are bounded by the value that is independent of the position of the point p_0. To do this, consider the increment of the function $V(x, y)$ along the arc p_2q_2 which is defined by

$$V(q_2) - V(p_2) = \int_0^{y_{q_2}} f(0,y)dy - \int_0^{k_1} q(x,0)dx - \int_0^{y_2} f(0,y)dy;$$

hence we find

$$\left|\int_{y_{q_2}}^{y_2} f(0,y)dy\right| \leq |V(q_2) - V(p_2)| + \left|\int_0^{k_1} g(x,0)dx\right|$$

$$\leq mk_1 + k_1 \sup_{x \in [0,k_1]} |g(x,0)|. \tag{5.5.27}$$

If the magnitude of $y_2 - y_{q_2}$ is not bounded above, then the left-hand side of relationship (5.5.27) is unbounded because of (5.5.21) while the right-hand side is a constant that is independent of the choice of point p_0.

We now evaluate the first two integrals on the right-hand side of (5.5.17):

$$\int_{q_1}^{r_1} W_2 dy = \int_{y_{q_1}}^{y_{r_1}} W_2(x,y)dy \leq \int_{y_{q_1}}^{y_{r_1}} m_2(y)dy$$

$$= \int_{y_3}^{y_2} m_2(y)dy - \int_{y_3}^{y_1} m_2(y)dy - \int_{y_1}^{y_{q_1}} m_2(y)dy - \int_{y_{r_1}}^{y_2} m_2(y)dy; \tag{5.5.28}$$

$$\int_{q_2}^{r_2} W_2(x,y)dy = -\int_{y_{r_2}}^{y_{q_2}} W_1(x,y)dy \leq -\int_{y_3}^{y_2} m_1(y)dy$$
$$+ \int_{y_3}^{y_{r_2}} m_1(y)dy + \int_{y_{q_2}}^{y_2} m_1(y)dy. \quad (5.5.29)$$

From (5.5.28), (5.5.29), and (5.5.17) we find

$$\int_{y_1}^{y_3} f(0,y)dy \leq \int_{y_3}^{y_2} (m_2(y) - m_1(y))dy + L + \int_{y_3}^{y_1} m_2(y)dy$$
$$- \int_{y_1}^{y_{q_1}} m_2(y)dy - \int_{y_{r_1}}^{y_2} m_2(y)dy + \int_{y_3}^{y_{r_2}} m_1(y)dy + \int_{y_{q_2}}^{y_2} m_1(y)dy. \quad (5.5.30)$$

The last five integrals on the right-hand side of (5.5.30) are bounded above by the number L_1 that is independent of their limits. Thus, from (5.5.30) we have

$$\int_{y_1}^{y_3} f(0,y)dy \leq \int_{y_3}^{y_2} (m_2(y) - m_1(y))dy + L + L_1. \quad (5.5.31)$$

Since the right-hand side of (5.5.31) is not bounded below as $y_2 \to +\infty$, then the assumption that the inequality $y_0 < y_2$ holds for all sufficiently large y_0 contradicts the condition

$$\int_{y_1}^{y_3} f(0,y)dy > 0.$$

Hence, (5.5.31) implies the existence of a number $y_0 > 0$ for which the inequality $y_2 \leq y_0$ holds, i.e. the point p_2 does not lie above the point p_0 on the positive semi-axis Oy. From Corollary to Theorem 91 we then conclude that system (5.5.1) has a periodic solution.

Note that an increment of the function V can be computed along the arc of integral curve $F(p_0, t)$ joining the points p_0 and p_2. The above statement then holds where conditions (5.5.24) and (5.5.25) are replaced by the conditions

$$-\int_a^b m_2(y)dy < m(n) < +\infty \text{ for } a < b \text{ and } b - a < n, \quad (5.5.32)$$

$$\int_a^b m_1(y)dy < m_1 < +\infty \text{ for } a < b. \quad (5.5.33)$$

Remark. The function V forms the basis of the above existence criterion for a periodic solution and can be replaced by any other function $V_1(x,y)$ having the properties as follows.

The function $V_1(x,y)$ is given on the plane (x,y), is real and continuously differentiable,

$$V_1(y + y_0) - V_1(y_0) \to +\infty \text{ as } y \to +\infty \quad (5.5.34)$$

uniformly in $y_0 > 0$,

$$V_1(y + y_0) - V_1(y_0) \to +\infty \text{ for } y \to -\infty \quad (5.5.35)$$

uniformly in $y_0 < 0$, and the quantities on the left-hand side of relationships (5.5.34) and (5.5.35) are positive for $y > y_0$, and $y < y_0$, respectively. The functions W_1 and W_2 are defined by V_1 in the following way:

$$W_1 = \frac{W}{f(x,y)}, \qquad W_2 = \frac{W}{g(x,y)},$$

where $W = \frac{dV_1}{dt}$; the functions m_1 and m_2 used for estimation of the functions W_1 and W_2 are given with regard to the fact that the integral curve $F(p_0, t)$ joining the points p_0 and p_3 stays in some bounded strip $x \in [a, b]$ which can be computed from the assignment of the point p_0 and conditions $A - D$. In other words, the arc of integral curve $F(p_0, t)$ joining the points p_0 and p_1 cannot lie beyond the strip $x \in [0, \xi]$, where ξ is the least root of the equation

$$\int_0^x \frac{f}{g} dx - y_1(x) = -y_0, \quad x > k_1.$$

It is also possible to select an estimate for the ordinate of the point p_1 and, with the help of this estimate, a number η such that the arc of integral curve $F(p_0, t)$ joining the points p_1 and p_2 does not leave the strip $x \in [\eta, 0]$. These considerations allow one to extend the criterion proposed here.

2. We now consider the uniqueness of a periodic solution to system (5.5.1) and derive conditions for its stability. First clarify the qualitative reasoning that is central to analytic criteria. System (5.5.1) has a unique stable limit cycle if and only if its every periodic solution is a stable limit cycle. Assume that this statement is not true. Then there are two stable limit cycles. The result of the presence of a unique equilibrium position is that one of the limit cycle is embedded into the other. The annular region confined to these limit cycles contains another periodic solution to system (5.5.1), since any integral curve $F(p, t)$ starting in this region stays there by the uniqueness for $t \in (-\infty, +\infty)$. As $t \to +\infty$, this integral curve may approach without bound one of the limit cycles which constitute the boundary of the annular region, and if $t \to -\infty$, then $F(p, t)$ cannot approach these limit cycles by the assumption of their stability. Hence, there exists the α-limit set of the trajectory $F(p, t)$ that is a periodic solution to system (5.5.1) since in the annular region there is no equilibrium for this system. The obtained periodic solution cannot be a stable limit cycle, since $F(p, t)$ approaches it without bound as $t \to -\infty$. The presence of such a periodic solution contradicts the assumption that every periodic solution to system (5.5.1) is a stable limit cycle.

The above reasoning implies some indications sufficient for every periodic solution of system (5.5.1) to be a stable limit cycle. Indeed, let us take on the semi-axis $y > 0$ two points p_0 and $\bar{p}_0$ related by $y_0 < \bar{y}_0$, one of these lying on the periodic solution. We shall construct the points p_2 and $\bar{p}_2$ that are the points of the first intersection of the integral curves $F(p_0, t), F(\bar{p}_0, t)$ and the positive semi-axis Oy for $t > 0$. Denote by y_2 and $\bar{y}_2$ the ordinates of these points. If the relationship

$$\bar{y}_2 - \bar{y}_0 < y_2 - y_0, \tag{5.5.36}$$

holds, then system (5.5.1) has a unique limit cycle. Indeed, if $y_2 = y_0$, then all integral curves $F(\bar{p}_0, t)$ for $\bar{y}_0 > y_0$ approach without bound the limit cycle of $F(p_0, t)$ as $t \to +\infty$, for there is the relationship $\bar{y}_2 = \bar{y}_0$, then from the above relationship it follows that all the integral curves $F(p_0, t)$ starting within the limit cycle approach it without bound as $t \to +\infty$, because the inequality $y_2 > y_0$ holds for $y_0 < \bar{y}_0$. By employing the sufficient test (5.5.36), we shall derive this relationship from the form of right-hand sides of (5.5.1), which in turn must be based on exact estimations.

We indicate another sufficient test for the existence of a unique limit cycle for symmetric systems. Assume that the right-hand sides of system (5.5.1) satisfy the additional conditions

$$f(-x, -y) = -f(x, y), \tag{5.5.37}$$

$$g(-x, -y) = -g(x, y). \tag{5.5.38}$$

If $F(q_0, t)$ is the integral curve of (5.5.1) satisfying (5.5.37), (5.5.38), then the following relationship holds

$$F(-q_0, t) = -F(q_0, t). \tag{5.5.39}$$

If $F(q_0, t)$ is the integral curve of (5.5.1), then, by (5.5.37) and (5.5.38), $-F(q_0, t)$ is also an integral curve of this system. The relationship (5.5.39) holds for $t = 0$, and by uniqueness, it is preserved for all t values while the integral curves are determined. Thus, when conditions (5.5.37), (5.5.38) are satisfied, the integral curves of system (5.5.1) starting for $t = 0$ at the points located symmetrically about the origin of coordinates, are arranged symmetrically about the same symmetry center during the motion. If, under these conditions, system (5.5.1) has a periodic solution passing through the point p_0, then $y_1 = -y_0$ and y_1 is the ordinate of the point, where this periodic solution intersects the negative semi-axis Oy. For symmetric systems, the criterion (5.5.36) can be represented in a slightly modified form. That is, for system (5.5.1) to have a unique stable limit cycle under conditions (5.5.37) and (5.5.38), it is sufficient that

$$\bar{y}_1 + \bar{y}_0 > y_1 + y_0, \tag{5.5.40}$$

where $\bar{y}_0 > y_0 > 0, \bar{y}_1 < y_1 < 0$. Recall that y_1 and $\bar{y}_1$ stand for the first points, where the respective integral curves $F(p_0, t)$ and $F(\bar{p}_0, t)$ intersect the negative semi-axis Oy for $t > 0$. Let $y_0 + y_1 = 0$. This means that the integral curve $F(p_0, t)$ is a periodic solution. It follows from the relationship (5.5.40) that $\bar{y}_1 > -\bar{y}_0$. Let us construct the arc of the integral curve $F(-\bar{p}_0, t)$ joining the points $-\bar{p}_0$ and $-\bar{p}_1$. In its passage through the point $\bar{p}_1$, the integral curve $F(\bar{p}_0, t)$ falls within the region bounded by the arcs of integral curves $F(\bar{p}_0, t)$ and $F(-\bar{p}_0, t)$ joining respectively the points $\bar{p}_0, \bar{p}_1$ and $-\bar{p}_0, -\bar{p}_1$, and by the portions of the ordinate axis $-\bar{p}_1\bar{p}_0$ and $-\bar{p}_0\bar{p}_1$. The integral curve involved cannot leave this region when time increases without bound. After $\bar{p}_0$, it will intersect for the first time the positive semi-axis Oy at the point $\bar{p}_2$ located below $-\bar{p}_1$, and hence below the point $\bar{p}_0$. In this case, the relationship $\bar{y}_0 > \bar{y}_2$ holds, whence it follows that the integral curves starting at $\bar{y}_0 > \bar{y}$ approach asymptotically a periodic solution $F(p_0, t)$ from outside. Likewise, if the condition (5.5.40) is satisfied, then all the integral curves of (5.5.1) starting in the region bounded by a periodic solution $F(p_0, t), 0 < \bar{y}_0 < y_0$ approach asymptotically this periodic solution from the inside of the above-mentioned region as $t \to +\infty$. Thus, if relationship (5.5.40) holds for system (5.5.1) subject to (5.5.37) and (5.5.38), then every periodic solution of this system is a stable limit cycle.

To find analytic criteria for the relationship (5.5.40) to be satisfied, introduce a function $V(x, y)$ such that $V(x, y)$ is given for $x \in (-\infty, +\infty), y \in (-\infty, +\infty)$, is real and continuously differentiable, and satisfies the equality

$$V(-x, -y) = V(x, y). \tag{5.5.41}$$

Calculate the total derivative of function V by system (5.5.1):

$$\frac{dV}{dt} = \frac{\partial V}{\partial x} f(x, y) + \frac{\partial V}{\partial y} g(x, y) = W(x, y). \tag{5.5.42}$$

The function $W(x, y)$ has the following properties. The equation $W(x, y) = 0$ defines two functions:

$$x = \varphi(y) \text{ and } x = -\varphi(y), \tag{5.5.43}$$

where $\varphi(y)$ is a continuous real function given for $y \in (-\infty, +\infty)$ whose plot $x = \varphi(y)$ on the plane xOy intersects the curve $y = y_1(x)$ once at the point $(x_0, y_0), x_0 > 0$. The function $\varphi(y)$ for $y \geq y_0$ decreases monotonically but remains positive, and for $y < y_0$ increases monotonically and also remains positive. The function $W(x, y)$ is non-negative for $x \in (-\varphi(y), \varphi(y))$ and does not vanish in the region $-\varphi(y) < x < \varphi(y)$ on any closed contour embracing the origin of coordinates and located in this region. The function

$$V(0, y) > 0 \text{ for } y > 0$$

and strictly monotonically increases; the function

$$W_1(x, y) = \frac{W(x, y)}{f(x, y)} \text{ for } y > y_1(x), \ 0 < x < \varphi(y),$$

is positive and decreases for any fixed x. For $y < y_1(x), 0 < x < \varphi(y)$ the function $W_1(x, y) < 0$ decreases if x is fixed. The function $W_2(x, y) = \dfrac{W(x, y)}{g(x, y)}$ increases with fixed y for $x \geq \varphi(y)$ and is non-negative in the region $x > \varphi(y)$.

Note that if system (5.5.1) under conditions (5.5.37), (5.5.38) has a periodic solution and there are the functions V and W possessing the above properties, then this periodic solution cannot lie entirely in the region $-\varphi(y) < x < \varphi(y)$, for otherwise the increment of the function V along this periodic solution will be nonzero. Consequently, the periodic solution (which is assumed to exist) $F(p_0, t)$ will intersect the plot of the curve $x = \varphi(y)$. We assume that all the integral curves intersect the curve $x = \varphi(y)$ in one direction. Suppose the integral curve $F(p_0, t)$ intersects the curve $x = \varphi(y)$ with $t > 0$ for the first time at the point q_1, for the second time at the point r_1, and then the negative semi-axis Oy at the point p_1. Similarly, suppose the integral curve $F(\bar{p}_0, t), \bar{y}_0 > y_0$ intersects the curve $x = \varphi(y)$ at the points q_2, r_2 and then the negative semi-axis Oy at the point $\bar{p}_1$; then we have

$$\int_{\bar{p}_0}^{\bar{p}_1} \frac{dV}{dt} dt = \int_{\bar{p}_0}^{\bar{p}_1} W\, dt = \int_{\bar{p}_0}^{q_2} W_1 dx + \int_{q_2}^{r_2} W_2 dy + \int_{r_2}^{\bar{p}_1} W_1 dx$$

$$< \int_{p_0}^{q_1} W_1 dx + \int_{q_1}^{r_1} W_2 dy + \int_{r_1}^{p_1} W_1 dx = \int_{p_0}^{p_1} \frac{dV}{dt} dt. \qquad (5.5.44)$$

From relationship (5.5.44) we find that

$$V(0, \bar{y}_1) - V(0, -\bar{y}_0) < V(0, y_1) - V(0, y_0),$$

or

$$V(0, \bar{y}_1) - V(0, -\bar{y}_0) < V(0, y_1) - V(0, -y_0).$$

From the strict monotonicity of the function $V(0, y)$ it follows that $\bar{y}_1 + \bar{y}_0 > y_1 + y_0$, i.e. relationship (5.5.40) is satisfied and therefore system (5.5.1) under conditions (5.5.37) and (5.5.38) has a unique stable limit cycle, if there are functions V and W possessing the above properties.

To complete the above reasoning, we will clarify the relationship (5.5.44). The quantity $\int_{\bar{p}_0}^{q_2} W_1 dx$ may be represented as $\int_0^{x_{q_2}} W_1 dx$, and $\int_{p_0}^{q_1} W_1 dx$ as $\int_0^{x_{q_1}} W_1 dx$. Since $x_{q_1} \geq x_{q_2}$ and, as y increases, the function W_1 decreases but remains positive, we have

$$\int_0^{x_{q_2}} W_1 dx \leq \int_0^{x_{q_1}} W_1 dx, \quad \int_{q_2}^{r_2} W_2 dy = \int_{q_2}^{\bar{q}_1} + \int_{\bar{q}_1}^{\bar{r}_1} + \int_{\bar{r}_1}^{r_2} W_2 dy,$$

where $\bar{r}_1$ and $\bar{q}_1$ are the points on the arc of the integral curve $\bar{p}_0\bar{p}_1$ having the same ordinates as the points q_1 and r_1; then

$$\int_{q_2}^{\bar{q}_1} W_2 dy < 0 \text{ and } \int_{\bar{p}_1}^{r_2} W_2 dy < 0.$$

As for the remaining integral, it can be evaluated by analogy with the first formula:

$$\int_{r_2}^{\bar{p}_1} W_1 dx < \int_{r_1}^{p_1} W_1 dx.$$

The above-mentioned qualitative and analytic criteria (or tests) for the existence of periodic solutions and the criteria for their stability underlie the principle which allows one to formulate some theorems for a two-equation system, or for one equation, describing the oscillations of an autonomous system with one degree of freedom that contain specific conditions for availability of periodic modes and their stability. This condition and the efficiency of the present reasoning may be illustrated by referring to some examples.

Example 1. Consider a two-equation system

$$\dot{x} = \sigma(y) - f(x), \qquad (5.5.45)$$

$$\dot{y} = -g(x).$$

Assume that the right-hand sides of system (5.5.45) satisfy the following conditions:

1) the functions σ, f and g are given and continuous in all finite values of their arguments and satisfy the Lipschitz condition in any finite region of their arguments;

2) $g = f = \sigma = 0$ when $x = y = 0$, here $g = 0$ only when $x = 0$, and $\sigma = 0$ only when $y = 0$;

3) $xg(x) > 0$ when $x \neq 0$, and $y\sigma(y) > 0$ when $y \neq 0$;

4) $\sigma(y)$ increases strict-monotonically from $-\infty$ to $+\infty$ as y varies from $-\infty$ to $+\infty$;

5) $xf(x) < 0$ when $x \neq 0$, where $|x|$ is sufficiently small.

Consider the function

$$V(x,y) = \int_0^y \sigma(y)dy + \int_0^x g(x)dx.$$

We will find the total derivative of this function in terms of system (5.5.45):

$$\frac{dV}{dt} = W = -g(x)f(x). \qquad (5.5.46)$$

The function V is positive-definite. By (5.5.45), its total derivative is non-negative in a sufficiently small neighborhood of the origin of coordinates. Therefore, the zero solution of system (5.5.45) is Lyapunov stable as $t \to -\infty$. We shall find conditions for the functions f, g, σ, under which all the integral curves of system (5.5.45) make spiral motions about an equilibrium as $t \to +\infty$. From Theorem 91 it follows that this occurs every time when conditions $A - D$ are satisfied. Condition A is satisfied if

$$\sup_{x>0} f(x) = +\infty,$$

and if there exists a finite limit

$$\overline{\lim_{x \to +\infty}} y_1(x) = c$$

and when

$$\int_0^{+\infty} \frac{-g(x)}{\sigma(\delta) - f(x)} dx = -\infty$$

for any choice of the constant $\delta > y_1(x)$, where y_1 is a solution to the equation $\sigma(y) = f(x)$ for $x > 0$. The function $\sigma(y) - f(x)$ may be represented as $\sigma(y_1 + \delta) - \sigma(y_1)$ and therefore it increases both for fixed x and for $y > y_1$. Hence the function $-\frac{g(x)}{\sigma(y) - f(x)}$ also increases for $y > y_1$. This suggests that condition A is satisfied.

Condition A is also satisfied if

$$\inf_{x>0} \left(\int_0^x \frac{-g(x)}{\sigma(\delta) - f(x)} dx - y_1(x) \right) = -\infty.$$

If $\inf_{x \leq 0} f(x) = -\infty$ or $\inf_{x \leq 0} \left(\int_0^x \frac{-g(x)}{\sigma(-\delta) - f(x)} dx - y_2(x) \right) = +\infty$ for any choice of a positive $\delta > -y_2$, then condition C is satisfied. We will show that, if conditions 1) — 5) hold for the right-hand side of system (5.5.45), conditions B and D are satisfied without further assumptions. Consider the strip

$$0 \leq x \leq x_q, -\infty < y \leq y_q,$$

Oscillations in Autonomous System

containing no points of the curve $y = y_1(x)$:

$$\int_{y_q}^{y} \frac{\sigma(y) - f(x)}{-g(x)} dy = \int_{y_q}^{y} \frac{-\sigma(y)}{g(x)} dy + \int_{y_q}^{y} \frac{f(x)}{g(x)} dy$$

$$\leq \frac{(-y + y_q)\sigma(y_q)}{\sup_{x \in [0, x_q]} g(x)} + (y - y_q) \inf_{x \in [0, x_q]} \frac{f(x)}{g(x)}. \tag{5.5.47}$$

The right-hand side of inequality (5.5.47) tends to $-\infty$ as $y \to -\infty$ if y_q is taken to be a negative number large enough in absolute value, and x_q is fixed, since $\sigma(y_q) \to -\infty$ as $y_q \to -\infty$. Fulfillment of condition D is demonstrated in the same manner.

Introduce the functions

$$W_1(x, y) = \frac{-g(x)f(x)}{\sigma(y) - f(x)} \text{ and } W_2(x, y) = f(x).$$

The function W_1 is bounded in any vertical strip $-k_2 \leq x \leq k_1$ for $y \in [-l_2, l_1]$, where k_1 and k_2 are any positive numbers, and l_1 and l_2 are sufficiently large positive numbers that are generally dependent on k_1 and k_2. Note that

$$\int_{y_0}^{y_0+y} \sigma(y) dy \to +\infty \text{ as } y \to +\infty$$

uniformly in $y_0 > 0$ and

$$\int_{y_0}^{y_0+y} \sigma(y) dy \to +\infty \text{ as } y \to -\infty$$

uniformly in $y_0 < 0$. The function $W_2(x, y)$ is not explicitly dependent on y, therefore the functions $m_1(y)$ and $m_2(y)$ are constants. Thus, system (5.5.45) has at least one periodic solution providing that for the right-hand sides of this system:

a) conditions 1) — 5) are satisfied;

b) $\sup_{x \geq 0} f(x) = +\infty$ or $\sup_{x \leq 0} \left(\int_0^x \frac{-g(x)}{\sigma(\delta) - f(x)} dx - y_1(x) \right) = -\infty$ for any $\delta > y_1(x)$;

c) $\inf_{x \leq 0} f(x) = -\infty$ or $\sup_{x \leq 0} \left(\int_0^x \frac{-g(x)}{\sigma(-\delta) - f(x)} dx - y_2(x) \right) = +\infty$ for any $\delta > -y_2(x)$,

where $y_1(x), y_2(x)$ are solutions to the equations $\sigma(y) = f(x)$ for $x \geq 0$ and $x < 0$, respectively;

d) there are two numbers $k_1 > 0$ and $k_2 > 0$ such that $f(x) \geq m_1$ for $x \geq k_1$, $f(x) \leq m_2$ for $x \leq -k_2$, where $m_1 > m_2$ are constants, $m_2 < 0$ or $m_1 > 0$.

Introduce sufficient conditions for the obtained periodic solution of system (5.5.45) to be a unique stable limit cycle. Suppose $f(x), g(x), \sigma(y)$ are odd functions and the function $f(x)$ vanishes at $x = x_0$, so that $f(x) < 0$ for $0 < x < x_0$, and for $x > x_0$ the function $f(x)$ increases monotonically. Using the equation $W(x, y) = 0$, the function $W(x, y)$ then determines two vertical straight lines $x = \pm x_0$ which can be taken as the functions $\varphi(y)$. Therefore, if in addition to conditions a) — d) the following conditions are satisfied

e) $f(-x) = -f(x)$, $g(-x) = -g(x)$, $\sigma(-y) = -\sigma(y)$;
f) $f(x_0) = 0$; $f(x) < 0$ for $0 < x < x_0$ and $f(x)$ increases monotonically for $x \geq x_0$, then system (5.5.45) has a unique stable limit cycle.

Note that system (5.5.45) incorporates all possible cases of oscillations of the autonomous system with one degree of freedom if friction is the function of the rate of motion only, and the restoration power is the function of motion. Indeed, if in system (5.5.45) the function $g(x)$ is linear, e.g., $g(x) = x$, then $\ddot{y} = \dot{x}$, and from (5.5.45) we have

$$\ddot{y} - f(-\dot{y}) + \sigma(y) = 0. \tag{5.5.48}$$

It is well known that Van der Pol's equation is a special case of (5.5.48), hence this example may be illustrated by directly referring to it.

Example 2. Consider a system of equations

$$\dot{x} = \sigma(y) - h_1(y)f(x), \tag{5.5.49}$$

$$\dot{y} = -g(x).$$

Assume that the functions $f(x), g(x)$ and $\sigma(y)$ have the same properties as given in Example 1. The function $h_1(y)$ is given for $y \in (-\infty, +\infty)$, and is positive and continuous, and satisfies the Lipschitz condition on every finite interval. The monotonicity condition for the function $\sigma(y)$ is replaced by the strict monotonicity condition for the function $\frac{\sigma(y)}{h_1(y)}$. This function is assumed to vary monotonically from $-\infty$ to $+\infty$ as y varies from $-\infty$ to $+\infty$. Take the function $V(x, y)$ to be

$$V(x, y) = \int_0^y \sigma(y) dy + \int_0^x g(x) dx.$$

Then, by system (5.5.49), the total derivative of the function V is

$$\frac{dV}{dt} = W(x, y) = -g(x)f(x)h_1(y).$$

Hence, the zero solution of (5.5.49) is Lyapunov stable as $t \to +\infty$.

Let $h_1(y)$ be a strictly monotonic function. We will impose such conditions on the right-hand sides of system (5.5.49) so that conditions $A - D$ are satisfied. Condition A is satisfied if

$$\sup_{x \geq 0} f(x) = +\infty \tag{5.5.50}$$

or

$$\inf_{x \geq 0} \left(\int_0^x \frac{-g(x)}{\sigma(\delta) - f(x)h_1(\delta)} dx - y_1(x) \right) = -\infty \tag{5.5.51}$$

for any $\delta > y_1(x)$. Condition C is satisfied when

$$\inf_{x \leq 0} f(x) = -\infty, \tag{5.5.52}$$

or

$$\sup_{x \leq 0} \left(\int_0^x \frac{-g(x)}{\sigma(-\delta) - f(x)h_1(-\delta)} dx - y_2(x) \right) = +\infty \tag{5.5.53}$$

Oscillations in Autonomous System

for any $\delta > -y_2$, where $y_1(x)$ is a solution to the equation $\sigma(y) = h_1(y)f(x)$ for $x \geq 0$, and y_2 is a solution to the same equation for $x < 0$.

If the function $f(x)$ under the integral sign in (5.5.51) is replaced by $\dfrac{\sigma(y_1)}{h_1(y_1)}$, then we get

$$\sigma(y) - f(x)h_1(y) = h(y)\left(\frac{\sigma(y)}{h_1(y)} - \frac{\sigma(y_1)}{h_1(y_1)}\right). \qquad (5.5.54)$$

The right-hand side of (5.5.54) with a fixed x is a strictly increasing function for $y > y_1$. Consequently, the same is true for the integrand in formula (5.5.51), and condition A in the case of (5.5.51) is satisfied. The same is with condition C. We will show that conditions B and D are also satisfied. Let us take a half-strip $0 \leq x \leq x_q$, $-\infty < y \leq y_q$, such that there are no points of the curve $y = y_1(x)$ therein. Then we get the relationships

$$\int_{y_q}^{y}\frac{\sigma(y) - f(x)h_1(y)}{-g(x)}dy = \int_{y_q}^{y}\frac{\sigma(y)}{-g(x)}dy + \int_{y_q}^{y}h_1(y)\frac{f(x)}{g(x)}dy$$

$$\leq a\int_{y_q}^{y}-\sigma(y)dy + b\int_{y_q}^{y}h_1(y)dy,$$

where

$$a = \inf_{0 \leq x \leq x_q}\frac{1}{g(x)}, \quad b = \inf_{0 \leq x \leq x_q}\frac{f(x)}{g(x)}.$$

Evaluate the expression on the right-hand side of the last inequality:

$$a\int_{y_q}^{y}-\sigma(y)dy + b\int_{y_q}^{y}h_1(y)dy = a\int_{y_q}^{y}h_1(y)\left(\frac{-\sigma(y)}{h(y)} + \frac{b}{a}\right)dy.$$

Since the function $\dfrac{\sigma_1(y)}{h_1(y)}$ tends monotonically to $-\infty$ as $y \to -\infty$, then y_q may be chosen to be a negative number so large in absolute value that

$$\frac{-\sigma(y)}{h_1(y)} + \frac{b}{a} > 0 \text{ for } y < y_q.$$

Hence, for such y_q we get

$$\int_{y_q}^{y}\frac{\sigma(y) - f(x)h_1(y)}{-g(x)}dx \to -\infty \text{ as } y \to -\infty,$$

which involves fulfillment of condition B. Fulfillment of condition D is established in the same way. Thus, the following assertion holds: if the conditions imposed on the functions f, g, σ, h_1 are satisfied and the relationships (5.5.50), (5.5.51), (5.5.52) or (5.5.53) are realized, then all the integral curves of system (5.5.49) spiral about an equilibrium as $t \to +\infty$.

We next set

$$W_1 = -\frac{f(x)g(x)h_1(y)}{\sigma(y) - f(x)h_1(y)}, \quad W_2 = f(x)h_1(y).$$

The function $W_1(x,y)$ is bounded in any region of the form $-k_2 \leq x \leq k_1, y \in [-l_2, l_1]$, where k_1, k_2 are arbitrary positive numbers, l_1 and l_2 are sufficiently large positive numbers that are dependent on k_1 and k_2. The function $W_2(x,y)$ may be estimated in terms of the functions $m_1(y)$ and $m_2(y)$. Assume that there are two positive constants k_1 and k_2 such that

$$f(x) > m_1 \text{ for } x \geq k_1, \qquad (5.5.55)$$

$$f(x) \leq m_2 \text{ for } x \leq k_2, \qquad (5.5.56)$$

where $m_2 < m_1, m_2 < 0$, or $m_1 > 0$. If the integral $\int_a^b h_1(y) dy$ is uniformly bounded above for $a < b, b - a < N < +\infty$ for any N, then system (5.5.49) has a periodic solution and spiral arrangement of integral curves provided that conditions (5.5.55), (5.5.56) are satisfied.

System (5.5.49) incorporates, in particular, some cases of oscillations in autonomous systems with one degree of freedom, where the frictional force may depend not only on the rate of motion, but also on the motion itself, and the restoration force depends on the motion only. For $g(x) = x$, system (5.5.49) is equivalent to an equation of the form $\ddot{y} - f(-\dot{y}) + h_1(y) + \sigma(y) = 0$. By analogy with system (5.5.49) we may consider a system of the form

$$\dot{x} = \sigma(y)k(x) - h_1(y)f(x), \qquad (5.5.57)$$

$$\dot{y} = -h_2(y)g(x).$$

In the case of system (5.5.57), we take function $V(x,y)$ to be

$$V(x,y) = -\int_0^y \frac{\sigma(y)}{h_2(y)} dy + \int_0^x \frac{g(x)}{k(x)} dx.$$

Then its total derivative is equal to

$$W(x,y) = -\frac{g(x)f(x)h_1(y)}{k(x)}$$

by (5.5.57). To be noted here is that the system (5.5.1) discussed at the beginning of this section, incorporates all autonomous systems with one degree of freedom. Indeed, oscillations in the autonomous system with one degree of freedom are described by the equation

$$\ddot{y} = f(y, \dot{y}). \qquad (5.5.58)$$

If $\dot{y} = -x$, then we have the system

$$\dot{x} = -f(y, -x) = \bar{f}(x,y), \qquad (5.5.59)$$

$$\dot{y} = -x.$$

System (5.5.59) is derived from (5.5.1) for $g(x,y) = -x$.

Oscillations in Autonomous System

Example 3. (relaxation oscillations) Suppose there is a system

$$\frac{\dot{x}}{\lambda} = \sigma(y) - f(x),$$

$$\dot{y} = -g(x), \qquad (5.5.60)$$

where $\lambda > 0$ is a parameter. We will focus on the development of oscillations in system (5.5.60) versus the parameter λ variations and, especially, as $\lambda \to +\infty$.

Assume that the right-hand side of system (5.5.60) have the same properties as those of system (5.5.45) and satisfy conditions a) — f) (see Example 1). For any $\lambda > 0$, system (5.5.60) then has a unique stable limit cycle. Take $V(x,y)$, to be the function

$$V(x,y) = \int_0^y \sigma(y) dy + \int_0^x \frac{g(x)}{\lambda} dx,$$

then

$$W(x,y) = \frac{dV}{dt} = -g(x)f(x), \quad W_1(x,y) = -\frac{g(x)f(x)}{\lambda(\sigma(y) - f(x))}, \quad W_2(x,y) = f(x)$$

and the parameter λ does not disturb the basic properties of functions employed in Example 1. Hence, for any $\lambda > 0$, system (5.5.60) has a unique stable limit cycle. Here a special feature is that the function W vanishes at $x = x_0$, where x_0 is independent of λ, the limit cycle for any λ comprises the point $x = y = 0$ and intersects both straight lines $x = \pm x_0$ for any $\lambda > 0$, and the direction of the field $\frac{dy}{dx}$ on the curve $\sigma(y) = f(x)$ is vertical. In any other point (x,y) of the plane, the field direction is arbitrarily close to horizontal for sufficiently large λ. Thus, the path of integral curves of system (5.5.60) for sufficiently large λ is determined within some limits by the equation $\frac{dy}{dx}$. On the curve

$$\sigma(y) = f(x) \qquad (5.5.61)$$

we find $\max y$ in the interval $[-x_0, 0]$. Suppose it is realized at the point $-x_1$ and equals η. The horizontal line (parallel to the Ox axis is drawn from the point $(-x_1, \eta)$ until it intersects the plot of the curve (5.5.61) at the point (x_2, ξ). Another line is drawn (parallel to the first) through the point $(x_1, -\eta)$ until it intersects the plot of the curve (5.5.61) at the point $(-x_2, -\xi)$. Let us construct a closed contour L composed of the line segments joining the points $(-x_1, \eta), (x_2, \xi), (x_1, -\eta)$ and $(-x_2, -\xi)$, and made up of two arcs of the curve (5.5.61) joining the points $(-x_2, -\xi), (-x_2, \eta)$ and $(x_1, -\eta), (x_2, \xi)$, respectively. The contour L is limit for the cycles of system (5.5.60) as $\lambda \to +\infty$. In other words, for any $\delta > 0$, it is always possible to select a number λ_δ such that for $\lambda > \lambda_\delta$ the limit cycles of system (5.5.60) lie entirely in the δ-neighborhood of the contour L.

3. Let us consider a system of two autonomous equations on the plane of a more general form than system (5.5.1):

$$\dot{x} = f_1(x,y), \qquad (5.5.62)$$

$$\dot{y} = f_2(x,y).$$

Assume that system (5.5.62) has a unique equilibrium $x = y = 0$, and consider the case where this equilibrium is asymptotically stable. We will study the problem of damped oscillations in system (5.5.62), clarify conditions under which the damped oscillations occur in system (5.5.62) for any initial deviations, and discuss the case involving the existence of a boundary which separates the region of damped oscillation from that of undamped oscillations. Since system (5.5.62) is the idealization of a physical system, whose oscillations are under study, then in the last the self-oscillations may occur in a sufficiently small neighborhood of the point $x = 0, y = 0$. In the general case, the undamped oscillations may occur, even though all oscillations in system (5.5.62) are damped. From the practical point of view, those cases may be considered equivalent. Assume that the right-hand sides of system (5.5.62) satisfy the following conditions:

1) functions f_1 and f_2 are given for $x \in (-\infty, +\infty)$, are real, continuous, and satisfy Lipschitz condition in any finite region;

2) equations $f_i = 0, i = 1, 2$ define two lines on the plane, each line breaking the plane into two regions where $f_1 > 0$ in one region and $f_1 < 0$ in the other, and respectively $f_2 > 0$ and $f_2 < 0$;

3) $f_1 = f_2$ only for $x = y = 0$;

4) the zero solution of system (5.5.62) is asymptotically stable in Lyapunov's sense;

5) there are no periodic solutions to system (5.5.62).

Suppose equation $f_1(x,y) = 0$ defines the continuous function $y = y(x)$ denoted by $y_1(x)$ for $x \geq 0$, and by $y_2(x)$ for $x \leq 0$. Moreover, assume that the equation $f_2(x,y) = 0$ defines, the continuous function $x = x(y)$ denoted by $x_1(y)$ for $y \geq 0$ and by $x_2(y)$ for $y \leq 0$. Also, we say that if $y > y_1(x)$, then $f_1 > 0$, and if $x > x_1(y)$, then $f_2 < 0$. Denote by G_1 the region confined between the curves $y_1(x)$ and $x_1(y)$, by G_2 the region between $y_1(x)$ and $x_2(y)$, by G_3 the region between $x_2(y)$ and $y_2(x)$, and by G_4 the region between $x_1(y)$ and $y_2(x)$. In the region G_1, along each integral curve the abscissa x increases and the ordinate y decreases since $f_1 > 0, f_2 > 0$ for $(x,y) \in G_1$; in the region G_2 the abscissa and ordinate of the integral curve decreases since $f_1 < 0, f_2 < 0$ for $(x,y) \in G_2$; in the region G_3 the abscissa on the integral curve decreases, and the ordinate increases since $f_1 < 0, f_2 > 0$ for $(x,y) \in G_3$; in the region G_4 we have $f_1 > 0, f_2 > 0$ and, consequently, its passage through the region G_4 the abscissa and ordinate of the integral curve increase. This behavior of integral curves in the regions $G_1 - G_4$ will form the basis for our investigations of the development of oscillations in system (5.5.62). If the right-hand sides of (5.5.62) satisfy conditions 1) — 5) and

$$\sup_{x \geq 0} y_1(x) = +\infty, \quad \inf_{x \leq 0} y_2(x) = -\infty,$$

$$\sup_{y \leq 0} x_2(y) = +\infty, \quad \inf_{y \geq 0} x_1(y) = -\infty,$$

then the zero solution of system (5.5.62) is asymptotically stable as a whole, that is, the oscillations in this system are damped under any initial deviations.

Every integral curve starting on the positive semi-axis Oy and not entering into the point $x = y = 0$ reintersects necessarilly the positive semi-axis Oy as time increases.

The second intersection point cannot lie above the first, otherwise there exists a periodic solution that is the α-limit set for the integral curve under consideration. Since, by the assumption, the system (5.5.62) has no periodic solutions, then the integral curve involved intersecting the positive semi-axis Oy below the starting point, and hence comes arbitrarily close to the point $x = y = 0$. We now prove the above assertion in detail. The positive semi-axis Oy lies at the union of regions G_1 and G_4, where the abscissa x of any integral curve of (5.5.62) monotonically increases. Therefore, all integral curves intersect the positive semi-axis Oy in one direction with time increase, passing from the left ($x < 0$) to the right ($x > 0$). We choose the point $p_0 = (0, y_0)$, $y_0 > 0$, then either $p_0 \in G_1$ or $p_0 \in G_4$. In the former case, the integral curve $F(p_0, t)$ stays in the region $x > 0, y < y_0, (x, y) \in G_1$ for $t > 0$ until it intersects the curve $y = y_1(x)$, or the straight line $y = y_0$ intersects the plot of the function $y = y_1(x)$. In the latter case, the integral curve $F(p_0, t)$ intersects the curve $x = x_1(y)$ and enters into the region G_1, since $\inf_{y \geq 0} x_1 = -\infty$ and hence there is a point $\bar{p}_0$ such that $x_1(\bar{y}_0) = 0$, $\bar{y}_0 > y_0$. From this it follows that the integral curve $F(p_0, t)$ stays in the region $y < \bar{y}_0, x > 0, (x, y) \in G_1$ and intersects the plot of the curve $y = y_1(x)$, entering into the region G_2. We say that the integral curve $F(p_0, t)$ does not tend to the point $x = 0, y = 0$ as $t \to +\infty$. Then it intersects the curve $x = x_2(y)$ and enters into the region G_2. Further it enters into the region G_3, then into G_4, and again intersects the positive axis Oy at the point $\tilde{p}_0 = (0, \tilde{y}_0)$. If $\tilde{y}_0 > y_0$, then the integral curve $F(p_0, t)$ stays for $t < 0$ in the bounded region containing the origin of coordinates, whose boundary is represented by the arc of the integral curve $F(p_0, t)$ joining the points $p_0, \tilde{p}_0$ and line segment $p_0, \tilde{p}_0$. Since among the α-limit points of the integral curve $F(p_0, t)$ there is no point $x = 0, y = 0$, then in the indicated region there is no periodic solution to (5.5.62), which is not possible. Hence, $\tilde{y}_0 < y_0$, then the integral curve $F(p_0, t)$ enters into a sufficiently small neighborhood of the point $x = 0, y = 0$ and tends to it as $t \to +\infty$ because of its stability. Any integral curve $F(p, t)$ of system (5.5.62) starting in any region G_i on their boundary and not bordering on the point $x = 0, y = 0$ as time increases, intersects the positive semi-axis Oy. This follows from the earlier considerations.

We now analyze the conditions under which the above assertion is valid. Proceeding from this analysis we then formulate more general conditions that are sufficient for stability in the global sense. From the above reasoning it follows that the condition $\sup_{x \geq 0} y_1(x) = +\infty$ is used only to ensure passage of the integral curve $F(p_1, t)$, starting in the region $G_1, x_p \geq 0$, to the region G_2. For this integral curve to pass from G_1 to G_2, it is necessary and sufficient that it would intersect the plot of the curve $y = y_1(x)$. This holds for any such integral curves if $\sup_{x \geq 0} y_1(x) = +\infty$. Let $x_p > 0$ and y_p be such that $p \in G_1$. Then for the integral curve $F(p, t)$ to intersect the plot of the curve $y = y_1(x)$, it is necessary and sufficient that the quantity

$$\int_{x_p}^{x} \frac{f_2(x, y)}{f_1(x, y)} dx - y_1(x)$$

would not be strictly bounded below by $-y_p, x \geq x_p$. As before, this condition will

be denoted by A. Let the point $q \in G_2$ be such that $y_q < 0$. Then, for the integral curve $F(q,t)$ to intersect the plot of the curve $x = x_2(y)$, it is necessary and sufficient that
$$\int_{y_q}^{y} \frac{f_1(x,y)}{f_2(x,y)} dy - x_2(y)$$
would not be strictly bounded below by $-x_q$ for $y \leq y_q$. Denote this condition by B. Let $r \in G_3, x_r < 0$; then for the integral curve $F(r,t)$ to intersect the plot of the curve $y = y_2(x)$, it is necessary and sufficient that the quantity
$$\int_{x_r}^{x} \frac{f_2(x,y)}{f_1(x,y)} dx - y_2(x)$$
would not be strictly bounded above by y_r for $x \leq x_r$. Denote this condition by C. Let the point $s \in G_4$ be such that $y_s > 0$. Then, for the integral curve $F(s,t)$ to intersect the plot of the curve $x = x_1(y)$, it is necessary and sufficient that the quantity
$$\int_{y_s}^{y} \frac{f_1(x,y)}{f_2(x,y)} dy - x_1(y)$$
would not be strictly bounded above by $-x_s$. Denote this condition by D. To be noted is that wherever conditions $A - D$ are constructed, we deal with the behavior of integral curves as t increases.

If conditions $A-D$ are satisfied and the right-hand sides of (5.5.62) have properties 1) — 5), then oscillations in this system are damped under any initial deviations. In other words, the zero solution of system (5.5.62) is global-asymptotically stable. From conditions $A-D$ it follows that any integral curve $F(p,t)$ intersects the positive semi-axis Oy at least twice and, consequently, tends to $x = 0, y = 0$ as $t \to +\infty$, because system (5.5.62) has no periodic solutions. This suggests that conditions $A - D$ are sufficient for the oscillations in system (5.5.62) with the properties 1) — 5) to be damped under any initial deviations.

Let us consider some properties of (5.5.62). From the assumption of the absence of periodic solutions in this system it follows that the investigation of the behavior of its integral curves may be started outside the circle of a sufficiently large radius, with its center at the origin of coordinates. Therefore, it suffices to assume that there exist constants $c_1 < 0$ and $c_2 > 0, d_1 < 0$ and $d_2 > 0$ such that if $x \bar{\in} [c_1, c_2]$, then $f_1 = 0$ defines the single-valued continuous functions $y_1(x), x \geq c_2$ and $y_2(x), x \leq c_1$, or this equation for $x > c_2$ and $x < c_1$ has no real solutions. The equation $f_2 = 0$ defines for $y \bar{\in} [d_1, d_2]$ two functions: $x_1(y)$ for $y \geq d_2$ and $x_2(y)$ for $y \leq d_1$, or has no real solutions. To be noted is that, in the general case, conditions $A - D$ are difficult to verify, because the integrals in these conditions are curvilinear and the integration is performed along the arc of integral curves. However, as shown, these conditions are satisfied under wide assumptions of the functions f_1 and f_2.

We shall clarify development of oscillations for the case where one of the conditions, say A, is not satisfied. If system (5.5.62) has properties 1) — 5), conditions $B-D$ are satisfied, and condition A is violated, then there is a unique integral curve breaking the plane into two regions, one of which contains the point $x = y = 0$ and is the attraction region of the equilibrium $x = y = 0$. In other words, this region has the property that all the oscillations of system (5.5.62), the initial deviations of which lying therein, are damped as time increases without bound. If condition A is not satisfied, then in G_1 there is no point $p = (x_p, y_p)$, through which the integral curve $F(p,t)$ passes the remaining at $t \geq 0$ in the region G_1. Along such an integral curve we obtain $x \to +\infty$ as $t \to +\infty$, y decreases monotonically as $t \to +\infty$. Denote by $\bar{p}$ the lower bound of the points p possessing the above property and lying on the straight line $x = x_p$. Such a point $\bar{p}$ exists, and for all points $y_p > y_1(x_p)$ the integral curve passing through the point $\bar{p}$ has the property $x \to +\infty$ as $t \to +\infty$, otherwise by the theorem of continuous dependence of solutions on initial data, the integral curve passing through the point p cannot leave G_1 because its further passage necessarily falls on the region G_2. This property is possessed by all integral curves starting in a sufficiently small neighborhood of the point $\bar{p}$. This is not possible because the integral curves possessing the property $x \to +\infty$ as $t \to +\infty$ start in an arbitrarily small neighborhood of the point $\bar{p}$. The integral curve $F(\bar{p}, t)$ is the desired boundary integral curve.

As time increases, any integral curve $F(\bar{p}, t), y_p < \bar{y}_p$ falls within the region G_2 and, therefore, either approaches $x = y = 0$ in G_2 or intersects the curve $x_2(y)$ within G_3, where it either approaches $x = y = 0$ or falls within G_4, after which this integral curve either falls within region G_1 or approaches $x = y = 0$ as $t \to +\infty$. Staying in the region G_1, this integral curve may now reintersect the straight line $x = x_p$ at a point p such that $y_p < \bar{y}_p$. From this it follows that in its further passage the integral curve approaches the point $x = y = 0$ as $t \to +\infty$, both in this and the other case. We shall draw the integral curve $F(\bar{p}, t)$ through the point $\bar{p}$. Its trajectory breaks the plane (x, y) into two regions. Denote by G the region containing the point $x = y = 0$. The region G is the attraction region of the zero solution $x = y = 0$. If system (5.5.62) has properties 1) — 5), conditions B and D are satisfied, and conditions A and C are violated, then there are not more than two integral curves bounding some region G which contains the point $x = y = 0$ and is the attraction region of the zero solution of system (5.5.62).

Let conditions A and C be violated. Then there exist points $p \in G_1$ and $r \in G_3$ such that along the integral curve $F(p, t)$ we have $x \to +\infty$ and y monotonically decreases as $t \to +\infty$, and along the integral curve $F(r, t)$ we have $x \to -\infty$ and y monotonically increases. Denote by $\bar{p}$ the lower bound of the points p possessing such a property on the straight line $x = x_p$, and by $\bar{r}$ the upper bound of the points r lying on the straight line $x = x_r$ and possessing the above property. We shall draw the integral curves $F(\bar{p}, t)$ and $F(\bar{r}, t)$ through $\bar{p}$ and $\bar{r}$ and construct the region G containing the point $x = y = 0$ whose boundary consists of one or both curves. It can be easily shown that this region is the attraction region of the zero solution. In fact, its construction can be described as follows. The integral curve $F(\bar{p}, t)$ breaks the plane into two regions. Denote by $\bar{G}$ one of the regions containing the point $x = y = 0$. The integral curve $F(\bar{r}, t)$ also breaks the plane into two regions. Denote one of the regions containing the point $x = y = 0$ by $\bar{\bar{G}}$. Then $G = \bar{G} \cap \bar{\bar{G}}$. If system (5.5.62) satisfies conditions 1) — 5) and $A-D$ are violated, then the attraction region of the zero solution of (5.5.62) has no more than four boundary integral curves.

If all conditions $A - D$ are violated, then there are the points p, q, r, s for which these conditions are exactly violated. As before, proceeding from these points we shall construct the points $\bar{p}, \bar{q}, \bar{r}, \bar{s}$. Let us draw the integral curves $F(\bar{p}, t)$, $F(\bar{q}, t)$, $F(\bar{r}, t)$, $F(\bar{s}, t)$ through these points. Using these curves, we shall construct the regions $\bar{G}, G', \bar{\bar{G}}, G''$ containing the point $x = y = 0$ as has been done for $\bar{G}$ and $\bar{\bar{G}}$. The region $G = \bar{G} \cap \bar{\bar{G}} \cap G' \cap G''$ is the zero solution attractor.

Below we give some examples illustrating application of the above theory for finding conditions under which the natural oscillations of system are damped for any initial deviations.

Example 1. Consider the system of equations

$$\dot{x} = \varphi_1(x) + Ay, \qquad (5.5.63)$$

$$\dot{y} = \varphi_2(x) + By,$$

where $\varphi_1(x), \varphi_2(x)$ are real continuously differentiable functions, given for $x \in (-\infty, +\infty)$. Assume that the curves $\varphi_1(x) + Ay = 0$ and $\varphi_2(x) + By = 0$ have a unique point of intersection $x = y = 0$, and that the equilibrium $x = y = 0$ is asymptotically stable, that is, the following inequalities are said to hold

$$\frac{d\varphi_1(0)}{dx} + B < 0, \quad B\frac{d\varphi_1(0)}{dx} - A\frac{\varphi_2(0)}{dx} > 0.$$

To apply the above ideas, we shall first establish the tests for which the system involved has no periodic solutions. This can always be done with the help of the Bendixson theorem stated as follows. In order for the system

$$\frac{dx}{dt} = f_1(x, y),$$

$$\frac{dy}{dt} = f_2(x, y)$$

to have no periodic solutions encompassing the point $x = y = 0$, it is sufficient that the quantity $\frac{\partial f_1}{\partial x} + \frac{\partial f_2}{\partial y}$ would have a constant sign and would not be identically zero in any simply connected region containing the point $x = y = 0$. For the system involved, the Bendixson theorem provides the absence condition of a periodic solution and is as follows: preserving its sign, the function $\frac{\partial \varphi_1}{\partial x} + B$ should not be identically zero in any of the intervals $[a, b]$ containing the point $x = 0$. Another absence condition of a periodic solution can be derived in the following manner. Let us take an arbitrary continuously differentable function $V = V(x, y)$ given on the entire plane. We shall construct its total derivative by the system

$$\frac{dx}{dt} = f_1(x, y),$$

$$\frac{dy}{dt} = f_2(x, y).$$

Then we have
$$\frac{dV}{dt} = \frac{\partial V}{\partial x} f_1(x,y) + \frac{\partial V}{\partial y} f_2(x,y) = W.$$

If the function $W(x,y)$ preserves its sign and is not identically zero on any closed continuous curve encompassing the point $x = y = 0$, then the system involved has no periodic solutions encompassing the point $x = y = 0$.

Suppose there exists such a periodic solution
$$x = x(t),\ y = y(t).$$

Let us compute the function W on it and denote its period by T. Then we have that
$$\int_0^T \frac{dV}{dt} dt = 0$$
and
$$\int_0^T W(t) dt \neq 0,$$
since the function W does not vanish at at least one point of the periodic solution. The obtained contradiction shows that the above statement is true. For our system, we take the function V to be
$$V(x,y) = \frac{(Bx - Ay)^2}{2} + \int_0^x (B\varphi_1(x) - A\varphi_2) dx.$$

Then, by the system under consideration, $\frac{dV}{dt}$ assumes the form
$$\frac{dV}{dt} = \frac{\partial V}{\partial x}(\varphi_1(x) + Ay) + \frac{\partial V}{\partial y}(\varphi_2(x) + By) = x^2 \sigma(x) \delta(x),$$
where
$$\sigma(x) = \frac{\varphi_1(x)}{x} + B,\ \delta(x) = B\frac{\varphi_1(x)}{x} - A\frac{\varphi_2(x)}{x},\ \sigma(x) \leq 0,\ \delta(x) \geq 0.$$

This suggests that the system has no periodic solutions if the functions $\sigma(x)$ and $\delta(x)$ are not identically zero in any interval $[a,b]$ containing the point $x = 0$. Assuming that the absence condition for periodic solutions has been satisfied, we shall replace in system (5.5.63) the desired functions by the formulas $\xi = x,\ \eta = Ay - Bx$:

$$\frac{d\xi}{dt} = \eta - f_1(\xi), \qquad (5.5.64)$$

$$\frac{d\eta}{dt} = f_2(\xi),$$

where
$$f_1(\xi) = -(\varphi_1(\xi) + B\xi),\ f_2(\xi) = A\varphi_2(\xi) - B\varphi_1(\xi).$$

Assume that $\xi f_2(\xi) < 0$ for $\xi \neq 0$. Then system (5.5.63) will have a unique global-asymptotically stable zero solution if

$$\sup_{\xi \geq 0} f_1(\xi) = +\infty, \qquad (5.5.65)$$

$$\inf_{\xi \leq 0} f_1(\xi) = -\infty.$$

In other words, under conditions (5.5.65) the oscillations in system (5.5.63) are damped for any initial deviations. Let, e.g., the first of conditions (5.5.65) be violated. Then, considering that $p \in G_1$, we get the inequality

$$\int_{\xi_p}^{\xi} \frac{f_2(\xi)}{\eta - f_1(\xi)} d\xi - f_1(\xi) \leq \int_{\xi_p}^{\xi} \frac{f_2(\xi)}{\eta_p - f_1(\xi)} d\xi - f_1(\xi). \qquad (5.5.66)$$

This inequality holds because $\eta - f_1(\xi) \leq \eta_p - f_1(\xi)$ and $f_2(\xi) < 0$ for $\xi \geq \xi_p \geq 0$. From (5.5.66) we have the following statement. The equilibrium of system (5.5.63) is global-asymptotically stable if the second condition of (5.5.65) and the additional relationship

$$\inf_{\xi \geq \xi_p} \int_{\xi_p}^{\xi} \frac{f_2(\xi)}{\eta_p - f_1(\xi)} d\xi - f_1(\xi) = -\infty \qquad (5.5.67)$$

hold under any choice of a positive number η_p satisfying the inequality $\eta_p \geq f_1(\xi)$ for $\xi \in [0, +\infty]$. Here it is implied that the first of conditions (5.5.65) has been violated. If the second condition in (5.5.65) is also violated, then for the zero solution of system (5.5.63) to remain global-asymptotically stable, it suffices to add the following condition to (5.5.67)

$$\sup_{\xi \leq \xi_r} \int_{\xi_r}^{\xi} \frac{f_2(\xi)}{\eta_r - f_1(\xi)} d\xi - f_1(\xi) = +\infty$$

for any negative number η_r satisfying the condition $\eta_r < f_1(\xi), \xi \in (-\infty, 0]$. Here it is assumed that the point $r = (\xi_r, \eta_r) \in G_3$.

Example 2. Consider the system

$$\frac{dx}{dt} = f_1(\varphi_1(x) + ay), \qquad (5.5.68)$$

$$\frac{dy}{dt} = f_2(\varphi_2(x) + by),$$

where $a > 0, b < 0$ are constants, and the functions $f_i(z)$ are given for $z \in (-\infty, +\infty)$, and are real and continuously differentiable, $z f_i(z) > 0$ for $z \neq 0$, $f_i(0) = 0$, $f_i'(z) > 0$ for $|z| > 0$. Assume that the functions $\varphi_1(x)$ and $\varphi_2(x)$ given for $x \in (-\infty, +\infty)$, are real and continuous, and

$$\sup_{x \geq 0}(-\varphi_1) = +\infty, \quad \inf_{x \leq 0}(-\varphi_1) = -\infty.$$

Oscillations in Autonomous System

If the zero solution of system (5.5.68) is asymptotically stable in Lyapunov's sense,

$$-\frac{\varphi_2(x)}{b} < -\frac{\varphi_1(x)}{a} \text{ for } x > 0, \quad -\frac{\varphi_2(x)}{b} > -\frac{\varphi_1(x)}{a} \text{ for } x < 0, \quad \varphi_1'(x) < 0 \text{ for all } x,$$

then the zero solution of system (5.5.68) will be global-asymptotically stable.

Thus, the zero solution of system (5.5.68) is asymptotically stable in Lyapunov's sense, the curves

$$y = -\varphi_1(x)/a, \quad y = -\varphi_2(x)/b,$$

on which the right-hand side of system (5.5.68) go to zero, satisfy all conditions of the above test and that of the Bendixson theorem. Hence, there are no periodic solutions. System (5.5.68) may be complicated if the functions $f_i(z)$ are also assumed to be dependent on x, y. Then the above-mentioned condition for global-asymptotic stability is preserved if the Bendixson test requires that the function

$$f(x,y) = \frac{\partial f_1(x,y,z_1)}{\partial x} + \frac{\partial f_1(x,y,z_1)}{\partial z} \frac{d\varphi_1}{dx} + \frac{\partial f_2(x,y,z_2)}{\partial y} + b\frac{\partial f_2(x,y,z_2)}{\partial z_2},$$

where $z_1 = \varphi_1(x) + ay$, $z_2 = \varphi_2(x) + by$, would not vanish in any simply connected region containing the point $x = y = 0$, and would retain its sign.

Example 3. Suppose we have a system

$$\dot{x} = f_1(x) + p(x)\sigma(y),$$

$$\dot{y} = f_2(x). \qquad (5.5.69)$$

Assume that $p(x) > 0$, $f_1(x), f_2(x), \sigma(y)$ are the continuous real functions given in all their variables, $xf_2(x) < 0$ for $x \neq 0$, $f_2(0) = 0$. The functions $f_1(x), p(x), \sigma(y)$ are said to be such that the equation $\sigma(y) = -\frac{f_1(x)}{p(x)}$ defines a single-valued continuous function $y = y(x)$ given for $x \in (-\infty, +\infty)$ and such that

$$\sup_{x \geq 0} y(x) = +\infty, \quad \inf_{x \leq 0} y(x) = -\infty, \quad \sigma(y) \xrightarrow[y \to \pm\infty]{} \pm\infty.$$

If system (5.5.69) has a unique equilibrium $x = y = 0$ that is asymptotically stable in Lyapunov's sense, and the function $\frac{df_1(x)}{dx}p(x) - \frac{dp(x)}{dx}f_1(x)$ retains its sign and is not identically zero in any interval $[a, b]$ containing the point $x = 0$, then the zero solution of system (5.5.69) is global-asymptotically stable.

Example 4. Consider the system

$$\frac{dx}{dt} = a_{11}x + a_{12}y + f_1(x), \qquad (5.5.70)$$

$$\frac{dx}{dt} = a_{21}x + a_{22}y,$$

where

$$f_1(x) = x \text{ for } x \in [l_2, l_1], \quad f_1(x) = l_1 \text{ for } x > l_1, \quad f_1(x) = l_2 \text{ for } x < l_2.$$

Moreover, $l_2 < 0 < l_1$. If $a_{11} + 1 < 0, a_{12} > 0, a_{21} < 0, a_{22} < 0, a_{22} < -a_{11}$, then the zero solution $x = y = 0$ of system (5.5.70) is global-asymptotically stable.

The above methods allow one to investigate the behavior of integral curves in the equation system of a more general form than (5.5.70), i.e. in a system of the form

$$\frac{dx}{dt} = f_1(x, y),$$

$$\frac{dy}{dt} = f_2(x, y),$$

where $f_i(x, y), i = 1, 2$ are piecewise-linear functions.

5.6 Analytical Test for Qualitative Behavior of Integral Curves on a Plane in the Neighborhood of Periodic Motion

1. Consider the system

$$\dot{x} = f_1(x, y), \qquad (5.6.1)$$

$$\dot{y} = f_2(x, y).$$

Assume that the system (5.6.1) has the properties:

a) the functions $f_i(x, y)$ in a region G of the xOy plane, are real and twice continuously differentiable in their variables;

b) there exist two continuously differentiable real functions

$$x = \varphi_1(t), \ y = \varphi_2(t), \qquad (5.6.2)$$

that give a solution to (5.6.1) and are periodic in t with period 2π, whose orbit M is contained in G.

By the structure of their neighborhood, the periodic solutions fall into two groups: isolated and nonisolated. We shall formulate the tests that allow one to distinguish a particular type of the qualitative behavior of integral curves, and establish the case of Lyapunov stability and instability of the periodic solution (5.6.2). Consider the construction of the fundamental system of equations. Let us draw normals through each point of the orbit M of periodic solution (5.6.2) and take the neighborhood $S(M, \delta) \subset G$ small enough for the line segments of the normals, contained therein and corresponding to distinct points on M, not to meet. The line segment of the normal passing through the point $(\varphi_1(t), \varphi_2(t))$ and contained in $S(M, \delta)$ will be denoted by N_t. By the theorem of solution continuity in initial data, it is possible to select a point $(x_0, y_0) \in N_0$ such that the solution of system (5.6.1) will be in $S(M, \delta)$ for $t \in [-T, T]$, where $T > 0$ is chosen arbitrarily. Fix the chosen point (x_0, y_0). Construct a line segment of the normal N_t corresponding to a sufficiently small $t > 0$, and denote by

$$\tau = \tau(t) \qquad (5.6.3)$$

Analytical Test for Qualitative Behavior

the first moment at which the solution trajectory corresponding to the fixed initial data x_0, y_0 intersects the normal N_t when moving in a positive direction $t \geq 0$. Introduce the functions

$$z_1(t) = x(\tau(t), x_0, y_0) - \varphi_1(t), \quad (5.6.4)$$

$$z_2(t) = y(\tau(t), x_0, y_0) - \varphi_2(t),$$

where $(x(t, x_0, y_0), y(t, x_0, y_0))$ are integral curves of the system (5.6.1), $x(0, x_0, y_0) = x_0$, $y(0, x_0, y_0) = y_0$. Consider the function

$$H(z_1, z_2, t) = z_1 f_1(t) + z_2 f_2(t), \quad (5.6.5)$$

where $f_i(t) = f_i(\varphi_1(t), \varphi_2(t))$. Since $(z_1(t), z_2(t))$ is a vector collinear with respect to N_t, then

$$H(z_1(t), z_2(t), t) \equiv 0.$$

Let us construct a differential equation whose solutions are functions (5.6.3) and (5.6.4)

$$\frac{dz_1}{dt} = \frac{d\tau}{dt} \frac{dx(\tau(t), x_0, y_0)}{d\tau} - \frac{d\varphi_1}{dt},$$

$$\frac{dz_2}{dt} = \frac{d\tau}{dt} \frac{dy(\tau(t), x_0, y_0)}{d\tau} - \frac{d\varphi_2}{dt}, \quad (5.6.6)$$

$$\frac{d}{dt} H(z_1(t), z_2(t), t) = \frac{dz_1}{dt} f_1(t) + \frac{dz_2}{dt} f_2(t) + z_1 \frac{df_1(t)}{dt} + z_2 \frac{df_2(t)}{dt} \equiv 0. \quad (5.6.7)$$

Using (5.6.1), (5.6.4), (5.6.6) and (5.6.7), we find

$$\frac{d\tau}{dt} = \frac{f_1^2(t) + f_2^2(t) - z_1 \frac{df_1(t)}{dt} - z_2 \frac{df_2(t)}{dt}}{f_1(t) f_1(z_1 + \varphi_1, z_2 + \varphi_2) + f_2(t) f_2(z_1 + \varphi_1, z_2 + \varphi_2)}; \quad (5.6.8)$$

$$\frac{dz_1}{dt} = \frac{d\tau}{dt} f_1(z_1 + \varphi_1, z_2 + \varphi_2) - f_1(t), \quad (5.6.9)$$

$$\frac{dz_2}{dt} = \frac{d\tau}{dt} f_2(z_1 + \varphi_1, z_2 + \varphi_2) - f_2(t).$$

The relationships (5.6.8) and (5.6.9) derived for the functions (5.6.3), (5.6.4), may be regarded as a system of differential equations intended to define the functions τ, z_1, z_2. The function $H(z_1, z_2, t)$ defined by relationship (5.6.5), is the integral of system (5.6.9) (see section 2.2). Consider a solution to system (5.6.9)

$$z_i = z_i(t, z_1^0, z_2^0, t_0), \quad i = 1, 2, \quad (5.6.10)$$

and a solution to equation (5.6.8)

$$\tau = \tau(t, z_1^0, z_2^0, t_0), \quad (5.6.11)$$

defined by the initial conditions

$$z_i = z_i^0 \text{ for } t = t_0, \quad \tau = t_0 \text{ for } t = t_0.$$

The function H computed on the solution (5.6.10) to system (5.6.9) remains constant.

This situation will be interpreted geometrically. Let us draw directions through the points M so that the direction ν_t passing through the point $(\varphi_1(t), \varphi_2(t))$ would make an angle ψ_t with the tangent such that

$$\cos \psi_t = \frac{H(z_1, z_2, t)}{\sqrt{f_1^2(t) + f_2^2(t)}\sqrt{z_1^2 + z_2^2}},$$

where z_1 and z_2 are determined from (5.6.10). The solution (5.6.10) of system (5.6.9) determines that

$$x(\tau - t_0, x_0, y_0) = z_1(t, z_1^0, z_2^0, t_0) + \varphi_1(t), \quad (5.6.12)$$

$$y(\tau - t_0, x_0, y_0) = z_2(t, z_1^0, z_2^0, t_0) + \varphi_2(t)$$

of system (5.6.1) with the initial data $x_0 = z_1^0 + \varphi_1(t_0), y_0 = z_2^0 + \varphi_2(t_0)$ at $t = t_0$ for system (5.6.1), intersects ν_t. Differentiating both parts of equalities (5.6.12) with respect to t and using equations (5.6.8) and (5.6.9), we find that the functions x and y, as functions of the argument $\tau - t$, satisfy system (5.6.1). In this case, (5.6.12) at $t = t_0$ passes through the point (x_0, y_0) lying on ν_{t_0}, and at the time t through the point $(x(t), y(t))$ lying on the direction ν_t. This assertion allows the order of system (5.6.8), (5.6.9) to be lowered by using its integral. Introduce the new desired functions by the formulas

$$\xi = z_1 f_2(t) - z_2 f_1(t), \quad \eta = z_1 f_1(t) + z_2 f_2(t). \quad (5.6.13)$$

Inversion of this transform yields

$$z_1 = \frac{\xi f_2(t) + \eta f_1(t)}{f_1^2(t) + f_2^2(t)},$$

$$z_2 = \frac{-\xi f_1(t) + \eta f_2(t)}{f_1^2(t) + f_2^2(t)}. \quad (5.6.14)$$

Differentiating (5.6.13) by system (5.6.8), (5.6.9), we get

$$\frac{d\xi}{dt} = \frac{d\tau}{dt}(f_1(z_1 + \varphi_1(t), z_2 + \varphi_2(t))f_2(t) - f_2(z_1 + \varphi_1(t),$$

$$z_2 + \varphi_2(t))f_1(t)) + \frac{z_1 df_2(t)}{dt} - \frac{z_2 df_1(t)}{dt}, \quad \frac{d\eta}{dt} = 0. \quad (5.6.15)$$

Considering that the quantity η is constant, we investigate the behavior of solutions of system (5.6.1) starting from N_0 and set $\eta \equiv 0$ without loss of generality. With this in mind, we eliminate from equations (5.6.8) and (5.6.15) the functions z_1 and z_2 defined by equalities (5.6.14):

$$\frac{d\tau}{dt} = \frac{f_1^2(t)f_2^2(t) - (\bar{a}_1 \frac{df_1(t)}{dt} + \bar{a}_2 \frac{df_2(t)}{dt})\xi}{f_1(t) + f_1(\bar{a}_1\xi + \varphi_1, \bar{a}_2\xi + \varphi_2) + f_2(t) + f_2(\bar{a}_1\xi + \varphi_1, \bar{a}_2\xi + \varphi_2)}; \quad (5.6.16)$$

Analytical Test for Qualitative Behavior

here

$$\bar{a}_1 = \frac{f_2(t)}{f_1^2(t) + f_2^2(t)}, \quad \bar{a}_2 = -\frac{f_1(t)}{f_1^2(t) + f_2^2(t)},$$

$$\frac{d\xi}{dt} = (f_1(\bar{a}_1\xi + \varphi_1(t), \bar{a}_2\xi + \varphi_2(t))f_2(t) - f_2(\bar{a}_1\xi + \varphi_1(t), \bar{a}_2\xi$$

$$+ \varphi_2(t))f_1(t))\frac{f_1^2(t) + f_2^2(t) - (\bar{a}_1\frac{df_1(t)}{dt} + \bar{a}_2\frac{df_2(t)}{dt})\xi}{f_1(t)f_1(\bar{a}_1\xi + \varphi_1, \bar{a}_2\xi + \varphi_2) + f_2(t)f_2(\bar{a}_1\xi + \varphi_1, \bar{a}_2\xi + \varphi_2)} +$$

$$+ (\bar{a}_1\frac{df_2(t)}{dt} - \bar{a}_2\frac{df_1(t)}{dt})\xi. \qquad (5.6.17)$$

Denote the right-hand side of (5.6.16) by $F(\xi, t)$, and the left-hand side of (5.6.17) by $G(\xi, t)$. In equation (5.6.16) we set $\tau = \Theta + t$. Then, to define the function Θ, we get the equation

$$\frac{d\Theta}{dt} = F(\xi, t) - 1^1. \qquad (5.6.18)$$

For the periodic solution of (5.6.1) to be stable in Lyapunov's sense, it is necessary and sufficient that the zero solution $\Theta = 0, \xi = 0$ of system (5.6.17), (5.6.18) be stable in Lyapunov's sense.

Remark. The closed curve M breaks its neighborhood $S(M, \delta)$ into two regions: S_1 and S_2. The problem may be stated with respect to Lyapunov stability of the periodic solution (5.6.2) to system (5.6.1) provided that the initial perturbations (x_0, y_0) belong to one of the regions: S_1 or S_2. Such a problem of conditional Lyapunov stability amounts to the issue of conditional Lyapunov stability of the zero solution to system (5.6.17), (5.6.18) on condition that $\xi > 0$ or $\xi < 0$. This statement follows from the reasoning given in section 2.2.

Any periodic solution of system (5.6.1) lying in a sufficiently small neighborhood of M corresponds to a 2π-periodic solution of equation (5.6.17). Moreover, equation (5.6.17) has no other periodic solutions lying in the corresponding neighborhood of the point $\xi = 0$.

Suppose the orbit of the periodic solution $(\psi_1(t), \psi_2(t))$ to system (5.6.1) with period T is in $S(M, \delta)$. Denote by (ψ_1^0, ψ_2^0) the point, at which the orbit of this solution intersects at N_0, and set

$$\xi_0 = f_2(0)(\psi_1^0 - \varphi_1(0)) - f_1(0)(\psi_2^0 - \varphi_2(0)).$$

We shall show that $\xi = \xi(t, \xi_0, 0)$ is a 2π-periodic solution to equation (5.6.17). The equation for the quantity τ corresponding to the periodic solution $(\psi_1(t), \psi_2(t))$ may be written as

$$\frac{d\tau}{dt} = f(\psi_1(\tau), \psi_2(\tau), t). \qquad (5.6.19)$$

Equation (5.6.19) is derived from (5.6.16) by replacing ξ therein

$$\xi = f_2(t)(\psi_1(\tau) - \varphi_1(t)) - f_1(t)(\psi_2(\tau) - \varphi_2(t)). \qquad (5.6.20)$$

---

[1] Section 2.2 relates the behavior of integral curves of the original system to that of the transformed one. The obtained results apply to the present case in full.

The right-hand side of equation (5.6.19) is a periodic function in t of period 2π and in τ of period T. Denote by $\tau(t)$ a solution of (5.6.19) with the initial condition $\tau = 0$ at $t = 0$. Geometrically, it is evident that $\tau(2\pi) = T$. Some simple verification shows that the function $\tau(t + 2\pi) - T$ is a solution of (5.6.19) with the same initial condition as $\tau(t)$. Hence, by uniqueness, we have $\tau(t + 2\pi) = \tau(t) + T$. The last relationship shows that the function ξ defined by formula (5.6.20) is 2π-periodic. If equation (5.6.17) is assumed to have a periodic solution, then one of the cases is possible here: either two successive values of ξ coincide in N_0, or there does not exist such a coincidence. In the former case, ξ is a 2π-periodic solution corresponding to a periodic solution of system (5.6.1), while in the latter case it describes a spiral and, therefore, cannot be a periodic motion to system (5.6.17).

2. Assume that the right-hand sides of system (5.6.1) are analytic with respect to x, y in a sufficiently small neighborhood of M. Then equations (5.6.17) and (5.6.18) can be represented as

$$\frac{d\Theta}{dt} = \sum_{k=1}^{\infty} a_k(t)\xi^k; \tag{5.6.21}$$

$$\frac{d\xi}{dt} = \sum_{k=1}^{\infty} b_k(t)\xi^k, \tag{5.6.22}$$

the series in (5.6.21) and (5.6.22) converging for $|\xi| < r, r > 0$. Computations show that

$$b_1(t) = \frac{\partial}{\partial x}f_1(\varphi_1(t), \varphi_2(t)) + \frac{\partial}{\partial y}f_2(\varphi_1(t), \varphi_2(t))$$

$$+ \frac{1}{2}\frac{d}{dt}\ln(f_1^2(t) + f_2^2(t)).$$

Set

$$G_1 = \int_0^{2\pi} b_1(t)dt = \int_0^{2\pi} \left(\frac{\partial}{\partial x}f_1(\varphi_1(t), \varphi_2(t)) + \frac{\partial}{\partial y}f_2(\varphi_1(t), \varphi_2(t))\right) dt.$$

It is well known (see section 2.2) that for $G_1 < 0$ the periodic solution (5.6.2) of system (5.6.1) is the Lyapunov stable limit cycle, for $G_1 > 0$ the unstable limit cycle, and it is Lyapunov stable for $t \to -\infty$. We shall find a solution for $G_1 = 0$. Let us make a replacement in equations (5.6.18) and (5.6.22) as follows:

$$\xi = \eta e^{\int_0^t b_1(t)dt}; \tag{5.6.23}$$

then

$$\frac{d\Theta}{dt} = \sum_{k=1}^{\infty} \bar{a}_k(t)\eta^k, \tag{5.6.24}$$

$$\frac{d\eta}{dt} = \sum_{k=2}^{\infty} \bar{b}_k(t)\eta^k. \tag{5.6.25}$$

We will seek a solution of (5.6.25) of the form

$$\eta = c + g_2(t)c^2 + \ldots + g_k(t)c^k + \ldots, \tag{5.6.26}$$

Analytical Test for Qualitative Behavior

where $g_k(t), k = 2, 3, \ldots$ are 2π-periodic functions to be defined, and c an arbitrary constant. Substituting (5.6.26) into (5.6.25) and equating the coefficients to the same power c, we get

$$\frac{dg_k}{dt} = r_k, \quad k = 2, 3, \ldots. \tag{5.6.27}$$

If $g_2, \ldots, g_{k-1}$ are defined to be periodic functions, then g_k is also a periodic function provided that

$$\frac{1}{2\pi}\int_0^{2\pi} r_k dt = G_k = 0.$$

Assume that for the first time $G_m \neq 0$ for $m \geq 2$. Let us fix this value of m. We shall seek a solution of equation (5.6.24) of the form

$$\Theta = \sum_{k=1}^{\infty} h_k(t)\eta^k. \tag{5.6.28}$$

Substituting (5.6.28) into (5.6.24) and equating the coefficients to the same power n, we get

$$\frac{dh_k}{dt} = P_k, \quad k = 1, 2, \ldots. \tag{5.6.29}$$

If $h_1, \ldots, h_{k-1}$ turn out to be periodic, then for

$$\frac{1}{2\pi}\int_0^{2\pi} P_k dt = F_k = 0$$

h_k is also periodic. Assume that there is a number $l \geq 1$ such that $F_l \neq 0$ for the first time. Let us fix this value of l.

Theorem 92 *1. If m is odd and $G_m < 0$, then the periodic solution (5.6.2) of (5.6.1) is a stable limit cycle. For $l + 1 > m$ this periodic solution is stable in Lyapunov's sense, and for $l + 1 \leq m$ it is unstable in Lyapunov's sense.*

2. If m is odd and $G_m > 0$, then the periodic solution (5.6.2) to (5.6.1) is an unstable limit cycle. For $l + 1 > m$ this periodic solution is stable in Lyapunov's sense as $t \to -\infty$, and for $l + 1 \leq m$ it is unstable in Lyapunov's sense.

3. If m is even, then the periodic solution (5.6.2) to (5.6.1) is a semi-stable limit cycle. Moreover, if $l + 1 > m$, then this periodic solution is conditionally Lyapunov stable in the direction in which the integral curves of system (5.6.1) approach M without bound. If, however, $l + 1 \leq m$, then such type of stability is nonexistent.

Proof: Let us make a replacement:

$$\eta = z + \sum_{k=2}^{m-1} g_k z^k + (g_m - G_m t)z^m,$$

$$\Theta = \sigma + \sum_{k=1}^{l-1} h_k \eta^k + (h_l - F_l t)\eta^l.$$

Then, to define the functions z and σ, we get

$$\frac{dz}{dt} = G_m z^m + \sum_{k=m+1}^{\infty} c_k(t) z^k, \tag{5.6.30}$$

$$\frac{d\sigma}{dt} = F_l z^l + \sum_{k=l+1}^{\infty} d_k(t) z^k. \qquad (5.6.31)$$

Without loss of generality, it may be assumed that the behavior of solutions can be investigated for $z > 0$.

$$\frac{dz}{dt} = G_m z^m, \quad \frac{d\sigma}{dt} = F_l z^l. \qquad (5.6.32)$$

Its direct integration leads to the formulas

$$z = \frac{z_0}{(1 + (1-m)G_m z_0^{m-1} t)^{\frac{1}{m-1}}},$$

$$\sigma = \sigma_0 + \frac{F_l}{(1-m)G_m z_0^{m-1}} \frac{(m-1)z_0^l}{(m-l-1)} \left((1 + (1-m)G_m z_0^{m-1} t)^{\frac{m-l-1}{m-1}} - 1 \right), \; l \neq m-1.$$

For $l = m - 1$

$$\sigma = \sigma_0 + \ln\left(1 + (1-m)G_m z_0^{m-1} t\right) \frac{F_l}{(1-m)G_m z_0^{m-1}}.$$

These formulas imply validity of the statements made in the theorem since the influence of the terms cancelled is immaterial.

Setting $z = z_0$ at $t = 0$, we derive from (5.6.30) and (5.6.31)

$$\sigma = \sigma_0 + \int_{z_0}^{z} z^{l-m} \frac{F_l + \ldots}{G_m + \ldots} dz. \qquad (5.6.33)$$

For all $z \leq r_0$, where r_0 is a sufficiently small positive number, the following inequality holds

$$a \leq \frac{F_l + \ldots}{G_m + \ldots} \leq b, \qquad (5.6.34)$$

where a and b are constant numbers $ab > 0$. From (5.6.33) and (5.6.34), for $z \leq r_0$ we have

$$\sigma_0 + \frac{a}{l-m+1}(z^{l-m+1} - z_0^{l-m+1}) \leq \sigma \leq \sigma_0 + \frac{b}{l-m+1}(z^{l-m+1} - z_0^{l-m+1}) \qquad (5.6.35)$$

for $l - m + 1 \neq 0$; for $l - m + 1 = 0$ we get

$$\sigma_0 + a \ln(z/z_0) \leq \sigma \leq \sigma_0 + b \ln(z/z_0). \qquad (5.6.36)$$

Additionally, the number r_0 can be so small that for $z \leq r_0$ there is

$$cz^m \leq G_m z^m + \ldots \leq dz^m, \; cd > 0. \qquad (5.6.37)$$

Integrating the inequality $cz^m \leq \frac{dz}{dt} \leq dz^m$, we get

$$\frac{z_0}{\left(1 + (1-m)dz_0^{m-1} t\right)^{\frac{1}{m-1}}} \leq z \leq \frac{z_0}{\left(1 + (1-m)cz_0^{m-1} t\right)^{\frac{1}{m-1}}}. \qquad (5.6.38)$$

Analytical Test for Qualitative Behavior

The inequality (5.6.38) holds as long as $z \leq r_0$. The inequalities (5.6.35), (5.6.36), (5.6.38) lead directly to the proof of the theorem in the general case. ∎

Theorem 93 *If $G_m = 0$ for any m, then a periodic solution of system (5.6.1) passes through each point in a sufficiently small neighborhood of the periodic solution (5.6.2) of system (5.6.1). Moreover, if $F_l = 0$ for any l, then the periodic solution (5.6.2) is stable in Lyapunov's sense. If, however, there is at least one l such that $F_l \neq 0$, then the periodic solution (5.6.2) is unstable in Lyapunov's sense.*

Proof: If $G_m = 0$ for $m \geq r$, then all the terms of series (5.6.26) are periodic functions. If $g_k = \int_0^t r_k dt$, then the series (5.6.26) converges (see section 2.2) for $|c| \leq c_0$ as a series in initial data. Hence, the series for each $|c| \leq c_0$ determines a periodic solution. In addition, equality (5.6.28) holds for $F_l = 0$, $l \geq 1$. In this inequality, $h_k(t)$ are 2π-periodic functions and the series on the right-hand side converges for $\eta \leq \eta_0$. Hence it follows that each periodic solution of system (5.6.1) lying in a sufficiently small neighborhood of M has a 2π period. If, however, $F_l \neq 0$ for a particular l, then we have Lyapunov stability of the periodic solution (5.6.2) to system (5.6.1). ∎

Theorem 94 *If the right-hand sides of system (5.6.1) are analytic with respect to x, y in a sufficiently small neighborhood of M, then the periodic solution (5.6.2) of system (5.6.1) is either a limit cycle, or that the periodic solution of system (5.6.1) passes through each point of its sufficiently small neighborhood $S(M, \delta)$.*

Proof follows directly from Theorems 92 and 93.

Remarks: 1. If the right-hand sides of system (5.6.1) are n times continuously differentiable in their variables, then such properties are possessed by the right-hand sides of (5.6.17) and (5.6.18). By applying Taylor's theorem to representation of the right-hand sides with subsequent elimination of the remainder terms, we get a system of equations to define Θ and ξ with polynomials on the right-hand side (the polynomials in ξ with periodic coefficients are on the right-hand side). If the application of Theorem 92 to these equations shows that $m < n$ and $l < n$, then all derivations in Theorem 94 hold for the original system (5.6.1).

2. Compared to familiar results, application of the principles contained in the proof of Theorem 92 makes it possible to obtain a deeper development for the systems involving a small (perturbation) parameter even in place where the solution (5.6.2) itself and a period of the system are independent of this parameter.

3. The general case may be covered by investigation of the roots of the equation $\psi(\xi_0) = 0$, where

$$\psi = \xi(2\pi, \xi_0, 0) - \xi_0, \quad |\xi_0| < \delta,$$

and $\delta > 0$ is sufficiently small.

3. We now discuss the behavior of integral curves of the system of differential equations

$$\frac{dx}{dt} = f_1(x, y) + r_1(t, x, y), \qquad (5.6.39)$$

$$\frac{dy}{dt} = f_2(x, y) + r_2(t, x, y)$$

in a neighborhood of the orbit M of the periodic solution (5.6.2) to system (5.6.1). Assume that the functions $r_1(t,x,y)$ and $r_2(t,x,y)$ given for $t \in (-\infty, +\infty), (x,y) \in S(M, \delta)$ are real and continuous in all their variables. Moreover, for fixed $(x,y) \in S(M, \delta)$ they are almost periodic functions with respect to t, and are continuous functions for x, y uniformly in $t \in (-\infty, +\infty)$ (see section 3.4). The results obtained in sections 2.5 and 3.5 fully apply to system (5.6.39). If the functions r_1 and r_2 are explicitly independent of time, and satisfy the Lipschitz condition

$$|r_i(\bar{x}, \bar{y}) - r_i(\bar{\bar{x}}, \bar{\bar{y}})| \leq aL\sqrt{(\bar{x} - \bar{\bar{x}})^2 + (\bar{y} - \bar{\bar{y}})^2}$$

and an inequality of the form

$$|r_i(x,y)| \leq am, i = 1, 2$$

in the region, where these functions are given, then, in the case where the periodic solution (5.6.2) is a self-oscillation of system (5.6.1), system (5.6.39) has a periodic solution for any $a, 0 \leq a \leq a_0$. This periodic solution lies in the region $S(M, \delta(a))$, where $\delta(a)$ is a positive (for $a \neq 0$) continuous function and is such that $\delta(a) \to 0$ as $a \to +0$. This result cannot be improved without further assumptions about functions r_1 and r_2 and functions f_1 and f_2. That is, we cannot select the number of such periodic solutions and the behavior of integral curves of system (5.6.39) in the neighborhood of these periodic solutions.

This may be illustrated by referring to the system

$$\dot{x} = -xf(x,y) - y + r_1(x,y), \tag{5.6.40}$$

$$\dot{y} = x - yf(x,y) + r_2(x,y).$$

Set $r_1 = -ag(x,y)x$, $r_2 = -ag(x,y)y$. For simplicity, assume that the functions f and g are given, real and continuous in a neighborhood of the unit circle $r^2 = x^2 + y^2 = 1$ and depend on $r = \sqrt{x^2 + y^2}$. If $f = 0$ for $r = 1, f < 0$ for $r < 1, f > 0$ for $r > 1$, then system (5.6.40) with $g = 0$ has the self-oscillation $x = \cos t, y = \sin t$. Any root $r = c$ of the equation

$$f + ag = 0 \tag{5.6.41}$$

yields a periodic solution $x = c\cos t, y = c\sin t$ to the system (5.6.40). The number of roots in equation (5.6.41) and the signs of the functions $f + ag$ in their neighborhood describe fully the behavior of integral curves of system (5.6.40).

Set, e.g.,

$$f = (r^2 - 1)^5, \ g = (r^2 - 1)^4 \sin\frac{1}{c^2 - 1}.$$

With such a choice of the functions f and g, equation (5.6.41) has, for any $a > 0$, an infinite number of periodic solutions in an arbitrarily small neighborhood of the set $r = 1$. These solutions correspond to both stable and unstable limit cycles; $r = 1$ corresponds to a nonisolated periodic solution, and for some values of a there may appear semistable limit cycles. An example may also be given for the case where the periodic solutions to system (5.6.39) fill the annular regions, with the spiral integral

Analytical Test for Qualitative Behavior 353

curves of this system lying between such regions. If the characteristic index λ_2 for the periodic solution (5.6.2) of system (5.6.1) is nonzero, then, in the case involved, system (5.6.39) has a unique periodic solution lying in $S(M, \delta(a))$ for any a, $0 < a \le a_0$. If $\lambda_2 < 0$, then this periodic solution is a self-oscillation of system (5.6.39).

The general case, where the functions r_1 and r_2 are explicitly dependent on time, was described for a system of differential equations in section 3.5. The result obtained there for the case of system (5.6.39), may be stated as follows. Let

$$\lambda_2 = \int_0^{2\pi} \left(\frac{df_1}{dx} + \frac{df_2}{dy} \right) dt < 0,$$

where the integrand is computed on the periodic solution (5.6.2). If the inequalities

$$\mid r_i(t, \bar{x}, \bar{y}) - r_i(t, \bar{\bar{x}}, \bar{\bar{y}}) \mid \le aL\sqrt{(\bar{x} - \bar{\bar{x}})^2 + (\bar{y} - \bar{\bar{y}})^2},$$

$$\mid r_i(t, x, y) \mid \le am, \ i = 1, 2$$

are satisfied, then there is a number $a_0 > 0$ such that for any a, $0 \le a \le a_0$ the system (5.6.39) has a unique integral manifold

$$y = \bar{y}(t, \varphi), \ x = \bar{x}(t, \varphi), \qquad (5.6.42)$$

lying in the region $S(M, \delta(a))$, where $\bar{x}(t, \varphi)$, $\bar{y}(t, \varphi)$ are 2π-periodic functions with respect to φ for any fixed t and are almost periodic functions with respect to t for any fixed φ. These functions satisfy the Lipschitz condition in φ, with the constant being independent of t over the entire range of variations in $\varphi \in (-\infty, +\infty)$. Following the ideas developed in section 3.5, we replace the desired functions in system (23.39) by the formulas

$$z_1 = x(t, x_0, t_0) - \varphi_1(\varphi), \ z_2 = y(t, x_0, t_0) - \varphi_2(\varphi),$$

where φ is the time function satisfying the equation

$$z_1 f_1(\varphi) + z_2 f_2(\varphi) = 0;$$

here $f_i(\varphi) = f_i(\varphi_1(\varphi), \varphi_2(\varphi))$. Following this replacement, the functions z_1 and z_2 undergo an orthogonal transformation of the form

$$z_1 = \frac{-f_2(\varphi)\xi + f_1(\varphi)\eta}{\sqrt{f_1^2(\varphi) + f_2^2(\varphi)}},$$

$$z_2 = \frac{\xi_1 f_1(\varphi) + \eta f_2(\varphi)}{\sqrt{f_1^2(\varphi) + f_2^2(\varphi)}}.$$

Inverting the transform and setting $\eta = 0$, for definition of functions φ and ξ we find the system of equations

$$\frac{d\varphi}{dt} = \frac{(f_1(h_1, h_2) + r_1(t, h_1, h_2))f_1(\varphi) + (f_2(h_1, h_2) + r_2(t, h_1, h_2))f_2(\varphi)}{f_1^2(\varphi) + f_2^2(\varphi) - (h_1 - \varphi_1(\varphi))\frac{\partial f_1(\varphi)}{\partial \varphi} - (h_2 - \varphi_2(\varphi))\frac{\partial f_2(\varphi)}{\partial \varphi}}$$

$$= f(\varphi,\xi) + \bar{r}_1(t,\varphi,\xi), \qquad (5.6.43)$$

$$\frac{d\xi}{dt} = \frac{-f_2(\varphi)(f_1(h_1,h_2) + r_1(t,h_1,h_2)) + f_1(\varphi)(f_2(h_1,h_2) + r_2(t,h_1,h_2))}{\sqrt{f_1^2(\varphi) + f_2^2(\varphi)}}$$

$$= g(\varphi,\xi) + \bar{r}_2(t,\varphi,\xi), \qquad (5.6.44)$$

where

$$h_1 = \frac{-f_2(\varphi)\xi}{\sqrt{f_1^2(\varphi) + f_2^2(\varphi)}} + \varphi_1(\varphi),$$

$$h_2 = \frac{f_1(\varphi)\xi}{\sqrt{f_1^2(\varphi) + f_2^2(\varphi)}} + \varphi_2(\varphi),$$

$\bar{r}_1$ and $\bar{r}_2$ are the linear combinations of functions r_1 and r_2 with the coefficients dependent on φ, ξ. Under the above-mentioned conditions, for the system (5.6.43) and (5.6.44) with any $a, 0 \le a \le a_0$ there is a unique function $\xi = \bar{\xi}(t,\varphi)$ such that $|\bar{\xi}(t,\varphi)| \le \delta(a)$, $\bar{\xi}(t,\varphi)$ is 2π-periodic for φ and almost periodic in t, and satisfies the Lipschitz condition in φ

$$|\bar{\xi}(t,\bar{\varphi}) - \bar{\xi}(t,\bar{\bar{\varphi}})| < l|\bar{\varphi} - \bar{\bar{\varphi}}|; \ t, \bar{\varphi}, \bar{\bar{\varphi}} \in (-\infty, +\infty) \text{ (see section 3.5)}.$$

Any pair of functions $\varphi = \varphi(t,\varphi_0,t_0), \xi = \bar{\xi}(t,\varphi(t,\varphi_0,t_0))$ constitutes a solution to system (5.6.43) and (5.6.44), where $\varphi(t,\varphi_0,t_0)$ is a solution to the equation (5.6.43), in which ξ is substituted by $\bar{\xi}(t,\varphi)$ with initial conditions $\xi = \xi_0$ at $t = t_0$. The integral manifold (5.6.42) of system (5.6.39) is defined by the function $\bar{\xi}(t,\varphi)$ from the formulas

$$\bar{x}(t,\varphi) = \varphi_1(\varphi) - \bar{\xi}(t,\varphi)\frac{f_2(\varphi)}{\sqrt{f_1^2(\varphi) + f_2^2(\varphi)}},$$

$$\bar{y}(t,\varphi) = \varphi_2(\varphi) + \bar{\xi}(t,\varphi)\frac{f_1(\varphi)}{\sqrt{f_1^2(\varphi) + f_2^2(\varphi)}}. \qquad (5.6.45)$$

The integral manifold implies that any integral curve starting on this manifold stays there throughout the motion. Here this condition is valid since any pair of functions $x = \bar{x}(t,\varphi(t,\varphi_0,t_0)), y = \bar{y}(t,\varphi(t,\varphi_0,t_0))$ constitutes a solution to system (5.6.39).

Consider the equation

$$\frac{d\varphi}{dt} = f(\varphi, \bar{\xi}(t,\varphi)) + \bar{r}(t,\varphi,\bar{\xi}(t,\varphi)). \qquad (5.6.46)$$

In the case where the functions r_1 and r_2 are periodic in t with period ω, the right-hand side of equation (5.6.46) is a periodic function in φ and t. The general theory for such equations was adequately developed by Poincaré and Denjoy. The developments made by these authors and Academician N.N.Bogolyubov, suggest that either each trajectory of this equation is everywhere dense on torus, or this equation has periodic motions that are approached indefinitely by the remaining motions from two sides. In the former case, any integral curve of system (5.6.39) lying on the integral manifold

Analytical Test for Qualitative Behavior 355

(5.6.42) is almost periodic. In the latter case, on the integral manifold (5.6.42) there are periodic motions of system (5.6.39), the other motions approaching them without bound. As shown in section 3.5, all the integral curves of system (5.6.39) starting in some neighborhood of the manifold (5.6.42) approach it without bound as $t \to +\infty$. We may also conclude that in the case of periodic functions r_1 and r_2 the integral curves approach indefinitely the stationary modes lying on the manifold (5.6.42). If $\lambda_2 = \lambda_1 = 0$, then all the statements made here cannot be derived without further assumptions about the functions r_1 and r_2. This generates a need for studies on the behavior of solutions to system (5.6.39) for the case $\lambda_1 = \lambda_2 = 0$.

Appendix
Theory of Rated Stability

1. Introduction and Basic Definitions

It is well known that in actual practice only stable modes of operation are implemented by the controlled engineering system. This raises the question as to how some of the desired modes can be stabilized in the system involved. This stabilization can be obtained at the sacrifice of their stability or asymptotic stability in Lyapunov's sense [Lyapunov (1950)]. A new challenge now is to stabilize the behavioral mode of the controlled system which is generally unimplementable, and hence is not a proper motion of the system involved. We shall investigate stability properties of the modes which, as distinct from proper motions, are called rated motions. Such issues have been discussed in detail [LaSalle and Rath (1964), Sebakhy (1975)] with reference to self-adapting automatic control systems. This appendix states the notions of stability, asymptotic stability, and instability as applied to rated motions in the engineering systems whose operational dynamics is described by nonlinear systems of ordinary differential equations. Further, based on the ideas of Lyapunov's direct method of investigating motion stability are the assertions providing the necessary and sufficient as well as the sufficient conditions for rated stability, rated asymptotic stability, and rated instability. The relations between the notions of rated stability and Lyapunov stability are also analyzed.

Suppose the operational dynamics of the system involved is described by the nonlinear system of ordinary differential equations

$$\frac{dx_i}{dt} = F_i(x_1, \ldots, x_n, t), i = 1, \ldots, n, \tag{1}$$

or, in vector form,

$$\frac{dx}{dt} = F(x, t), x = (x_1, \ldots, x_n)^*, F(x, t) = (F_1(x, t), \ldots, F_n(x, t))^*.$$

Here $*$ indicates transposition. Let us introduce the Euclidean vector norm $\|x\| = (x_1^2 + \ldots + x_n^2)^{1/2}$. In what follows the n-dimensional vector function $F(x, t)$ is taken to be given in the set $\Omega = \{(x, t) : \|x\| < R, t \geq 0\}$, where R is a positive real number or symbol $+\infty$. Suppose the function $F(x, t)$ satisfies in Ω the conditions which ensure the existence, uniqueness, and continuability in t, from $t = t_0 \geq 0$ to $+\infty$, for all the solutions of system (1) whose graphs are in the set Ω. By the graph of the solution $x(t)$ to system (1) is meant the collection of all pairs $(x(t), t), t \geq t_0$ such that the solution $x(t)$ is defined at time t and satisfies the inequality $\|x(t)\| < R, t \geq t_0 \geq 0$. In a case $R = +\infty$ these conditions (see, e.g., [Coddington and Levinson (1955), Hale (1969), Cesari (1959)]) are: the continuity of function $F(x, t)$ in the set Ω in all its independent variables, the boundedness of function $F(x, t)$ in the set Ω, and the local Lipschitz condition for $F(x, t)$ in x with respect to all components of vector x in the set Ω. Note that $F(x, t)$ satisfies the local Lipschitz condition in x in the set Ω, if for any subset $\omega \subset \Omega$ there exists a constant $L_\omega, L_\omega > 0$ such that the inequality $\|F(x, t) - F(y, t)\| \leq L_\omega \|x - y\|$ holds for all pairs $(x, t) \in \omega$, $(y, t) \in \omega$. Let $x(t, t_0, x_0)$ be a solution of system (1) with initial condition (t_0, x_0), i.e. this is a solution of system (1) that is defined for all $t \geq t_0 \geq 0$ and is such that the equality $x(t_0, t_0, x_0) = x_0$ holds. Note that, under the above assumptions of all pairs $(x_0, t_0) \in \Omega$, for any $t \in [t_0, \bar{t}\,]$, where $\bar{t}$ is merely a number and $\bar{t} > t_0$ or $\bar{t}$ is merely a symbol $+\infty$, we get $\|x(t, t_0, x_0)\| < R$. In what follows we will study the behavior of a scalar function $V(x, t)$ that is given in the set Ω on the solutions of system (1) possessing the above property, i.e. those graphs are contained in the set Ω.

Denote $V_0(t, t_0, x_0) = V(x(t, t_0, x_0), t)$ and consider $V_0(t, t_0, x_0)$ to be the function of t for all $t \geq t_0$ as long as the pair $(x(t, t_0, x_0), t) \in \Omega$. We shall take a property of the function $V(x, t)$ computed along the solutions of system (1) to mean a suitable property of the function $V_0(t, t_0, x_0)$.

We shall investigate the behavior of solutions to system (1) in the neighborhood of the motion $x(t) \equiv 0$ which can or cannot be a solution to system (1) (the latter is possible, say, where the identity $F(0, t) \equiv 0$ holds for all $t \geq 0$). Thus, the motion $x = 0$ for system (1) is a rated motion. Note that the investigation of properties of solutions of system (1) in the neighborhood of the motion given by an arbitrary n-dimensional differentiable vector function $\varphi(t)$, whose graph belongs to the set Ω, is limited to this case. Consideration of a system in deviations from the motion involved investigation of the behavior of solutions to the system of differential equations

$$\frac{dy}{dt} = F(y + \varphi(t), t) - \frac{d\varphi(t)}{dt}, y = x - \varphi(t) \qquad (2)$$

in the neighborhood of the rated motion $y(t) \equiv 0$. Therefore, the investigation of the dynamics of system (1) in the neighborhood of the rated motion $x = 0$ is in no way restricts the generality of the resulting conditions which ensure its rated stability, since the above passage to the system in deviations (2) makes it possible to study the behavior of solutions to system (1) in the neighborhood of any motion $x = \varphi(t)$ provided that the n-dimensional vector function $\varphi(t)$ is differentiable in t and its graph is in the set Ω.

The stability of solutions to system (1) will be defined in Lyapunov's sense.

Definition 1. The solution $x(t, t_0, x_0)$ of system (1) is called Lyapunov stable if the following conditions are satisfied:

1) the solution $x(t, t_0, x_0)$ is defined for all $t \geq t_0$;

Introduction and Basic Definitions

2) for any numbers $\epsilon > 0$ and $t'_0 \geq t_0$ there is $\delta = \delta(\epsilon, t'_0) > 0$ such that for any n-dimensional vector x'_0 satisfying the inequality $\|x'_0 - x(t'_0, t_0, x_0)\| < \delta$ all the solutions $x(t, t'_0, x'_0)$ of system (1) are continuable in t from $t = t'_0$ to $+\infty$;

3) the inequality $\|x(t, t'_0, x'_0) - x(t, t_0, x_0)\| < \epsilon$ holds for such a quantity $\delta(\epsilon, t'_0)$ and the selected initial values of x'_0 for all $t \geq t'_0$.

Furthermore, if for any number $t'_0 \geq t_0$ there exists a quantity $\Delta = \Delta(t'_0) > 0$, $\Delta(t'_0) < \delta(\epsilon, t'_0)$ such that for all n-dimensional vectors x'_0 satisfying the inequality $\|x'_0 - x(t'_0, t_0, x_0)\| < \Delta(t'_0)$ there is a limit relationship $\|x(t, t'_0, x'_0) - x(t, t_0, x_0)\| \to 0$ as $t - t'_0 \to +\infty$, then the solution $x(t, t_0, x_0)$ is called asymptotically Lyapunov stable.

Definition 2. The Lyapunov stable solution $x(t, t_0, x_0)$ of system (1) is called uniformly stable in the initial moment t_0 if the above quantity $\delta(\epsilon, t'_0)$ can be chosen to be independent of t'_0, i.e. $\delta(\epsilon, t'_0) = \delta(\epsilon) > 0$.

In what follows it is precisely the uniform stability in Lyapunov's sense or simply the uniform stability that is taken to mean stability of this kind.

The asymptotically Lyapunov stable solution $x(t, t_0, x_0)$ of system (1) that is stable uniformly in the initial moment t_0 is called uniform-asymptotically stable if the above limit relationship $\|x(t, t'_0, x'_0) - x(t, t_0, x_0)\| \to 0$ as $t - t'_0 \to +\infty$ holds uniformly for all numbers $t'_0 \geq t_0$ and for any n-dimensional vector x'_0 satisfying the inequality $\|x'_0 - x(t'_0, t_0, x_0)\| < \Delta$, where the constant $\Delta > 0$.

Definition 3. The solution $x(t, t_0, x_0)$ of system (1) is called Lyapunov unstable if at least one of the following requirements is satisfied:

1) the solution $x(t, t_0, x_0)$ is noncontinuable in t from $t = t_0$ to $+\infty$;

2) for any number $\bar{\epsilon} > 0$ there exists a $t'_0 \geq t_0$ and an n-dimensional vector x'_0 satisfying the condition $\|x'_0 - x(t'_0, t_0, x_0)\| < \bar{\epsilon}$ such that the solution $x(t, t'_0, x'_0)$ of system (1) is noncontinuable in t from $t = t'_0$ to $+\infty$;

3) for any number $t'_0 \geq t_0$ there exists $\delta = \delta(t'_0) > 0$ such that for any n-dimensional vector x'_0 satisfying the condition $\|x'_0 - x(t'_0, t_0, x_0)\| < \delta$ the corresponding solution $x(t, t'_0, x'_0)$ of system (1) are continuable in t from $t = t'_0$ to $+\infty$ and there is, in addition, a number $\bar{\epsilon} > 0$ such that for any values $\bar{t}_0 \geq t_0$ and $\bar{\delta}, 0 < \bar{\delta} < \bar{\epsilon}$ it is possible to choose $\bar{t}_0 = \bar{t}_0(\bar{\delta}) > \bar{t}_0$ and an n-dimensional vector $\bar{x}_0 = \bar{x}_0(\bar{\delta})$ such that the inequalities $\|\bar{x}_0 - x(\bar{t}_0, t_0, x_0)\| < \bar{\delta}$ and $\|x(\bar{t}_0, \bar{t}_0, \bar{x}_0) - x(\bar{t}_0, t_0, x_0)\| \geq \bar{\epsilon}$ hold.

Let us describe the properties of the function $V(x, t)$ to be used in the investigation of the behavior of solutions to system (1) in the neighborhood of the motion $x = 0$. Introduce the appropriate notions.

Definition 4. The scalar function $V(x, t)$ given in the set Ω is called positive-definite if there exists a scalar function $V_1(x)$ that is continuous for $\|x\| < R$ and is such that $V_1(0) = 0, V_1(x) > 0$ for $0 < \|x\| < R$ and the inequality $V(x, t) \geq V_1(x)$ holds for all pairs $(x, t) \in \Omega$. In this case the identity $V(0, t) \equiv 0$ is not taken to hold for all $t \geq 0$.

Definition 5. We say that, by system (1), the function $V(x, t)$ has a derivative with respect to t if the function $V_0(t, t_0, x_0)$ for all pairs $(x_0, t_0) \in \Omega$ is differentiable in t for those values of t for which it is defined.

If the function $V(x, t)$ has partial derivatives in all its independent variables that are continuous in the set Ω, then we get

$$\frac{dV_0(t,t_0,x_0)}{dt} = \frac{\partial V(x',t)}{\partial t} + \sum_{i=1}^{n}\frac{\partial V(x',t)}{\partial x'_i}\frac{dx'_i}{dt} = \frac{\partial V(x',t)}{\partial t} + \sum_{i=1}^{n}\frac{\partial V(x',t)}{\partial x'_i}F_i(x',t),$$

where $x' = x(t,t_0,x_0)$ represents the solution of system (1) along which the derivative is computed.

Definition 6. The function

$$\frac{\partial V(x,t)}{\partial t} + \sum_{i=1}^{n}\frac{\partial V(x,t)}{\partial x_i}F_i(x,t)$$

considered without regard for solutions to the system involved is called the total derivative of function $V(x,t)$ with respect to t by system (1) and is denoted as $\frac{dV}{dt}$.

2. An Investigation of Rated Stability by Lyapunov's Direct Method

Let us introduce the notion of stability of a rated motion $x = 0$ for the system of differential equations (1) and analyze the relation between the rated stability of the motion $x = 0$ of system (1) and the stability in Lyapunov's sense for the case where the rated motion $x = 0$ is a solution to the system involved. Further, using the ideas of Lyapunov's direct method for investigating motion stability we will establish the sufficient as well as the necessary conditions for the rated stability of the motion $x = 0$.

Definition 7. The motion $x = 0$ of system (1) is called rated stable if for any number $\epsilon > 0$ there are $T(\epsilon) \geq 0$ and $\delta(\epsilon) > 0$ such that for any value $t_0 \geq T(\epsilon)$, and for any n-dimensional vector x_0 satisfying the condition $\|x_0\| < \delta(\epsilon)$ the inequality $\|x(t,t_0,x_0)\| < \epsilon$ holds for all $t \geq t_0$.

Theorem 1 *Let the motion $x = 0$ of system (1) be rated stable and suppose, in Definition 7, it is possible to set $T(\epsilon) = T_0 = const \geq 0$ for all sufficiently small values $\epsilon > 0$. Then, for all $t \geq T_0$, the system involved realizes the motion $x = 0$.*

Proof: Suppose the opposite is true, i.e., for $t \geq T_0$ the motion $x = 0$ is not a solution to system (1). Then there exist numbers t_0, t_1 such that the inequalities $t_1 > t_0 \geq T_0$ and $\|x(t_1,t_0,0)\| > 0$ are satisfied. Let $\epsilon = \frac{1}{2}\|x(t_1,t_0,0)\|$. From the rated stability of the motion $x = 0$ it follows that the inequality $\|x(t,t_0,0)\| < \epsilon$ holds for all $t \geq t_0$ and, in particular, for $t = t_1$; hence we get $2\epsilon < \epsilon$, which is a contradiction. Consequently, $x(t,t_0,0) \equiv 0$ for all $t \geq t_0 \geq T_0$ and $F(0,t) \equiv 0$ for all $t \geq T_0$. This completes the proof of Theorem 1.

Thus, under the above restriction on the quantity $T(\epsilon)$, the rated stability of the motion $x = 0$ of system (1) entails realization of this motion for the system involved, and uniform stability of the solution $x = 0$ in Lyapunov's sense. The opposite is also true.

Theorem 2 *If the system of equations (1) is autonomous or has the solution $x = 0$, then the rated stability of the motion $x = 0$ of this system is equivalent to uniform Lyapunov stability.*

Proof: If system (1) is autonomous, i.e. it is independent of the variable t, then the initial time t_0 can be chosen in an arbitrary way and the Lyapunov stability becomes uniform, coinciding with the rated stability. Here it is possible to set $T(\epsilon) = 0$ for all values $\epsilon > 0$. From Theorem 1 it follows that system (1) has the solution $x = 0$. If system (1) is not autonomous and has the solution $x = 0$, then by the uniqueness of solutions and their continuous dependence on the initial conditions (t_0, x_0) on a finite time interval (see, e.g., [Hale (1969)], [Cesari (1959)], [Coddington and Levinson (1955)]), we have that the rated stability of the motion $x = 0$ of system (1) implies uniform Lyapunov stability. In fact, by the chosen properties of solutions to system (1), for any number $\epsilon > 0$ and the corresponding $T(\epsilon) \geq 0$ there exists a value $\delta_1(\epsilon), 0 < \delta_1(\epsilon) < \delta(\epsilon)$ such that the inequalities $\|x(\tau, T(\epsilon), x_0)\| < \epsilon$ and $\inf_{0 \leq \tau \leq T(\epsilon)} \|x(\tau, T(\epsilon), x_0)\| = \delta_2(\epsilon) > 0$ hold for any n-dimensional vector x_0 satisfying the condition $\|x_0\| < \delta_1(\epsilon)$ and for all values $\tau \in [0, T(\epsilon)]$. Then the quantity $\delta_2(\epsilon)$ is the required quantity for defining the uniform Lyapunov stability of the solution $x = 0$ to system (1), since the inequality $\|x(t, t_0, x_0)\| < \epsilon$ holds for any number $t_0 \geq 0$ and for any n-dimensional vector x_0 satisfying the inequality $\|x_0\| < \delta_2(\epsilon)$ for all $t \geq t_0$. The reverse follows from the fact that in Definition 7 it is possible to set $T(\epsilon) = 0$. This completes the proof of Theorem 2.

We shall derive conditions under which the motion $x = 0$ is the rated stable motion of system (1).

Theorem 3 *For the motion $x = 0$ of system (1) to be rated stable, it is necessary and sufficient that there exists a function $V(x, t)$ given in the set Ω such that the following conditions are satisfied:*

1) $V(x, t)$ is positive-definite;

2) $V(x, t) \to 0$ as $\|x\| \to 0, t \to +\infty$;

3) $V_0(t, t_0, x_0)$ is a nonincreasing function in the variable t for all $t \geq t_0$.

Proof: *Necessity.* Let $V(x_0, t_0) = \sup_{\tau \geq t_0} \|x(\tau, t_0, x_0)\|$. The thus introduced function $V(x, t)$ is given in the set Ω and is positive-definite, which follows from the inequality $V(x, t) \geq \|x\|$. Since the motion $x = 0$ of system (1) is rated stable, for a number $\epsilon > 0$ it is possible to select $T(\epsilon) \geq 0$ and $\delta(\epsilon) > 0$ such that the inequality $\|x(t, t_0, x_0)\| < \epsilon$ holds for any n-dimensional vector x_0 satisfying the inequality $\|x_0\| < \delta(\epsilon)$, and for any value $t_0 \geq T(\epsilon)$ with all $t \geq t_0$, and hence $V(x_0, t_0) \leq \epsilon$. Since the inequality $V(x, t) \geq 0$ is valid, this implies the limit relationship $V(x, t) \to 0$ as $\|x\| \to 0, t \to +\infty$. Let the numbers t_1, t_2 be such that the inequalities $t_2 > t_1 \geq t_0$ hold and the function $V_0(\tau, t_0, x_0)$ is defined for $\tau \in [t_0, t_2]$. Since the relationships $V_0(t, t_0, x_0) = \sup_{\tau \geq t} \|x(\tau, t_0, x_0)\|, \sup_{\tau \geq t_2} \|x(\tau, t_0, x_0)\| \leq \sup_{\tau \geq t_1} \|x(\tau, t_0, x_0)\|$ are valid, the inequality $V_0(t_2, t_0, x_0) \leq V_0(t_1, t_0, x_0)$ is also valid. Thus, all the conditions of Theorem 3 are satisfied and necessity is proved.

Sufficiency. Let $V(x, t)$ be the function satisfying all the conditions of Theorem 3.

Then, by condition 1) and Definition 4, we take the corresponding continuous scalar function $V_1(x)$ and define the quantity $\lambda = \inf\limits_{\|x\|=\epsilon} V_1(x) > 0$ for any number ϵ satisfying the inequalities $0 < \epsilon < R$. By condition 2), for a given number ϵ it is possible to select quantities $T(\epsilon) \geq 0$ and $\delta(\epsilon), 0 < \delta(\epsilon) < \epsilon$ such that we obtain $V(x_0, t_0) < \lambda$ for all values $t_0 \geq T(\epsilon)$ and for all n-dimensional vectors x_0 satisfying the condition $\|x_0\| < \delta(\epsilon)$. Then, by condition 3), the inequalities $V_0(t, t_0, x_0) \leq V(x_0, t_0) < \lambda$ hold for all those values $t \geq t_0$ for which the function $V_0(t, t_0, x_0)$ is defined. The inequality $\|x(t, t_0, x_0)\| < \epsilon$ holds for all values $t \geq t_0$. Indeed, if the opposite is true, then for a number $t_1 > t_0$ there are the relationships $\|x(t_1, t_0, x_0)\| = \epsilon$, $V_0(t_1, t_0, x_0) \geq V_1(x(t_1, t_0, x_0)) \geq \lambda$ which yield a contradiction. Consequently, when all the conditions of Theorem 3 are satisfied, the motion $x = 0$ is rated stable, which proves the sufficiency. This completes the proof of Theorem 3.

Condition 3) of Theorem 3 is boundless, since it is stated for some set of solutions to system (1). Sufficient conditions for the rated stability of the motion $x = 0$ of system (1) which seem to be more constructive are given below.

Theorem 4 Let the scalar functions $V(x, t)$ and $W(x, t)$ with the following properties be given in the set Ω:
1) $V(x, t)$ is positive-definite;
2) $V(x, t) \to 0$ as $\|x\| \to 0, t \to +\infty$;
3) there exists $\frac{dV}{dt} = -W(x, t), W(x, t) \geq 0$.
Then the motion $x = 0$ of system (1) is rated stable.

Proof: The conditions of Theorem 3 are satisfied when the conditions of Theorem 4 are satisfied. This implies the validity of Theorem 4 which is analogous to Lyapunov's stability theorem [Lyapunov (1950)].

Theorem 5 Suppose the scalar functions $V(x, t)$ and $W(x, t)$ given in the set Ω are such that the following conditions are satisfied:
1) $V(x, t) \to V_1(x)$ as $t \to +\infty$ uniformly in $x, \|x\| < R$, the function $V_1(x)$ is continuous in x for $\|x\| < R, V_1(0) = 0, V_1(x) > 0$ for all $x, 0 < \|x\| < R$;
2) there exists $\frac{dV}{dt} = -W(x, t), W(x, t) \geq 0$.
Then the motion $x = 0$ of system (1) is rated stable.

Proof: Let ϵ be an arbitrary number satisfying the inequalities $0 < \epsilon < R$. For the number ϵ we define the quantity $\inf\limits_{\|x\|=\epsilon} V_1(x) = \lambda$. By the properties of the function $V_1(x)$, we obtain $\lambda > 0$. Then, by condition 1), there exists a quantity $T_1 \geq 0$ such that the inequality $|V(x, t) - V_1(x)| \leq \frac{1}{2}\lambda$ holds for all n-dimensional vectors x satisfying the condition $\|x\| < R$ and for all values $t \geq T_1$. From this it follows that, under the condition $\|x\| = \epsilon$, the inequality $V(x, t) \geq \frac{1}{2}\lambda$ holds for all values $t \geq T_1$. By condition 1), it is possible find quantities $T(\epsilon) \geq T_1$ and $\delta(\epsilon), 0 < \delta(\epsilon) < \epsilon$ such that for any n-dimensional vectors satisfying the inequality $\|x_0\| < \delta(\epsilon)$ and for any values $t_0 \geq T(\epsilon)$ we get $V(x_0, t_0) < \frac{1}{2}\lambda$. For such initial conditions (t_0, x_0) with all $t \geq t_0$ the inequality $\|x(t, t_0, x_0)\| < \epsilon$ holds, for otherwise the relationships $\|x(t_1, t_0, x_0)\| = \epsilon$ and $V_0(t_1, t_0, x_0) = V(x(t_1, t_0, x_0), t_1) \geq \frac{1}{2}\lambda$ hold for some $t_1 > t_0$, which contradicts the lack of increase in the function $V_0(t, t_0, x_0)$, the latter being subject to condition 2). Thus, the motion $x = 0$ of system (1) is rated stable. This

An Investigation of Rated Stability by Lyapunov's Direct Method

completes the proof of Theorem 5.

We shall now introduce the notion of rated instability for the motion $x = 0$ of system (1) and then, using the ideas of Lyapunov's direct method for investigating motion stability, establish the sufficient as well as the necessary and sufficient conditions for the rated instability of the motion $x = 0$ of system (1).

Definition 8. The motion $x = 0$ is called the rated unstable motion of system (1) if there exists a number $\bar{\epsilon}, 0 < \bar{\epsilon} < R$ such that for any $T \geq 0$ and $\delta, 0 < \delta < R$ it is possible to select an n-dimensional vector x_0 satisfying the condition $\|x_0\| < \delta$, and $t_0 \geq T$ such that the inequality $\|x(t_1, t_0, x_0)\| \geq \bar{\epsilon}$ holds for some $t_1 > t_0$.

If the system of equations (1) is autonomous or has a zero solution ($F(0,t) \equiv 0$ for all $t \geq 0$), then the rated instability of the motion $x = 0$ is equivalent to the property that is opposite to the uniform Lyapunov stability of the solution $x = 0$ to system (1). The validity of this statement follows from Theorem 2. Let us derive the rated instability conditions for the motion $x = 0$ of system (1).

Theorem 6 *For the motion $x = 0$ of system (1) to be rated unstable, it is necessary and sufficient that there exist two scalar functions $V(x,t)$ and $W(x,t)$ given in the set Ω and such that the following conditions are satisfied:*

1) for any numbers $T \geq 0$ and $\delta, 0 < \delta < R$ it is possible to select an n-dimensional vector x_0 satisfying the condition $\|x_0\| < \delta$, and $t_0 \geq T$ such that the inequality $V(x_0, t_0) > 0$ holds;

2) there exists $\frac{dV}{dt} = \lambda V + W$, where the functions $\lambda = \lambda(x,t)$ and $W(x,t)$ are continuous in Ω, and the inequalities $W(x,t) \geq 0, \lambda(x,t) > 0$ hold for all $(x,t) \in \Omega$;

3) there exists a number $\epsilon, 0 < \epsilon < R$ such that for any values $t_0 \geq 0$ and $\delta, 0 < \delta < \epsilon$, and for any n-dimensional vector function $y(t)$ satisfying the conditions $\|y(t_0)\| < \delta, V(y(t_0), t_0) > 0, \|y(t)\| < \epsilon$ for all $t \geq t_0$ the inequality

$$|V(y(t),t)| \geq V(y(t_0), t_0) \exp\left(\int_{t_0}^{t} \lambda(y(\tau), \tau) d\tau\right)$$

does not remain valid for any $t \geq t_0$.

Theorem 7 *For the motion $x = 0$ of system (1) to be rated unstable, it is necessary and sufficient that there exist two scalar functions $V(x,t)$ and $W(x,t)$ given in the set Ω and possessing the properties:*

1) $V(x,t)$ is bounded in Ω;

2) for any numbers $T \geq 0$ and $\delta, 0 < \delta < R$ there exist an n-dimensional vector x_0 satisfying the condition $\|x_0\| < \delta$, and $t_0 \geq T$ such that the inequality $V(x_0, t_0) > 0$ holds;

3) there exists $\frac{dV}{dt} = \lambda V + W$, where the number $\lambda > 0$, and the function $W(x,t)$ is continuous in Ω such that for all $(x,t) \in \Omega$ there is $W(x,t) \geq 0$.

Proof of Theorems 6 and 7 : Necessity. Let the motion $x = 0$ of system (1) be rated unstable. Then there exists a number $\bar{\epsilon}, 0 < \bar{\epsilon} < R$ such that for any $T \geq 0$ and $\delta, 0 < \delta < \bar{\epsilon}$ it is possible to select an n-dimensional vector x_0 satisfying the condition $\|x_0\| < \delta$, and a value $t_0 \geq T$ such that the inequality $\|x(t, t_0, x_0)\| < \bar{\epsilon}$ fails to hold for all $t \geq t_0$. Let $t(t_0, x_0)$ be the first time after t_0 when this inequality is disturbed. The set $\omega = \{(x,t) : \|x\| < \bar{\epsilon}, t \geq 0\}$ is made up of pairs of two types. Those pairs (x_0, t_0) for which the inequality $\|x(t, t_0, x_0)\| < \bar{\epsilon}$ holds with all $t \geq t_0$ fall into the first type, while those pairs for which this inequality fails to hold fall into the second type. Setting $V(x_0, t_0) = 0$ for pairs of the first type and $V(x_0, t_0) = \exp(t_0 - t(t_0, x_0))$ for pairs of the second type we obtain the function satisfying the conditions of Theorems 6 and 7. Show this.

Indeed, the inequality $0 \leq V(x, t) < 1$ holds for all $(x, t) \in \Omega$, which follows from construction of the function $V(x, t)$. Construction of the function also implies that for any numbers $T \geq 0$ and $\delta, 0 < \delta < \bar{\epsilon}$ there exists an n-dimensional vector x_0 satisfying the inequality $\|x_0\| < \delta$, and a quantity $t_0 \geq T$ such that the condition $V(x_0, t_0) > 0$ is satisfied. The identity $\frac{dV}{dt} = V$ holds along the solutions of system (1) whose graphs are in the set ω, i.e., for the corresponding functions we get $W(x, t) \equiv 0, \lambda(x, t) \equiv 1$. The inequality $|V(y(t), t)| \geq V(y(t_0, t_0)) \exp(t - t_0)$ fails to hold for all $t \geq t_0$, since $V(y(t_0), t_0) > 0$ and the function $V(x, t)$ is bounded in the set ω. This proves the necessity parts of Theorems 6 and 7.

Sufficiency. Suppose conditions 1) – 3) of Theorem 6 are satisfied and the motion $x = 0$ of system (1) is rated stable. Then for any number $\epsilon, 0 < \epsilon < R$ there exist quantities $T(\epsilon) \geq 0$ and $\delta(\epsilon), 0 < \delta(\epsilon) < \epsilon$ such that the inequality $\|x(t, t_0, x_0)\| < \epsilon$ holds for any n-dimensional vector x_0 satisfying the condition $\|x_0\| < \delta(\epsilon)$ and for any value $t_0 \geq T(\epsilon)$ with all $t \geq t_0$. Hence the function $V_0(t, t_0, x_0)$ is defined for all $t \geq t_0$. Suppose an n-dimensional vector x_0 and a value t_0 are taken to be such that the inequality $V(x_0, t_0) > 0$ holds. Then the equality $\frac{dV_0(t, t_0, x_0)}{dt} = \lambda_0(t, t_0, x_0) V_0(t, t_0, x_0) + W_0(t, t_0, x_0)$ holds for all $t \geq t_0$ and, since the inequality $W_0(t, t_0, x_0) \geq 0$ holds for all $t \geq t_0$, the inequality

$$V_0(t, t_0, x_0) \geq V(x_0, t_0) \exp\left(\int_{t_0}^{t} \lambda_0(\tau, t_0, x_0) d\tau\right)$$

also holds for all $t > t_0$. This last inequality contradicts condition 3) of Theorem 6, where $y(t) = x(t, t_0, x_0)$. If, however, conditions 1) – 3) of Theorem 7 are satisfied, then the obtained inequality $V_0(t, t_0, x_0) \geq V(x_0, t_0) \exp(\lambda(t - t_0))$ which holds for all $t \geq t_0$ contradicts the boundedness of the function $V(x, t)$. Thus, the assumption of rated stability for the motion $x = 0$ of system (1) contradicts the conditions of Theorems 6 and 7, and hence sufficiency is proved. This completes the proof of Theorems 6 and 7.

Theorem 8 *Let the scalar functions $V(x, t)$ and $W(x, t)$ be given in the set Ω and suppose they satisfy the conditions:*

1) $V(x, t) \to 0$ as $\|x\| \to 0, t \to +\infty$;

2) for any numbers $T \geq 0$ and $\delta, 0 < \delta < R$ it is possible to select an n-dimensional vector x_0 satisfying the condition $\|x_0\| < \delta$, and $t_0 \geq T$ such that the inequality $V(x_0, t_0) < 0$ holds;

3) there exists $\frac{dV}{dt} = -W$ such that the function $W(x, t)$ is positive-definite. Then the motion $x = 0$ of system (1) is rated unstable.

Proof: Suppose the motion $x = 0$ of system (1) is rated stable. Then for any number $\epsilon, 0 < \epsilon < R$, it is possible to select $T(\epsilon) \geq 0$ and $\delta(\epsilon), 0 < \delta(\epsilon) < \epsilon$, such that the inequality $\|x(t, t_0, x_0)\| < \epsilon$ holds for any n-dimensional vector x_0 satisfying the

An Investigation of Rated Stability by Lyapunov's Direct Method 365

condition $\|x_0\| < \delta(\epsilon)$, and for any value $t_0 \geq T(\epsilon)$ when all $t \geq t_0$. By condition 1), a number ϵ and a quantity $T(\epsilon)$ can be taken to be such that the inequality $|V(x_0, t_0)| < 1$ holds for all n-dimensional vectors x_0 satisfying the condition $\|x_0\| < \epsilon$, and for all values of $t_0, t_0 \geq T(\epsilon)$. In this case, by condition 2), the n-dimensional vector x_0 and the value of t_0 can be chosen to be such that the inequality $V(x_0, t_0) < 0$ holds. Therefore, by condition 3), we have that the function $V_0(t, t_0, x_0)$ is given for all $t \geq t_0$ and does not increase, hence the inequalities $V_0(t, t_0, x_0) \leq V(x_0, t_0) < 0$ hold for all $t \geq t_0$. There is no sequence $t_n \to +\infty$ as $n \to \infty$ such that $\|x(t_n, t_0, x_0)\| \to 0$ as $n \to \infty$, for otherwise $V_0(t_n, t_0, x_0) \to 0$ as $n \to \infty$, which is impossible by the above inequalities. Therefore, there exist numbers $\bar{\epsilon}, 0 < \bar{\epsilon} < \epsilon$ and $T_1 \geq t_0$ such that the inequality $\|x(t, t_0, x_0)\| \geq \bar{\epsilon}$ holds for all values $t \geq T_1$. Consequently, the following inequalities hold for all values $t \geq T_1$:

$$\frac{dV}{dt} \leq -\inf_{\bar{\epsilon} \leq \|x\| \leq \epsilon} W_1(x) = -\alpha < 0, V_0(t, t_0, x_0) \leq V_0(T_1, t_0, x_0) - \alpha(t - T_1).$$

From this it follows that $V_0(t, t_0, x_0) \to -\infty$ as $t \to +\infty$, but this contradicts the boundedness of the function $V(x, t)$. Thus, the motion $x = 0$ of system (1) is rated unstable. This completes the proof of Theorem 8.

Theorems 7 and 8 are analogous to Lyapunov's first and second stability theorems, respectively [Lyapunov (1950)].

Let us introduce the notion of asymptotic rated stability for the motion $x = 0$ of system (1) and, using the ideas of Lyapunov's direct method, establish the sufficient as well as the necessary conditions for asymptotic rated stability of the motion $x = 0$ of the system involved.

Definition 9. The motion $x = 0$ of system (1) is called asymptotically rated stable if it is rated stable and, additionally, for any n-dimensional vector x_0 satisfying the condition $\|x_0\| < \delta(\epsilon)$ and for any $t_0 \geq T(\epsilon)$, where the quantities $\delta(\epsilon), T(\epsilon)$ are selected for the number $\epsilon, 0 < \epsilon < R$, the limit relationship $\|x(t, t_0, x_0)\| \to 0$ holds as $t \to +\infty$ by the definition of rated stability.

Let us introduce conditions under which the motion $x = 0$ is the asymptotically rated stable motion of system (1).

Theorem 9 *For the motion $x = 0$ of system (1) to be asymptotically rated stable, it is necessary and sufficient that there exists a scalar function $V(x, t)$ given in the set Ω and such that conditions 1) – 3) of Theorem 3 and the condition below hold for it:*

4) it is possible to select quantities $T_0 \geq 0$ and $\Delta, 0 < \Delta < R$ such that for any n-dimensional vector x_0 satisfying the condition $\|x_0\| < \Delta$ and for any $t_0 \geq T_0$ the function $V_0(t, t_0, x_0)$ is defined for all $t \geq t_0$, and $V_0(t, t_0, x_0) \to 0$ as $t \to +\infty$.

Proof: *Necessity.* Suppose the motion $x = 0$ of system (1) is asymptotically rated stable, then it is rated stable. Let $V(x, t)$ be the function constructed in the proof of Theorem 3. Fix some number $\epsilon_1, 0 < \epsilon_1 < R$, and let an arbitrary n-dimensional vector x_0 and an arbitrary t_0 satisfy the conditions $\|x_0\| < \Delta = \delta(\epsilon_1)$ and $t_0 \geq T_0 = T(\epsilon_1)$, respectively. Here the quantities $\delta(\epsilon_1)$ and $T(\epsilon_1)$ are taken from the definition of rated stability for the motion $x = 0$ and correspond to a fixed number ϵ_1. Then we have that for all $t \geq t_0$ the inequality $\|x(t, t_0, x_0)\| < \epsilon_1$ holds, and $\|x(t, t_0, x_0)\| \to 0$ as $t \to +\infty$. The proof of Theorem 3 gives the equality $V_0(t, t_0, x_0) = \sup_{\tau \geq t} \|x(t, t_0, x_0)\|$ which holds for all $t \geq t_0$; it follows that condition 4) is valid. Theorem 3 implies that the other conditions are satisfied. This completes the proof of necessity.

Sufficiency. From Theorem 3 it follows that the motion $x = 0$ is rated stable. As in the proof of necessity, fix a number $\epsilon_1, 0 < \epsilon_1 < R$, and define the corresponding $T_0 = T(\epsilon_1), \Delta = \delta(\epsilon_1)$. Suppose there exists an n-dimensional vector x_0 satisfying the condition $\|x_0\| < \Delta$ and there exists a value $t_0 \geq T_0$ such that the limiting relationship $\|x(t, t_0, x_0)\| \not\to 0$ holds as $t \to +\infty$. Then there exists a number $\bar{\epsilon}, 0 < \bar{\epsilon} < \epsilon_1$, and a sequence $t_n \to +\infty$ as $n \to \infty$ such that the inequalities $\|x(t_n, t_0, x_0)\| \geq \bar{\epsilon}_1$ hold for all $n = 1, 2, \ldots$. Hence the inequalities $\bar{\epsilon} \leq \|x(t_n, t_0, x_0)\| < \epsilon_1, V_0(t_n, t_0, x_0) \geq \inf_{\bar{\epsilon} \leq \|x\| \leq \epsilon_1} V_1(x) > 0$ hold for all $n = 1, 2, \ldots$, but this contradicts condition 4). Thus, the motion $x = 0$ is an asymptotically rated stable motion of system (1) if it is taken to be in the set $\omega = \{(x, t) : \|x\| < \Delta, t \geq T_0\} \subset \Omega$. This proves sufficiency and the proof of Theorem 9 is complete.

Condition 4) of Theorem 9 is boundless, since it is stated for some set of solutions to system (1). We shall provide more constructive sufficient conditions for asymptotic rated stability of the motion $x = 0$ of system (1).

Theorem 10 *Let the scalar functions $V(x, t)$ and $W(x, t)$ given in the set Ω satisfy conditions 1) – 3) of Theorem 4 and the condition:*

4) the function $W(x, t)$ is positive-definite.

Then the motion $x = 0$ of system (1) is asymptotically rated stable.

Proof: Theorem 4 implies that the motion $x = 0$ of system (1) is rated stable. As in the proof of Theorem 9, fix a number $\epsilon_1, 0 < \epsilon_1 < R$ and define the corresponding $T_0 = T(\epsilon_1), \Delta = \delta(\epsilon_1)$. Then the inequality $\|x(t, t_0, x_0)\| < \epsilon_1$ holds for any n-dimensional vector x_0 satisfying the condition $\|x_0\| < \Delta$, and for any value $t_0 \geq T_0$ when all $t \geq t_0$. There exists a sequence $t_n \to +\infty$ as $n \to \infty$ such that $\|x(t_n, t_0, x_0)\| \to 0$ as $n \to \infty$. In fact, otherwise it is possible to select values $T_1 \geq T_0$ and $\bar{\epsilon}, 0 < \bar{\epsilon} < \epsilon_1$, such that for all $t \geq T_1$ there is the inequality $\|x(t, t_0, x_0)\| \geq \bar{\epsilon}$, and hence the inequality $\frac{dV}{dt} \leq - \inf_{\bar{\epsilon} \leq \|x\| \leq \epsilon_1} W_1(x) = -\alpha < 0$ is valid. It follows that for all $t \geq T_1$ there is the inequality $V_0(t, t_0, x_0) \leq V_0(T_1, t_0, x_0) - \alpha(t - T_1)$, and hence $V_0(t, t_0, x_0) \to -\infty$ as $t \to +\infty$, which contradicts condition 1). Thus the chosen sequence exists. By the rated stability of the motion $x = 0$, for any number $\epsilon, 0 < \epsilon < \epsilon_1$, it is possible to select $T(\epsilon) \geq 0$ and $\delta(\epsilon), 0 < \delta(\epsilon) < \Delta$, such that for any n-dimensional vector x_0' satisfying the condition $\|x_0'\| < \delta(\epsilon)$, and for any value $t_0' \geq T(\epsilon)$ the inequality $\|x(t, t_0', x_0')\| < \epsilon$ holds for all $t \geq t_0'$. Then for all $t \geq T_2(\epsilon)$, the inequality $\|x(t, t_0, x_0)\| < \epsilon$ holds for all $t \geq t_0'$. Then for all $t \geq T_2(\epsilon)$ there is $\|x(t, t_0, x_0)\| < \epsilon$, where $x_0' = x(t_n, t_0, x_0), T_2(\epsilon) = t_0' = t_n$ for a natural number n such that the conditions $\|x(t_n, t_0, x_0)\| < \delta(\epsilon)$ and $t_n \geq T(\epsilon)$ are satisfied. This means that for any n-dimensional vector x_0 subject to the condition $\|x_0\| < \Delta$ and for any $t_0 \geq T_0$ the limit relationship $\|x(t, t_0, x_0)\| \to 0$ holds as $t \to +\infty$. It follows that the motion $x = 0$ is the asymptotically rated stable motion for system (1) if it is taken

An Investigation of Rated Stability by Lyapunov's Direct Method 367

to be in the set $\omega = \{(x,t) : \|x\| < \Delta, t \geq T_0\} \subset \Omega$. This completes the proof of Theorem 10.

Theorem 10 is analogous to Lyapunov's asymptotic stability theorem [Lyapunov (1950)].

Theorem 11 *Suppose the scalar functions $V(x,t)$ and $W(x,t)$ are given in the set Ω and satisfy conditions 1) and 2) of Theorem 5 and condition*

3) $W(x,t) \to W_1(x)$ as $t \to +\infty$ uniformly in x, $\|x\| < R$, the function $W_1(x)$ is continuous in x for $\|x\| < R$, $W_1(0) = 0$, $W_1(x) > 0$ for all x, $0 < \|x\| < R$.

Then the motion $x = 0$ of system (1) is asymptotically rated stable.

Proof: Theorem 5 implies that the motion $x = 0$ of system (1) is rated stable. Show that for any n-dimensional vector x_0 satisfying the condition $\|x_0\| < \Delta$ and for any value $t_0 \geq T_0$ there is the limit relationship $\|x(t,t_0,x_0)\| \to 0$ as $t \to +\infty$, where the variables $T_0 \geq 0$ and $\Delta > 0$ are chosen in much the same way as in the proof of Theorem 9. There exists a sequence $t_n \to +\infty$ as $n \to \infty$ such that we get $\|x(t_n,t_0,x_0)\| \to 0$ as $n \to \infty$; otherwise it is possible to select values $T_1 \geq T_0$ and $\bar{\epsilon}, 0 < \bar{\epsilon} < \epsilon_1$ such that the inequalities $\bar{\epsilon} \leq \|x(t,t_0,x_0)\| < \epsilon_1$ hold for all $t \geq T_1$. Let us choose a quantity $T_2 \geq T_1$ such that for all $t \geq T_2$ and all n-dimensional vectors x satisfying the condition $\|x\| < R$ there is the inequality $|W(x,t) - W_1(x)| < \frac{1}{2}\alpha$, where $\alpha = \inf_{\bar{\epsilon} \leq \|x\| \leq \epsilon_1} W_1(x)$. Then the inequalities $W_0(t,t_0,x_0) \geq \frac{1}{2}\alpha$ and $\frac{dV}{dt} \leq -\frac{1}{2}\alpha$ hold for all $t \geq T_2$. Consequently, the inequality $V_0(t,t_0,x_0) \leq V_0(T_2,t_0,x_0) - \frac{1}{2}(t - T_2)\alpha$ holds for all $t \geq T_2$, whence it follows that $V_0(t,t_0,x_0) \to -\infty$ as $t \to +\infty$, but this contradicts the boundedness of the function $V(x,t)$ implied by condition 1). Thus, the above sequence exists. Further, by the argument given in the proof of Theorem 10, we conclude that $\|x(t,t_0,x_0)\| \to 0$ as $t \to +\infty$. Thus, the motion $x = 0$ of system (1) is asymptotically rated stable. This completes the proof of Theorem 11.

Theorem 12 *Let the scalar functions $V(x,t)$ and $W(x,t)$ be given in the set Ω, satisfying conditions 1) and 2) of Theorem 5 and condition*

3) $W(x,t) \geq u(V(x,t))$, where the scalar function $u(\theta)$ of the real argument θ is continuous for $0 \leq \theta < +\infty$, $u(0) = 0$ and $u(\theta) > 0$ for $\theta > 0$.

Then the motion $x = 0$ of system (1) is asymptotically rated stable.

Proof: Theorem 5 implies that the motion $x = 0$ of system (1) is rated stable. Fix a number $\epsilon_1, 0 < \epsilon_1 < R$, and choose, as in the proof of Theorem 9, the corresponding quantities $T_0 \geq 0$ and $\Delta, 0 < \Delta < \epsilon_1$. Then for any n-dimensional vector x_0 satisfying the condition $\|x_0\| < \Delta$ and for any value $t_0 \geq T_0$ the inequality $\|x(t,t_0,x_0)\| < \epsilon_1$ holds for all $t \geq t_0$, and hence the function $V_0(t,t_0,x_0)$ is defined for all $t \geq t_0$. From conditions 1) and 2) it follows that the function $V_0(t,t_0,x_0)$ does not increase, is bounded, and therefore has a non-negative limit as $t \to +\infty$. Let $\lim_{t \to +\infty} V_0(t,t_0,x_0) = \alpha > 0$. Then, with α_1 as some constant, the inequalities $V_0(t,t_0,x_0) \geq \alpha$ and $0 < \alpha_1 \leq u(V_0(t,t_0,x_0))$ hold for all $t \geq t_0$. Further, by employing the argument given in the proofs of Theorems 10 and 11, we obtain $V_0(t,t_0,x_0) \to -\infty$ as $t \to +\infty$, which yields a contradiction. Hence $V_0(t,t_0,x_0) \to 0$ as $t \to +\infty$. From this, by employing the argument given in the proof of sufficient conditions of Theorem 9, it is concluded that the limit relationship $\|x(t,t_0,x_0)\| \to 0$ holds as $t \to +\infty$. Thus, the motion $x = 0$ of system (1) is asymptotically rated stable. This completes the proof of Theorem 12.

Rated stability is closely related to stability in Lyapunov's sense and is conceptually associated with it. This will be illustrated in the proof of the assertions below.

Theorem 13 *Let the n-dimensional vector function $F(x,t)$ be defined and continuous in the set Ω and let the limit relationship $F(x,t) \to f(x)$ hold as $t \to +\infty$ uniformly in x, $\|x\| < R$. Suppose that $f(0) = 0$ and the solution $z = 0$ of the system of differential equations*

$$\frac{dz}{dt} = f(z) \qquad (3)$$

is stable (asymptotically stable) in Lyapunov's sense. Denote

$$G(y,\tau) = \begin{cases} F(y, -\ln|\tau|), & 0 < |\tau| \le 1, \\ f(y), & \tau = 0. \end{cases}$$

Then $G(0,0) = 0$ and the function $G(y,\tau)$ is continuous in all its independent variables for $\|y\| < R, |\tau| < 1$. The motion $x = 0$ of system (1) is said to be rated stable (asymptotically rated stable) if and only if the solution $y = 0, \tau = 0$ of the autonomous system of differential equations

$$\frac{dy}{dt} = G(y,\tau), \qquad \frac{d\tau}{dt} = -\tau \qquad (4)$$

is stable (asymptotically stable) in Lyapunov's sense.

Proof: First we show that the function $G(y,\tau)$ is continuous in all (y,τ) for all $\|y\| < R, |\tau| < 1$. For $\|y\| < R, 0 < |\tau| < 1$, this follows from the continuity of the function $F(x,t)$ in the set Ω. Let us see that the function $G(y,\tau)$ is continuous in all (y,τ) in the set $\omega_1 = \{(y,\tau) : \|y\| < R, \tau = 0\}$. To this end, we first establish the continuity of the function $f(x)$ in x for $\|x\| < R$. Indeed, if the opposite is true, then there exists a number $\bar{\epsilon} > 0$ and an n-dimensional vector x_0 satisfying the condition $\|x_0\| < R$, and there exists a sequence of n-dimensional vectors x_n satisfying the condition $\|x_n\| < R$ and the limit relationship $\|x_n - x_0\| \to 0$ as $n \to \infty$ such that the inequalities $\|f(x_n) - f(x_0)\| \ge \bar{\epsilon}$ hold for all $n = 1, 2, \ldots$. Since $F(x,t) \to f(x)$ as $t \to +\infty$ uniformly in $x, \|x\| < R$, there exists $T_0 \ge 0$ such that the inequality $\|F(x,t) - f(x)\| < \frac{\bar{\epsilon}}{4}$ holds for any $t > T_0$ and any n-dimensional vector x satisfying the condition $\|x\| < R$. Therefore, the inequality $\|F(x_n,t) - F(x_0,t)\| \ge \frac{1}{2}\bar{\epsilon}$ holds for all $t > T_0$ and all $n = 1, 2, \ldots$, which contradicts the continuity of the function $F(x,t)$ in the set Ω. Thus, for any number $\epsilon > 0$ it is possible to select $\delta = \delta(\epsilon) > 0$ such that the inequality $\|f(x') - f(x'')\| < \frac{1}{2}\epsilon$ holds for any two n-dimensional vectors x' and x'' satisfying the conditions $\|x'\| < R, \|x''\| < R$ and $\|x' - x''\| < \delta$. Since $F(x,t) \to f(x)$ as $t \to +\infty$ uniformly in $x, \|x\| < R$, there exists $T_1 \ge 0$ such that the inequality $\|F(x'',t) - f(x'')\| < \frac{1}{2}\epsilon$ holds for any $t > T_1$ and any n-dimensional vector x'' satisfying the condition $\|x''\| < R$. Note that in $G(y,\tau)$ the variable τ is related to the variable t in the function $F(x,t)$ by the formula $t = -\ln|\tau|$, and therefore the condition $t > T_1$ is equivalent to the condition $|\tau| < \exp(-T_1)$. In view of this relation, it follows from the above two inequalities that the inequality $\|G(x'', \tau) - G(x', 0)\| < \epsilon$ holds when the condition $\|x' - x''\|^2 + \tau^2 < \min\{\delta^2(\epsilon), \exp(-2T_0)\}$ is satisfied. This proves the continuity of the function $G(y, \tau)$ in the set ω_1, i.e., for $\|y\| < R, \tau = 0$. Thus, the function $G(y, \tau)$ is continuous in the set $\omega_2 = \{(y, \tau) : \|y\| < R, |\tau| < 1\}$, and hence the existence conditions hold for solutions to the system of differential equations (4) (see, e.g., [Coddington and Levinson (1955)], [Hale (1969)], [Cesari (1959)]). Consequently, for any initial conditions (y_0, τ_0) from the set ω_2 the solution $(y(t, t_0, y_0, \tau_0), \tau(t, t_0, \tau_0))$ of system (4) is defined for all $t \geq t_0$ such that the inequalities $\|y(t, t_0, y_0, \tau_0)\| < R, |\tau(t, t_0, \tau_0)| < 1$ hold. Note that system (1) is autonomous, and therefore the initial time t_0 can be chosen in an arbitrary way. Then we set $t_0 = 0$. In what follows the solution of system (4) will be denoted as $(y(t, y_0, \tau_0), \tau(t, \tau_0)) = (y(t), \tau(t))$.

We shall now prove the statements of Theorem 13. First prove the direct statement. Let the motion $x = 0$ of system (1) be rated stable, i.e., for any number $\epsilon, 0 < \epsilon < \min\{R, 1\}$ there exist $T(\epsilon) \geq -\ln \epsilon$ and $\delta(\epsilon), 0 < \delta(\epsilon) < \epsilon$, such that for any $t_0 \geq T(\epsilon)$ and for any n-dimensional vector x_0 satisfying the condition $\|x_0\| < \delta(\epsilon)$ the inequality $\|x(t, t_0, x_0)\| < \epsilon$ holds for all $t \geq t_0$. Since the system of equations (3) is autonomous, for solutions of this system we set the initial time $t_0 = 0$ and denote the solution as $z(t, z_0) = z(t), z_0 = z(0)$. Since the solution $z(t)$ of system (3) is stable in Lyapunov's sense, for a number ϵ it is possible to select $\Delta(\epsilon), 0 < \Delta(\epsilon) < \epsilon$, such that for any n-dimensional vector z_0 satisfying the condition $\|z_0\| < \Delta(\epsilon)$ for all $t \geq 0$ the inequality $\|z(t)\| < \epsilon$ holds. Let $\delta_0(\epsilon) = \min\{\delta(\epsilon), \Delta(\epsilon), \exp(-T(\epsilon))\}$. Show that $\delta_0(\epsilon)$ is the required quantity for defining Lyapunov stability of the solution $y = 0, \tau = 0$ for system (4).

We say that the initial conditions (t_0, x_0) of the solution $x(t, t_0, x_0)$ of system (1) are related to the initial conditions (y_0, τ_0) of the solution $(y(t), \tau(t))$ of system (4), namely, $x_0 = y_0$ and $t_0 = -\ln |\tau_0|$ with $\tau_0 \neq 0$. This means that for the choice of the value of τ_0, $0 < |\tau_0| < 1$ and n-dimensional vector y_0, $\|y_0\| < R$ in system (4) taken for all $t \geq 0$ there is the choice of $x_0 = y_0$ in system (1) taken for all $t \geq t_0$. Thus, with $\tau_0 \neq 0$ the identity $y(t) \equiv x(t + t_0, t_0, x_0)$ holds for all $t \geq 0$. In the case $\tau_0 = 0$, for all $t \geq 0$ we get $\tau(t) \equiv 0$, which corresponds to the solution $z(t)$ of system (3) for $z_0 = y_0$. In other words, with $\tau_0 = 0$, the identity $y(t) \equiv z(t), \tau(t) \equiv 0$ holds for all $t \geq 0$. Let the initial conditions (y_0, τ_0) satisfy the inequality $\|y_0\|^2 + \tau_0^2 < \delta_0^2(\epsilon)$, then the following inequalities hold: $\|x_0\| < \delta(\epsilon), t_0 > T(\epsilon), \|z_0\| < \Delta(\epsilon)$. Therefore, by the choice of $T(\epsilon), \delta(\epsilon)$ and $\Delta(\epsilon)$, we have that the inequality $\|x(t, t_0, x_0)\| < \epsilon$ holds for all $t \geq t_0$ and that the inequalities $\|z(t)\| < \epsilon$ and $\tau^2(t) \leq \tau_0^2 < \exp(-2T(\epsilon))$ hold for all $t \geq 0$. It follows that the inequality $\|y(t)\|^2 + \tau^2(t) < \epsilon^2 + \exp(-2T(\epsilon)) < 2\epsilon^2$ holds for all $t \geq 0$. And this means that the solution $y = 0, \tau = 0$ for system (4) is Lyapunov stable. If the motion $x = 0$ of system (1) is asymptotically rated stable, and the solution $z = 0$ of system (3) is asymptotically Lyapunov stable, then the quantities $\delta(\epsilon)$ and $\Delta(\epsilon)$ can be chosen such that for all values of $t_0 \geq T(\epsilon)$ and for all n-dimensional vectors x_0 and z_0 satisfying the conditions $\|x_0\| < \delta(\epsilon)$ and $\|z_0\| < \Delta(\epsilon)$ the respective limit relationships hold: $\|x(t, t_0, x_0)\| \to 0$ as $t \to +\infty$ and $\|z(t)\| \to 0$ as $t \to +\infty$. Show that with the above choice of $\delta_0(\epsilon)$ for all initial conditions (y_0, τ_0) satisfying the inequality $\|y_0\|^2 + \tau_0^2 < \delta_0^2(\epsilon)$ the limit $\|y(t)\|^2 + \tau^2(t) \to 0$ holds as

$t \to +\infty$. In fact, when this inequality holds the conditions $\|x_0\| < \delta(\epsilon)$, $\|z_0\| < \Delta(\epsilon)$ also hold and, by the above argument, the limit $\|y(t)\| \to 0$ holds as $t \to +\infty$. Since $\tau(t) = \tau_0 \exp(-t) \to 0$ as $t \to +\infty$, the limit $\|y(t)\|^2 + \tau^2(t) \to 0$ holds as $t \to +\infty$. Thus, the solution $y = 0, \tau = 0$ of system (4) is asymptotically Lyapunov stable.

We shall now prove the reverse statement of Theorem 13. Let the solution $y = 0, \tau = 0$ of system (4) be Lyapunov stable. Then for any number $\epsilon, 0 < \epsilon < \min\{R, 1\}$, it is possible to select a quantity $\delta_0(\epsilon), 0 < \delta_0(\epsilon) < \epsilon$, such that for any initial conditions (y_0, τ_0) satisfying the inequality $\|y_0\|^2 + \tau_0^2 < \delta_0^2(\epsilon)$ the inequality $\|y(t)\|^2 + \tau^2(t) < \epsilon^2$ holds for all $t \geq 0$, and so does the inequality $\|y(t)\|^2 < \epsilon^2 - \tau^2(t) \leq \epsilon^2$, i.e., $\|y(t)\| < \epsilon$ for all $t \geq 0$. In view of this, select the values of τ_0 in the initial conditions so that the inequality $0 < \tau_0 < \frac{\delta_0(\epsilon)}{\sqrt{2}}$ holds. Note that this corresponds to selecting the value $t_0 \geq \ln\sqrt{2} - \ln \delta_0(\epsilon)$ on examination of the solution $x(t, t_0, x_0)$ of system (1). Take the quantities $T(\epsilon) = \ln\sqrt{2} - \ln \delta_0(\epsilon), \delta(\epsilon) = \frac{\delta_0(\epsilon)}{\sqrt{2}}$. Then for all $t \geq 0$ the above inequality $\|y(t)\| < \epsilon$ holds for any $t_0 \geq T(\epsilon)$ and for any n-dimensional vector x_0 satisfying the condition $\|x_0\| < \delta(\epsilon)$, because the inequality $\|y_0\|^2 + \tau_0^2 < \delta_0^2(\epsilon)$ is valid. Since the identity $y(t) \equiv x(t + t_0, t_0, x_0)$ holds for any $t \geq 0$, the inequality $\|x(t, t_0, x_0)\| < \epsilon$ holds for the same initial conditions (t_0, x_0) at all $t \geq t_0$, and the motion $x = 0$ of system (1) is rated stable.

If the solution $y = 0, \tau = 0$ of system (4) is asymptotically Lyapunov stable, the above variable $\delta_0(\epsilon)$ can be chosen such that for all initial conditions (y_0, τ_0) satisfying the inequality $\|y_0\|^2 + \tau_0^2 < \delta_0^2(\epsilon)$ there is additionally the limit relationship $\|y(t)\|^2 + \tau^2(t) \to 0$ as $t \to +\infty$. Since $\tau(t) = \tau_0 \exp(-t) \to 0$ as $t \to +\infty$, this implies the limit relationship $\|y(t)\| \to 0$ as $t \to +\infty$. Since the identity $y(t) \equiv x(t + t_0, t_0, x_0)$ holds for any $t \geq 0$, selecting the initial conditions (t_0, x_0) by the above procedure yields the limit relationship $\|x(t, t_0, x_0)\| \to 0$ as $t \to +\infty$. This means that the motion $x = 0$ of system (1) is asymptotically rated stable. We have thus proved Theorem 13.

Remark. Under the conditions of Theorem 13, the motion $x = 0$ of system (1) is rated unstable if and only if the solution $y = 0, \tau = 0$ of system (4) is Lyapunov unstable. The validity of this assertion follows from Theorem 13.

Theorem 14 *Suppose $\varphi(t)$ is a solution of system (1) which is defined for all $t \geq 0$ and its graph is contained in the set Ω. Let $\|\varphi(t)\| \to 0$ as $t \to +\infty$. It is stated that:*

1) if the solution $\varphi(t)$ is uniformly Lyapunov-stable, then the motion $x = 0$ of system (1) is rated stable.

2) if the solution $\varphi(t)$ is also asymptotically stable, then the motion $x = 0$ of system (1) is asymptotically rated stable;

3) if the solution $\varphi(t)$ is not uniformly stable, then the motion $x = 0$ of system (1) is rated unstable.

Proof: 1) Since the solution $\varphi(t)$ is uniformly stable, then for all $t \geq t_0$ the inequality $\|x(t, t_0, x_0) - \varphi(t)\| < \epsilon$ holds for any $t_0 \geq 0$ and for any n-dimensional vector x_0 satisfying the condition $\|x_0 - \varphi(t_0)\| < \delta(\epsilon)$. Since $\|\varphi(t)\| \to 0$ as $t \to +\infty$, it is possible to select $T(\epsilon) \geq 0$ such that the inequality $\|\varphi(t)\| < \frac{1}{2}\delta(\epsilon) < \frac{1}{2}\epsilon$ holds for

An Investigation of Rated Stability by Lyapunov's Direct Method

all $t \geq T(\epsilon)$. Let the initial conditions (t_0, x_0) be such that $t_0 \geq T(\epsilon), \|x_0\| < \frac{1}{2}\delta(\epsilon)$. Then the inequality $\|x_0 - \varphi(t_0)\| \leq \|x_0\| + \|\varphi(t_0)\| < \delta(\epsilon)$ holds. By the stability of the solution $\varphi(t)$, it follows that the inequality $\|x(t, t_0, x_0) - \varphi(t)\| < \epsilon$ holds for all $t \geq t_0$. Therefore, the inequality $\|x(t, t_0, x_0)\| \leq \|x(t, t_0, x_0) - \varphi(t)\| + \|\varphi(t)\| < \frac{3}{2}\epsilon$ holds for all $t \geq t_0$, which implies that the motion $x = 0$ of system (1) is rated stable.

2) In addition, if the solution $\varphi(t)$ is asymptotically stable, then $\|x(t, t_0, x_0) - \varphi(t)\| \to 0$ as $t \to +\infty$. Since $\|\varphi(t)\| \to 0$ as $t \to +\infty$, the inequality $\|x(t, t_0, x_0)\| \leq \|x(t, t_0, x_0) - \varphi(t)\| + \|\varphi(t)\|$ implies the validity of the limit relationship $\|x(t, t_0, x_0)\| \to 0$ as $t \to +\infty$. Consequently, the motion $x = 0$ of system (1) is asymptotically rated stable.

3) If the solution $\varphi(t)$ is not uniformly stable, then there exists a number $\bar{\epsilon} > 0$ such that for any $T \geq 0$ and $\delta > 0$ it is possible to select initial conditions (t_0, x_0) such that $t_0 \geq T, \|x_0 - \varphi(t_0)\| < \delta$ and the inequality $\|\varphi(t_1) - x(t_1, t_0, x_0)\| \geq \bar{\epsilon}$ holds for some $t_1 > t_0$. Since $\|\varphi(t)\| \to 0$ as $t \to +\infty$, there exists $T_0 = T_0(\delta)$ such that the inequality $\|\varphi(t)\| < \min\{\delta, \frac{1}{2}\bar{\epsilon}\}$ holds for all $t \geq T_0$. By the foregoing, we may choose $T \geq T_0(\delta)$, and let (t_0, x_0) be the corresponding initial conditions, i.e. the inequalities $t_0 \geq T, \|x_0 - \varphi(t_0)\| < \delta$ and $\|x(t_1, t_0, x_0) - \varphi(t_1)\| \geq \bar{\epsilon}$ hold for some $t_1 > t_0$. Then the inequalities $\|x_0\| \leq \|x_0 - \varphi(t_0)\| + \|\varphi(t_0)\| < 2\delta$ and $\|x(t_1, t_0, x_0)\| = \|x(t_1, t_0, x_0) - \varphi(t_1) + \varphi(t_1)\| \geq \|x(t_1, t_0, x_0) - \varphi(t_1)\| - \|\varphi(t_1)\| > \bar{\epsilon} - \frac{1}{2}\bar{\epsilon} = \frac{1}{2}\bar{\epsilon}$ are valid. Consequently, for any quantities $\delta > 0$ and $T \geq T_0(\delta)$ we have constructed the rule of selecting initial conditions (t_0, x_0) such that $t_0 \geq T, \|x_0\| < 2\delta$ and the inequality $\|x(t_1, t_0, x_0)\| > \frac{1}{2}\bar{\epsilon}$ holds for some value $t_1 > t_0$, which implies the rated stability of the motion $x = 0$ of system (1). This completes the proof of Theorem 14.

Theorem 15 *Let system (1) have a solution $x = 0$ which is stable (asymptotically stable) in Lyapunov's sense, the stability being uniform. Then for any n-dimensional vector function $\varphi_1(t)$ (whose graph belongs to the set Ω), which is given for all $t \geq 0$ and is such that $\|\varphi_1(t)\| \to 0$ as $t \to +\infty$, we have that the motion $x = \varphi_1(t)$ of system (1) is rated stable (asymptotically rated stable).*

Proof: Let $\epsilon \in (0, R)$ be an arbitrary number. Then, by the uniform stability of the solution $x = 0$ of system (1), there exists a quantity $\delta(\epsilon), 0 < \delta(\epsilon) < \frac{1}{2}\epsilon$ such that for all $t \geq t_0$ the inequality $\|x(t, t_0, x_0)\| < \frac{1}{2}\epsilon$ holds for any initial conditions (t_0, x_0) satisfying the inequalities $t_0 \geq 0$ and $\|x_0\| < \delta(\epsilon)$. Since $\|\varphi_1(t)\| \to 0$ as $t \to +\infty$, it is possible to select $T(\epsilon) \geq 0$ such that the inequality $\|\varphi_1(t)\| < \frac{1}{2}\delta(\epsilon) < \frac{1}{2}\epsilon$ holds for all $t \geq T(\epsilon)$. Consequently, if the initial conditions (t_0, x_0) satisfy the inequalities $t_0 \geq T(\epsilon), \|x_0 - \varphi_1(t_0)\| < \frac{1}{2}\delta(\epsilon)$, then by the above inequalities, we get $\|x_0\| \leq \|\varphi_1(t_0)\| + \|x_0 - \varphi_1(t_0)\| < \delta(\epsilon)$, and the inequality $\|x(t, t_0, x_0) - \varphi_1(t)\| \leq \|x(t, t_0, x_0)\| + \|\varphi_1(t)\| < \epsilon$ holds for all $t \geq t_0$. Thus, the motion $x = \varphi_1(t)$ of system (1) is rated stable. In the case of asymptotic stability of the solution $x = 0$ of system (1) we have in addition the limit relationship $\|x(t, t_0, x_0)\| \to 0$ as $t \to +\infty$. Since $\|\varphi_1(t)\| \to 0$ as $t \to +\infty$, the inequality $\|x(t, t_0, x_0) - \varphi_1(t)\| \leq \|x(t, t_0, x_0)\| + \|\varphi_1(t)\|$ implies the validity of the limit relationship $\|x(t, t_0, x_0) - \varphi_1(t)\| \to 0$ as $t \to +\infty$. Consequently, in this case the motion $x = \varphi_1(t)$ of system (1) is asymptotically rated stable. This completes the proof of Theorem 15.

Remark. The vector function $\varphi_1(t)$ in the formulation of Theorem 15 may be a nondifferentiable and even a discontinuous function. Thus, the passage to a system as (2) is found to be impossible for cases of this kind.

Theorem 16 Let system (1) have a Lyapunov-stable (asymptotically stable) solution $x = 0$ such that the stability is uniform. Let $\psi(t)$ be a differentiable n-dimensional vector function, with its graph in the set Ω, which is given for all $t \geq 0$. Also, suppose $\|\psi(t)\| \to 0$ holds as $t \to +\infty$. Then for the system of differential equations

$$\frac{dy}{dt} = f(y - \psi(t), t) + \frac{d\psi(t)}{dt} \tag{5}$$

the motion $y = 0$ is rated stable (asymptotically rated stable). Here $y = (y_1, \ldots, y_n)$ is the n-dimensional vector.

Proof: It can be easily shown that the existence and uniqueness conditions of the solution for system (1) also hold for the system of equations (5), but in the set $\Omega_1 = \{(y,t) : \|y\| < R_1, t \geq 0\}$, R_1 is a constant, $0 < R_1 < R$. Denoting $x = y - \psi(t)$ we have that if $y(t, t_0, y_0)$ is a solution of system (5), then $x(t, t_0, x_0) = y(t, t_0, y_0) - \psi(t)$ is the corresponding solution of system (1) such that $x_0 = y_0 - \psi(t_0)$. System (5) has a solution $y(t, t_0, \psi(t_0)) = \psi(t)$ which corresponds to the solution $x = 0$ of system (1). Therefore, the solution $\psi(t)$ of system (5) is Lyapunov-stable (asymptotically stable), the stability being uniform. This immediately follows from the stability properties of the solution $x = 0$ of system (1). Since the limit relationship $\|\psi(t)\| \to 0$ holds as $t \to +\infty$, by Theorem 14, the motion $y = 0$ of system (5) is rated stable (asymptotically rated stable). This completes the proof of Theorem 16.

Of interest is the invertibility of Theorem 16. Let us state the following problem. Suppose the motion $y = 0$ of a system of differential equations

$$\frac{dy}{dt} = \Phi(y, t) \tag{6}$$

is rated stable (asymptotically rated stable). We say that the assumptions of the existence and uniqueness of the solution made for system (1) also hold for system (6). We shall derive the conditions under which there exists a differentiable n-dimensional vector function $\psi(t)$ with its graph in the set Ω which is given for all $t \geq 0$ and is such that the limit relationship $\|\psi(t)\| \to 0$ holds as $t \to +\infty$, and the right-hand sides of system (6) coincide with those of system (5). In this case the corresponding system as (1) has a Lyapunov-stable (asymptotically stable) solution $x = 0$ such that the stability is uniform.

In the case of asymptotically rated stability of the motion $y = 0$ of system (6), the stated problem can be approached as follows. By the asymptotically rated stability of the motion $y = 0$ of system (6), for any number ϵ, $0 < \epsilon < R$, there are $T(\epsilon) \geq 0$ and $\delta(\epsilon), 0 < \delta(\epsilon) < \epsilon$ such that for all $t \geq t_0$ the inequality $\|y(t, t_0, y_0)\| < \epsilon$ holds for any initial conditions (t_0, y_0) satisfying the inequalities $t_0 \geq T(\epsilon)$ and $\|y_0\| < \delta(\epsilon)$, and, in addition, $\|y(t, t_0, y_0)\| \to 0$ as $t \to +\infty$. Thus, there exists a family of the desired functions $\psi(t) = y(t, t_0, y_0)$. From this family we may choose a function $\psi(t)$

such that the solution $x = 0$ of the linear system of differential equations

$$\frac{dx}{dt} = \left.\frac{\partial \Phi(y,t)}{\partial y}\right|_{y=\psi(t)} \cdot x \tag{7}$$

is asymptotically Lyapunov-stable. Here $x = (x_1, \ldots, x_n)$ is an n-dimensional vector, while the matrix on the right-hand side of system (7) is the functional matrix of components of the vector $\Phi(y,t)$ in components of the vector y which is computed on the corresponding solution $y(t, t_0, y_0)$. Then the desired vector of the right-hand sides of system as (1) is the vector $F(x,t) = \Phi(x + \psi(t), t) - \Phi(\psi(t), t)$, and the system of linear equations (7) is the first approximation system for system as (1). Since the solution $x = 0$ of system (7) is asymptotically Lyapunov-stable, under some additional assumptions of the right-hand sides of systems (6) and (7) [Cesari (1959), Hirsch and Smale (1974)] it is possible to make inferences about asymptotic stability of the solution $x = 0$ to system as (1). However, in the case of rated stability (not asymptotic) of the motion $y = 0$ for system (6) the issue remains open.

We shall now investigate the rated stability of the motion $x = 0$ of system (1) in its simplest form – a nonhomogeneous linear system of ordinary differential equations. First prove the auxiliary assertion below.

Lemma 1 *Let λ be an arbitrary real number, $\lambda < 0$. Assume that the real-valued function $g(t)$ which is given and continuous for all $t \geq 0$ satisfies one of the relationships*

$$g(t) \to 0 \quad \text{as} \quad t \to +\infty, \tag{8}$$

$$\left|\int_0^{+\infty} g(t)dt\right| < +\infty. \tag{9}$$

Then we get

$$e^{\lambda t} \int_0^t e^{-\lambda \tau} g(\tau) d\tau \to 0 \quad \text{as} \quad t \to +\infty. \tag{10}$$

Proof: Let condition (8) be satisfied. From the inequality

$$0 \leq |e^{\lambda t} \int_0^t e^{-\lambda t} g(\tau) d\tau| \leq e^{\lambda t} \int_0^t e^{-\lambda \tau} |g(\tau)| d\tau$$

it follows that relationship (10) is valid if it holds for the function $|g(t)|$. Since the inequality $\exp(-\lambda t)|g(t)| \geq 0$ holds for all $t \geq 0$, one of the following two situations is realized:

1) there exists a constant $L > 0$ such that the following inequality holds for all $t \geq 0$

$$0 \leq \int_0^t e^{-\lambda \tau} |g(\tau)| d\tau \leq L,$$

then we get

$$0 \leq e^{\lambda t} \int_0^t e^{-\lambda \tau} |g(\tau)| d\tau \leq L e^{\lambda t} \to 0 \quad \text{as} \quad t \to +\infty,$$

i.e. relationship below is valid;
2) the limit relationship below holds

$$\int_0^t e^{-\lambda\tau}|g(\tau)|d\tau \to +\infty \quad \text{as} \quad t \to +\infty,$$

then, applying the L'Hospital rule to the fraction

$$\frac{\int_0^t e^{-\lambda\tau}|g(\tau)|d\tau}{e^{-\lambda t}},$$

we derive from condition (8) the validity of relationship (10).

Let condition (9) be satisfied, which is equivalent to the following

$$\int_t^{+\infty} g(\tau)d\tau \to 0 \quad \text{as} \quad t \to +\infty. \tag{11}$$

Let us use the transform

$$e^{\lambda t}\int_0^t e^{-\lambda\tau}g(\tau)d\tau = -e^{\lambda t}\int_0^t e^{-\lambda\tau}d(\int_\tau^{+\infty} g(\xi)d\xi),$$

and take this integral by parts. This yields

$$e^{\lambda t}\int_0^t e^{-\lambda\tau}g(\tau)d\tau = -\int_t^{+\infty} g(\tau)d\tau + e^{\lambda t}\int_0^{+\infty} g(\tau)d\tau - \lambda e^{\lambda t}\int_0^t e^{-\lambda\tau}\int_\tau^{+\infty} g(\xi)d\xi d\tau.$$

By condition (11), the foregoing implies that each summand on the right-hand side of the equation derived tends to zero as $t \to +\infty$, i.e. relationship (10) is valid. This completes the proof of Lemma 1.

Remark 1. The assertion of Lemma 1 is true when λ is a complex number, $\operatorname{Re}\lambda < 0$, while the function $g(t)$ takes complex values. In this case, relationships (8) – (11) hold for the real and the imaginary parts of the above expressions.

Remark 2. Conditions (8) and (9) are not mutually exclusive. For example, the function $g(t) = h(t)\exp(-\alpha t)$ satisfies these conditions simultaneously; here the constant $\alpha > 0$, the function $h(t)$ is given, continuous and bounded for all $t \geq 0$.

We shall give examples of the functions subject to conditions (8) and (9).

1) $g_1(t) = (1+t)^{-\alpha}$, the constant $\alpha > 0$. This function satisfies condition (8) at all times, and condition (9) for $\alpha > 1$. For $0 < \alpha \leq 1$ the function $g_1(t)$ does not satisfy condition (9).

2) $g_2(t) = 3\cos(1+t)^3 - \dfrac{\sin(1+t)^3}{(1+t)^2}$.

3) $g_3(t) = 4(1+t)\cos(1+t)^4 - \dfrac{\sin(1+t)^4}{(1+t)^2}$.

The functions $g_2(t), g_3(t)$ satisfy condition (9), but fail to satisfy condition (8). Note that the functions $g_2(t), g_3(t)$ are strongly oscillating depending on t, while $g_3(t)$ is unbounded for $t \geq 0$.

Consider a nonhomogeneous linear system of the form

$$\frac{dx}{dt} = \mathbf{A}x + f(t), \tag{12}$$

where $x = (x_1, \ldots, x_n)^*$, $f(t) = (f_1(t), \ldots, f_n(t))^*$, $\mathbf{A}$ is a constant real $(n \times n)$ matrix, the functions $f_j(t), j = 1, \ldots, n$, are defined, continuous, and take real values for any $t \geq 0$.

Theorem 17 *Let all eigenvalues of the matrix $\mathbf{A}$ of system (12) have negative real parts, and the functions $f_j(t)$, $j = 1, \ldots, n$, simultaneously satisfy condition (8) or (9). Then the motion $x = 0$ of system (12) is asymptotically rated stable.*

Proof: It is well known [Hirsch and Smale (1974)] that there exists a constant nonsingular $(n \times n)$ matrix $\mathbf{S}$ for which the matrix $\mathbf{J} = \mathbf{SAS}^{-1}$ is a Jordan form of the matrix $\mathbf{A}$. In this case, the elements of the matrix $\mathbf{S}$ can be complex numbers. Let us make in system (12) a nonsingular linear transformation $x = \mathbf{S}^{-1}y$, $y = \mathbf{S}x$, where the components of the vector $y = (y_1, \ldots, y_n)$ are generally complex numbers. Then the system of equations (12) becomes

$$\frac{dy}{dt} = \mathbf{J}y + \mathbf{S}f(t). \tag{13}$$

Denote $\mathbf{S}f(t) = h(t) = (h_1(t), \ldots, h_n(t))^*$. Under the above assumptions of $f(t)$, we have that the functions $h_j(t)$, $j = 1, \ldots, n$, are defined, continuous for all $t \geq 0$, and simultaneously satisfy condition (8) or condition (9). The system of equations (13) decomposes into subsystems of smaller dimension, which correspond to Jordan boxes in Jordan form $\mathbf{J}$ of the matrix $\mathbf{A}$. To prove the assertion of Theorem 17 it suffices to prove the corresponding assertion for each subsystem of system (13); then Theorem 17 will hold for the entire system (13), and hence for the original system (12).

Let us prove the assertion of Theorem 17 for a subsystem for which there is a Jordan box of order k $(1 \leq k \leq n)$:

$$\frac{dz_j}{dt} = \lambda z_j + z_{j+1} + h_j(t), \ j = 1, \ldots, k-1,$$

$$\frac{dz_k}{dt} = \lambda z_k + h_k(t). \tag{14}$$

Here λ is the eigenvalue of the matrix $\mathbf{A}$ corresponding to the chosen Jordan box. In vector notation, system (14) becomes

$$\frac{dz}{dt} = \mathbf{\Lambda}z + h(t), \tag{15}$$

where $z = (z_1, \ldots, z_k)^*$, $h(t) = (h_1(t), \ldots, h_k(t))^*$, $\mathbf{\Lambda}$ is a $(k \times k)$ matrix, the Jordan box of order k corresponding to system (14). Note that the functions $h_j(t), j = 1, \ldots, k$, are defined, continuous, and simultaneously satisfy condition (8) or condition

(9), taking generally complex values. The solution of system (15) for all $t \geq t_0 \geq 0$ is representable with the use of the exponential of matrix Λ [Hirsch and Smale (1974)]:

$$z(t, t_0, z_0) = e^{\Lambda(t-t_0)} z_0 + e^{\Lambda t} \int_{t_0}^{t} e^{-\Lambda \tau} h(\tau) d\tau. \tag{16}$$

Since $\text{Re}\lambda < 0$, there exists [Hirsch and Smale (1974)] a constant $M > 0$ such that the inequality $\|e^{\Lambda t}\| \leq M$ holds for all $t \geq 0$. Furthermore, $\|e^{\Lambda t}\| \to 0$ as $t \to +\infty$. It is taken to mean the matrix norm fitted to the Euclidean vector norm. Denote by $H(t, t_0)$ the addend on the right-hand side of (16). For the vector function $H(t, t_0)$ we formulate the following assertions:

A. For any fixed number $t_0 \geq 0$ the limit relationship $\|H(t, t_0)\| \to 0$ holds as $t \to +\infty$.

B. For any number $\epsilon > 0$ there is a quantity $T(\epsilon) \geq 0$ such that the inequality $\|H(t, t_0)\| < \epsilon$ holds for all $t \geq t_0 \geq T(\epsilon)$.

Let us prove these assertions with the use of the scalar form (14) of system (15). The solution of system (14) for all $t \geq t_0 \geq 0$ can be written as follows:

$$z_j(t, t_0, z_0) = e^{\lambda(t-t_0)} z_{j0} + e^{\lambda t} \int_{t_0}^{t} e^{-\lambda \tau} (z_{j+1}(\tau, t_0, z_0) + h_j(\tau)) d\tau, \quad j = 1, \ldots, k-1,$$

$$z_k(t, t_0, z_0) = e^{\lambda(t-t_0)} z_{k0} + e^{\lambda t} \int_{t_0}^{t} e^{-\lambda \tau} h_k(\tau) d\tau, \tag{17}$$

where $z(t, t_0, z_0) = (z_1(t, t_0, z_0), \ldots, z_k(t, t_0, z_0))$, $z_0 = (z_{10}, \ldots, z_{k0})$. The vector function $H(t, t_0)$ is independent of components of the vector z_0 and therefore it is necessary to select in relationships (17) those summands which contain no quantities $z_{10}, \ldots, z_{k0}$. Transforming and relabelling relationships (17) we obtain

$$z_j(t, t_0, z_0) = e^{\lambda(t-t_0)} \left(z_{j0} + \sum_{\ell=1}^{k-j} \frac{(t-t_0)^\ell z_{j+\ell 0}}{\ell!} \right) + g_j(t), \quad j = 1, \ldots, k-1,$$

$$z_k(t, t_0, z_0) = e^{\lambda(t-t_0)} z_{k0} + g_k(t), \tag{18}$$

$$g_j(t) = e^{\lambda t} \int_{t_0}^{t} e^{-\lambda \tau} (g_{j+1}(\tau) + h_j(\tau)) d\tau, \quad j = 1, \ldots, k-1,$$

$$g_k(t) = e^{\lambda t} \int_{t_0}^{t} e^{-\lambda \tau} h_k(\tau) d\tau, \tag{19}$$

Since $\text{Re}\lambda < 0$, relationships (18) imply the properties of the matrix $e^{\Lambda t}$ formulated above. The components of the vector $H(t, t_0)$ are the functions $g_1(t), \ldots, g_n(t)$ given in (19).

Let us prove assertion A. Since the function $h_k(t)$ satisfies condition (8) or condition (9), Lemma 1 implies that $g_k(t) \to 0$ as $t \to +\infty$. By relationships (19), the function $g_{k-1}(t)$ is representable as the sum

$$g_{k-1}(t) = e^{\lambda t} \int_{t_0}^{t} e^{-\lambda \tau} g_k(\tau) d\tau + e^{\lambda t} \int_{t_0}^{t} e^{-\lambda \tau} h_{k-1}(\tau) d\tau,$$

where, by what has been proved, $g_k(t)$ satisfies condition (8), while $h_{k-1}(t)$ satisfies condition (8) or condition (9). Hence, by Lemma 1, we have $g_{k-1}(t) \to 0$ as $t \to +\infty$. The same procedure can be applied to the remaining functions $g_j(t)$. This proves assertion A.

Let us now prove assertion B. We find an estimate for the functions $g_j(t), j = 1, \ldots, k$, given by formulas (19). Estimate the quantities

$$e^{\lambda t}\int_{t_0}^{t} e^{-\lambda \tau} h_j(\tau)d\tau = e^{\lambda(t-t_0)}\int_{0}^{t-t_0} e^{-\lambda \tau} h_j(\tau + t_0)d\tau, \quad j = 1, \ldots, k.$$

Denote $\mu_0 = 1, \mu_j = \min\{1, |\operatorname{Re}\lambda|^j\}, j = 1, \ldots, k$. It can be easily established that $\mu_j \le 1, j = 1, \ldots, k, \frac{\mu_j}{|\operatorname{Re}\lambda|} \le \mu_{j-1}, j = 1, \ldots, k$. Let the functions $h_j(t), j = 1, \ldots, k$, satisfy condition (8). Then for any number $\epsilon > 0$ there exists a quantity $T(\epsilon) \ge 0$ such that for all $t \ge T(\epsilon)$ there are inequalities $|h_j(t)| < \frac{\epsilon \mu_k}{2\sqrt{k}}, j = 1, \ldots, k$. Hence, with any $t \ge t_0 \ge T(\epsilon)$, for all $j = 1, \ldots, k$ we have

$$|e^{\lambda(t-t_0)}\int_{0}^{t-t_0} e^{-\lambda\tau}h_j(\tau+t_0)d\tau| \le e^{\operatorname{Re}\lambda(t-t_0)}\int_{0}^{t-t_0} e^{-\operatorname{Re}\lambda\tau}|h_j(\tau+t_0)|d\tau$$

$$< \frac{\epsilon\mu_k}{2\sqrt{k}}e^{\operatorname{Re}\lambda(t-t_0)}\int_{0}^{t-t_0} e^{-\operatorname{Re}\lambda\tau}d\tau = \frac{\epsilon\mu_k}{2\sqrt{k}|\operatorname{Re}\lambda|}e^{\operatorname{Re}\lambda(t-t_0)}(e^{-\operatorname{Re}\lambda(t-t_0)} - 1)$$

$$\le \frac{\epsilon\mu_{k-1}}{2\sqrt{k}}(1 - e^{\operatorname{Re}\lambda(t-t_0)}) < \frac{\epsilon\mu_{k-1}}{2\sqrt{k}} \le \frac{\epsilon}{2\sqrt{k}}.$$

In particular, with $j = k$ these relationships imply that the inequality $|g_k(t)| < \frac{\epsilon\mu_{k-1}}{2\sqrt{k}} \le \frac{\epsilon}{2\sqrt{k}}$ holds for all $t \ge t_0 \ge T(\epsilon)$. In a similar manner, we may form an estimate for the quantities

$$|e^{\lambda t}\int_{t_0}^{t} e^{-\lambda\tau} g_{j+1}(\tau) d\tau| < \frac{\epsilon\mu_{j-1}}{2\sqrt{k}} \le \frac{\epsilon}{2\sqrt{k}}, \quad j = 1, \ldots, k-1.$$

From the established inequalities it follows that for any $t \ge t_0 \ge T(\epsilon)$ the inequalities $|g_j(t)| < \frac{\epsilon}{\sqrt{k}}$ hold for all $j = 1, \ldots, k$, and so does the inequality $\|H(t, t_0)\| < \epsilon$. Thus, in the case of condition (8) assertion B is proved. Now, let the functions $h_j(t), j = 1, \ldots, k$, satisfy condition (9). The proof of Lemma 1 has made use of the identity which, after a simple transformation, becomes

$$e^{\lambda t}\int_{t_0}^{t} e^{-\lambda\tau}h_j(\tau)d\tau = -\int_{t}^{+\infty} h_j(\tau)d\tau + e^{\lambda(t-t_0)}\int_{t_0}^{+\infty} h_j(\tau)d\tau$$

$$-\lambda e^{\lambda(t-t_0)}\int_{0}^{t-t_0} e^{-\lambda\tau}\int_{\tau+t_0}^{+\infty} h_j(\xi)d\xi d\tau. \qquad (20)$$

By the above procedure, we estimate the quantities appearing on the right-hand side of identity (20). Since the functions $h_j(t), j = 1, \ldots, k$, satisfy condition (9), and hence condition (11), for any number $\epsilon > 0$ there exists a quantity $T(\epsilon) \ge 0$ such that for all $t \ge T(\epsilon)$ there are the inequalities.

$$|\int_{t}^{+\infty} h_j(\tau)d\tau| < \frac{\epsilon\mu_{k-1}}{6\sqrt{k}}, \quad j = 1, \ldots, k.$$

Similarly, estimating the right-hand side of identity (20) we have that, for any $t \geq t_0 \geq T(\epsilon)$, the inequalities

$$|e^{\lambda t} \int_{t_0}^{t} e^{-\lambda \tau} h_j(\tau) d\tau| < \frac{\epsilon \mu_{k-1}}{2\sqrt{k}} \leq \frac{\epsilon}{2\sqrt{k}}$$

hold for all $j = 1, \ldots, k$. Further, applying the established estimates to the functions $g_1(t), \ldots, g_k(t)$ we have that, with any $t \geq t_0 \geq T(\epsilon)$, the inequalities $|g_j(t)| < \frac{\epsilon}{\sqrt{k}}$ hold for all $j = 1, \ldots, k$. Therefore, the inequality $\|H(t,t_0)\| < \epsilon$ is also valid, which proves assertion B in the case of condition (9). This completes the proof of assertion B.

Using assertions A and B it can be easily shown that Theorem 17 holds for the system of equations (15). Let $\epsilon > 0$ be an arbitrary number. In relation to the properties of the matrix $e^{\Lambda t}$ and vector function $H(t,t_0)$, take the quantities $T_1(\epsilon) = T(\frac{\epsilon}{2})$ and $\delta(\epsilon) = \frac{\epsilon}{2M}$. Then, with all $t \geq t_0$, the inequality $\|z(t,t_0,z_0)\| < \epsilon$ holds for any initial conditions (t_0, z_0) satisfying the inequalities $t_0 \geq T_1(\epsilon)$ and $\|z_0\| < \delta(\epsilon)$. This follows from equality (16) and from the estimates: $\|z(t,t_0,z_0)\| = \|e^{\Lambda(t-t_0)} z_0 + H(t,t_0)\| \leq \|e^{\Lambda(t-t_0)}\| \cdot \|z_0\| + \|H(t,t_0)\| < M\delta(\epsilon) + \frac{\epsilon}{2} = \epsilon$. Furthermore, by the properties of the matrix $e^{\Lambda t}$ and vector function $H(t,t_0)$, for the selected initial conditions (t_0, z_0) there is the limit relationship $\|z(t,t_0,z_0)\| \to 0$ as $t \to +\infty$. Thus, the motion $z = 0$ of system (15) is asymptotically rated stable. Consequently, the motion $y = 0$ of system (13) is also asymptotically rated stable, since for each of its subsystems, as (15), the corresponding zero motions are asymptotically rated stable. Because of the nonsingularity of the transformation, which permitted the passage from system (12) to system (13), we have that the motion $x = 0$ of system (12) is asymptotically rated stable. This completes the proof of Theorem 17.

Theorem 18 *Let all eigenvalues of the matrix* $\mathbf{A}$ *of system (12) have negative or zero real parts for which there are prime elementary divisors. Suppose the functions* $f_j(t), j = 1, \ldots, n$, *simultaneously satisfy the condition*

$$\int_0^{+\infty} |f_j(t)| dt < +\infty, j = 1, \ldots, n. \tag{21}$$

Then the motion $x = 0$ *of system (12) is rated stable.*

Proof: Note that condition (21) leads to satisfaction of condition (9), therefore we will use the procedure as in the proof of Theorem 17. Passing to system (13) with the use of linear nonsingular transformation we have that the corresponding Jordan boxes fall into two types. One type corresponds to negative real parts of the eigenvalues of the matrix $\mathbf{A}$. This case has already been discussed in the proof of Theorem 17. The other type of Jordan boxes corresponds to zero real parts of the eigenvalues of the matrix $\mathbf{A}$ such that, by condition of Theorem 18, the order of these boxes is 1. Thus, for Jordan boxes of the second type in system (14) $k = 1$ and, as system (15), we obtain the differential equation

$$\frac{dz}{dt} = \lambda z + h(t), \tag{22}$$

where $\text{Re}\lambda = 0$, the function $h(t)$ is defined and continuous for all $t \geq 0$ and satisfies condition (21). Show that the motion $z = 0$ of equation (22) is rated stable. Equation (22) integrates to the following expression

$$z(t, t_0, z_0) = e^{\lambda(t-t_0)} z_0 + e^{\lambda t} \int_{t_0}^{t} e^{-\lambda \tau} h(\tau) d\tau. \tag{23}$$

The addend in equality (23) is a scalar counterpart of the function $H(t, t_0)$ in the proof of Theorem 17, and assertion B also holds for it. Indeed, form an estimate:

$$\left| e^{\lambda t} \int_{t_0}^{t} e^{-\lambda \tau} h(\tau) d\tau \right| = \left| \int_{t_0}^{t} e^{-\lambda \tau} h(\tau) d(\tau) \right| \leq \int_{t_0}^{t} |h(\tau)| d\tau \leq \int_{t_0}^{+\infty} |h(\tau)| d\tau. \tag{24}$$

Condition (21) is equivalent to the limit relationship

$$\int_{t}^{+\infty} |h(\tau)| d\tau \to 0 \quad \text{as} \quad t \to +\infty.$$

Hence, for any number $\epsilon > 0$ there is $T(\epsilon) \geq 0$ such that for all $t_0 \geq T(\epsilon)$ there is

$$\int_{t_0}^{+\infty} |h(\tau)| d\tau < \epsilon.$$

Choosing $T_1(\epsilon) = T(\frac{\epsilon}{2})$, $\delta(\epsilon) = \frac{\epsilon}{2}$ and using (24) we then have that, for all $t \geq t_0$, the inequality $|z(t, t_0, z_0)| < \epsilon$ holds for all initial conditions (t_0, z_0) satisfying the inequalities $t_0 \geq T(\epsilon)$ and $|z_0| < \delta(\epsilon)$. Thus, the motion $z = 0$ of equation (22) is rated stable. Rated stability of the motion $z = 0$ of system (15) for Jordan boxes of the first type has been established in the proof of Theorem 17. We have thus obtained the rated stability of zero motions from subsystems of system (13). This also implies the rated stability of the motion $y = 0$ for the entire system (13) as well as the rated stability of the motion $x = 0$ for system (12). This completes the proof of Theorem 18.

Consider the linear homogeneous and nonhomogeneous systems

$$\frac{dx}{dt} = \mathbf{P}(t)x, \tag{25}$$

$$\frac{dy}{dt} = \mathbf{P}(t)y + f(t), \tag{26}$$

where $x = (x_1, \ldots, x_n)^*$, $y = (y_1, \ldots, y_n)^*$, $f(t) = (f_1(t), \ldots, f_n(t))^*$, the $(n \times n)$ matrix $\mathbf{P}(t)$ has the elements $p_{ij}(t)$, $i = 1, \ldots, n$, $j = 1, \ldots, n$ that are defined, continuous, bounded for all $t \geq 0$ and take real values, while the functions $f_j(t), j = 1, \ldots, n$ are defined, continuous for all $t \geq 0$, and also take real values.

Theorem 19 *Let the vector function $f(t)$ on the right-hand side of system (26) be such that this system has a solution $\psi(t)$ possessing the property: $\|\psi(t)\| \to 0$ as $t \to +\infty$. The assertion is:*

1) if the solution $x = 0$ of system (25) is Lyapunov-stable (asymptotically stable) and such that the stability is uniform, then the motion $y = 0$ of system (25) is rated stable (asymptotically rated stable);

2) *if the solution $x = 0$ of system (25) is not uniformly stable, then the motion $y = 0$ of system (25) is rated unstable.*

Proof: 1). It follows from Theorem 15 that the motion $x = \psi(t)$ of the system of equations (25) is rated stable (asymptotically rated stable). Setting up a system of differential equations for the deviation $y = \psi(t) - x$ we have that the y satisfies the system of equations (26). In this case, the motion $x = \psi(t)$ of system (25) corresponds to the motion $y = 0$ of system (26). Hence, the motion $y = 0$ of system (26) is rated stable (asymptotically rated stable).

2). Note that the system of differential equations for the deviation $y = \psi(t) - x$ has the solution $y = \psi(t)$ which corresponds to the solution $x = 0$ of system (25) and hence is not uniformly stable. By Theorem 14, we then derive rated stability for the motion $y = 0$ of system (26). This completes the proof of Theorem 19.

3. Analysis of Rated Stability by Lyapunov's First Method

Lyapunov's first method [Lyapunov (1950)] of investigating motion stability is used, along with the direct (second) method, to analyze stability conditions and investigate the behavior of solutions of the systems of differential equations describing the dynamics of the engineering systems under study. The first method is based on the approach of constructing the solutions of systems of ordinary differential equations which consists of constructing series in integral positive powers of arbitrary constants with some time-dependent coefficients. In particular, these arbitrary constants can be the components of initial values for solutions of the systems of equations under study. Of interest is the application of Lyapunov's first method to the study of the behavior of solutions to systems of differential equations in the neighborhood of rated motions that are not solutions of the systems involved. In what follows Lyapunov's conditional stability theorem [Lyapunov (1950)] will be used to establish asymptotic properties of solutions to the systems of ordinary differential equations with holomorphic right-hand terms in the neighborhood of the rated motion $x = 0$.

First, for convenience of further presentation, we will provide some knowledge and assertions of Lyapunov's first method to be used. In particular, the Lyapunov notion of the characteristic number of function [Lyapunov (1950)].

Definition 10. Let $h(t)$ be the continuous scalar function given for all $t \geq 0$. The characteristic number of this function is called λ such that the following relationships hold for all arbitrarily small $\epsilon > 0$

$$\overline{\lim_{t \to +\infty}} h(t)e^{(\lambda+\epsilon)t} = +\infty, \quad \lim_{t \to +\infty} |h(t)|e^{(\lambda-\epsilon)t} = 0. \tag{27}$$

The characteristic number of the function $h(t)$ is denoted as $\lambda = \chi(h)$. It is easy to establish that if $\chi(h)$ is finite, then

$$\chi(h) = \overline{\lim_{t \to +\infty}} \frac{1}{t} \ln |h(t)|, \quad h(t) \neq 0. \tag{28}$$

By formula (28), it is possible to define the notion of a characteristic number itself, and then relationships (27) describe the properties of this notion. Note that the limit in (28) can be infinite, and then $\chi(h)$ symbolizes $\pm\infty$. We shall now present the properties of characteristic numbers established by Lyapunov [Lyapunov (1950)].

1) The characteristic number of the sum of two functions is equal to the least of the characteristic numbers of these functions and not less than they when they are equal.

2) The characteristic number of the product of two functions is not less than the sum of their characteristic numbers, and hence

$$\chi(h \cdot \frac{1}{h}) = 0 \geq \chi(h) + \chi(\frac{1}{h}). \tag{29}$$

3) For the sum of the characteristic numbers of functions h and $1/h$ to be zero, i.e. for inequality (29) to become an equality, it is necessary and sufficient that there be a finite limit

$$\lim_{t \to +\infty} \frac{1}{t} \ln |h(t)|.$$

4) If the sum of the characteristic numbers of functions h and $1/h$ is zero, then for an arbitrary function g the characteristic number of the product $h \cdot g$ is equal to the sum of characteristic numbers of factors

$$\chi(h \cdot g) = \chi(h) + \chi(g).$$

5) The characteristic number of the integral is no less than the characteristic number of the integrand, so that the following integrals are considered:

$$\int_{t_0}^{t} h(\tau)d\tau, \ \chi(h) \leq 0; \quad \int_{t}^{+\infty} h(\tau)d\tau, \ \chi(h) > 0.$$

Further, we will consider the homogeneous system of linear differential equations

$$\frac{dx_s}{dt} = p_{s1}(t)x_1 + \ldots + p_{sn}(t)x_n, \ s = 1, \ldots, n, \tag{30}$$

where the functions $p_{si}(t), s = 1, \ldots, n, i = 1, \ldots, n$, are taken to be real-valued, continuous and bounded for all $t \geq 0$. For the system of equations (30) the following system of linear homogeneous differential equations

$$\frac{dy_j}{dt} = -p_{1j}(t)y_1 - \ldots - p_{nj}(t)y_n, \ j = 1, \ldots, n, \tag{31}$$

is called conjugate.

Definition 11. The fundamental solutions of the system of equations (30) will be any collection of n linearly independent solutions $z_{ij}(t), i = 1, \ldots, n, j = 1, \ldots, n$, such that the vector functions $z_j(t) = (z_{1j}(t), \ldots, z_{nj}(t))^*, j = 1, \ldots, n$, are linearly independent for all $t \geq 0$; so the $(n \times n)$ matrix $\mathbf{Z}(t) = \{z_{ij}(t)\}$ is nonsingular for all $t \geq 0$ and linearly independent solutions of system (30) are located in its columns.

In what follows ∗ indicated transposition.

Definition 12. Let $x(t) = (x_1(t), \ldots, x_n(t))^*$ be a nonzero solution to system (30). The least of the characteristic numbers $\chi(x_s), s = 1, \ldots, n$, will be called the characteristic number of this solution $\chi(x) = \chi(x_1, \ldots, x_n)$.

We shall now provide the properties of the characteristic numbers of solutions of system (30) established by Lyapunov [Lyapunov (1950)].

1) Every nonzero solution $x(t) = (x_1(t), \ldots, x_n(t))^*$ to system (30) has a finite characteristic number.

2) Let $z_{ij}(t), i = 1, \ldots, n, j = 1, \ldots, n$, be the fundamental solutions for the system of equations (30), and assume that $\lambda_j = \chi(z_{1j}, \ldots, z_{nj})$ is the characteristic number of the solution $(z_{1j}(t), \ldots, z_{nj}(t))^*, j = 1, \ldots, n$, with all $\lambda_1, \ldots, \lambda_n$ distinct. Then the characteristic number of the solution $x_s = \sum_{\ell=1}^{k} c_\ell z_{s\ell}$ with constant c_ℓ is equal to the least of the characteristic numbers $\lambda_1, \ldots, \lambda_n$. Hence, if $\lambda_1, \ldots, \lambda_n$ are distinct, then the characteristic number of any nonzero solution to system (30) is one of the numbers $\lambda_1, \ldots, \lambda_n$. Thus, system (30) may have no more than n solutions with distinct characteristic numbers.

Suppose that among the characteristic numbers $\lambda_1, \ldots, \lambda_n$ of some fundamental system of solutions to system (30) there are equal numbers. Then the characteristic number of the solution that is a linear combination of solutions with the same characteristic number λ_j of the fundamental system involved may prove to be greater than λ_j. In this case, the new solution is incorporated into the fundamental system instead of one of the combined solutions. If the new fundamental system again has solutions with the same characteristic number, whose linear combination yields a solution with a greater characteristic number than that of the solutions combined, then this system can be transformed in much the same way as the preceding one. Since the quantity of distinct characteristic numbers is finite, this results in the fundamental solutions, where any linear combination of solutions has a characteristic number equal to the characteristic number of one of the solutions combined.

Definition 13. The fundamental solutions to the system of equations (30) possessing the above property is called normal. The characteristic numbers $\lambda_1, \ldots, \lambda_n$ of the normal system (these may include the same numbers) are called characteristic numbers of the system of differential equations (30).

3) If all the characteristic numbers $\lambda_1, \ldots, \lambda_n$ of some fundamental system are distinct, then this system is normal.

4) Any fundamental solutions, for which the sum of characteristic numbers $\sum_{j=1}^{n} \lambda_j$ of all its solutions is at a maximum, are normal.

5) The sum of characteristic numbers $\sum_{j=1}^{n} \lambda_j$ of system (30) does not exceed the characteristic number of the function

$$\exp\left(\int_0^t \sum_{j=1}^{n} p_{jj}(\tau)d\tau\right). \tag{32}$$

Let us prove this assertion. Let $\mathbf{Z}(t) = \{z_{ij}(t)\}$ be an $(n \times n)$ matrix of the fundamental solutions $z_{ij}(t), i = 1,\ldots,n, j = 1,\ldots,n$ for system (30). Denote by $\Delta(t) = \det \mathbf{Z}(t)$ the determinant of the matrix $\mathbf{Z}(t)$. By Liouville's formula [Hale (1969)], [Coddington and Levinson (1955)], we then obtain

$$\Delta(t) = \Delta(0)\exp(\int_0^t \sum_{j=1}^n p_{jj}(\tau)d\tau).$$

Utilizing the properties of the characteristic numbers of the sum and product of the functions we see that the following inequality is valid

$$\chi(\exp(\int_0^t \sum_{j=1}^n p_{jj}(\tau)d\tau)) = \chi(\Delta) \geq \sum_{j=1}^n \lambda_j. \tag{33}$$

6) If for some fundamental solutions the sum of its characteristic numbers is equal to the characteristic number of function (32), then this system is normal, since (33) implies that the characteristic number of function (32) is the upper limit for the sum of characteristic numbers of system (30).

Definition 14. The homogeneous system of linear differential equations (30) is called regular, if the sum of its characteristic numbers $\lambda_1,\ldots,\lambda_n$ satisfies the relationship

$$\sum_{j=1}^n \lambda_j = -\chi(\exp(-\int_0^t \sum_{j=1}^n p_{jj}(\tau)d\tau)). \tag{34}$$

Note that if all the coefficients $p_{ij}(t), i = 1,\ldots,n, j = 1,\ldots,n$ in the system of differential equations (30) are constant or are periodic functions with the same period, then system (30) is regular. The following equality holds for the regular system

$$\chi(\exp(\int_0^t \sum_{j=1}^n p_{jj}(\tau)d\tau)) = -\chi(\exp(-\int_0^t \sum_{j=1}^n p_{jj}(\tau)d\tau)).$$

Consider a nonhomogeneous linear system of differential equations

$$\frac{dx_s}{dt} = p_{s1}(t)x_1 + \ldots + p_{sn}(t)x_n + f_s(t), \quad s = 1,\ldots,n, \tag{35}$$

which is distinguished from system (30) by the presence of the vector function $f(t) = (f_1(t),\ldots,f_n(t))^*$ on the right-hand side; here the functions $f_s(t), s = 1,\ldots,n$ are defined and continuous for any $t \geq 0$. The functions $p_{sj}(t), s = 1,\ldots,n, j = 1,\ldots,n$ are assumed, as before, to be given, continuous, and bounded for all $t \geq 0$. Let $x(t, x_0)$ be a solution to system (35) satisfying the initial condition $x(0, x_0) = x_0$, where $x(t, x_0) = (x_1(t, x_0),\ldots,x_n(t, x_0))^*$, $x_0 = (x_{01},\ldots,x_{0n})^*$. By the Cauchy formula [Coddington and Levinson (1955)], we then obtain

$$x(t, x_0) = \mathbf{Z}(t)\mathbf{Z}^{-1}(0)x_0 + \mathbf{Z}(t)\int_0^t \mathbf{Z}^{-1}(\tau)f(\tau)d\tau. \tag{36}$$

Here $\mathbf{Z}(t)$ is $(n \times n)$ matrix whose columns contain the solutions $(z_{1j}(t),\ldots,z_{nj}(t))^*$, $j = 1,\ldots,n$ of the fundamental solution system for the homogeneous system of equations (30), while $f(t)$ is the n-dimensional vector function on the right-hand side of the nonhomogeneous system of equations (35) [Coddington and Levinson (1955), Hale (1969)]. Denoting n-dimensional vector of arbitrary constants $c = \mathbf{Z}^{-1}(0)x_0 = (c_1, \ldots, c_n)^*$ we rewrite formula (36)

$$x_s(t) = \sum_{j=1}^{n} z_{sj}(t)c_j + \sum_{i=1}^{n}\sum_{j=1}^{n} z_{sj}(t) \int_0^t \frac{\Delta_{ij}(\tau)}{\Delta(\tau)} f_i(\tau)d\tau, \quad s = 1, \ldots, n. \quad (37)$$

Here, as before, the determinant $\Delta(t) = \det \mathbf{Z}(t)$, while $\Delta_{ij}(t)$ is a cofactor of the element $z_{ij}(t)$ of this determinant.

Let us see that the collections of the functions

$$\frac{\Delta_{1j}(t)}{\Delta(t)}, \ldots, \frac{\Delta_{nj}(t)}{\Delta(t)}, \quad j = 1, \ldots, n,$$

satisfy the linear homogeneous system of equations conjugated to the system of equations (30). In other words, we need to establish the validity of the following relationships:

$$\frac{d}{dt}\left(\frac{\Delta_{ij}(t)}{\Delta(t)}\right) = -\frac{\Delta_{1j}(t)}{\Delta(t)} p_{1i}(t) - \ldots - \frac{\Delta_{nj}(t)}{\Delta(t)} p_{ni}(t), \quad i = 1, \ldots, n, \quad j = 1, \ldots, n \quad (38)$$

for all $t \geq 0$. Prove relationships (38), e.g., for $j = 1$. For other values of j the argument is similar. Immediate differentiation yields

$$\frac{d}{dt}\left(\frac{\Delta_{i1}(t)}{\Delta(t)}\right) = \frac{\Delta'_{i1}(t)\Delta(t) - \Delta'(t)\Delta_{i1}(t)}{\Delta^2(t)},$$

and since by Liouville's formula [Coddington and Levinson (1955), Hale (1969)]

$$\Delta'(t) = \Delta(t) \sum_{s=1}^{n} p_{ss}(t),$$

the following relationship is valid

$$\frac{d}{dt}\left(\frac{\Delta_{i1}(t)}{\Delta(t)}\right) = \frac{\Delta'_{i1}(t)}{\Delta(t)} - \frac{\Delta_{i1}(t)}{\Delta(t)} \sum_{s=1}^{n} p_{ss}(t). \quad (39)$$

It remains to obtain an expression for $\Delta'_{i1}(t)$. This derivative can be represented as a sum of $n-1$ determinants of the form $\Delta_{i1}(t)$ just that the k-th row is differentiated in each summand, $k = 1, \ldots, n, k \neq i$, i.e., the k-th components of the vector functions $z_2(t), \ldots, z_n(t)$ in this row are replaced by their derivatives:

$$\Delta'_{i1}(t) = \sum_{k=1, k \neq i}^{n} \Delta^k_{i1}(t). \quad (40)$$

The derivatives of the corresponding components in each such row are expressed in terms of equation (30) as follows:

$$\frac{dz_{k\ell}(t)}{dt} = \sum_{s=1}^{n} p_{ks}(t) z_{s\ell}(t), \quad k = 1, \ldots, n, \quad k \neq i, \quad \ell = 2, \ldots, n. \quad (41)$$

Therefore, substituting (41) into $\Delta_{i1}^k(t)$, we obtain

$$\Delta_{i1}^k = p_{kk}(t)\Delta_{i1}(t) - p_{ki}(t)\Delta_{k1}(t), \qquad (42)$$

because the other determinants vanish. The minus sign appears in front of the second addend in (42) because substituting (41) into $\Delta_{i1}^k(t)$ for $s = i$ yields a determinant which is different from $\Delta_{k1}(t)$ by the locations of two rows. The first term corresponds to $s = k$. Thus, (40), (42) imply the validity of the equality

$$\Delta_{i1}'(t) = \Delta_{i1}(t)\sum_{s=1}^n p_{ss}(t) - \Delta_{i1}(t)p_{ii}(t) - \sum_{k=1, k\neq i}^n p_{ki}(t)\Delta_{ki}(t),$$

whence, by (39), follows relationship (38).

Lemma 2 *Suppose that in the homogeneous linear system of equations (35) the functions $p_{si}(t), s = 1, \ldots, n, i = 1, \ldots, n$, are continuous and bounded for $t \geq 0$, while the functions $f_s(t)$ continuous for $t \geq 0$ are such that*

$$\left|\int_t^{+\infty} f_s(\tau)d\tau\right| \leq A\exp(-\alpha t), \quad s = 1, \ldots, n, \qquad (43)$$

for all $t \geq 0$, where the constants $A > 0$ $\alpha > 0$. Suppose the homogeneous system (30) is regular. Then system (35) has the solution with the characteristic number λ such that $\lambda \geq \alpha$.

Proof: Let, as before, $\mathbf{Z}(t)$ be an $(n \times n)$ matrix for the normal system of solutions to the system of equations (30) whose columns correspond to linearly independent solutions of system (30). Let the j-th column $(z_{1j}(t), \ldots, z_{nj}(t))^*$ represent the j-th solution with the characteristic number $\lambda_j, j = 1, \ldots, n$. Show that arbitrary constants c_j in formula (37) can be chosen such that the inequalities $\chi(x_s) \geq \alpha, s = 1, \ldots, n$ hold. To this end, take the integral in (37) by parts. We have

$$\int_0^t \frac{\Delta_{ij}(\tau)}{\Delta(\tau)} f_i(\tau)d\tau = -\int_0^t \frac{\Delta_{ij}(\tau)}{\Delta(\tau)} d\left(\int_\tau^{+\infty} f_i(\theta)d\theta\right) = -\left(\frac{\Delta_{ij}(\tau)}{\Delta(\tau)} \int_\tau^{+\infty} f_i(\theta)d\theta\right)\Big|_0^t$$

$$+ \int_0^t \int_\tau^{+\infty} f_i(\theta)d\theta d\left(\frac{\Delta_{ij}(\tau)}{\Delta(\tau)}\right) = \frac{\Delta_{ij}(0)}{\Delta(0)}\int_0^{+\infty} f_i(\theta)d\theta - \frac{\Delta_{ij}(t)}{\Delta(t)}\int_t^{+\infty} f_i(\theta)d\theta$$

$$- \int_0^t \sum_{k=1}^n \frac{\Delta_{kj}(\tau)}{\Delta(\tau)} p_{ki}(\tau) \int_\tau^{+\infty} f_i(\theta)d\theta d\tau, \quad i = 1, \ldots, n, \ j = 1, \ldots, n.$$

The last equality holds by the relationships (38) proved above. Finally we obtain

$$x_s(t) = \sum_{j=1}^n z_{sj}(t)\left(c_j + \sum_{i=1}^n \frac{\Delta_{ij}(0)}{\Delta(0)}\int_0^{+\infty} f_i(\theta)d\theta\right) - \sum_{i=1}^n \int_t^{+\infty} f_i(\theta)d\theta \sum_{j=1}^n z_{sj}(t)\frac{\Delta_{ij}(t)}{\Delta(t)}$$

$$- \sum_{j=1}^n z_{sj}(t)\int_0^t \sum_{k=1}^n \frac{\Delta_{kj}(\tau)}{\Delta(\tau)} \sum_{i=1}^n p_{ki}(\tau) \int_\tau^{+\infty} f_i(\theta)d\theta d\tau \qquad (44)$$

$$= -\sum_{j=1}^{n} z_{sj}(t) \int \sum_{k=1}^{n} \frac{\Delta_{kj}(t)}{\Delta(t)} \sum_{i=1}^{n} p_{ki}(t) \int_{t}^{+\infty} f_i(\theta) d\theta dt - \int_{t}^{+\infty} f_s(\theta) d\theta, \quad s = 1, \ldots, n.$$

Here the indefinite integral symbolizes the presence of the arbitrary constants c_j. Since the functions $p_{ki}(t)$ are continuous and bounded for $t \geq 0$, by the properties of the characteristic numbers of functions and by relationships (43) we get

$$\chi\left(\sum_{i=1}^{n} p_{ki}(t) \int_{t}^{+\infty} f_i(\theta) d\theta\right) \geq \alpha. \qquad (45)$$

By hypothesis, system (30) is regular, and hence the following inequality holds

$$\chi\left(\frac{\Delta_{kj}}{\Delta}\right) = \chi(\Delta_{kj}) + \chi\left(\frac{1}{\Delta}\right) \geq \sum_{i=1, i \neq j}^{n} \lambda_i + \chi\left(\frac{1}{\Delta}\right) = -\lambda_j. \qquad (46)$$

Indeed, regularity of system (30) implies that the following relationships hold

$$\chi(\Delta) + \chi\left(\frac{1}{\Delta}\right) = 0, \quad \chi\left(\frac{1}{\Delta}\right) = -\sum_{i=1}^{n} \lambda_i.$$

Hence, from the properties of the characteristic numbers of functions we derive

$$\chi\left(\frac{\Delta_{kj}}{\Delta}\right) = \chi(\Delta_{kj}) + \chi\left(\frac{1}{\Delta}\right), \quad \chi(\Delta_{kj}) \geq \sum_{i=1, i \neq j}^{n} \lambda_i.$$

From the preceding four relationships follows inequality (46). Thus, from the properties of the characteristic numbers of functions and from relationships (45), (46) it follows that $\chi(R_j) \geq \alpha - \lambda_j$, where $R_j(t)$ is the integrand in (44). Then we have $\chi(\int R_j(t)dt) \geq \alpha - \lambda_j$, when it is integrated from t to $+\infty$ for $\chi(R_j) > 0$ and from 0 to t otherwise. In fact, if $\chi(R_j) > 0$, then by selecting

$$c_j = \int_{0}^{+\infty} R_j(t) dt - \sum_{i=1}^{n} \frac{\Delta_{ij}(0)}{\Delta(0)} \int_{0}^{+\infty} f_i(\theta) d\theta$$

we ensure that $\chi(\int R_j(t)dt) \geq \alpha - \lambda_j$. For $\chi(R_j) \leq 0$ the constant c_j is arbitrary. In both cases $\chi(z_{sj}(t) \int R_j(t)dt) \geq \alpha$, and hence $\chi(x_s) \geq \alpha$. Consequently, the characteristic number of the solution $x_s(t), s = 1, \ldots, n$, obtained by the above procedure for system (35) is at least α. This completes the proof of Lemma 2.

Remark. If, under the other conditions of Lemma 2, system (30) is not regular and the inequality

$$\sigma = -\left(\sum_{i=1}^{n} \lambda_i + \chi\left(\frac{1}{\Delta}\right)\right) > 0$$

holds, then system (35) has the solution with a characteristic number λ such that $\lambda \geq \alpha - \sigma$. The proof of this assertion is the same as the proof of Lemma 2 except that inequality (46) is replaced by an inequality of the form

$$\chi\left(\frac{\Delta_{kj}}{\Delta}\right) \geq \chi(\Delta_{kj}) + \chi\left(\frac{1}{\Delta}\right) \geq \sum_{i=1, i \neq j}^{n} \lambda_i + \chi\left(\frac{1}{\Delta}\right) = -\sigma - \lambda_j.$$

By properly selecting constants c_j we have, as in the proof of Lemma 2, that the relationships $\chi(x_s) \geq \alpha - \sigma, s = 1, \ldots, n$ are valid. If $\alpha > \sigma$, it follows that $\chi(x_s) > 0, s = 1, \ldots, n$.

We shall give examples of the functions satisfying conditions (43) in the formulation of Lemma 2:

a) $f_s(t) = \alpha g_s(t) \exp(-\alpha t)$, where the functions $g_s(t)$ are continuous and bounded for all $t \geq 0$;

b) $f_s(t) = \alpha a_s \cos(\exp(\alpha t))$;

c) $f_s(t) = (\alpha + \beta) a_s \exp(\beta t) \cos(\exp((\alpha + \beta)t))$, where $\beta > 0$, a_s are constants.

These examples show that the functions satisfying an integral estimate of the form (43) can be strongly oscillating relative to time t and even unbounded for all $t \geq 0$.

We shall now examine the asymptotic behavior of solutions to a system of ordinary differential equations as $t \to +\infty$ in the neighborhood of the rated motion $x = 0$. Consider the system

$$\frac{dx_s}{dt} = p_{s1}(t)x_1 + \ldots + p_{sn}(t)x_n + X_s + cf_s(t), \quad s = 1, \ldots, n, \tag{47}$$

where X_s are holomorphic real-valued functions of the real variables $x_1, \ldots, x_n$ with the coefficients that are continuous and bounded for all $t \geq 0$

$$X_s = \sum_{m_1+\ldots+m_n > 1} P_s^{(m_1,\ldots,m_n)}(t) x_1^{m_1}, \ldots, x_n^{m_n}, \quad s = 1, \ldots, n. \tag{48}$$

Series (48) are taken to be absolutely and uniformly convergent in $t \geq 0$ for $|x_s| \leq A_s$, where the constants $A_s > 0, s = 1, \ldots, n$.

Theorem 20 *Suppose that in the system of differential equations (47) the functions $p_{si}(t), s = 1, \ldots, n, i = 1, \ldots, n$ are continuous and bounded for $t \geq 0$, while the functions $f_s(t)$ are continuous for $t \geq 0$ and are such that*

$$\left| \int_t^{+\infty} f_s(\tau) d\tau \right| \leq A \exp(-\alpha t), \quad s = 1, \ldots, n, \tag{49}$$

where the constants $A > 0, \alpha > 0$. Now suppose that the linear homogeneous system of differential equations corresponding to system (47)

$$\frac{dx_s}{dt} = p_{s1}(t)x_1 + \ldots + p_{sn}(t)x_n, \quad s = 1, \ldots, n \tag{50}$$

is regular, and its normal system of solutions has k ($0 \leq k \leq n$) solutions with positive characteristic numbers.

Then for every arbitrarily small $\epsilon > 0$ there is $\delta > 0$ such that if the initial values $a_s = x_s(0)$ and the parameter c satisfy the inequalities $|a_s| < \delta, |c| < \delta, s = 1, \ldots, n$ and the equations

$$F_\ell(c, a_1, \ldots, a_n) = 0, \quad \ell = 1, \ldots, n - k, \tag{51}$$

then for all $t \geq 0$ the inequalities $|x_s(t)| < \epsilon, s = 1, \ldots, n$ hold for the corresponding solution $x_s(t), s = 1, \ldots, n$ of system (47). Moreover, any such solution $x_s(t), s =$

$1, \ldots, n$ of system (47) tends to a zero motion as $t \to +\infty$, i.e. $x_s(t) \to 0$ as $t \to +\infty$, $s = 1, \ldots, n$. In relationships (51) F_ℓ are holomorphic real-valued functions of real independent variables that are given for all sufficiently small $|c|, |a_s|$ and vanish when they are all zero. In this case, the functions F_ℓ are such that $a_s, s = 1, \ldots, n$ can be expressed as holomorphic real-valued functions of the parameter c and some k independent real variables.

Proof: Applying Lyapunov's first method we will construct the solution of system (47) as series in integral positive powers of the parameter c and arbitrary constants $c_1, \ldots, c_k$:

$$x_s = x_s^{(1)} + x_s^{(2)} + \ldots, \quad s = 1, \ldots, n, \tag{52}$$

where $x_s^{(q)}$ is the q-th power function that is homogeneous in $c, c_1, \ldots, c_k$. Further, the sum $x_s^{(1)} + \ldots + x_s^{(q)}$ is called the q-th approximation ($q \geq 1$) of the desired solution $x_s, s = 1, \ldots, n$. Set $x_s^{(1)} = cz_s + \sum_{r=1}^{k} c_r z_{sr}, s = 1, \ldots, n$ where the solutions $(z_{1r}, \ldots, z_{nr})^*, r = 1, \ldots, k$ are taken from the normal system for the system of equations (50) and have characteristic numbers $\lambda_r > 0, r = 1, \ldots, k$, while $(z_1, \ldots, z_n)^*$ is the solution of the linear nonhomogeneous system of differential equations

$$\frac{dz_s}{dt} = p_{s1}(t)z_1 + \ldots + p_{sn}(t)z_n + f_s(t), \quad s = 1, \ldots, n, \tag{53}$$

which has the characteristic number $\lambda \geq \alpha > 0$. From Lemma 2 it follows that, under the conditions of Theorem 20, such first approximation can be constructed.

In order to construct $x_s^{(q)}$, we introduce the $(q-1)$ approximation into series (48) and select in the resulting expressions the homogeneous functions of $c, c_1, \ldots, c_k$ to the power of q that are denoted by $R_s^{(q)}$. We take $x_s^{(q)}$ to be a solution to the linear nonhomogeneous system

$$\frac{dx_s^{(q)}}{dt} = p_{s1}(t)x_1^{(q)} + \ldots + p_{sn}(t)x_n^{(q)} + R_s^{(q)}, \quad s = 1, \ldots, n.$$

Let $(z_{1s}(t), \ldots, z_{ns}(t))^*, s = 1, \ldots, n$, be the fundamental system of solutions for the system of equations (50), $\Delta(t)$ the determinant of matrix for this fundamental system, and $\Delta_{ij}(t)$ the cofactor of the element $z_{ij}(t)$ of the determinant $\Delta(t)$. Then, by the Cauchy formula [Coddington and Levinson (1955), Hale (1969)], we obtain

$$x_s^{(q)} = \sum_{i=1}^{n} \sum_{j=1}^{n} z_{sj}(t) \int \frac{\Delta_{ij}(\tau)}{\Delta(\tau)} R_i^{(q)}(\tau) d\tau, \quad s = 1, \ldots, n.$$

Denote

$$\int \frac{\Delta_{ij}(\tau)}{\Delta(\tau)} R_i^{(q)}(\tau) d\tau = \sum_{m+m_1+\ldots+m_k=q} c^m c_1^{m_1} \ldots c_k^{m_k} \int T_{ij}^{(m,m_1,\ldots,m_k)}(\tau) d\tau.$$

Each of these integrals is taken from t to $+\infty$ if the integrand has a positive characteristic number; otherwise we set

$$\int T_{ij}^{(m,m_1,\ldots,m_k)}(\tau) d\tau = \int_0^t T_{ij}^{(m,m_1,\ldots,m_k)}(\tau) d\tau + C_{ij}^{(m,m_1,\ldots,m_k)}.$$

In this case, for the constants $C_{ij}^{(m,m_1,\ldots,m_k)}$ there are defined values that are independent of $c, c_1, \ldots, c_k$. We have thus constructed series (52) formally satisfying the system of equations (47). These series (52) can be represented as follows:

$$x_s = \sum_{m+m_1+\ldots+m_k>0} L_s^{(m,m_1,\ldots,m_k)}(t) c^m c_1^{m_1} \ldots c_k^{m_k} \exp(-t(\lambda m + \sum_{r=1}^{k} \lambda_r m_r)), \quad s=1,\ldots,n,$$
(54)

where $\lambda_r = \chi(z_{1r}, \ldots, z_{nr}) > 0, r = 1, \ldots, k, \lambda = \chi(z_1, \ldots, z_n) \geq \alpha > 0$, and the functions $L_s^{(m,m_1,\ldots,m_k)}(t)$ are defined by expressions of the form

$$L_s^{(m,m_1,\ldots,m_k)}(t) = \exp(t(\lambda m + \sum_{r=1}^{k} \lambda_r m_r)) \sum_{i=1}^{n} \sum_{j=1}^{n} z_{sj}(t) \int T_{ij}^{(m,m_1,\ldots,m_k)}(\tau) d\tau, \quad s=1,\ldots,n.$$

It can be readily seen that the characteristic numbers of all functions $L_s^{(m,m_1,\ldots,m_k)}(t)$ are non-negative. By the conditional stability theorem from Lyapunov's first method [Lyapunov (1950)], the series (54) with sufficiently small $|c|, |c_r|, r = 1, \ldots, k$, are absolutely and uniformly convergent for all $t \geq 0$ and hence do represent a solution of the system of differential equations (47) with suitable initial values. Setting in (54) $t = 0$ we obtain

$$a_s = \sum_{m+m_1+\ldots+m_k>0} L_s^{(m,m_1,\ldots,m_k)}(0) c^m c_1^{m_1} \ldots c_k^{m_k}, \quad s=1,\ldots,n.$$
(55)

Let us select from equalities (55) k expressions whose functional determinant is nonzero for $c_r, r = 1, \ldots, k$, with $c = c_1 = \ldots = c_k = 0$. This is possible, because the functions $z_{1r}, \ldots, z_{nr}, r = 1, \ldots, k$, are linearly independent for all $t \geq 0$. Denote by $s_1, \ldots, s_k$, the indices of the above k expressions from equalities (55). For all sufficiently small $|c|, |a_{s_r}|, r = 1, \ldots, k$ we have thus defined the holomorphic functions specifying $c_r, r = 1, \ldots, k$ as series in $c, a_{s_1}, \ldots, a_{s_k}$ and vanishing at $c = a_{s_1} = \ldots = a_{s_k} = 0$. Substituting the resulting series into (55) for the remaining a_s, we obtain the sought-for equations (51) for the initial values of the solution of system (47). From the construction it follows that the functions F_ℓ satisfy all requirements of Theorem 20. The form of the resulting series (54) implies that $x_s(t) \to 0$ as $t \to +\infty, s = 1, \ldots, n$. Moreover, Lyapunov's conditional stability theorem implies that all quantities $|x_s(t)|, s = 1, \ldots, n$ remain arbitrarily small for all $t \geq 0$ only if $|c|, |a_s|, s = 1, \ldots, n$ are sufficiently small and $c, a_s, s = 1, \ldots, n$ satisfy equations (51).

Note that, with $k = 0$, equations (51) automatically follow from relationships (55). This completes the proof of Theorem 20.

Remark 1. Suppose the system of equations (50) is not regular and the remaining conditions of Theorem 20 are all satisfied. Then Theorem 20 remains valid under the additional assumption that the constant α from conditions (49) satisfies the inequality $\alpha > 2\sigma > 0$ and the normal system of solutions for the system of equations (50) has k $(0 \leq k \leq n)$ solutions with characteristic numbers $\lambda_r > \sigma > 0, r = 1, \ldots, k$. Here $\sigma = -(S + \mu) > 0, S$ is the sum of characteristic numbers of solutions from the normal system for (50), $\mu = \chi(\frac{1}{\Delta}), \Delta(t)$ is the determinant of matrix for this normal system of solutions. From the remark of Lemma 2, the proof of this assertion is the same as the proof of Theorem 20 except that the construction of the first approximation $x_s^{(1)}, s = 1, \ldots, n$ utilizes only the solutions $(z_1, \ldots, z_n)^*$ of system (53) and $(z_{1r}, \ldots, z_{nr})^*, r = 1, \ldots, k$ from the normal system of solutions for equations (50) whose characteristic numbers satisfy the inequalities $\lambda > \sigma$, $\lambda_r > \sigma, r = 1, \ldots, k$.

Remark 2. If the relation $f_s(t) \not\equiv 0$ for all $t \geq 0$ holds for at least one $s, s = 1, \ldots, n$, then the rated motion $x_s = 0, s = 1, \ldots, n$ is not realized by the system of equations (47), for it is not a solution to this system of equations.

Consider the system

$$\frac{dx_s}{dt} = p_{s1}(t)x_1 + \ldots + p_{sn+k}(t)x_{n+k} + X_s, \quad s = 1, \ldots, n,$$

$$\frac{dx_{n+r}}{dt} = X_{n+r}, \quad r = 1, \ldots, k, \tag{56}$$

$$X_\ell = \sum_{m_1+\ldots+m_n>1} P_\ell^{(m_1,\ldots,m_n)}(t) x_1^{m_1} \ldots x_n^{m_n}, \quad \ell = 1, \ldots, n+k,$$

where the functions $P_\ell^{(m_1,\ldots,m_n)}(t)$ are continuous and bounded for any $t \geq 0$. The series are taken to be absolutely and uniformly convergent in $t \geq 0$ for $|x_s| \leq A_s$, where the constants $A_s > 0, s = 1, \ldots, n$.

Theorem 21 *Suppose that in the system of differential equations (56) the functions $p_{si}(t), s = 1, \ldots, n, i = 1, \ldots, n$ are continuous and bounded for $t \geq 0$, and the functions $p_{sn+r}(t), s = 1, \ldots, n, r = 1, \ldots, k$ are continuous for $t \geq 0$ and are such that*

$$|\int_t^{+\infty} p_{sn+r}(\tau)d\tau| \leq A\exp(-\alpha t), \quad s = 1, \ldots, n, \quad r = 1, \ldots, k,$$

where the constants $A > 0, \alpha > 0$. We assume that the homogeneous linear system of differential equations

$$\frac{dx_s}{dt} = p_{s1}(t)x_1 + \ldots + p_{sn}(t)x_n, \quad s = 1, \ldots, n \tag{57}$$

is regular and its normal system of solutions $(z_{1s}, \ldots, z_{ns})^*, s = 1, \ldots, n$, has $m, 0 \leq m \leq n$, solutions with positive characteristic numbers.

Then the zero solution $x_\ell = 0, \ell = 1, \ldots, n+k$ of system (56) is conditionally stable. That is, the initial values $a_\ell = x_\ell(0), \ell = 1, \ldots, n+k$, must satisfy the equations

$$F_j(a_1, \ldots, a_{n+k}) = 0, \quad j = 1, \ldots, n-m. \tag{58}$$

These equations are such that $a_\ell, \ell = 1, \ldots, n+k$ can be expressed as functions of some $m+k$ independent variables. Furthermore, for sufficiently small $|a_\ell|, \ell = 1, \ldots, n+k$ as $t \to +\infty$, we obtain

$$x_s(t) \to 0, \quad s = 1, \ldots, n,$$

$$x_{n+r}(t) \to c_{m+r}, \quad r = 1, \ldots, k, \tag{59}$$

where the variables $|c_{m+r}|, r = 1, \ldots, k$, are arbitrarily small if $|a_\ell|, \ell = 1, \ldots, n+k$, are sufficiently small.

Analysis of Rated Stability by Lyapunov's First Method

Proof: Following Lyapunov, we construct a solution to system (56) as a series in integral positive powers of arbitrary constants $c_1, \ldots, c_{m+k}$

$$x_\ell = x_\ell^{(1)} + x_\ell^{(2)} + \ldots, \quad \ell = 1, \ldots, n+k, \qquad (60)$$

where $x_\ell^{(q)}$ denotes the q-th function that is homogeneous in $c_1, \ldots, c_{m+k}$. The sum $x_\ell^{(1)} + \ldots + x_\ell^{(q)}$ is called the q-th approximation ($q \geq 1$) of the sought-for solution x_ℓ. Let

$$x_s^{(1)} = c_1 z_{s1} + \ldots + c_m z_{sm} + c_{m+1} z_{sn+1} + \ldots + c_{m+k} z_{sn+k},$$
$$x_{n+r}^{(1)} = c_{m+r}, \quad s = 1, \ldots, n, \quad r = 1, \ldots, k,$$

where the solutions $(z_{1s}, \ldots, z_{ns})^*, s = 1, \ldots, m$ are taken from the normal system of solutions of the system of equations (57) and have positive characteristic numbers, while $(z_{1n+r}, \ldots, z_{nn+r})^*$ are the solutions of nonhomogeneous linear systems of differential equations of the form

$$\frac{dz_{sn+r}}{dt} = p_{s1}(t) z_{1n+r} + \ldots + p_{sn}(t) z_{nn+r} + p_{sn+r}(t), \quad s = 1, \ldots, n, \quad r = 1, \ldots, k, \qquad (61)$$

which also have positive characteristic numbers. Lemma 2 implies that, under conditions of Theorem 21, such first approximation does exist. The construction of the higher approximations and the convergence proof of the resulting series (60) are an exact replica of the argument in Lyapunov's conditional stability theorem. The assertions of Theorem 21 can be derived in much the same way as in the presentation of Lyapunov's first method [Lyapunov (1950)]. Note that the series (60) converge absolutely and uniformly in $t \geq 0$ for all sufficiently small values $|a_\ell|, \ell = 1, \ldots, n+k$, satisfying the equations (58). These series do represent a solution to the system of differential equations (56). In this case, the functions $F_j(a_1, \ldots, a_{n+k}), j = 1, \ldots, n-m$, are series in integral positive powers of $a_\ell, \ell = 1, \ldots, n+k$ with some constant coefficients. These series converge in all sufficiently small $|a_\ell|, \ell = 1, \ldots, n+k$, and vanish when $a_\ell = 0, \ell = 1, \ldots, n+k$. From the form of the resulting series (60) we may conclude about stability of a zero solution to the system of differential equations (56) and about asymptotic behavior (59) of solutions to this system. This completes the proof of Theorem 21.

Remark 1. Suppose the homogeneous linear system of differential equations (57) is not regular and the other conditions of Theorem 21 are satisfied. Then Theorem 21 holds under the additional assumption that the constant α from conditions of Theorem 21 satisfies the inequality $\alpha > 2\delta > 0$, while the normal system of solutions to system (57) has $m, 0 \leq m \leq n$ solutions with characteristic numbers $\lambda_i > \sigma > 0, i = 1, \ldots, m$. Here $\sigma = -(S + \mu) > 0$, S is the sum of characteristic numbers of solutions from the normal system for system (57), $\mu = \chi(1/\Delta)$, $\Delta(t)$ is the determinant of matrix for this normal system of solutions. From the remark of Lemma 2, the proof of this assertion differs from the proof of Theorem 21 only in that the construction of the first approximation $x_\ell^{(1)}, \ell = 1, \ldots, n+k$, utilizes only the solutions $(z_{1n+r}, \ldots, z_{nn+r})^*, r = 1, \ldots, k$ for the systems of nonhomogeneous linear equations (61) and the solutions $(z_{1i}, \ldots, z_{ni})^*, i = 1, \ldots, m$ from the normal system of solutions for the system of homogeneous linear equations (57), whose characteristic numbers satisfy the inequalities $\lambda_r > \sigma, r = 1, \ldots, k, \lambda_i > \sigma, i = 1, \ldots, m$.

Remark 2. If the relation $p_{sn+r}(t) \not\equiv 0$ holds for at least one $s, s = 1, \ldots, n$, and for one $r, r = 1, \ldots, k$, with all $t \geq 0$, then the motion (59) that is limiting as $t \to +\infty$ for solutions of the system of differential equations (56) (i.e. the motion $x_s = 0, s = 1, \ldots, n, x_{n+r} = c_{m+r}, r = 1, \ldots, k$) is the rated motion of the system of equations under study. This motion is not realized by the system of equations (56), for it is not a solution to this system.

Remark 3. If, under all conditions of Theorem 21, there is $m = n$, then the zero solution $x_\ell = 0, \ell = 1, \ldots, n + k$, for the system of differential equations (56) is stable in Lyapunov's sense. From the limit relationships (59) it follows that no asymptotic stability may occur here. Moreover, in this case, equations (58) are nowhere to be seen.

Bibliography

Aizerman, M. A., *Theory of Automatic Control*, p. 452. 3-d ed., Nauka, Moscow, 1966.

Andronov, A. A., *Collected Works*, p. 538. USSR Acad. of Sci., Moscow, 1956.

Andronov, A. A., Witt, A. A. and Khaikin, S. E., *Theory of Oscillations*, p. 915. Fizmatgiz, Moscow, 1959. Available in German: *Theorie der Schwingungen*, vol. 1, p. 492, vol. 2, p. 414. Akademie-Verlag, Berlin, 1965 (vol. 1), 1969 (vol. 2).

Barbashin, E. A., *Math. Proceedings*, vol. 29, no. 2, 1951.

Birkhoff, G., *Dynamical Systems*, p. 295. AMS, New York, 1927.

Birkhoff, G., *Ann. de L'Institut H. Poincaré*. Vol. 3, 1932.

Cesari, L., *Asymptotic Behavior and Stability Problems in Ordinary Differential Equations*, p. 271. Springer-Verlag, Berlin, 1959.

Chetaev, N. G., *Stability of Motion*, p. 207. 3rd ed., Nauka, Moscow, 1965.

Coddington, E. and Levinson, N., *Theory of Ordinary Differential Equations*, p. 429. McGraw-Hill, New York, 1955.

Erougin, N. P., *Reducible Systems*, p. 96. Steklov Math. Inst. USSR Acad. of Sci., Moscow – Leningrad, 1946.

Faddeev, D. K. and Faddeeva, V. N., *Computational Methods of Linear Algebra*, p. 734. 2nd ed., Fizmatgiz, Moscow – Leningrad, 1963. Available in German: *Numerische Methoden der Linearen Algebra*, p. 771. Deutscher Verlag der Wissenschaften, Berlin, 1970.

Fichtengol'c, G. M., *A Course in Differential and Integral Calculus*, vol. 1, p. 607, vol. 2, p. 800. Nauka, Moscow, 1969 (vol. 1), 1970 (vol. 2). Available in German: *Differential– und Integralrechnung*, vol. 1, p. 836, vol. 2, p. 572. Deutscher Verlag der Wissenschaften, Berlin, 1970 (vol. 1), 1971 (vol. 2).

Hadamard, J., *Bull. Soc. Math. de France*, vol. 29, 1901.

Hale, J., *Ordinary differential equations*, p. 332. Wiley–Interscience, New York, 1969.

Bibliography

Hirsch, M. and Smale, S., *Differential equations, dynamical systems, and linear algebra*, p. 358. Academic Press, New York – London, 1974.

Kantorovič, L. V. and Akilov, G. P., *Functional Analysis*, p. 742. 2nd ed., Nauka, Moscow, 1977. Available in German: *Funktionalanalysis in Normierten Räumen*, p. 622. Akademie-Verlag, Berlin, 1964. Available in French: *Analyse Fonctionnelle*, vol. 1, p. 490, vol. 2, p. 343. Mir, Moscow, 1981.

Krasovskii, N. N., *Appl. Math. Mech.*, vol. 16, no. 5, 1952.

LaSalle, J. and Rath, R., *Proc. of the 2nd Cong. of the Int. Fed. of Autom. Contr. (IFAC)*. Butterworths – Oldenbourg, London – Munich, vol. 1, 1964.

Letov, A. M., *Stability of Controlled Nonlinear Systems*, p. 483. 2nd ed., Fizmatgiz, Moscow, 1962. Available in English: *Stability in Nonlinear Control Systems*, p. 315. Princeton Univ. Press, Princeton, NJ, 1961.

Lur'ye, A. I., *Some Nonlinear Problems in the Theory of Automatic Control*, p. 216. Gostekhteoretizdat, Moscow – Leningrad, 1951. English translation available: Her Majesty's Stationery Office, 1957.

Lyapunov, A. M., *The General Problem of Stability of Motion*, p. 472. Gostekhteoretizdat, Moscow – Leningrad, 1950. Available in French: *Problème Général de la Stabilité du Mouvement*, pp. 203–474. Annales de la Faculté des Sciences, Toulouse, 2nd ser., vol. 9, 1907. Reprinted in Annals of Mathematics Studies, no. 17. Princeton Univ. Press, Princeton, NJ, 1949.

Malkin, I. G., *Appl. Math. Mech.*, vol. 18, no. 2, 1954.

Malkin, I. G., *Some Problems in the Theory of Nonlinear Oscillations*, p. 491. Gostekhizdat, Moscow, 1956.

Malkin, I. G., *Theory of Stability of Motion*, p. 530. 2nd ed., Nauka, Moscow, 1966. Available in German: *Theorie der Stabilität einer Bewegung*, p. 402. Akademie – Verlag, Berlin, 1959. Available in English: AEC translation 3352, Dept. of Commerce USA, 1958.

Nemyckij, V. V. and Stepanov, V. V., *Qualitative Theory of Differential Equations*, p. 550. 2nd ed., Gostekhteoretizdat, Moscow – Leningrad, 1949. Available in English: translation of 2nd ed., p. 523. Princeton Univ. Press, Princeton, NJ, 1960.

Perron, O., *Mathem. Zeitschrift.*, Bd. 31, 1930.

Popov, E. P., *Dynamics of Automatic Control Systems*, p. 799. Gostekhizdat, Moscow, 1954. Available in German: *Dynamik Automatischer Regelsysteme*, p. 780. Akademie-Verlag, Berlin, 1958.

Popov, E. P. and Pal'tov, I. P., *Approximate Methods for Investigation of Nonlinear Automatic Systems*, p. 792. Fizmatgiz, Moscow, 1960. Available in German: *Näherungsmethoden zur Untersuchung Nichtlinearer Regelungssysteme*, p. 786. Akademie-Verlag — Gesst & Portig, Leipzig, 1963.

Rumjancev, V. V., *USSR Acad. of Sci., Mech. and Machine Building*, no. 6, 1963.

Sebakhy, O., *IEEE Trans. on Autom. Contr.*, vol. AC–20, no. 10, 1975.

Tsypkin, Ya. Z., *Automatic Relay Systems*, p. 575. Nauka, Moscow, 1974.

Zubov, V. I., *Oscillations in Nonlinear and Controlled Systems*, p. 631. Sudpromgiz, Leningrad, 1962.

Zubov, V. I., *Mathematical Methods for the Study of Automatic Control Systems*, p. 327. Pergamon Press, New York, 1962.

Zubov, V. I., *Methods of A. M. Lyapunov and Their Application*, p. 263. Noordhoff, Groningen, 1964.

Zubov, V. I., *Theory of Optimal Control*, p. 352. Sudostroyeniye, Leningrad, 1966.

Zubov, V. I., *Analytical Dynamics of Gyro Systems*, p. 318. Sudostroyeniye, Leningrad, 1970.

Zubov, V. I., *Lectures on Control Theory*, p. 496. Nauka, Moscow, 1975. Available in French: *Théorie de la Commande*, p. 469. Mir, Moscow, 1978.

Index

almost periodic Bohr function, 199, 212
almost periodic mode, 15
almost periodic motion, 15, 76, 78
almost periodic oscillation, 19, 20, 24, 25
almost periodic solution, 16, 19, 167, 172, 173, 175, 177, 179, 180, 185, 190, 192, 193, 252–254, 257
Andronov's theorem, 89
Arzela theorem, 1, 2, 42

Bendixson theorem, 340, 343
Birkhoff theorem, 76

case of several complex roots with unity moduli, 100
　general case, 106
　special case, 106
characteristic equation, 21–23, 88–90, 98, 100, 108, 152, 161, 167, 258, 261
　of the first approximation system, 89
characteristic function of the set, 202, 205
characteristic index, 22–25, 58, 117, 126, 157, 158, 161, 190, 194, 247, 248, 252, 353
characteristic number
　of function, 26, 27, 29, 30, 34, 44, 52, 380–383
　of solution, 27–29, 382, 385–387
characteristic numbers
　of matrix, 13
　of system, 28–31, 45–47, 52, 53, 59, 382, 383, 389–391
characteristic polynomial, 151, 152

differentiability of function in terms of system, 62
dynamical system in E_n, 70
　invariant set, 70
　　Lyapunov asymptotically stable, 73
　　Lyapunov stable, 73
　　Lyapunov unstable, 73
　　minimal, 76
　motion, 70
　　α-limit point, 72
　　ω-limit point, 72
　　almost periodic, 76
　　central, 74
　　periodic, 73
　　Poisson-stable point, 73, 74
　　Poisson-stable point as $t \to +\infty$, 73
　　Poisson-stable point as $t \to -\infty$, 73
　　recurrent, 76
　　recurrent t, ϵ-almost period, 76
　negative semitrajectory, 70
　positive semitrajectory, 70
　representative point, 70
　rest point, 71
　　isolated, 72
　　Lyapunov asymptotically stable, 72
　　Lyapunov stable, 72
　　Lyapunov unstable, 72
　　nonisolated, 72
　trajectory, 70
　wandering point, 74

eigenvalues of matrix, 13, 22
equations of characteristics, 47, 223

Euler formula, 101
Euler polygonal line, 3

group property, 70, 75

integral curve, 3, 5, 78
  almost periodic, 173, 174
  asymptotic, 78
  doubly asymptotic, 78
  periodic, 173, 174
irregular system, 30, 44, 45, 52, 60, 386, 389, 391
irregularity coefficient, 44, 60

Jordan
  box, 266, 375, 378, 379
  canonical form of matrix, 12, 15, 22, 55, 58, 88, 258, 263, 266, 375

Lyapunov function, 62
Lyapunov theorem
  of asymptotic stability, 64
  of instability, 64
  of stability, 62

matrix equation by Lyapunov, 68, 291, 293
matrix of Lyapunov type, 28, 30, 31

negative-definite
  function, 61, 66, 110, 127, 181, 182, 186, 188, 302, 303, 307
    for components of vector, 295, 296, 307
  homogeneous function, 93, 95, 110, 195
  matrix, 68
  quadratic form, 66, 95
non-resonance
  case, 17, 58, 153, 157, 248
  system, 17
nonperturbed motion, 23, 60, 61
  asymptotically Lyapunov stable, 61
  Lyapunov stable, 60, 61
normal fundamental system, 28–31, 35, 47, 48, 382, 383, 385, 387–391

Peano theorem, 3
periodic boundary conditions, 247
periodic dynamical system in E_n, 132, 133
  motion
    periodic, 134, 135, 143
periodic mode, 15, 16, 189, 257
periodic motion, 15, 73, 76, 78, 134, 143, 188, 189, 231
periodic oscillation, 18, 19, 24, 25, 167
periodic self-oscillation, 77, 83, 84, 89, 93, 95, 116, 126, 127
periodic solution, 16–18, 20, 23, 25, 77, 78, 81–87, 89, 91–93, 95, 96, 98, 99, 105, 108, 110–112, 115–118, 121, 122, 124–129, 146–161, 167, 172–175, 180, 181, 188–190, 193–197, 231, 248–250, 252, 253, 257
  Lyapunov stable, 77
  orbit-asymptotically stable, 77
  orbitally stable, 77
  stable under persistent perturbations, 115, 116
Perron theorem, 30
perturbed motion, 60, 61
phase plane, 309, 310
Picard theorem, 4, 6, 239
positive-definite
  function, 61–64, 66, 67, 95, 110, 127, 181, 186, 188, 189, 302, 307, 330, 359, 361, 362, 364, 366
    for components of vector, 295, 296
  homogeneous function, 93, 110, 195
  matrix, 68
  quadratic form, 66, 195, 227, 291
reducible system, 30, 88
regular system, 30, 32, 34, 47, 53, 59, 383, 385–387, 390
resonance
  case, 17, 19, 24, 153, 156–158, 248
  eigenvalues, 153, 154, 156, 157

Index   399

    system, 17

self-oscillation, 278, 280, 314, 315, 336
set of functions
    equicontinuous, 1, 3, 42, 177, 178, 187
    uniformly bounded, 1, 4
stability problem, 66, 161, 169, 257, 262–264
stability under persistent perturbations, 115, 116
stationary field
    electric, 306
    magnetic, 296, 297, 303, 305
stationary mode, 20, 24, 265, 267, 268, 270, 271, 273, 274, 276, 277, 310, 355
stationary motion, 230, 231, 270, 271, 274, 313
stationary oscillations, 15, 17, 76, 268
    "linear", 267
    "nonlinear", 267
    forced, 16, 17
    natural, 15, 16, 24, 267, 270, 271
sweep method, 231

the general solution of the nonhomogeneous system, 12, 91, 248
theorem
    of integral continuity, 71, 322
    of solution continuity in initial data, 8, 181, 285, 311, 312, 339, 344
    of solution continuously differentiability in initial data, 11, 151
total derivative of function in terms of system, 63–66, 95, 110, 111, 127, 181, 186, 291, 295, 328, 330, 332, 346, 359, 360

unity roots of characteristic equation, 87
    general case, 91
    nonsingular case, 89
    singular cases, 89

    special case, 91
zero solution
    asymptotically Lyapunov stable, 64, 66–68
    Lyapunov stable, 61–64
    Lyapunov unstable, 64, 65
    uniformly Lyapunov stable, 64, 65